STP 924

Basic Questions in Fatigue: Volume I

Jeffrey T. Fong and Richard J. Fields, editors

ASTM
1916 Race Street
Philadelphia, PA 19103

Library of Congress Cataloging-in-Publication Data

Basic questions in fatigue / Jeffrey T. Fong . . . [et al.].

(STP; 924)
"ASTM publication code number (PCN) 04-924002-30."
Papers presented at the Symposium on Fundamental Questions and Critical Experiments on Fatigue held 23–24 Oct. 1984 in Dallas, Tex. and sponsored by Committee E-9 on Fatigue.
Includes bibliographies and indexes.
ISBN 0-8031-0925-3
1. Materials—Fatigue—Congresses. 2. Metals—Fatigue—Congresses. I. Fong, J. T. (Jeffrey Tse-wei), 1934– II. Symposium on Fundamental Questions and Critical Experiments on Fatigue (1984; Dallas, Tex.) III. American Society for Testing Materials. Committee E-9 on Fatigue. IV. Series: ASTM special technical publication; 924.
TA418.38.B375 1988 88-14555
620.1'123—dc19 CIP

NOTE

The Society is not responsible, as a body,
for the statements and opinions
advanced in this publication.

Peer Review Policy

Each paper published in this volume was evaluated by three peer reviewers. The authors addressed all of the reviewers' comments to the satisfaction of both the technical editor(s) and the ASTM Committee on Publications.

The quality of the papers in this publication reflects not only the obvious efforts of the authors and the technical editor(s), but also the work of these peer reviewers. The ASTM Committee on Publications acknowledges with appreciation their dedication and contribution of time and effort on behalf of ASTM.

Printed in Ann Arbor, MI
August 1988

Foreword

The papers in this publication, *Basic Questions in Fatigue, Volume I,* contain papers presented at the symposium on Fundamental Questions and Critical Experiments on Fatigue held 22–23 October 1984 in Dallas, Texas. The symposium was sponsored by Committee E-9 on Fatigue. Jeffrey T. Fong, National Bureau of Standards, and Richard J. Fields, National Bureau of Standards, are editors of this volume.

Fatigue of Steels—Mechanical Loading

Fatigue of Aluminum and Other Alloys—Mechanical Loading

Overview

Structural fatigue, or simply, *fatigue,* has been of interest to civil and mechanical engineers, materials scientists, applied mathematicians, plant managers, and the public for a long time. In 1978, at the ASTM International Symposium on Fatigue Mechanisms held at Kansas City, Missouri, an *ad hoc* estimate was made of the annual world-wide cost of fatigue testing and research at about one billion 1978 U.S. dollars (*see* pp. 730–731, *ASTM STP 675*).

Undoubtedly, most of that effort each year is on fatigue testing with perhaps only a few percent of that effort on research. Nevertheless, the total effort on fatigue research over a period of, say, 20 to 30 years, may be looked upon as a sizable investment by both the private and the public sectors to the tune of many thousands of person-years. At that level of effort, members of the public, the technical community, and the next generation of engineers and materials scientists about to enroll in a course on fatigue, have a right to ask some obvious questions on the state of fatigue research such as:

(*a*) Has the concept of fatigue evolved over the past 30 years from an empirical subject of engineering practice to a well-defined discipline of materials science?
(*b*) Are the methodologies of fatigue research sufficiently scientific to yield a core of knowledge known as "fatigue science?"
(*c*) Are the current procedures for predicting the fatigue lives of structures in ordinary and severe environments based on sound theories and credible experiments?

In an attempt to shed some light on this and to ascertain whether there indeed existed a "scientific basis" of fatigue, the ASTM Committee E-9 on Fatigue initiated as early as 1982 the planning of a unique 5-day international symposium entitled:

"Fundamental Questions and Critical Experiments on Fatigue."

The symposium was held in October 1984 at Dallas, Texas, and was attended by over 250 researchers, engineers, and managers from 14 countries. Co-sponsoring the symposium were the ASTM Committee E-24 on Fracture and the U.S. National Bureau of Standards (NBS).

The symposium consisted of a 3-day workshop (18–20 Oct.) at Arlington, a suburb of Dallas, and a 2-day conference (22–23 Oct.) at Dallas, Texas, during a scheduled Committee Week of ASTM. Of the 43 contributed papers that were presented, 37 manuscripts were eventually submitted for inclusion in the two-volume proceedings. The papers in these two volumes represent the bulk of deliberations by some of the most distinguished and knowledgeable researchers from the international fatigue community.

To appreciate the significance of these papers, it is useful to recall some of the statements made in the original Call for Papers. In that document, which was released in the summer of 1983, potential contributors were advised that the symposium was a *new forum* designed for researchers to meet and exchange *not necessarily* their recent results, as would normally be expected of them at traditional symposia, but rather their *burning questions* on some aspects of fatigue so long as the questions were "basic" and were aimed toward a better understanding of fatigue.

To guide the contributors in preparing their abstracts, the Call for Papers stipulated that each abstract must contain the following items:

1. A clear statement of the fundamental question and its importance.
2. A well-defined critical experiment to answer, unequivocally, the question posed.
3. Measurements to be made in the proposed critical experiment.

The goals of the symposium, as stated in the Call for Papers, were:

Goal 1—To Advance the Understanding of Fatigue

By emphasizing the coupling of fundamental questions with critical experiments, the multidisciplinary nature of fatigue may be brought into sharper focus in order to accelerate the understanding of fatigue in the following four subareas (for both metals and nonmetals):

1. Nucleation of fatigue damage.
2. Transition between nucleation and propagation.
3. Propagation of fatigue damage.
4. Environmental effects.

Goal 2—To Lay the Foundation for a Scientific Basis of Fatigue

Invited researchers from around the world will contribute open questions and critical experiments, including new results, for an intensive discussion and debate, thereby providing the framework for a scientific basis.

Goal 3—To Mold a Consensus on Research Priorities

Leading experts and practicing engineers will discuss and debate on the merits of a list of open questions and ideas for experiments. It is expected that a consensus on research priorities may be reached in time for inclusion in the symposium proceedings to guide the research direction of major fatigue laboratories.

To achieve the goals of the symposium, the Program Committee adopted a 3-part format for each accepted paper, namely: (i) presentation, (ii) invited official discussion, and (iii) general discussion. To preserve a continuity in technical discussion and debate leading to a consensus on a scientific basis of fatigue, the Committee also adopted a policy of not scheduling any parallel sessions.

Both the format and the single-session policy placed a severe restriction on the number of papers that could be scheduled in the final program. For a 10-session symposium lasting a total of 5 days, the upper bound of that number was somewhere between 40 and 50. This was about half of the 96 questions submitted to the Program Committee from authors of 14 countries (Austria, Canada, China, Finland, France, F.R. Germany, Italy, Japan, Korea, Sweden, Switzerland, U.K., U.S.A., and U.S.S.R.).

Fortunately, a good number of researchers were still able to contribute as invited official discussors. This led to a new activity of the Program Committee by introducing the concept of an *open preview*, where the extended abstract of every accepted paper was reviewed by two or more invited discussers, and both the abstracts and the discussers' comments were sent to all pre-registrants one month before the meeting in the form of a 426-page symposium preview. This had the advantage that by the time the symposium opened, most of the participants had already digested the pros and cons of the relevance of each fundamental question and were able to zero in on the central issues of each paper as soon as it was presented.

So much for the background of the symposium. The two-volume proceedings is divided into eight sections of which five are in Volume I and three in Volume II. In the section on Introductory Remarks, we include the "Historical Account of the Symposium" by Dr. J. T. Cammett, then Chairman of ASTM Committee E-9. There are also two other remarks, one on "The International Role of ASTM" by Dr. D. R. Johnson of NBS, then a member of the Board of Directors of ASTM, and the other on "Experimentation and Measurement" by Dr. H. H. Ku, then Chief of the NBS Statistical Engineering Division.

In the next two sections, the questions of nucleation of fatigue damage in single crystals and polycrystals are addressed, with principal emphasis on the observation of damage at the microstructural level. Following these are sections dealing primarily with the role of mechanical variables on fatigue crack growth in ferrous and nonferrous alloys, and focus attention at the continuum level. This completes the contents of Volume I with three opening remarks and 20 contributed papers.

The remaining 17 contributed papers appear in Volume II, where the topics of research are more complicated, and the state of knowledge is very much in the formative stage. In the first section, the complex interactions associated with combined fatigue and creep damage are considered. The next section, by far the largest group of papers, deals with the questions of environmental effects. The last section contains the only papers that address fatigue of nonmetals. (It is recognized that fatigue research on nonmetals is customarily reported through a different forum.)

The 5-day symposium was well received and enthusiastically attended. The open preview concept, the single-session policy, and the 3-part presentation format, were most often cited as the principal factors in keeping everyone interested in the debate. The three questions posed earlier in this review were addressed throughout the symposium, and the final consensus appeared to be:

(*a*) The concept of fatigue did evolve over the last 30 years from an empirical subject of engineering practice to a well-recognized topic of materials science research, but the evolution fell short of reaching the goal of a mature discipline.

(*b*) The methodologies of fatigue research vary among researchers with some sufficiently scientific but others less so. A core of knowledge known as "fatigue engineering" already exists, but what may pass as "fatigue science" is yet to emerge.

(*c*) The current procedures for predicting the fatigue lives of structures in ordinary and severe environments are still based on a combination of empirical data, plausible theories, and experts' judgment. The day of making predictions from sound theories, credible experiments, and operational data, is still very much in the future.

The real value of the symposium lies in the searching questions and the critical experiments that are carefully laid out in these two volumes for the next generation of fatigue researchers to take advantage of. It is hoped that in a decade or two when we meet again to take stock of what we know, most of the questions listed in this book would be either fully or partially answered to yield a truly scientific basis of fatigue.

It gives us great pleasure in acknowledging the tremendous help and cooperation we received from hundreds of researchers all over the world in making this symposium a reality.

We are also indebted to the following organizations in the United States for financial grants, without which many United States and foreign-based authors would not have been able to attend:

Aluminum Company of America–Alcoa Laboratories
Boeing Commercial Airplane Company
Exxon Research & Engineering Company
General Dynamics Corporation
General Electric Corporate R & D
Lockheed California Company
McDonnell Douglas Corporation
MTS Systems Corporation
U.S. Army Research Office
U.S. National Bureau of Standards
U.S. Office of Naval Research

Jeffrey T. Fong
Symposium chairman and principal editor, Volume I.

Robert P. Wei
Symposium co-chairman and principal editor, Volume II.

Richard J. Fields
Co-editor, Volume I.

Richard P. Gangloff
Co-editor, Volume II.

Introductory Remarks

Historical Account of the Symposium

It is not my duty to inform you in detail of the objectives and specific format of the Symposium; I will leave that to the Organizing Committee and others. For now, just let it suffice for me to tell you that the objectives are different and the format is different from any other Fatigue Symposium which you may have attended or experienced.

Regardless of the specific objectives, however, the underlying purpose of this and any symposium is interchange of ideas, and we trust that this purpose will be served in exemplary fashion. As part of my message to you today, I wish to trace the history of ASTM Committee E-9 on Fatigue, which is the sponsor of the symposium.

The Committee, as it is now constituted, has been in existence for nearly 40 years; however, its roots trace back much farther than that. In 1928, the ASTM Research Committee on Fatigue was formed under the leadership of Professor H. F. Moore. This Committee served the Society until 1946 at which time it was dissolved and its duties were transferred to the ASTM Committee E-9 on Fatigue, which was formed at the ASTM Annual Meeting 1946 in Buffalo, New York.

The membership of the first Advisory Committee, now called the Executive Subcommittee, was comprised of individuals whose names will be familiar to all of you who are familiar with the fatigue literature. The Chairman was R. E. Peterson; the Secretary was O. J. Horger; other members were M. A. Grossman, J. M. Wassels, H. F. Moore, and R. L. Templin. From an initial membership of 53, Committee E-9 has now grown to its current membership of 378; this is a very current figure, among which are included many international members.

Since 1946, the Committee which meets twice annually has been sponsor or co-sponsor of 60 or more fatigue symposia, as well as many, many workshops and less formal technical meetings. Sponsorship of this current symposium and other fatigue symposia is quite consistent with the overall scope of the Committee. Bear with me while I read verbatim the scope of our Committee E-9 on Fatigue.

"The formulation of methods for the determination of the fatigue characteristics of simple and composite materials, the promotion of research leading to the determination of the nature of fatigue, and to methods for determining fatigue characteristics and the coordination of society activities in these areas conducted by other committees."

Now, I underscore the words, *Promotion of Research Leading to Determination of the Nature of Fatigue* because this is exactly the underlying purpose of this Symposium. Fulfillment of this purpose moreover is entirely consistent with the other elements of the Committee's scope. In particular, we view the maintenance of a continuing research forum within the Committee as a most instrumental and necessary precursor to the creation and advancement of standard methods for fatigue testing and evaluation, this being the consummate function of our Committee.

Now, the credit for creating this Symposium and the immediately-preceding workshop, which was held in Arlington, Texas, last week, must go to the Symposium Organizing Committee, the Program Committee, and the International Advisory Board. There are many names—too numerous to cite here, however, I would be remiss not to mention the names of Dr. Jeffrey Fong, Professor Robert Wei, and Dr. Louis Coffin, for their considerable efforts in this regard.

I also cite the generous financial contributions of the following organizations which have provided travel grants which have helped many of you to attend this Symposium. They are the U.S. Army Research Office; the U.S. Office of Naval Research; Alcoa Laboratories of the Aluminum Company of America; the Boeing Commercial Airplane Company; Exxon Research and Engineering Company; General Dynamics Corporation; Corporate Research and Development Laboratory of the General Electric Company; Lockheed California Company; McDonnell Douglas, and MTS Systems Corporation.

Unquestionably, I must also cite the considerable contributions of the National Bureau of Standards in terms of personnel staff time in the organization of this Symposium. I also must cite the personal commitment to this field of fatigue by all of you, and your employers in providing the funds to send you here because without you, there would be no Symposium.

John T. Cammett
Vice president, Metcut Research Associates, Inc., Cincinnati, Ohio; past chairman of ASTM Committee E-9 on Fatigue.

The International Role of ASTM

The topic I will address with you this morning is the general issue of standards organizations in the international arena. For the past several years, we have all struggled with economic problems; it certainly has been true in the United States, and I believe it is true in every one of the countries that is represented at this gathering. Here in the United States there has been mounting concern over the last several years about our balance of trade with other countries. There has been also concern about the fact that the dollar is extraordinarily strong. I am sure that any of you who have come into this country in the last few days and had to exchange your money for dollars are aware of the problem. It is a problem that affects us in many ways. The strong dollar makes it more difficult to sell American products abroad because the costs associated with the manufacture of those products in the United States, by the time they are converted into other currency, cause our products to be priced high. On the other hand, it tends to make the United States a lucrative marketplace for our colleagues from other countries.

I think all of us in the standards business have wondered what we can do to help. That has certainly been the case at the National Bureau of Standards (NBS) and, in fact, in nearly every standards organization that I know of. We deal with many standards organizations at NBS and nearly every one is considering what an appropriate role should be in the international arena. Certainly, ASTM is no exception to that. There has been considerable discussion for an extended period of time within ASTM about international activities and about the proper role of ASTM. The issue of determining the proper role for an organization like ASTM is really what I want to address. It is not an easy matter, and I think in the end, not everyone is going to agree that the right direction is chosen.

I think that it has become clear from the discussions within ASTM and within many other standards organizations in the United States, that there are certain basic facts that must be established. To use Jeffrey Fong's terms, there are certain questions that must be answered. Questions such as, what kind of credibility does ASTM have in the international arena? You can substitute other words for credibility here: What kind of recognition . . . what kind of clout does ASTM have? What we are after is some assessment of the ability of the organization to follow through with what it proposes to do. Perhaps credibility is the best word to choose, to get a sense of ASTM's goals. How much recognition does the organization have? Does it have name recognition in Europe, for example? Can it compete favorably with the other well-established standards writing organizations in Europe?

Yet another question, and a question that is a fundamental one is, what is the purpose of ASTM and who does it serve? You might say that answer is simple. The purpose of ASTM is to write voluntary standards. ASTM is a voluntary standards or concensus standards writing organization. I think that is generally well understood and well accepted. But when one begins to think about alternative activities in the international arena, one begins to expand upon that writing of standards theme and add other things. I think one has to focus sharply on what the purpose of the organization is.

You might ask questions such as, who are the customers of ASTM? How can you define the customers of ASTM? Who is ASTM here to serve? ASTM is a private sector organization in the United States. It is clearly not here to serve the U.S. Federal Government. From the discussions we have had at the ASTM Board of Directors, I think the general concensus is that the primary clientele, as represented by the majority of membership, is North American industry; not just U.S industry, but Canadian industry and Mexican industry as well. On the other hand, ASTM is also well represented by members from Europe and Asia, and certainly, they have interests that must be considered. So, I think the issue of, who does ASTM serve?, is not an easy one; and it is an issue that is going to require considerable discussion.

The last issue that I feel needs to be addressed is the question of what resources are available to implement any programs that are planned in the international arena? Here is an area where ASTM as an organization has great riches. There is a strong management team at ASTM; a good, strong headquarters staff located in Philadelphia. ASTM is a healthy organization financially. The annual budget for 1984 will be approximately $15 million. The sources of income for ASTM are strong and well established. The membership of ASTM is strong; nearly 30 000 members this year. So, it is an organization with great resources, which can be brought to bear on the international problems.

With these questions in mind, a thorough examination of the whole subject of ASTM in the international arena was initiated about 1980, begun with an international symposium, where the general issues were aired. Then a special committee was established. That special committee was chaired by a long-term ASTM member, Frank LaQue, and was charged with the responsibility of trying to determine what was really fact and what was just folklore as far as ASTM's reputation was concerned. The LaQue Committee carried out a number of studies. Some of those studies were carried out on contract for the Committee; other studies carried out by the ASTM staff.

There is one study that I would like to highlight for you, as an example. It is a study carried out by Keith Gorton from Humberside College in England, and the use of ASTM standards in Europe. This is called the Humberside Study and has been reported in great detail in the *Standardization News*, the ASTM magazine. If you have not read the reports in Standardization News, I urge you to do so. But now I will just touch on some of the basic facts that are involved.

The study involves three commodities (*a*) iron and steel, (*b*) plastics and resins, and (*c*) bulk chemicals. It involves four countries: Great Britain, West Germany, the Netherlands, and Sweden. It was carried out with a total of 202 so-called belly-to-belly interviews, direct interviews of people who would be involved in the use of ASTM standards. These are people who run industrial organizations that use standards; they are people who are involved in the purchasing and the procurement side of the standards business.

I am just going to summarize a couple of the key results because I want to make a point here. There is a great deal of detail on how the countries were selected; how the companies within those countries were selected. I also should make it clear that at the beginning it was understood that the interviewers were to address the questions in a general sense and to not divulge the fact that this was an ASTM sponsored investigation. We were interested here in getting as even-handed and balanced a view as possible.

I think there are two results among the many that probably indicate most clearly how ASTM is viewed. One of those came from the first question that was asked by the interviewers: what kinds of books of standards do you keep on your premises? The results are shown there: DIN, ASTM, BSI, and ISO standards are kept by the vast majority of the people interviewed. The other standards systems are used significantly less. It is clear just in terms of the standards kept on the shelf that the top four overshadow the others. If we explore further and we ask, which standards are used? ASTM Standards has been normalized at 100, and again, you will notice, that ASTM, BSI, DIN, and ISO win out. You can explore more detailed questions here and ask, why particular standards are being used? You can ask, who makes the choice of a standard? And I think you can reach some general conclusions.

First of all, concensus standards are a preferred means of communication in world markets. Essentially, every individual who was interviewed in this study made that point. Concensus standards are the common meeting ground in foreign trade. They form the basic understanding of any kind of contractual arrangement. A conclusion that came from the study that I think is perhaps most significant is that the customer is the one who usually specifies the standard to be used; not the seller, but the buyer. That means that if you are an exporter you must have available to you several different books of standards so that you can meet whatever terms your customers specify.

We have found in this study that customers specify ASTM standards as frequently as they do British Standards or DIN Standards in the European system, perhaps more frequently than ISO. You might ask, why more frequently than ISO? The usual answer to that question is that ASTM standards are more quickly updated; they are more quickly modified than ISO standards and they tend to be more readily available. The last point here, which is one of the major issues that we were exploring, is that ASTM is viewed as a major supplier of standards worldwide. The conclusion one has to reach is that ASTM is credible; that it does have clout in the standards writing arena; it is important to world trade.

The next questions to ask are, what can we do about that? What kind of program can we implement at ASTM which will provide an expanded role in the international arena?

There are several ways that we could go. One of the major operating functions of ASTM is to publish books of standards. Certainly, wider distribution of those standards documents would further goals in the international arena. ASTM is in the process of developing a technology training program, aimed at training individuals in the use of ASTM standards; not necessarily training ASTM members, but rather training individuals who use ASTM standards at the bench—technicians, draftsmen, procurement agents, and the like. Certainly, taking training programs of that kind into other countries would further the use of ASTM standards in those countries.

There is also the issue of laboratory accreditation. Accrediting laboratories so that those laboratories can certify products that will be sold internationally is another worthwhile goal. All three of those areas could have significant impact on the posture of ASTM internationally.

The remaining issues that are now being heavily debated at the Board are related to implementation. How should we go about dealing with each of the three areas mentioned previously? One can certainly take extremes. For example, if you view the results of the Humberside Study to mean that ASTM has a well recognized name and reputation in Europe, then one might assert that the organization has a responsibility to provide a certain level of service to our European clients and colleagues. This assertion would lead to a view of ASTM as a worldwide organization responsible for the standards needs of people in all countries. Logically one would then aim a distribution program at getting the widest circulation of ASTM documents possible.

On the other hand, one could take the opposite extreme, an extreme that Jeffrey Fong tells me is a protectionist view. You could assert that the purpose of ASTM is to write voluntary standards; the clientele of ASTM is North American industry and that any distribution of documents in the international market should be focused on those areas where North American industry wants to market. For example, if the plastics industry in the United States wants to sell plastic pipe in Argentina, it would be appropriate for ASTM to distribute standards documents related to ASTM Standards for plastic pipe as completely as possible in Argentina. One could expect those documents to contribute to specifications in requests for bids, allowing U.S. companies an advantage in sale in that market. That is a protectionist view and one that is certainly extreme.

I think you can see that there are arguments that can be made either way: the United Nations of International Voluntary Standards on one hand and the extreme protectionist view on the other. One can make the same kind of arguments with technology training and laboratory accreditation. These are the kinds of issues that are being debated by the ASTM Board of Directors at the present time.

It is clear that we are reaching decision points; we are looking at implementation of programs, but the final strategy, the final approach has not been decided upon. My guess, and I am giving a personal comment now, is that we will end up somewhere in the middle with some moderate position, which tries to take the balanced view that I think is appropriate for this organization.

Donald R. Johnson

Director of National Measurement Laboratory, National Bureau of Standards, Gaithersburg, Maryland; past member of the Board of Directors at ASTM.

EXPERIMENTATION AND MEASUREMENT

The title of this remark, "Experimentation and Measurement," was that of a book written by Dr. William J. Youden in 1962. That book was one of a series of Vistas of Science books published by the National Science Teachers Association for the purpose of the "improvement of the teaching of science." The original version has been out-of-print for many years. Recently, we at the National Bureau of Standards have reprinted a limited number of copies for use in our statistics-related workshops and courses.[1]

[1] *Experimentation and Measurement*, W. J. Youden, NBS Special Publication 672, National Bureau of Standards, Gaithersburg, MD, March 1984.

Here we see the standard deviation of each difference is σ, and each result is used three times. If by chance, none of the variables has an effect, we can calculate a grand average where the higher level and lower level are equally represented. This cannot be done with the first arrangement. Moreover, we see that there are exactly two points on each of the six surfaces of the cube—balanced and satisfying to the aesthetic minded.

This simple design is a one half replication of three factors each at two levels, and belongs to the class of fractional factorial designs. When the number of factors and levels increase, the pattern gets complicated, yet the general principal is the same as in this example—to get the maximum useful information by carefully selecting the experimental points.

One of the main problems in the design of experiments is what to do with the large number of variables or factors that may have a bearing on the results. Take, for example, the life test of saucer-type springs in clutches in a car, we may have

A. Shape: 3 levels
B. Hole ratio: 2 levels
C. Coining: 2 levels
D. Tension stress: 3 levels
E. Comp. stress: 3 levels
F. Shot peening: 3 levels
G. Outside perimeter planning: 3 levels

There are altogether $2^2 \times 3^4 = 324$ combinations. To find out an optimal combination requires a tremendous effort, even without replication.

To concentrate only on a selected few factors is like drilling 100 holes within a square mile area to find oil, rather than drilling 1 hole each in 100 square miles.

A solution to this type of problem has been advanced by Genichi Taguchi of Japan using what are called "Orthogonal Array" designs. These designs are a class of super fractional factorial which Taguchi has used successfully in Japan. In 27 experiments using three springs each he pin pointed that A_3 C_2E_1 D_2 F_2 and A_3 C_2E_1 D_3 F_2 an optimal conditions, and give confidence limits to the estimates of average life.

I must admit that I know little about Taguchi's work since most of his publications are in Japanese. Taguchi teaches his design and analysis to engineers and uses signal to noise ratios as response. According to him, a 20 000 employee company, Nippon Denso, used the orthogonal array design 2700 times in 1976. You cannot argue with success!

Last May, a Conference on "Frontiers of Industrial Experimentation" was held in New York, sponsored by Bell Labs Quality Assurance Group. Taguchi was invited, and a number of papers were presented on his methods. A most readable reprint was by M. S. Phadke et al.[2] of Quality Assurance Center, Bell Labs, Holmdel, New Jersey.

How good is Taguchi's method? I cannot give you a definitive answer. Obviously some information in the full experimentation has been neglected. I like to make three points though. First, if one cannot possibly do the full experimentation, it would be better to follow a systematic approach like Taguchi's than select a subset based on instinct. Second, is the lost information really important for your purpose? If not, nothing is lost. Thirdly, the result of such an "exploratory" experiment can be always verified by confirmatory experiments.

[2] M. S. Phadke, R. N. Kackar, D. V. Speeney, and M. J. Grieco, "Off-Line Quality Control in Integrated Circuit Fabrication Using Experimental Design," *The Bell System Technical Journal*, Vol. 62, No. 5, May-June, 1983.

It appears to me that the design advocated by Taguchi can be of use to experimenters in fatigue test to deal with the larger number of factors and variables. It is worth looking into and trying out. I myself would like to learn more about it and study its properties.

H. H. Ku
Chief, Statistical Engineering Division, National Bureau of Standards, Gaithersburg, Maryland.

Fatigue of Metal Single Crystals

N. Y. Jin[1] *and A. T. Winter*[1]

The Role of Cross-Slip of Screw Dislocations in Fatigue Behavior of Copper Single Crystals

REFERENCE: Jin, N. Y. and Winter, A. T., **"The Role of Cross-Slip of Screw Dislocations in Fatigue Behavior of Copper Single Crystals,"** *Basic Questions in Fatigue: Volume I, ASTM STP 924,* J. T. Fong and R. J. Fields, Eds., American Society for Testing and Materials, Philadelphia, 1988, pp. 17–25.

ABSTRACT: Differently oriented copper single crystals were fatigued, where the Schmid factor for primary screw dislocations moving on the cross-slip plane differed markedly while that for primary slip remained more or less constant. The experimental results show that the cyclic hardening rate depends on crystal orientation. The saturation properties, the saturation stress, and the plastic strain in the persistent slipbands, are orientation independent. It seems likely that cross-slip is important only in the early stage of cyclic deformation, but not in the saturation.

KEY WORDS: cyclic stress-strain response, copper single crystals, crystal orientation, persistent slip bands, cross-slip, dislocation microstructure

Recently, in the field of metal fatigue, much work has been devoted to studying cyclic deformation and its microscopic mechanisms. This is not only because new methods of designing against fatigue failure have been developed on the basis of cyclic deformation, but also because cracks have been found to nucleate preferentially in the persistent slipbands (PSBs). Similar cyclic behavior, in particular the formation of PSBs and their characteristic dislocation ladder structure [*1*], has been observed in various materials of diverse crystal structures [*2–7*]. This suggests that the dislocation phenomenon represented in the PSBs is of universal significance. It is hoped that, in the long run, the study of cyclic deformation and its mechanisms will elucidate some fundamental problems in fatigue fracture.

It is generally accepted that cyclic hardening is due mainly to the accumulation of edge dislocation dipoles laid down by irreversible motion of screw dislocations via cross-slip. In the saturation stage, a dynamic equilibrium between multiplication and annihilation occurs [*8–10*]. According to some authors, such as Brown [*9*], the saturation stress may be viewed as the stress at which screw dislocation dipoles become unstable and annihilate by mutual cross-slip. Therefore, in cases favoring cross-slip we would expect a high initial hardening rate, lower saturation stress, and smaller plastic strain amplitude in the PSBs.

One great attraction of assigning an important role to cross-slip is that being a thermally activated process, it might be able to account for the very strong temperature dependence

[1] Cavendish Laboratory, Madingley Road, Cambridge CB3 OHE, England. The present address of Dr. Jin is: Materials Science and Engineering Department, Shanghai Jiao Tong University, Shanghai, China.

TABLE 1—*Crystallographic characteristics of the specimens used.*

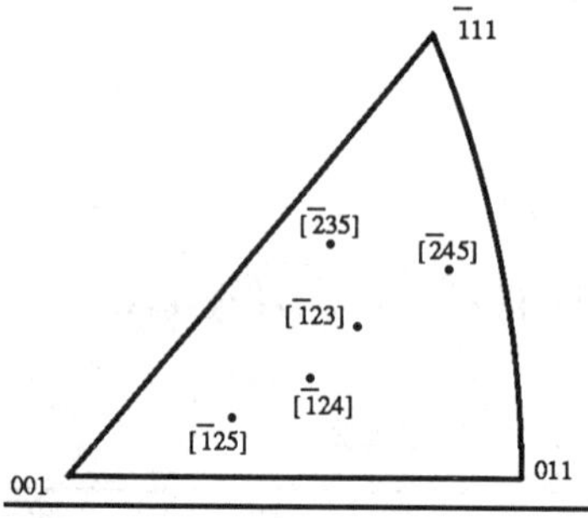

Tensile Axis	Schmid Factors: Primary Slip	Schmid Factors: Cross Slip	Ratio of Schmid Factors on the Two Most Highly Stressed Systems Q
$[\bar{1}23]$	0.467	0	0.749
$[\bar{1}25]$	0.490	+0.163	0.888
$[\bar{2}45]$	0.445	−0.064	0.856

of fatigue properties. For copper, a higher temperature (that is, easier cross-slip) leads to a higher initial hardening rate and lower saturation stress [*11,12*].

In an attempt to throw more light on this subject, differently oriented copper single crystals were fatigued, where the Schmid factors for primary screw dislocations moving on the cross-slip plane, λ_c, differed markedly while those for primary slip, λ_p, remained more or less constant. Such investigation was expected to provide data for clarifying dislocation mechanisms in cyclic deformation, since the ease of cross-slip should be different in these crystals.

Experiments and Results

The crystallographic characteristics of $[\bar{1}23]$, $[\bar{1}25]$, and $[\bar{2}45]$ crystals employed in this work are listed in Table 1, with a figure of the poles of their tensile axes in the standard triangle. The crystals were grown from high-purity (0.9999) copper by a Bridgman technique, with reduced section of length 18 mm and cross section 5 mm by 7 mm. The orientations were checked by X-ray Laue diffraction to be within ±1 deg of the required orientations.

All crystals were fatigued at constant plastic strain amplitudes within the range 7×10^{-4} to 3.5×10^{-3} with zero mean strain. The strain was measured by a clip-on extensometer and was converted to the plastic strain using a control circuit. Consistent test conditions, in particular identical strain rate ($1.2 \times 10^{-2}\,s^{-1}$) in all tests, promised reliability of comparisons

TABLE 2—*Experimental parameters and results.*

Tensile Axis	Specimen No.	Plastic Shear Strain Amplitude, $e_p/10^{-3}$	Saturation Shear Stress Amplitude, σ_s/MPa (±0.05)	Initial Hardening Rate, θ/MPa
$[\bar{1}23]$	J127	0.78	31.3	9.1
	J130	1.45	31.2	12.5
	J128	1.95	30.9	15.0
	J129	3.02	29.7	25.9
$[\bar{1}25]$	J313	0.80	32.2	9.1
	J319	1.35	32.3	13.2
	J318	1.93	31.4	21.6
	J312	3.15	28.4	30.6
$[\bar{2}45]$	J513	0.87	32.6	9.1
	J512	1.45	32.4	11.7
	J507	1.97	32.1	13.1
	J510	3.00	32.2	20.5

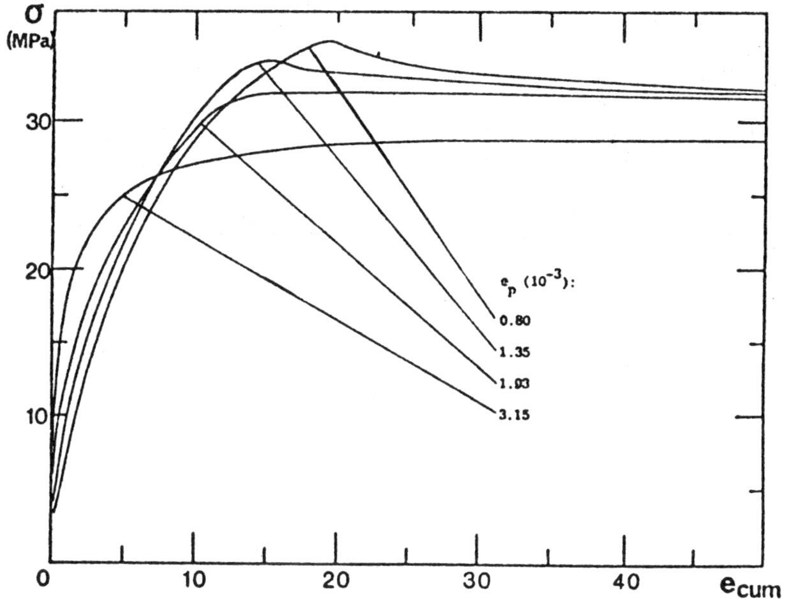

FIG. 1—*Cyclic hardening curves of [$\bar{1}$25] crystals fatigued at various plastic strain amplitudes.*

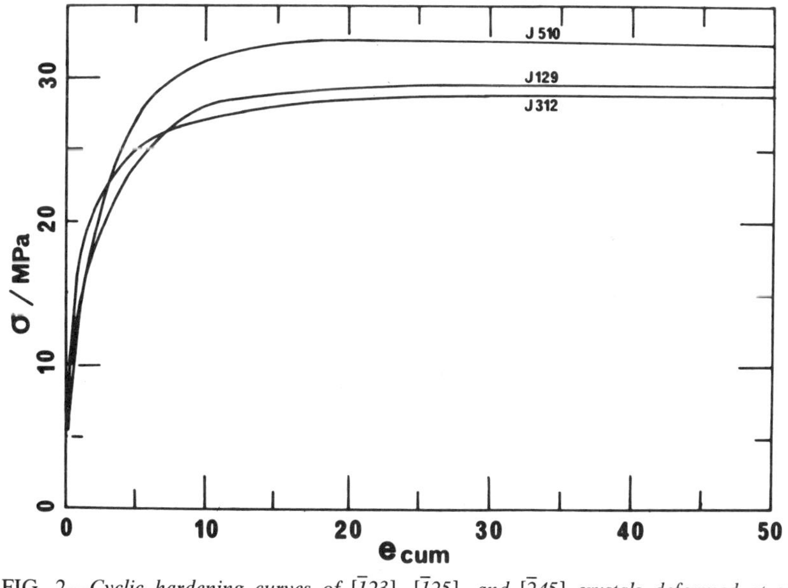

FIG. 2—*Cyclic hardening curves of [$\bar{1}$23], [$\bar{1}$25], and [$\bar{2}$45] crystals deformed at* $e_p \sim 3 \times 10^{-3}$. *J129: [$\bar{1}$23] crystal; J312: [$\bar{1}$25] crystal; J510: [$\bar{2}$45] crystal.*

between crystals. Tests were stopped at a cumulative strain $e_{cum} \sim 200$ ($e_{cum} = 4Ne_p$, where N is the cycle number and e_p the plastic strain amplitude). At the end of each test, the measurement of the fractional area fraction occupied by the PSBs was carried out at a magnification of about ×130. A separation of 2 μm between micro-PSBs or micro-PSBs of a width of 2 μm were easily measured at this magnification. Etching and transmission electron microscopy (TEM) were employed for examining the dislocation structures.

The main experimental parameters and results are summarized in Table 2. At all plastic strain amplitudes the peak stresses in compression were greater than in tension by a few percent; the saturation stresses listed here were taken as the averages of tension and compression. The initial hardening rate was defined as the value of $\theta = d\sigma/de_{cum}$ at $e_{cum} = 0.1$.

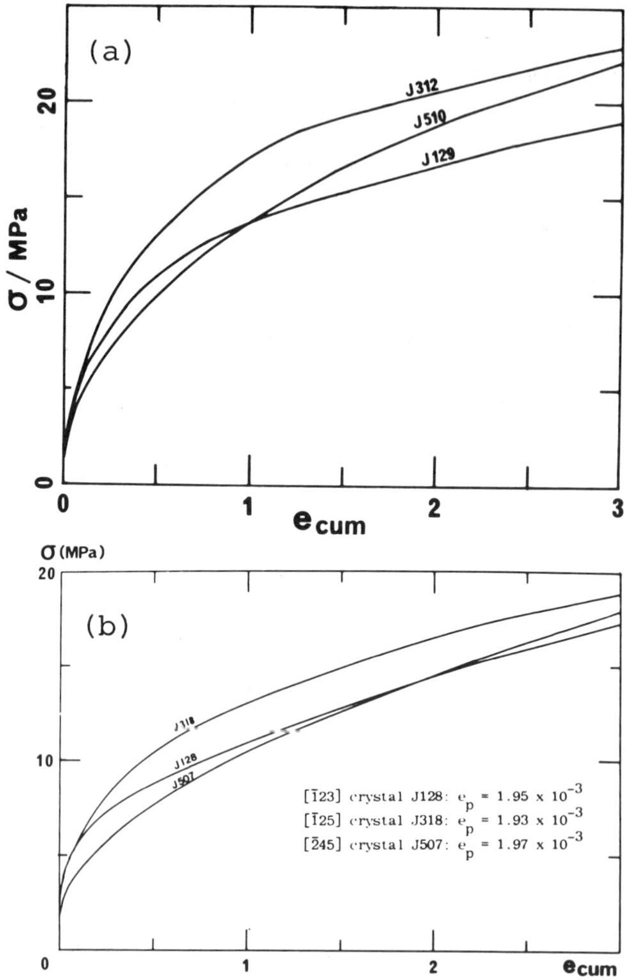

FIG. 3—*Beginning parts of cyclic hardening curves, showing orientation effect of the initial hardening rate. Note the effect is more pronounced in* (a) $e_p \sim 3 \times 10^{-3}$ *than in* (b) $e_p \sim 2 \times 10^{-3}$.

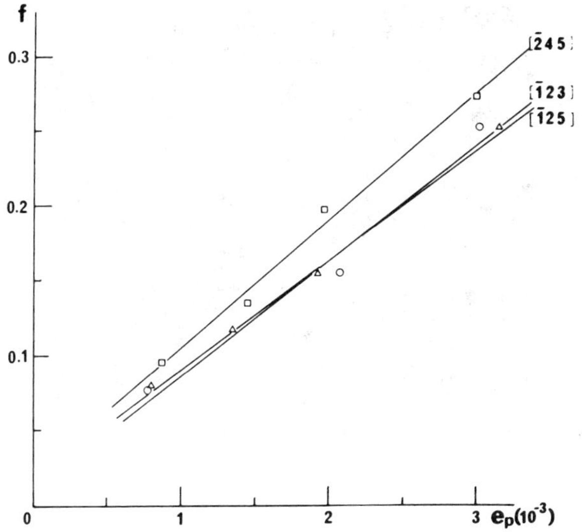

FIG. 4—*Volume fractions of PSBs in* [$\bar{1}$23], [$\bar{1}$25], *and* [$\bar{2}$45] *crystals.* ○ [$\bar{1}$23] *crystals;* △ [$\bar{1}$25] *crystals;* □ [$\bar{2}$45] *crystals.*

Cyclic Hardening

All crystals showed rapid hardening at the initial stages followed by saturation. For $e_p \leq 2 \times 10^{-3}$ the stress amplitude approached from above after overshooting. As examples, cyclic hardening curves of [$\bar{1}$25] crystals tested at various strain amplitudes are shown in Fig. 1. The overshooting phenomenon was revealed to be most prominent in [$\bar{2}$45] crystals. As always observed, crystals hardened faster at higher strain amplitude, that is, θ was an increasing function of e_p (Table 2).

Cyclic hardening curves for $e_p = 3 \times 10^{-3}$ of crystals with various orientations are presented in Fig. 2 and their values of θ are given in Table 2. Orientation can be seen to have significant effects on early hardening. At a given e_p, the initial hardening rate was highest in a [$\bar{1}$25] crystal and lowest in a [$\bar{2}$45] crystal. Such effects held for all e_p, but were more pronounced at higher e_p (Fig. 3). For example, $\theta_{[\bar{1}25]}/\theta_{[\bar{1}23]}$ was about 1.2 at $e_p = 3 \times 10^{-3}$, but only 1.06 at $e_p = 1.5 \times 10^{-3}$.

The cumulative strain for establishing a saturation state varied in different e_p tests, being higher for lower e_p, as can be seen in Fig. 1. However, crystals always saturated at similar cumulative strain at a given e_p with no significant difference from one orientation to another.

Saturation Stress

Saturation stress in these three types of crystals, as listed in Table 2, did not scatter widely; it was approximately 30 MPa. The higher value compared with the results from other authors is attributed to the higher strain rate employed in the present work. Although some [$\bar{1}$25] crystals showed slightly higher values, this is not statistically significant. The [$\bar{2}$45] specimens achieved slightly higher stresses, and the variation between crystals was within 5%. This suggests that, unlike initial hardening, the saturation stress is very weakly (if at all) affected by crystal orientation, as long as the orientation remains for single slip. This agrees with the experimental observations by Cheng and Laird [*13*].

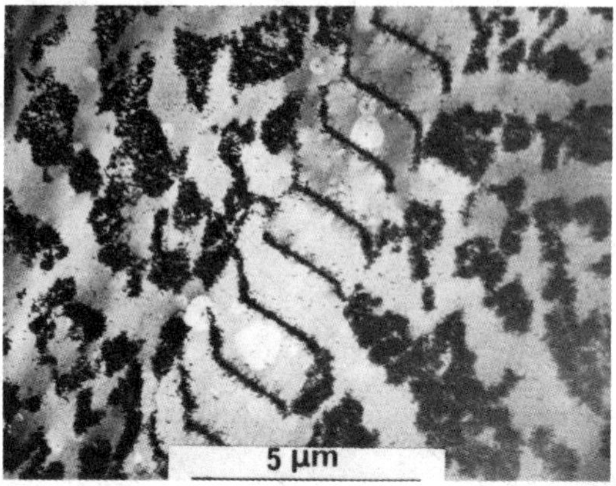

FIG. 5—*A TEM micrograph of a saturated* [$\bar{1}25$] *crystal, showing the typical PSB ladder structure and the matrix vein structure in crystals oriented for single slip. The foil plane is* ($1\bar{1}1$).

Plastic Strain Amplitude in the PSBs

According to the two-phase model, the plastic strain amplitude localized in the PSBs, e_b, can be deduced as the reciprocal of the slope of a f-e_p line, where f is the volume fraction of the PSBs obtained from the optical observation on the gage surfaces of a saturated crystal. The results are presented in Fig. 4. There seems to be no significant difference between the

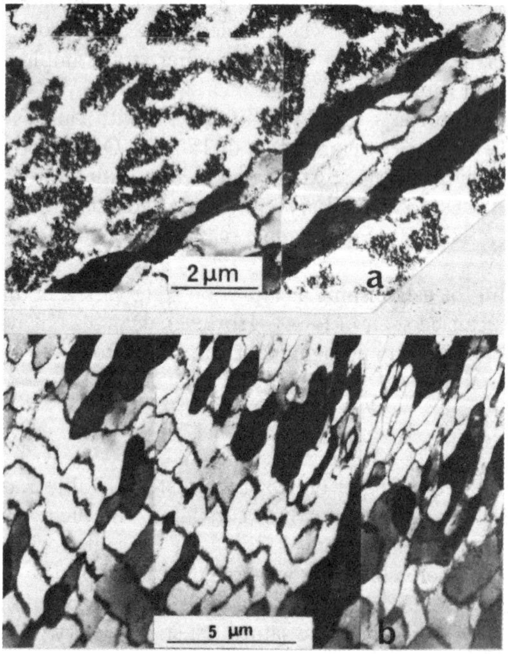

FIG. 6—*Dislocation structures in the saturated* [$\bar{2}45$] *and* [$\bar{1}22$] *crystals. Dislocations with all three Burgers vectors on the* (*111*) *plane formed the misoriented cells.* (a) [$\bar{2}45$] *crystal, the foil plane is* ($1\bar{1}1$); (b) [$\bar{1}22$) *crystal, the foil plane is* ($1\bar{1}3$).

three orientations; all data appear to scatter within experimental errors. Although, using regression, one may obtain f-e_p lines with slightly different slopes, as also drawn in the figure, the orientation effect on the plastic strain in the PSBs in respect of λ_c is, however, not large.

Dislocation Microstructures

The results from saturated [$\bar{1}$23] and [$\bar{1}$25] crystals were in accordance with earlier publications, namely, the typical PSB wall or ladder structure and the matrix vein structure [*1,2,4*] (Fig. 5). The primary dislocations were overwhelmingly predominant, and the density of secondary dislocations was several orders or magnitude lower than the primaries.

In saturated [$\bar{2}$45] crystals, however, there existed dense [$\bar{1}$10] dislocations, indicating the activity of secondary slip during cyclic deformation. In some matrix regions, veins consisted

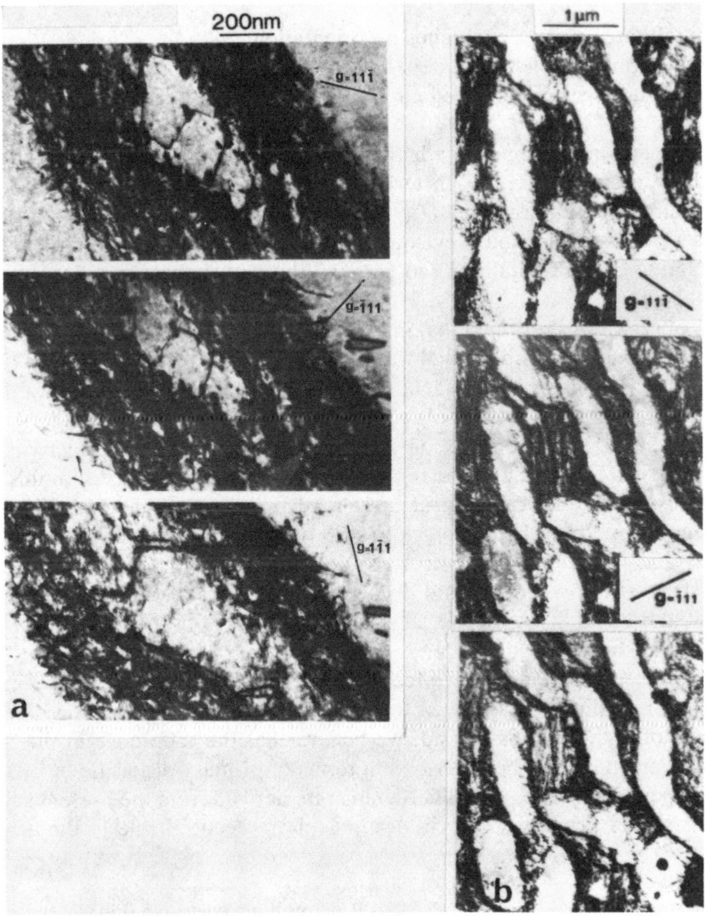

FIG. 7—*Comparison between the dislocation structures of* [$\bar{2}$45] *crystals fatigued into different stages at the same* e_p *of* 3×10^{-3}. *The foil planes are* (*111*). (a) *The cyclic hardening stage,* $e_{cum} \sim 10$. *Secondary* [$\bar{1}$*10*] *dislocations, which are in residual contrast at* g = [*11*$\bar{1}$], *coexist with primary* [$\bar{1}$*01*] *dislocations, which are in residual contrast at* g = [*1*$\bar{1}$*1*], *in the veins.* (b) *The saturation stage,* $e_{cum} \sim 210$. *Veins consist of* [$\bar{1}$*01*] *dislocations in their centers and* [$\bar{1}$*10*] *dislocations on their edges.*

of $[\bar{1}01]$ dislocations in their centers and $[\bar{1}10]$ dislocations on their edges. The PSBs contained both $[\bar{1}01]$ dense dislocation walls and loosely packed dislocations with all three Burgers vectors on the (111) plane, and these constructed misoriented cells (Fig. 6), similar to what was observed in a $[\bar{1}22]$ crystal for double slip [*14*].

Dislocation configurations in the $[\bar{2}45]$ crystal deformed into the stage when the PSBs emerged on the surfaces have also been examined. Secondary $[\bar{1}10]$ dislocations were found to distribute in the veins together with $[\bar{1}01]$ primary dislocations (Fig. 7). This implies the activity of secondary slip in the hardening stage and the redistribution of the different types of dislocations before saturation.

Discussion

The present work shows, for the cyclic deformation of copper single crystals oriented for single slip, that

1. The initial hardening rate depends on orientation.
2. The saturation stress does not depend on orientation.
3. The local plastic strain amplitude in the PSBs does not depend on orientation.

Points 1 and 2 are in agreement with the experimental work of Cheng and Laird [*13*]. However, their interpretation was in terms of secondary slip during the hardening stage. Cheng and Laird used a parameter Q (<1), the ratio of Schmid factors on the two most highly stressed slip systems, and they concluded that the rate of rapid hardening increased as Q rose. None of our orientations had $Q > 0.9$, but yet showed significant differences of hardening behavior.

Our TEM observations revealed that in $[\bar{1}23]$ crystals ($Q \sim 75\%$) and $[\bar{1}25]$ crystals ($Q \sim 89\%$), almost all dislocations belonged to the primary system. In contrary, in $[\bar{2}45]$ crystals ($Q \sim 86\%$), dislocations of vector $[\bar{1}10]$ were found with a density comparable to that of $[\bar{1}01]$. This suggests that the activity of secondary slip is not simply related to the parameter Q. The driving stress on the dislocations in a slip system other than primary is the sum of several components, for example, the resolved part of the applied stress on this system, the long-range stress of the dislocation structure already formed at that stage of hardening, and the short-range stress from their interaction with other dislocations or obstacles or both. In fact, in $[\bar{1}22]$ and $[\bar{1}12]$ double slip crystals, Q-values were the same as 100%; however, their cyclic stress-strain responses and dislocation structures were very different. Interestingly, the initial hardening rate of a $[\bar{1}22]$ crystal was much lower than that of a $[\bar{1}23]$ crystal, and its saturation stress was higher [*14*].

Considering these observations, we feel that the behavior of $[\bar{2}45]$ crystals may be complicated by multiple slip, perhaps even from the earliest stage of cyclic hardening. Having left the results of $[\bar{2}45]$ crystals behind, we believe that the difference in the behaviors of $[\bar{1}23]$ and $[\bar{1}25]$ crystals must be described in terms of primary dislocations.[2]

Strong λ_c dependence of the initial hardening rate derived from the present study suggests that the cross-slip of primary screw dislocations plays a central role in the early stages of

[2] We later tested $[\bar{1}24]$ and $[\bar{2}35]$ crystals; λ_c is 0.097 in the former and 0 in the latter. The results show that their saturation stresses are very similar to those of $[\bar{1}23]$ and $[\bar{1}25]$ crystals, and that at the same e_p the initial hardening rate of a $[\bar{1}24]$ crystal is the intermediate of those of $[\bar{1}23]$ and $[\bar{1}25]$ crystals, while that of a $[\bar{2}35]$ crystal is the same as that of the $[\bar{1}23]$ crystal. These fit well with the results reported in this paper, and hence support the hypothesis of different control mechanisms for cyclic hardening and saturation.

cyclic hardening. This is coincident with the current theories of cyclic hardening. At low e_p, the glide of dislocations on the slip plane is sufficient to carry the applied strain; therefore the cross-slip occurs much less frequently than in high-e_p cases. This leads to a slow accumulation of edge dipoles and hence a slow initial hardening rate. Obviously, the effect of λ_c on hardening, caused by the irreversible cross-slip of screw dislocations, is also diminished, as was observed in the present work.

The observation of orientation independence of saturation properties is not easy to reconcile with the theories of saturation based on cross-slip. This distinction between mechanisms in saturation and in the early stages of hardening was also found in Cu-5Al [*4*]. There, the initial hardening rate in the alloy was much lower than that in pure copper. This is consistent with the idea that the reduced stacking fault energy in the alloy makes cross-slip more difficult. However, the saturation stress of copper-aluminum alloy was not much different from that in pure copper. This agrees with our result that cross-slip is less important in saturation than in the early stages.

It seems likely to us that the cross-slip of screw dislocations plays a central role in the early stages of cyclic hardening but not in the saturation. However, in order to confirm the results, an investigation on the behavior of crystals with $\lambda_c = 0$, but with orientation other than $[\bar{1}23]$, is necessary. We are working on this (see footnote 2). We also hope to develop the modeling of cross-slip to give a stricter comparison with experiments. The insignificance of λ_c on the saturation properties might, however, be explained in other ways, such as if λ_c does not really represent the ease of cross-slip. For example, the effect of crystal orientation might diminish if the cross-slip occurred along a non-crystallographic path. The hypothesis of different control mechanisms for cyclic hardening and saturation could be tested with other experiments, for example, cyclic deformation of alloys, in which differing stacking fault energies would affect cross-slip. At present, the cyclic deformation data of various face-centered cubic materials give support to such an hypothesis.

Acknowledgment

The authors thank Dr. L. M. Brown, FRS, and Dr. M. K. Surappa for useful discussions. N. Y. Jin is grateful to the Chinese Education Ministry for the research studentship.

References

[*1*] Winter, A. T., *Philosophical Magazine,* Vol. 30, 1974, p. 719.
[*2*] Mughrabi, H., Achermann, F., and Herz, K., *Fatigue Mechanism, ASTM STP 675*, American Society for Testing and Materials, Philadelphia, 1979.
[*3*] Kwadjo, R. and Brown, L. M., *Acta Metallurgica,* Vol. 26, 1978, p. 1117.
[*4*] Woods, P. J., *Philosophical Magazine,* Vol. 28, 1973, p. 155.
[*5*] Wilhelm, M. and Everwin, P., *Proceedings,* 5th Conference on the Strength of Metals and Alloys, Vol. 2, 1979, p. 1089.
[*6*] Mecke, K., *Crystal Research Technology,* Vol. 17, 1982, p. 1557.
[*7*] Pohl, K., Mayr, P., and Macherauch, E., *Scripta Metallurgica,* Vol. 14, 1981, p. 1167.
[*8*] Mughrabi, H., *Materials Science and Engineering,* Vol. 33, 1978, p. 207.
[*9*] Brown, L. M. in *Dislocation Modelling of Physical Systems,* M. F. Ashby et al., Eds., Gainesville, FL, 1980, p. 51.
[*10*] Kuhlmann-Wilsdorf, D. and Laird, C., *Materials Science and Engineering,* Vol. 46, 1980, p. 209.
[*11*] Basinski, Z. S., Korbel, A. S., and Basinski, S. J., *Acta Metallurgica,* Vol. 28, 1980, p. 191.
[*12*] Hancock, J. R. and Grosskreutz, J. C., *Acta Metallurgica,* Vol. 17, 1969, p. 77.
[*13*] Cheng, A. S. and Laird, C., *Materials Science and Engineering,* Vol. 51, 1981, p. 111.
[*14*] Jin, N. Y. and Winter, A. T., *Acta Metallurgica,* Vol. 32, 1984, p. 989.

Axel Hunsche[1] and Peter Neumann[2]

Crack Nucleation in Persistent Slipbands

REFERENCE: Hunsche, A. and Neumann, P., **"Crack Nucleation in Persistent Slipbands,"** *Basic Questions in Fatigue: Volume I, ASTM STP 924,* J. T. Fong and R. J. Fields, Eds., American Society for Testing and Materials, Philadelphia, 1988, pp. 26–38.

ABSTRACT: The development of persistent slipband topography, crack nucleation, and crack growth in fatigued copper single crystals was studied by an new section technique, which reveals surface topographies with a resolution of 20 nm. Experiments were performed in air, oxygen, hydrogen, water vapor, high vacuum, and ultrahigh vacuum in order to study the influence of the environment on these processes. The development of persistent slipband topography, including intrusions, extrusions, and the protrusion of the whole persistent slipband, does not depend on environment. Crack nucleation and early propagation, however, do strongly depend on the environment. Intrusions can also be distinguished from crack nuclei by their finite vertex angle of approximately 30 deg. Quantitative data of the average protrusion growth as well as crack growth are presented. Chains of large voids were found in high vacuum and ultrahigh vacuum along the trace of the primary slip plane and are interpreted as remnants of rewelded Stage I cracks.

KEY WORDS: copper, single crystal, fatigue, fatigue crack nucleation, fatigue crack propagation, intrusion, extrusion, persistent slipband, environment, sectioning technique

Two fundamental questions form the basis of this paper:

1. Is crack nucleation in persistent slipbands identical to surface roughening due to slip or is it an independent process?
2. How does the environment influence surface roughening, crack nucleation, and the growth of small cracks?

In materials which have no extended flaws, crack initiation occurs by slip localization in so-called persistent slipbands (PSBs) [*1*]. The mechanism of this crack nucleation is not yet understood quantitatively. It has been assumed since the work of May [*2*] that repeated compressive and tensile slip produces a random roughening of the surface. With increasing cumulative strains the valleys between the peaks grow deeper statistically. The deepest valleys may eventually be called cracks when they have reached such a depth that—due to the change in geometry—the strains concentrate at the bottoms of these grooves, thus producing an accelerated growth.

On the other hand it is well known that the fatigue life is usually extended by one order of magnitude if the test is performed in high vacuum [*3–6*]. There are at least two ways in which this result can come about: First, the environment changes the mechanical properties of the surface, thus producing a different surface roughening, and second, the environment influences only the process of crack nucleation and early crack growth. These questions could not be answered reliably because the rough surface structure of the slipbands hinders

[1] Research assistant and director, Max-Planck-Institut für Eisenforschung, 4006 Düsseldorf, F.R.G. Present address Brankamp System, 4006 Erkrath, F.R.G.
[2] Director at the Max-Planck-Institut für Eisenforschung, 4006 Düsseldorf, F.R.G.

reliable observation of crack nuclei from outside the specimen. Therefore crack nucleation can be followed only by means of destructive techniques such as by sectioning the specimen.

Before the scanning electron micrograph (SEM) was available, attempts were made to resolve the fine details of the surface topography with the help of optical microscopy plus a purely geometrical enlargement of the surface profile by using small taper sectioning angles (about 10 deg) [*7*]. In this way qualitative information about the change of the surface topography during fatigue was obtained. The low taper sectioning angle, however, can easily produce artifacts in the observed profiles; for example, surface-connected pits may be reproduced as isolated holes.

Thus a sectioning technique involving the so-called micromilling was developed. This method, which is described elsewhere [*8*], allows the preparation of cross sections perpendicular to the specimen surface without measurable distortion of the resulting edge [*9*]. Similar results were obtained by Basinski and Basinski [*10,11*] by conventional polishing techniques. In this way the topography of PSBs can be measured in the SEM with a resolution of about 20 nm. Thus crack nucleation and the early stages of crack propagation can be observed. In order to determine the effect of the environment on these processes, fatigue tests were performed with copper single crystals under different environments.

Experimental Details

The crystals were grown by the Bridgman technique from 99.99% oxygen-free, high-conductivity (OFHC) copper. A gage length with a square 4 by 4 mm^2 cross section was machined into the specimen by spark erosion and electropolishing. All specimens were oriented for single slip. One pair of the side surfaces contained the primary Burgers vector.

All crystals were cyclically hardened very gradually under stress amplitude control up to the saturation stress of 32 MPa. During the hardening, the rate of shear stress amplitude increase was controlled to be constant, 1 kPa/cycle, and the frequency used was 40 Hz.

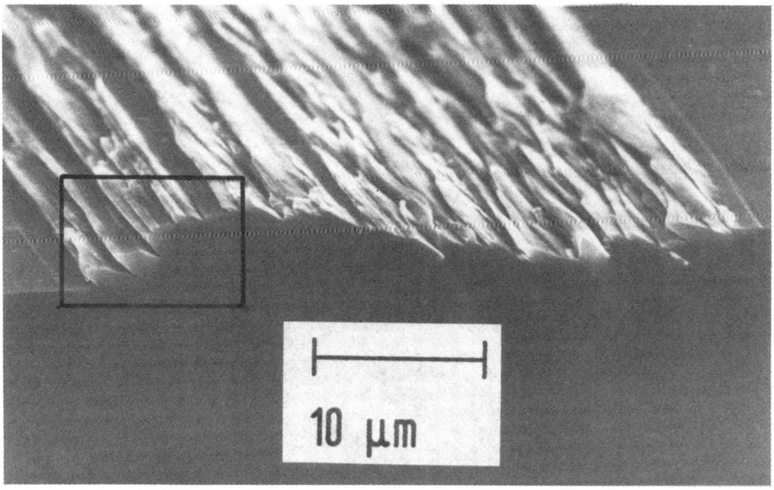

FIG. 1—*90-deg section (no taper magnification) through a persistent slipband after 3 × 10^4 cycles (with γ_{pl} = 2 × 10^{-3}, 20 Hz) at saturation stress (32 MPa). Final failure occurs after 0.9 × 10^5 more cycles. The marked area is shown in Fig. 2. At both sides of the PSB the flat surface of the matrix area is visible. Thus the protrusion of the PSB is obvious.*

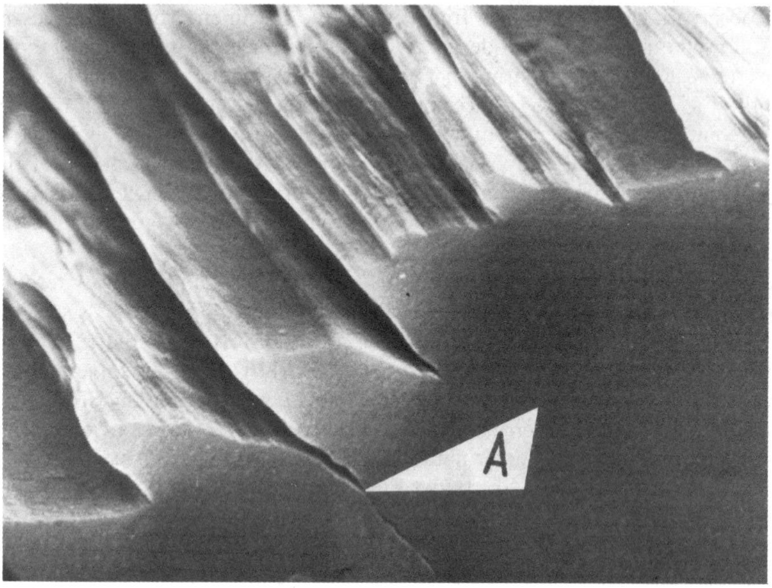

FIG. 2—*Magnified view of the marked area of Fig. 1, demonstrating the edge resolution of the sectioning technique (20 nm). "A" points at the vertex of an intrusion, at which a crack nucleus has formed.*

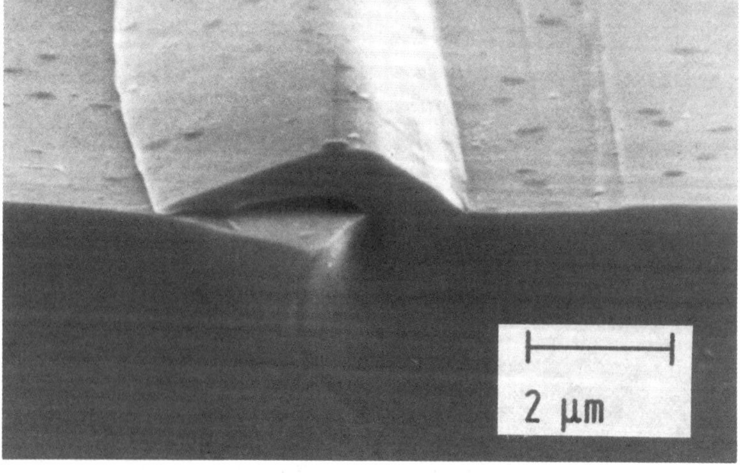

FIG. 3—*90-deg section (no taper magnification) through an isolated extrusion after* 6×10^4 *cycles (with* $\gamma_{pl} = 2 \times 10^{-3}$*, 20 Hz) at saturation stress (32 MPa). The "balcony" underneath the extrusion is a polishing artefact due to this peculiar geometry.*

After the hardening, at the saturation stress, the plastic shear strain amplitude was as small as 8×10^{-5} and no PSBs had formed. Subsequent to the hardening, the specimens were subjected to a constant plastic shear strain fatigue test at a frequency of 20 Hz. The plastic shear strain amplitude had values ranging from 2×10^{-4} to 2×10^{-3}. At $\gamma_{pl} = 2 \times 10^{-3}$, PSBs formed within less then 1000 cycles.

Due to the identical hardening procedure the PSBs always formed in the same matrix structure and at the same stress amplitude—independent of the value of the constant plastic shear strain amplitude chosen after hardening. Consequently it is very likely that all PSBs have the same local plastic shear strain amplitude. This is concluded from the identical behavior of individual PSBs and the very slight dependence of life on the applied plastic shear strain amplitude [*8,9*], but was not verified by local measurements at individual PSBs. The situation is different in tests with a constant plastic shear strain amplitude from the very beginning. Then PSBs form already during hardening, that is, before saturation. Cheng and Laird [*12*] have shown by local strain measurements that there are large differences in the local strains in individual PSBs. Correspondingly, the life depends on the applied plastic shear strain amplitude. We feel, however, that experiments with specimens with nearly identical PSBs can be interpreted more readily and thus should be preferred for the study of crack nucleation. From the applied plastic shear strain amplitude and the volume fraction of the PSBs, the localized plastic shear strain in the PSB, $\gamma_{pl}^{PSB} = 0.01$, was obtained, consistent with the results of many authors [*13–16*].

All SEMs in the following show views from a direction 10 deg off the normal of the sectioning plane. Thus the sectioning plane and the surface profile are shown with a negligible perspective contraction by a factor of cos 10 deg = 0.98, whereas the original specimen surface is contracted in the *y*-direction by the factor of cos 80 deg = 0.17. This is in contrast to the so-called "taper sections," which exaggerate the profile by a factor of 10 to 20. During the work with the SEM, the edge of the sections was observed from various directions, thus removing any doubt in interpreting the topography. Further details of the experimental procedure and the sectioning technique can be found in Refs *8* and *9*.

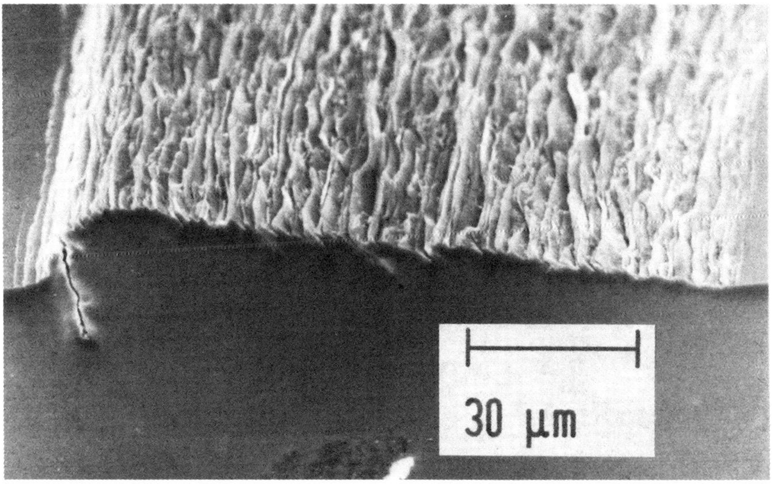

FIG. 4—*90-deg section (no taper magnification) through a persistent slipband after* 6×10^4 *cycles (with* $\gamma_{pl} = 2 \times 10^{-3}$, *20 Hz) at saturation stress (32 MPa). A Stage II crack develops at the edge of the persistent slipband. At both sides of the PSB the flat surface of the matrix area is visible. Thus the protrusion of the PSB is obvious.*

FIG. 5—*90-deg section (no taper magnification) through a persistent slipband after 10^7 cycles of $\gamma_{pl} = 2 \times 10^{-3}$ at 20 Hz in ultrahigh vacuum.*

Results and Discussion

Geometrical Difference Between Intrusions and Crack Nuclei

Since the sectioning technique is destructive, the development of individual PSBs cannot be followed. If a large number of sections are studied, however, and the frequency of the features of interest is evaluated, statistically relevant information can be obtained. In this way the evolution of PSBs was studied in [*8,9*] and the main results are quoted here as far as is necessary for the purpose of this paper.

The extrusions form with a rate of roughly 1 nm/cycle up to a height of about 3 μm. The profile of the extrusions within the PSBs is approximately triangular with a base width of 2 μm and a height of 3 μm. After such an extrusion has formed the flanks do not change much and stay rather smooth. Within a PSB the extrusions are closely proximate and thus

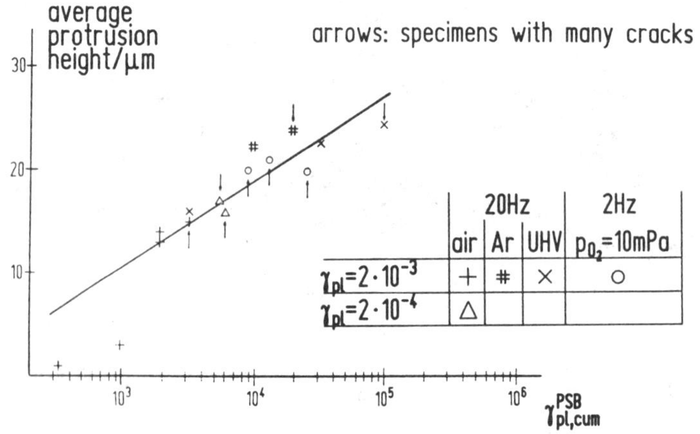

FIG. 6—*PSB protrusion height averaged over all PSBs in the specimen as a function of the accumulated plastic shear strain in the PSB for various environments.*

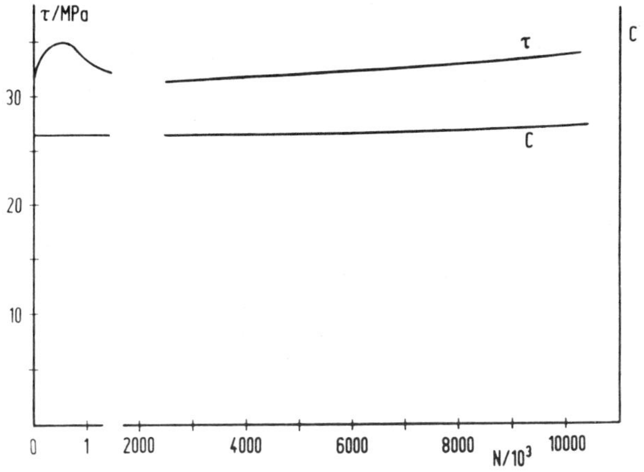

FIG. 7—*Secondary hardening and increase of the compliance of the specimen at large numbers of cycles* ($\gamma_{pl} = 2 \times 10^{-3}$, *20 Hz, vacuum*).

form valleys between each other. These valleys are usually called intrusions. Figure 1 shows these details in a section through a typical PSB after 3×10^4 cycles (at an accumulated plastic shear strain in the PSBs $\gamma_{cum,pl}^{PSB} = 1200$). Figure 2 shows the left-hand side of Fig. 1 at a higher magnification. This micrograph demonstrates the sharpness of the section edges, that is, the resolution of the technique. Occasionally, very narrow PSBs are observed. They may consist of only one single extrusion which frequently is bent over like the example of Fig. 3. It is obvious that observing the free surface is insufficient to reveal the real topography.

The sawtooth-like profile of the PSBs is quite regular and has a dominant wavelength of 2 μm, the base width of the extrusions. This profile is not statistical at all and thus cannot be explained by a statistical egress of dislocations [*17*]. Superimposed on this sawtooth-like profile there is a protrusion of the PSBs as a whole; the vertices of the intrusions are above the level of the surrounding matrix. Figure 4 shows a typical example. More details about the protrusion development are given in Ref *8*.

The intrusions always have a finite vertex angle on the order of 30 deg. Crack nuclei may form at the vertices of these intrusions as shown at A in Fig. 2. The intrusions near the interface between PSBs and the matrix are preferred sites for crack nucleation [*8,9*]. These crack nuclei are truly closed cracks with unresolvably small crack tip angles. Therefore intrusions and crack nuclei are easily distinguishable by the value of their vertex angle. Furthermore, the effect of the environment on both processes is different as shown in the following subsection. It must be concluded, therefore that crack nucleation is not a simple continuation of intrusion formation. This clearly answers the first question: Crack nucleation is a process distinct, indeed, from the formation of surface topography including the formation of intrusions. In the conclusions different mechanisms will thus be proposed for both processes.

Effect of the Environment on the Formation of PSBs, Intrusions, and Cracks

Most experiments in air were performed with a constant plastic shear strain amplitude of 0.002. This results in a life of about 1.4×10^5 cycles. If the tests were performed under ultrahigh vacuum conditions the hardening behavior as well as the development of the surface topography in the first 10^5 cycles was undistinguishable from that found in air. In particular, the appearance of the extrusions and PSBs was not changed by performing the test in an inert environment. The only difference, which is responsible for the prolongation of the life, is a pronounced retardation of crack nucleation and growth. Thus much larger accumulated plastic shear strains can be reached before failure of the specimen. In spite of the large number of cycles, most tests did not result in a real fatigue failure, but had to be stopped because of buckling after extremely large applied accumulated shear strains of $4 \times 0.002 \times 10^7 = 8 \times 10^4$.

The resulting shape of a PSB after 10^7 cycles is shown in Fig. 5. If this is compared with Fig. 4, the profile looks more serrated, but this is entirely due to the difference in the applied accumulated plastic shear strain, which is 160 times larger in Fig. 5 than in Fig. 4.

The amount of PSB-protrusion was found to be larger in vacuum than in air. But again this is due only to the larger accumulated plastic shear strains which can be reached in vacuum because of retarded crack nucleation. Figure 6 shows the protrusion height—averaged over all PSBs within the specimen—as a function of the applied accumulated plastic shear strain found in experiments in different environments. It is obvious that the average protrusion height is a function of the applied accumulated plastic shear strain only and does not depend explicitly on the environment.

Secondary Hardening in Inert Environments

Most of the experiments in this work were performed with an applied plastic shear strain amplitude of $\gamma_{pl} = 2 \times 10^{-3}$, which produces PSBs with a volume fraction of about 20%. The volume fraction remains constant for the whole life if the experiment is performed in air. In vacuum or ultrahigh vacuum, on the other hand, the life is extended considerably. Then secondary hardening, described by Abel [*18*] and Wang [*19*], takes place in the PSBs (Fig. 7). This leads to an increasing stress in the matrix, which therefore fills with PSBs.

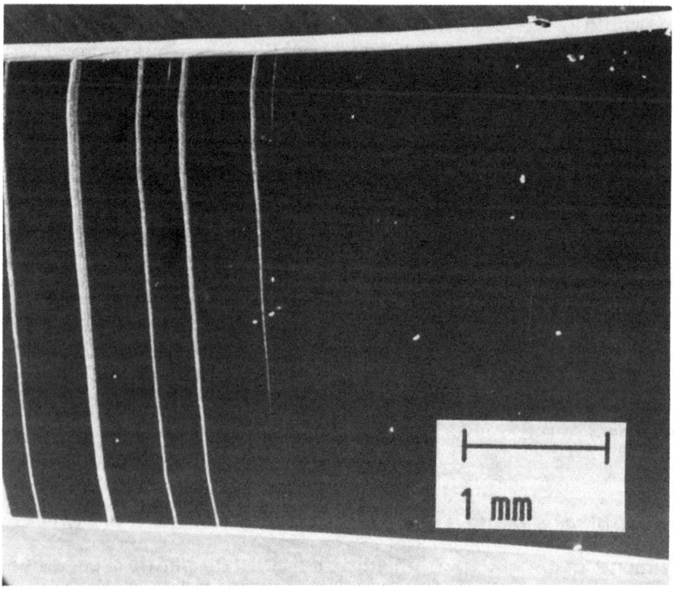

FIG. 8—*Surface of a specimen showing PSBs after 10^5 cycles in ultrahigh vacuum. The volume fraction of the PSBs is also typical for a specimen fatigued in air under otherwise identical conditions.*

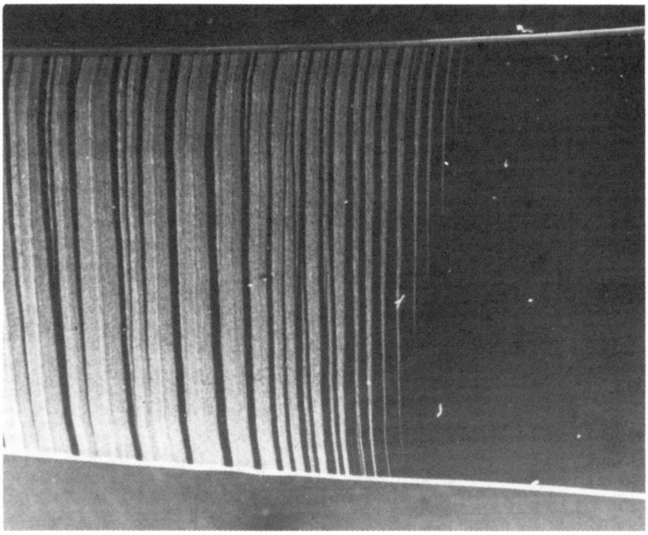

FIG. 9—*Same area as in Fig. 8 but after 10^6 cycles, demonstrating the increase of the volume fraction of the PSBs due to secondary hardening. The lack of PSBs at the left is due to the gentle increase of specimen thinkness near the end of the gage length.*

The resulting surface is shown in Figs. 8 and 9. This homogenization of slip in vacuum, which has been observed by Verkin and Grinberg [*20*] as well as Alekseev [*21*], certainly is a consequence and not the cause of the prolonged life in an inert environment.

Because of this secondary hardening the whole gage length is covered with extrusions, intrusions, and partly rewelded slipband cracks. These cracks are believed to be responsible for the gentle rise of the elastic compliance during the very long experiments in vacuum (Fig. 7).

Effect of Environment on Early Crack Growth

The retardation of crack nucleation in vacuum was quantified by measuring the lengths of a large number of microcracks to establish a statistically significant determination of average crack length as a function of $\gamma_{cum,pl}^{PSB}$. Since almost all intrusions contain very small crack nuclei at the later stages of life, the average crack size was calculated by dividing the sum of the lengths of all observed cracks by the number of intrusions. The results are shown in Fig. 10. Early crack propagation down to average crack lengths of about 0.1 μm are delayed by a factor of 20 to 100 when an inert environment is compared with air. These findings answer Question 2: The environment does not influence the development of the surface topography; however, it strongly affects crack nucleation and growth.

The reason for the reduction in growth rate may be found in Fig. 11. If a specimen is tested in vacuum, voids are observed close to the surface but not in contact with the surface. They are often elongated and are aligned along the trace of the primary slip plane. Because they are found only in the vicinity of the surface (<50 μm) it is unlikely that they are due to the condensation of vacancies produced by slip, since the high slip activity of the PSBs extends throughout the specimen. Thus the voids are believed to be the remnants of imperfectly rewelded cracks. This agrees with indications of cold welding of clean fatigue cracks under compression in vacuum found previously by Fuhlrott and Neumann [*22*].

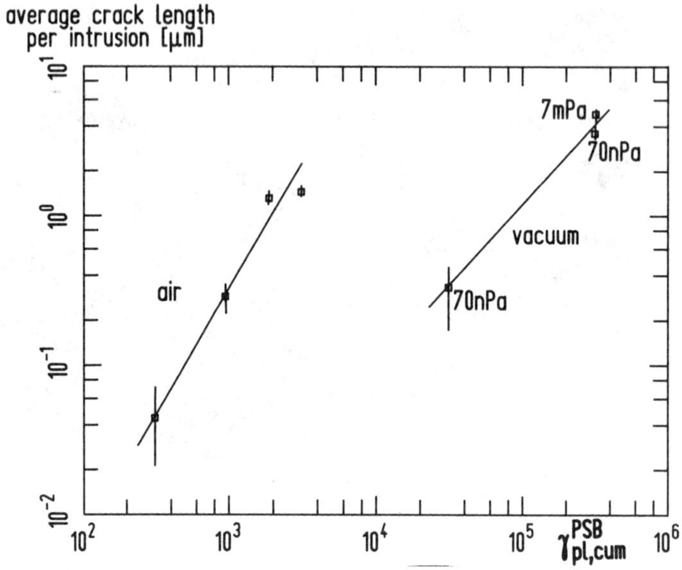

FIG. 10—*Average crack length per intrusion as a function of the accumulated plastic shear strain in the PSB,* $\gamma_{cum,pl}^{PSB}$, *in air, high vacuum, and ultrahigh vacuum.*

FIG. 11—*90-deg section (no taper magnification) through a specimen fatigued for* 10^7 *cycles (with* $\gamma_{pl} = 2 \times 10^{-3}$, *20 Hz) at saturation stress (32 MPa). Elongated voids along the primary slip plane traces are visible in the near-surface region.*

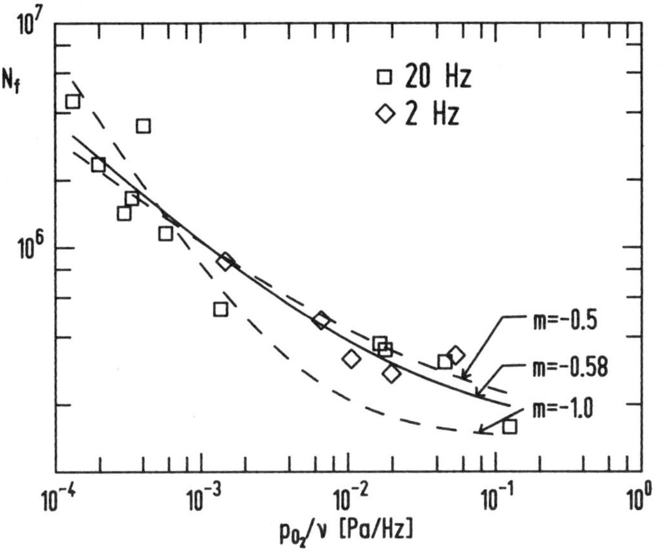

FIG. 12—*Cycles to failure as a function of the ratio of oxygen partial pressure and frequency. The dashed curves are optimal fits of the function* $N_f = N_f^{air} + c \cdot p_{O_2}{}^m/\nu$ *with* $N_f^{air} = 1.4 \times 10^5$ *for* $m = -0.5$ *and* $m = -1$. *The solid curve was obtained with* m *as a second free fit parameter resulting in the optimal* m-*value of* $m = -0.58$. *The values of* c *and related standard deviations are found in Table 1.*

The observed mean crack lengths are the same in a high vacuum of 7 mPa as well as in an ultrahigh vacuum of 70 nPa after identical cumulative shear strains (Fig. 10). Thus at a frequency of 20 Hz the environment is irrelevant for crack growth at pressures of 7 mPa and below. On the other hand, it is obvious that cracks are still nucleating and growing under these conditions. Therefore it must be concluded that the observed crack growth at 70 nPa (about 1% of that in air) has occurred without the help of the environment and is due to slip irreversibilities. The actual crack growth due to slip irreversibilities may in fact be larger because of the possibility of rewelding.

Cycles to Failure as a Function of Oxygen (O_2) Pressure

In order to identify the active component in the air environment, experiments were performed in water (H_2O) vapor and in hydrogen (H_2) gas of 0.4 Pa. No fatigue fracture occurred under otherwise identical conditions within 2.5×10^6 cycles. In O_2 of the same pressure, however, fracture occurred after 3.3×10^5 cycles. Therefore oxygen was identified as the active gas for fatigue damage of copper specimens in agreement with Thompson et al. [*1*] and Wadsworth and Hutchings [*23*]. In order to complete the picture, the fatigue life was measured for various O_2 partial pressures and two loading frequencies. In Fig. 12 these data are plotted against the ratio of the O_2 partial pressure and the loading frequency ν. They fall on a common line, which means that there is a frequency effect as strong as the effect of partial pressure. If it is assumed that the concentration of chemisorbed oxygen at the crack tip is the parameter which controls the crack propagation, it must be concluded from the frequency effect that under the experimental conditions the reaction is far from equilibrium. Since the crack nuclei are very narrow, it is even possible that the flow of O_2 molecules towards the crack tip may be rate limiting; even at a freely exposed plane surface

TABLE 1—*Values of the fit parameters and of the standard deviations of the data for the fit curves in Fig. 12.*

c	m	Standard Deviation of Log $10(N_f)$
29 000 ± 3000	−0.5	0.303
17 000 ± 7000	−0.58 ± 0.06	0.268
700 ± 140	−1.0	0.764

the time for complete coverage, assuming a sticking coefficient of 1 at a typical pressure of 20 mPa, is 20 ms, which is already of the order of the cycle time.

If the gas flow toward the crack tip is rate limiting, the rate should be proportional to the pressure itself or $N_f - N_f^{air} = 1/p_{02}$. If the reaction is in equilibrium according to the Langmuir isotherm of dissociative adsorbtion, a $p_{02}^{-0.5}$ dependence is expected. Figure 12 shows that the experimental data can be described reasonably well by either law. The fit is optimal for an exponent $m = -0.58$. The fits were performed with the logarithm of the data (as plotted); that is, a constant error was assumed in the log-log plot. Table 1 contains all values of the fitted parameters together with their standard deviation and the standard deviations of the log $10(N_f)$ data in the fits. A more intricate analysis of the data does not seem to be appropriate because of the complexity of the underlying situation.

Conclusions

Intrusions and crack nuclei are different in two respects:

1. Their geometrical shapes are different (vertex angle 30 deg or 0 deg, respectively).

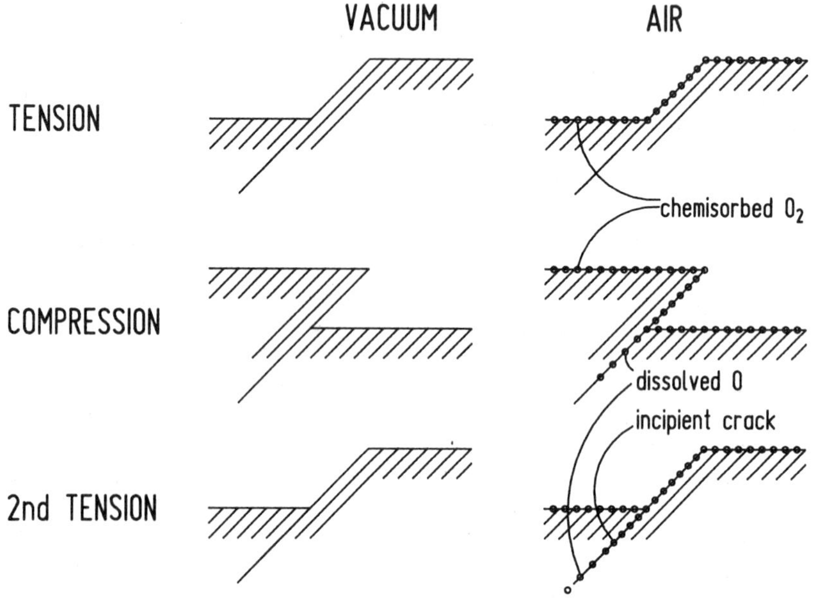

FIG. 13—*Mechanism of environmentally assisted Stage I crack growth after Ref* 1.

2. Intrusion formation does *not* depend on the environment, whereas crack formation and growth do *strongly* depend on the environment.

It is suggested therefore that both phenomena are produced by two different mechanisms:

• Intrusions are formed as valleys between extrusions by the egress of dislocation dipoles [*9,24*]. However, the triangular shape of the extrusions is not yet fully understood. The only available explanation for this shape is due to Ref *25*. But the authors of Ref *25* required for their explanation a hypothetical triangular shape of the underlying ladder structure, which could not yet be verified experimentally.

• The crack nuclei form and grow most likely according to the mechanism proposed by Thompson et al. [*1*], which is schematically shown in Fig. 13. This mechanism agrees—qualitatively at least—with all the known experimental facts about crack nucleation and growth.

In detail, the following conclusions can be drawn from the experiments described in this paper:

1. A new sectioning technique was developed to obtain the profile of PSBs, extrusions, intrusions, and crack nuclei with an edge resolution of 20 nm.
2. There is a distinct geometrical difference between intrusions and crack nuclei; the former have a finite vertex angle of about 30 deg, whereas the latter are closed cracks (vertex angle about zero).
3. In vacuum, crack nucleation and growth is retarded up to two orders of magnitude but otherwise the PSB topography is not affected if compared at identical accumulated plastic shear strains.
4. At large accumulated strains, elongated voids are observed along near-surface slip planes, which look like remnants of rewelded cracks.
5. The retardation of crack nucleation and growth in vacuum allows PSBs to be observed at much larger accumulated strains and thus reveals more extreme topographies (higher PSB protrusion). Furthermore, it allows a new effect, secondary hardening, with the consequence of PSB widening.
6. It has been confirmed that oxygen is the active component of the air environment, which is responsible for the reduction of fatigue life of copper if compared with that in vacuum.
7. Evidence has been found for crack growth due to slip irreversibilities to an extent of 1% of the crack growth in air. However, this value is a lower bound only, because of the possibility of rewelding.

Acknowledgment

The authors gratefully acknowledge financial support from the Deutsche Forschungsgemeinschaft under Contract No. 193/6 and 193/8-2.

References

[*1*] Thompson, N., Wadsworth, N. J., and Louat, N., *Philosophical Magazine*, Vol. 1, 1956, p. 113.
[*2*] May, A. N., *Nature*, 1960, Vol. 185, p. 303.
[*3*] Gough, H. J. and Sopwith, D. C., *Journal of the Institute of Metals*, Vol. 49, 1932, p. 93.

[4] Gough, H. J. and Sopwith, D. C., *Journal of the Institute of Metals*, Vol. 72, 1946, p. 415.
[5] Munz, D., Schwalbe, K., and Mayr, P., *Dauerschwingverhalten Metallischer Werkstoffe*, Vieweg, Braunschweig, Germany, 1971.
[6] Laird, C. and Duquette, D. J. in *Proceedings*, Second International Conference on Corrosion Fatigue, National Association of Corrosion Engineers, Houston, TX, 1972, p. 88.
[7] Wood, W. A., *Philosophical Magazine*, Vol. 3, 1958, p. 692.
[8] Hunsche, A. and Neumann, P., *Acta Metallurgica*, Vol. 34, 1986, p. 207.
[9] Neumann, P. in *Physical Metallurgy*, R. W. Cahn and P. Haasen, Eds., North-Holland, Amsterdam, 1983.
[10] Basinski, Z. J. and Basinski, S. J. in *Proceedings*, Sixth International Conference on the Strength of Metals and Alloys, Vol. 2, 1983, p. 819.
[11] Basinski, Z. J. and Basinski, S. J. in *Proceedings*, Eshelby Memorial Symposium, Sheffield, U.K., Cambridge University Press, Cambridge, 1985, p. 583.
[12] Cheng, A. S. and Laird, C., *Materials Science and Engineering*, Vol. 51, 1981, pp. 55 and 111.
[13] Roberts, W. N., *Philosophical Magazine*, Vol. 20, 1969, p. 675.
[14] Winter, A. T., *Philosophical Magazine*, Vol. 30, 1974, p. 719.
[15] Finney, J. M. and Laird, C., *Philosophical Magazine*, Vol. 31, 1975, p. 339.
[16] Mughrabi, H., *Materials Science and Engineering*, Vol. 33, 1978, p. 207.
[17] Essmann, U. and Mughrabi, H., *Philosophical Magazine*, Series A, Vol. 40, 1979, p. 731.
[18] Abel, A., *Materials Science and Engineering*, Vol. 36, 1978, p. 117.
[19] Wang, R., Mughrabi, H., McGovern, S., and Rapp, M., *Materials Science and Engineering*, Vol. 65, 1984, pp. 219 and 235.
[20] Verkin, B. I. and Grinberg, N. M., *Materials Science and Engineering*, Vol. 41, 1979, p. 149.
[21] Alekseev, A. I., *International Journal of Fatigue*, Vol. 3, 1981, p. 129.
[22] Fuhlrott, H. and Neumann, P. in *Proceedings*, Conference on the Mechanisms of Environment Sensitive Cracking of Materials, The Metals Society, London, 1977, p. 375.
[23] Wadsworth, N. J. and Hutchings, J., *Philosophical Magazine*, Vol. 3, 1958, p. 1154.
[24] Partridge, P. G., *Acta Metallurgica*, Vol. 13, 1965, p. 517.
[25] Essmann, U., Goesele, U., and Mughrabi, H., *Philosophical Magazine*, Series A, Vol. 44, 1981, p. 405.

Yukitaka Murakami,[1] *Toshio Mura,*[2] *and Masaki Kobayashi*[1]

Change of Dislocation Structures and Macroscopic Conditions from Initial State to Fatigue Crack Nucleation

REFERENCE: Murakami, Y., Mura, T., and Kobayashi, M., **"Change of Dislocation Structures and Macroscopic Conditions from Initial State to Fatigue Crack Nucleation,"** *Basic Questions in Fatigue: Volume 1, ASTM STP 924,* J. T. Fong and R. J. Fields, Eds., American Society for Testing and Materials, Philadelphia, 1988, pp. 39–63.

ABSTRACT: Dislocation structures in fatigue were investigated with special consideration given to the compatibility of the macroscopic observation by an optical microscope and the microscopic observation by a transmission electron microscope (TEM). First, previous works are reviewed from this point of view, that is, the compatability of macro- and microscopic observations. It is postulated that this viewpoint is very important for both the elucidation of fatigue mechanism and the construction of crack nucleation models. Second, the experiments are planned and carried out, and the results are discussed in this direction. Copper polycrystalline sheets with a thickness of 0.1 mm are used for the preparation of specimens for fatigue testing. The specimen contains an artificial hole of 200 μm diameter. The fatigue processes at the edge of the hole were observed successively with an optical microscope. In the early stage of the fatigue test, slipbands initiated at the hole edge and subsequently cracks nucleated after stress cycles of $N \cong 8 \times 10^5$. The foil specimens for TEM observation were prepared from the sheet specimens that were tested for various numbers of cycles, say, $N = 2 \times 10^5$, 4×10^5, and 8×10^5. Every effort was made to show the correspondence between the site of TEM observation and that of macroscopic observation. Since it was very difficult to thin the edge of a hole sufficiently, only a few specimens were available for TEM observation though more than 100 sheet specimens were tested. The first part of the paper poses five fundamental questions; especially considered in the direction described above are (1) the relationship between vein structures and persistent slipbands (PSBs), (2) the relationship between crack nucleation and PSBs, and (3) dislocation structure at the stress concentration or stress gradient.

Conclusions may be summarized as follow: (1) Vein structures and ladder structures in PSBs are closely related. (2) In the fatigued copper polycrystal specimens with no hole, PSBs with ladder structure can develop in certain grains of the specimen. (3) At the edge of a hole of the specimens which contain a hole, however, PSBs with ladder structure are unlikely to be formed. (4) The crack nucleation sites at the edge of a hole are PSBs without ladder structure.

KEY WORDS: fatigue, dislocation, crack nucleation, vein structure, persistent slipbands, ladder structure, copper, small artificial hole, stress concentration, observation by optical microscope, observation by transmission electron microscope

Recently many transmission electron microscope (TEM) observations of dislocation structures during fatigue have been directed to the discussion of the evolution of dislocation structure from the early stage of fatigue to crack nucleation. Most of these investigations have treated the dislocation structures in single-crystal specimens or in plain polycrystal specimens under uniform cyclic stress or strain. Therefore, the question remains whether

[1] Professor and research associate, respectively, Department of Mechanics and Strength of Solids, Faculty of Engineering, Kyushu University, Higashi-ku, Fukuoka, 812 Japan.

[2] Professor, Department of Civil Engineering, The Technological Institute, Northwestern University, Evanston, IL 60201.

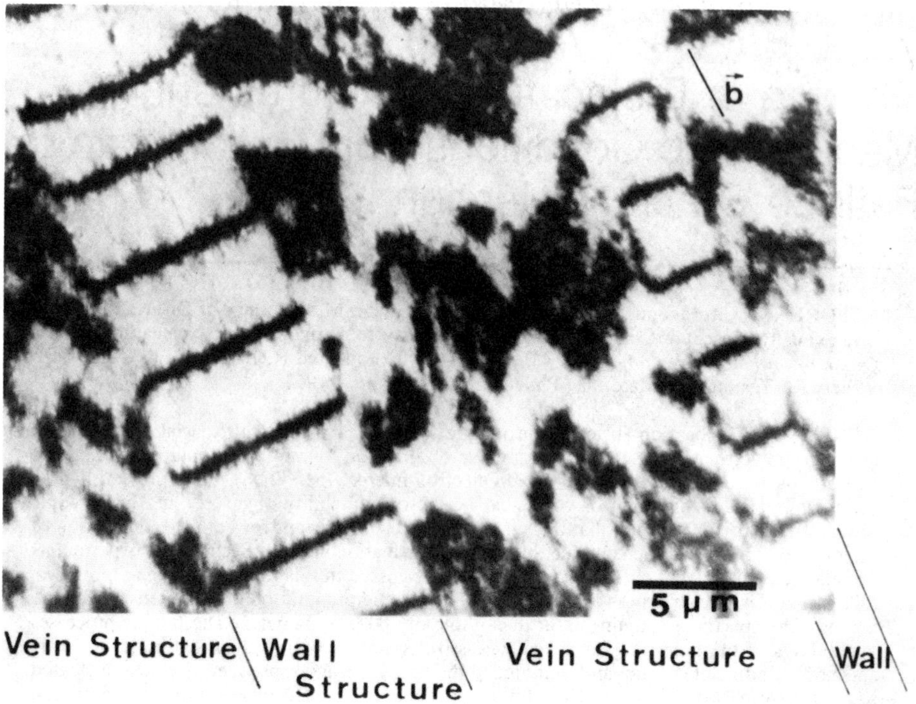

FIG. 1—*Ladder structures and vein structures in fatigued copper single crystal. Courtesy of Imura and Yamamoto* [26].

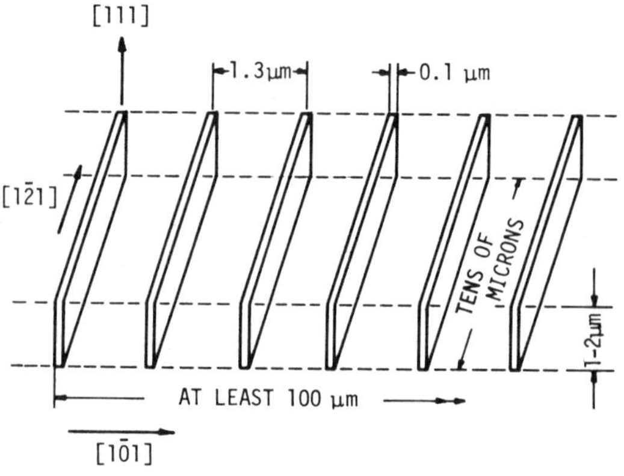

FIG. 2*a*—*Three-dimensional model of PSB; from Ref* 16.

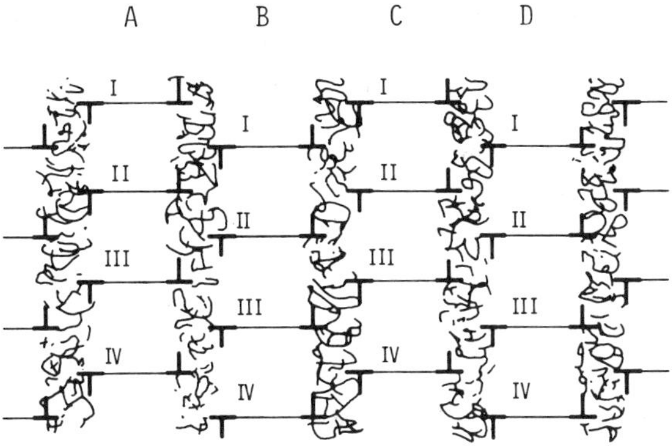

FIG. 2*b*—*Idealized schematic representation of dipolar walls within a PSB; from Ref* 36.

the experimental results obtained in this way may or may not be applied directly to the problems of stress concentrations or stress gradients at a notch or a hole. Moreover, in these investigations the evolution of dislocation structure was not observed continuously at the same point. In other words, since the continuous observation of the site of crack nucleation has never been carried out, the mechanism of crack nucleation still remains unclear.

Although it is very interesting to know how the dislocation structures evolve at the origin of a fatal crack, it is impossible to perform the continuous observation by TEM from the early stage to crack nucleation because the fatigue test must be stopped for the preparation of foil specimens. An equivalent and available method may be as follows: Prepare and test many specimens; stop the fatigue test at various numbers of cycles; find by macroscopic observation (observation by optical microscope) the site where a fatal crack will probably nucleate, and carry out microscopic observation (observation by TEM) on the same site.

The mutually correlated macroscopic and microscopic observation of this kind is very important for understanding the crack nucleation mechanism. The results of TEM observations on foil specimens prepared randomly from fatigue specimens may not necessarily offer representative information on the fatigue process of a particular specimen and accordingly such results sometimes may be the cause of misunderstandings and inadequate fatigue models.

In the present study, special emphasis is placed on the compatibility between macroscopic and microscopic observations on fatigue crack nucleation.

First, previous work on typical dislocation structures in fatigue will be briefly reviewed. A few models already proposed for crack nucleation will also be reviewed from the viewpoint of the compatibility between macroscopic and microscopic observations. In the course of the discussion, reference will be made to the available experimental results on copper and 70/30 brass.

Second, the results of fatigue test carried out using sheet specimens of copper polycrystal will be discussed. The specimens contain a very small artificial hole, and the progress of fatigue process at the hole edge is investigated by macroscopic and microscopic observations.

Consequently the *Fundamental Questions* selected in the present study are as follows:

1. How does dislocation structure during fatigue change from initial state to the nucleation of a crack?
2. Does a crack always initiate along persistent slipbands (PSBs)?

3. Are the dislocation structures of PSBs very near the surface the same as those in the interior of a specimen?
4. Do PSBs follow vein structure? Is vein structure the preceding structure of PSBs?
5. How do stress concentration and stress gradient influence the dislocation structure and the crack nucleation along PSBs? Are the dislocation structures under stress concentration and stress gradient the same as those of single-crystal specimens under uniform cyclic strain?

Review of Previous Work

Introduction

The results of dislocation behavior observed at the level of optical microscopic observation in fatigue show the appearance of slipbands, extrusions and intrusions, microcracks along slipbands, and microcracks along grain boundaries (for example, see Ref *1* and its references).

In the past two decades many investigations have been made of the dislocation structures in fatigue by the transmission electron microscope [*2–31*]. The typical dislocation structures in the range of high-cycle fatigue (low-strain fatigue) are vein structures and persistent slipbands [*8,16,22,23,28,29*]. In the range of low-cycle fatigue (high-strain fatigue) cell structures become predominant [*9,16,22,23,28,29*].

Persistent Slipbands

Since Laufer and Roberts [*2,3*] and Klesnil, Lukas, Krejci, and Rys [*4–6*] observed the dislocation structures of PSBs in fatigued copper by TEM, many other investigators have recognized similar dislocation structures. The common results of the observation of structures of PSBs in copper is a ladder-like structure on the primary slip plane. Examples are shown in Fig. 1. Figure 1 shows two PSBs in vein structures of a fatigued copper single crystal [*26*]. It is known that PSBs can develop also in other metals [*22,30,31*].

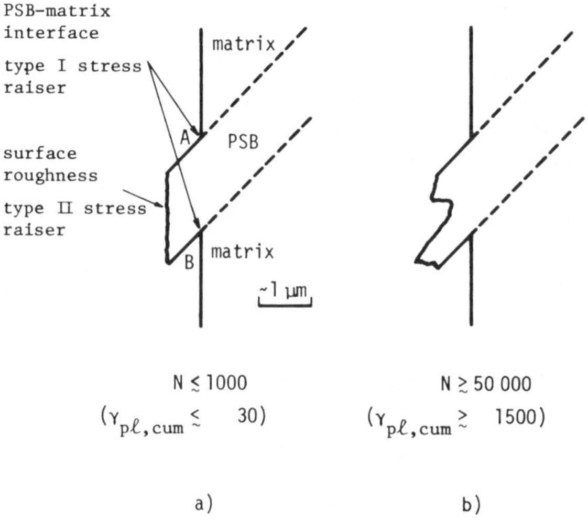

FIG. 3—*Surface profile of PSB in EGM model* [24,38].

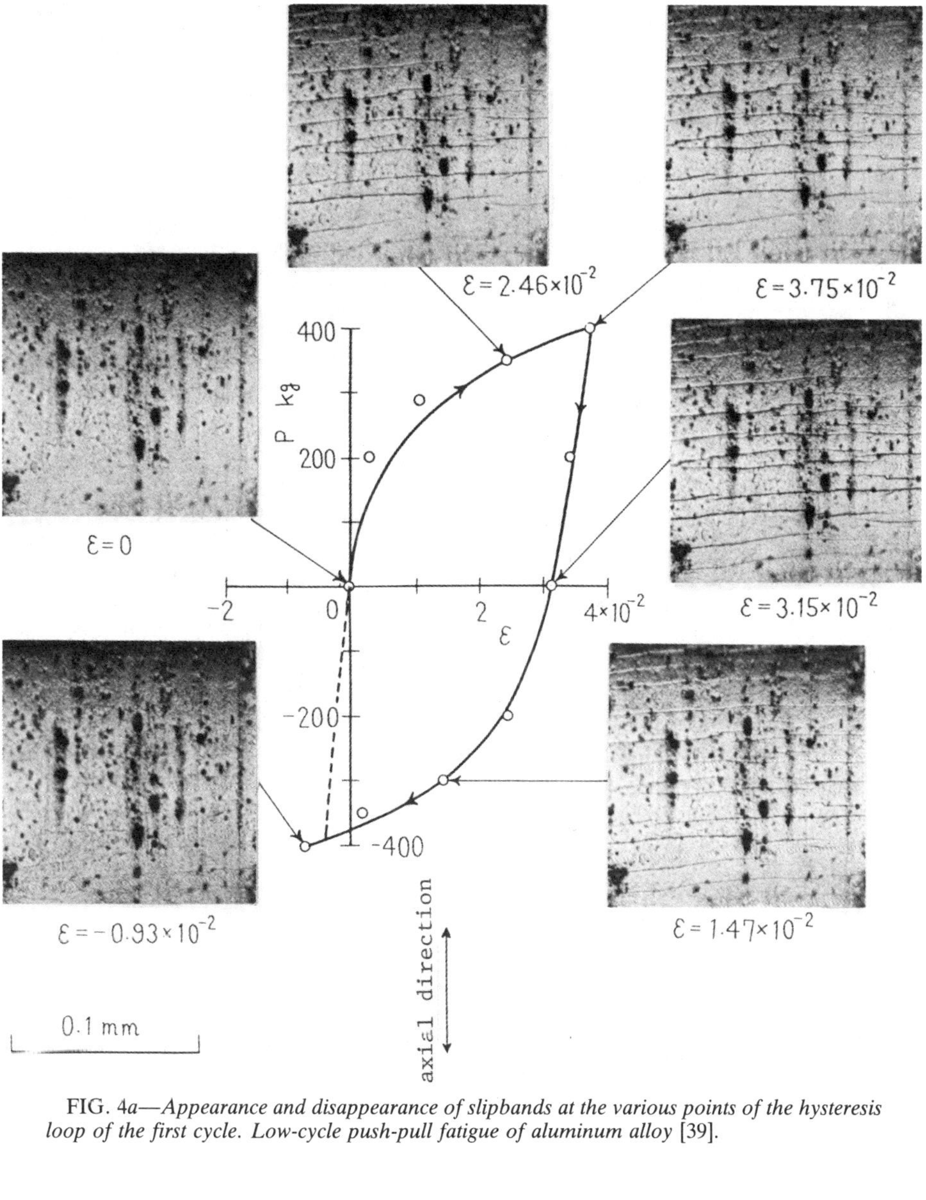

FIG. 4*a*—*Appearance and disappearance of slipbands at the various points of the hysteresis loop of the first cycle. Low-cycle push-pull fatigue of aluminum alloy* [39].

Recent detailed investigations have revealed the three-dimensional structures of PSBs fairly well [*16,23,26,27*]. It is said that the rungs of a ladder consist of condensed edge dislocation dipoles [*15–18,22,24*]. Finney and Laird [*16*] proposed a model as shown in Fig. 2*a*. It is well known that the wall spacings (the distance between two rungs) have surprisingly uniform distance. In the three-dimensional model by Finney and Laird [*16*], mobile dislocations emerge on a primary slip plane from wall to neighboring wall by the motion analogous to the Frank–Read source. The experiment by Imura and Yamamoto [*26*] supports this model. Finney and Laird [*16*] estimated the structural dimensions of PSBs as in Fig. 2*a* and

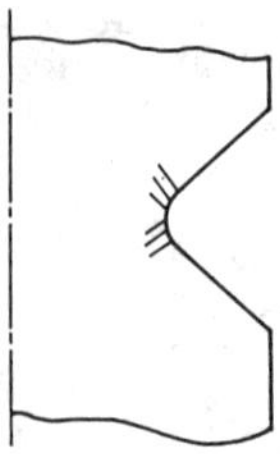

FIG. 4*b*—*Schematic illustration of the sectioned view of Fig. 4*a.

Mughrabi [*24*] also estimated similar dimensions. But it should be noted that the PSBs which exist in the same neighborhood sometimes have different wall heights as shown in Figs. 1.

Extrusions and Intrusions

Although many models of intrusions and extrusions have been proposed since the studies by Forsyth and Stubbington [*32*], Cottrell and Hull [*33*], and Mott [*34*], the recent near-surface observations of PSBs by TEM revealed that an extrusion was usually caused by a single PSB rather than the mechanism of intersecting slips [*33*].

The dislocation structure of PSBs illustrated in Fig. 2 [*16,18,36*] is often used as the basic idea for making the model of extrusions and intrusions. Although the exact mechanism of the formation of extrusions and intrusions still remains unclear, many investigators have emphasized the importance of the elucidation of the mechanism because extrusions and intrusions are considered to be strongly responsible for the crack nucleation along PSBs [*1,20,21,22,36*].

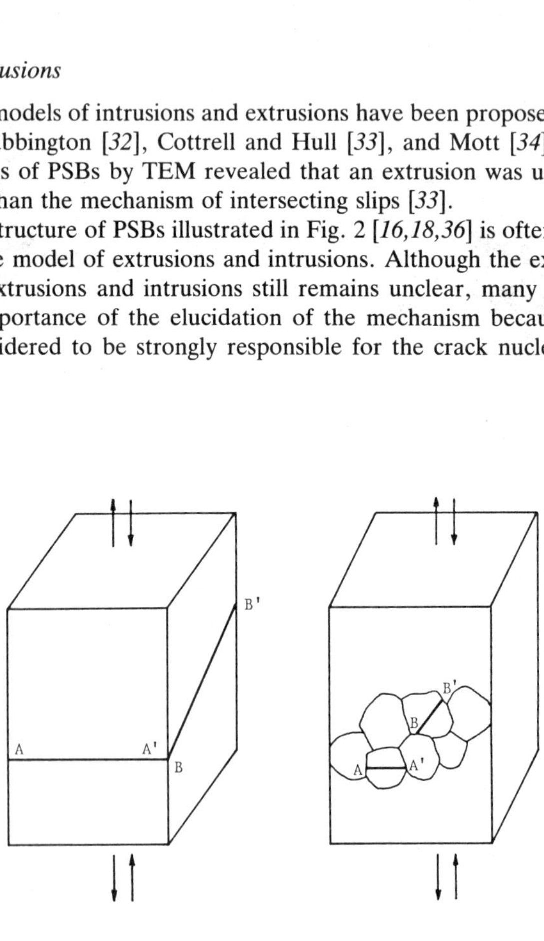

FIG. 5—*Appearance of slipbands at surface. Extrusions or intrusion may be formed along A-A'. Along B-B' the surface roughening may not form.*

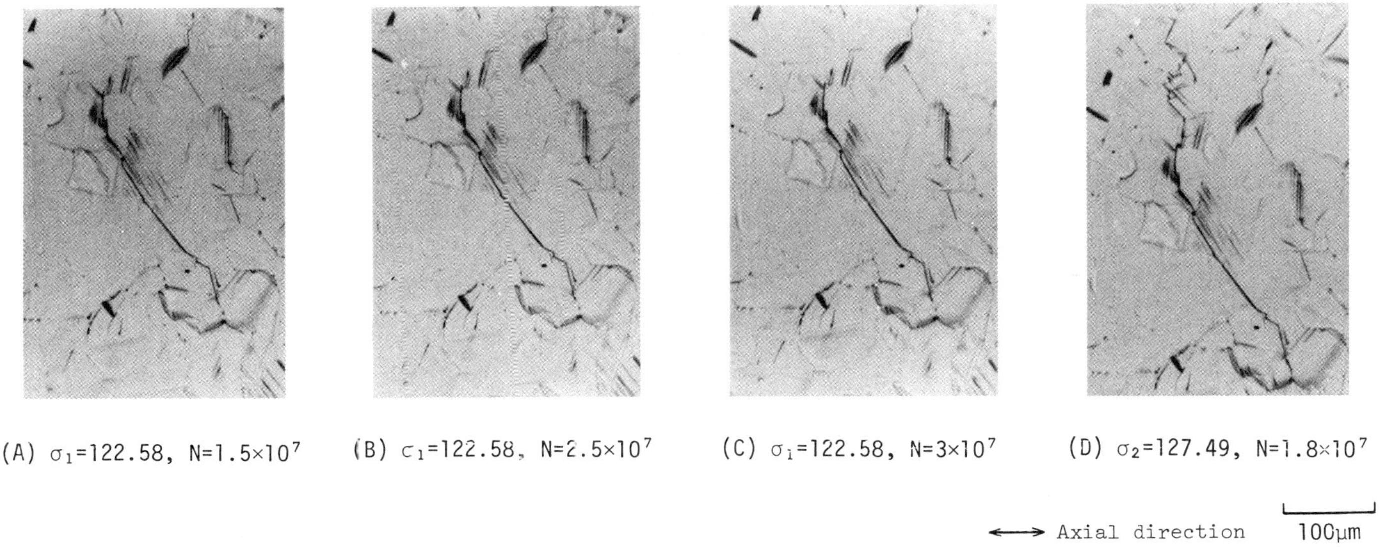

FIG. 6—*Nucleation of a transgranular crack of the type B-B′ in Fig. 5: Rotating bending high-cycle fatigue of 70/30 brass; from Ref* 40. *Units of stress in MPa. After* N = 3×10^7 *the stress was increased from* σ_1 = *122.58 MPa to 127.49 MPa as shown in* (D).

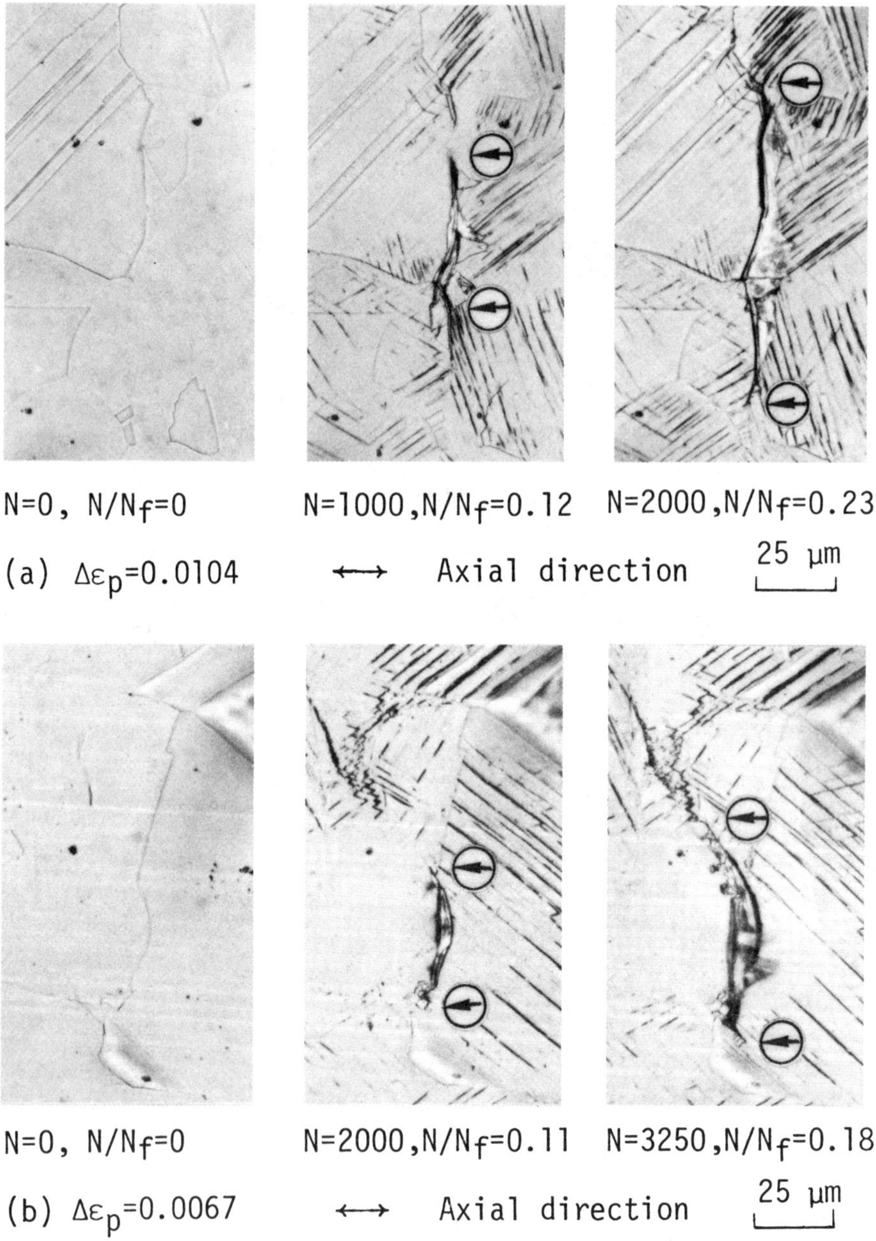

FIG. 7—*Nucleation of intergranular cracks due to PSB-GB interaction. Push-pull low-cycle fatigue of 70/30 brass.* N_f *is the cycle to failure. The same locations were monitored by the plastic replica method from the early stage of fatigue. The fins of replica which were made by the penetration of replica into the crack interrupt clear sight of the grain boundaries* (*from Ref* 43).

Crack Nucleation at PSBs

In the model proposed by Brown [*35*] and Antonopoulos, Brown, and Winter [*17*] cracks nucleate at the roots of extrusions based on the criterion that elastic misfit strain energy caused by the increase in dipole density is converted into surface energy (see also Ref *37*).

Considering the experimental results of extrusion formation at PSBs [*6*], Essman, Goesele, and Mughrabi (EGM model) [*24,38*] carried out computer stimulation of surface roughening by the random irreversible slip process in planar- and wavy-slip materials and concluded that there are two types of stress raisers at an extrusion (Fig. 3). One is the surface steps at the PSB-matrix interface which is formed in the early stage of fatigue. The other is the notch-like profile which is formed within an extrusion in the later stage of fatigue. In the EGM model, cracks are expected to nucleate at either one of these two stress raisers (Fundamental Questions 5). The models just described are those for low-strain fatigue of a single crystal.

Mughrabi et al. [*24,38*] suggested that the EGM model can be also applied to the nucleation of a transgranular Stage I crack of polycrystals if the hill-and-valley surface roughening due to random slip becomes predominant in a grain. They proposed also another model for the nucleation of intergranular cracks of polycrystal. In the model, cracks are expected to nucleate along grain boundaries due to the stress concentration caused by the interaction between PSBs and grain boundaries (PSB-GB crack). However, the present authors doubt that the microscopic stress concentration induced by PSBs is the most important factor in fatigue crack nucleation, because we have similar microscopic stress concentration induced

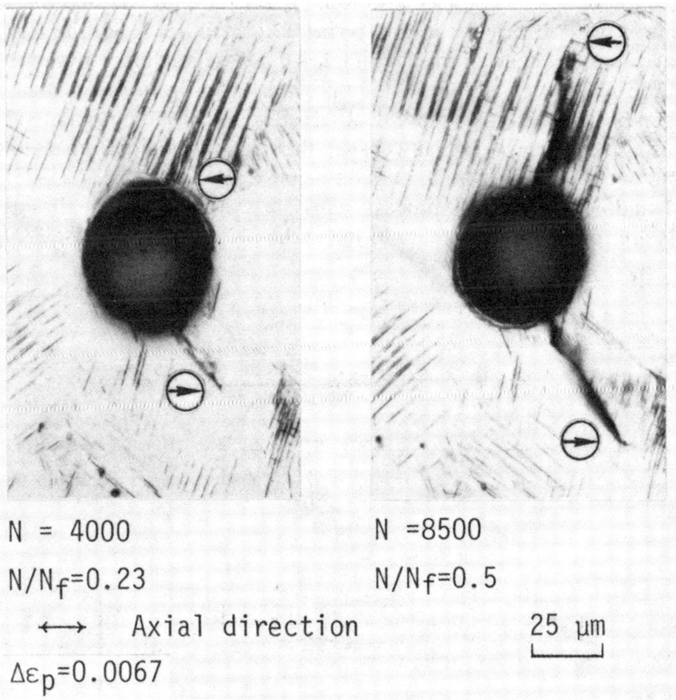

FIG. 8—*Nucleation of slipband cracks at the edge of a very small artificial hole. Low-cycle fatigue of 70/30 brass. The same specimen as that of Fig.* 7b. *All cracks nucleated at grain boundaries except for the edge of the hole.*

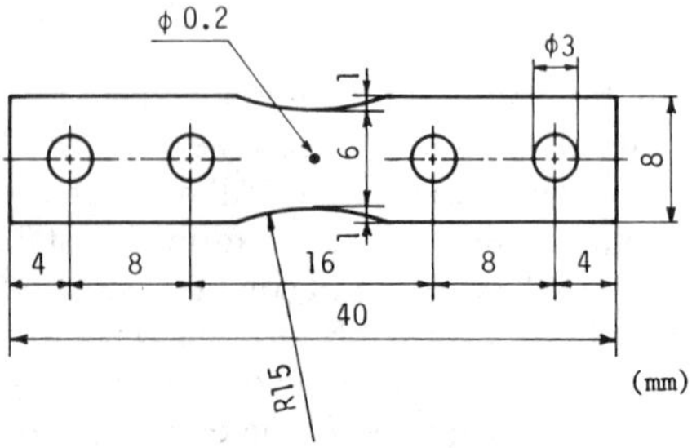

FIG. 9—*Copper sheet specimen (thickness = 0.1 mm).*

by slipbands in static deformation also, and usually there is no crack nucleation at these sites.

Mechanism of Ladder-like PSBs Formation

The mechanism of formation of a ladder-like structure in PSBs is not well known compared with the details of its final structure. Some results of TEM observations show that the wall spacings are ~1.3 ~ 1.4 μm in copper at room temperature and the wall height h_b are 1 ~ 2 μm [*16*]. However, as shown in Fig. 1, PSBs with different wall heights are observed on the same section of a specimen (see also Refs *30,31*). This implies that the wall height h_b is independent of the structures of matrix (vein). On the other hand, the fact that two PSBs in Fig. 1 have nearly equal wall spacings implies the close relation between the wall spacings and structural dimensions of veins. Many previous investigations have shown that ladder structures can be observed in commonly observed veins or veins which are almost like cells. It is thought also that veins gradually change their structure under repeated stress from so-called young to old veins and are arranged to form PSBs [*16,18,23,24,26–28*] (Fundamental Question 1 and 4). During this process veins are cyclically hardened and, subsequently, softer PSBs are formed to contribute to the cyclic plastic strain [*16,18,23,24,26–28*]. How-

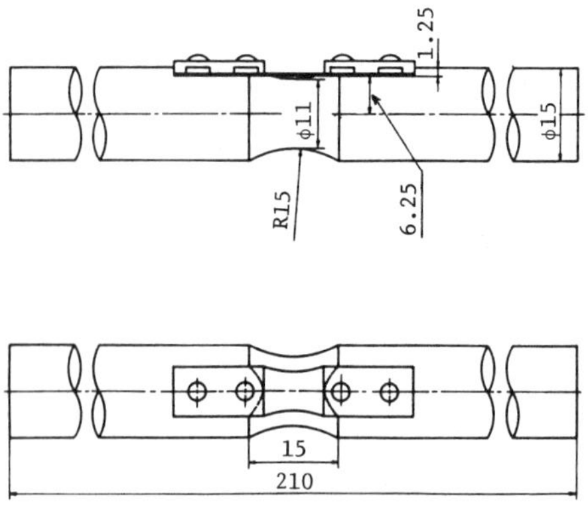

FIG. 10—*A sheet specimen attached on a rotating bending specimen of steel.*

ever, the questions of why the wall spacings are surprisingly uniform and why the dislocations are condensed in only very narrow regions still remain unsolved.

Importance of the Compatibility of Macroscopic and Microscopic Observations

The reason that single-crystal specimens are often used for the observation of dislocation structure in fatigue is that we know the resolved shear stress and strain and also expect the analogous development of dislocation structures in individual grains of polycrystals. However, in observing fatigued polycrystalline specimens with an optical microscope, we recognize that fatigue damage changes from grain to grain. Indeed, it may be often observed that cracks nucleate in particular grains, while some other grains appear to have no damage. This may be due to the difference of crystal orientations and the constraint effects of surrounding grains. Therefore, in the preparation of foil specimens for TEM observation from polycrystalline specimens, we cannot necessarily thin the exact portion where we wanted to observe; that is, the observed sites are not necessarily representative of the fatigued conditions of the specimen. Because of the statistical irregularities of individual grains, the fatigue damages vary from place to place in the specimen. The ideal method for clarifying the fatigue mechanism is to trace the location of a fatal crack nucleation macroscopically and microscopically from the early stage of fatigue to crack nucleation. It is impossible, however, to use this method because we cannot continue the fatigue test once we have prepared foil specimens from the fatigued specimen. In the present study the substitutional method will be adopted as explained later under "Test Procedure." The one-to-one correspondence of the macroscopic and microscopic observations of the same site of specimens is very important for studying the crack nucleation along PSBs, as well as the contribution of extrusions and intrusions to crack nucleation and also the mechanism of intergranular and transgranular crack nucleation.

Figure 4*a* shows the slip deformation at particular positions of the hysteresis loop which was obtained in the first loading cycle of a push–pull low-cycle fatigue test of a circumferentially notched specimen of 6061-T6 aluminum alloy (minimum diameter 8.0 mm, notch depth 1.0 mm, and notch root radius 0.4 mm) [*39*]. The notch root of the specimen was observed continuously with an optical microscope. Then the sectioned view of the slip deformation may be like that illustrated in Fig. 4*b*. Since the grain size of this material is smaller than 0.1 mm in the transverse direction, the slipbands in Fig. 4*a* are not necessarily confined to one grain. In the macroscopic views, all slip steps in Fig. 4*a* tend to vanish at the bottom of the hysteresis loop where the plastic strain at the observed area becomes approximately zero. However, this does not necessarily mean the complete reversibility of the slip process. Microscopically this process must contain the irreversible slip, because with increasing stress cycles the slip steps became macroscopically visible at any point of the hysteresis loop and cracks nucleated at the slip steps. Since the number of slip steps is a nonlinear function of the plastic strain range $\Delta\epsilon_p$, the increase in plastic strain range is accommodated by the increase in both the number and the size of slip steps [*39*]. The

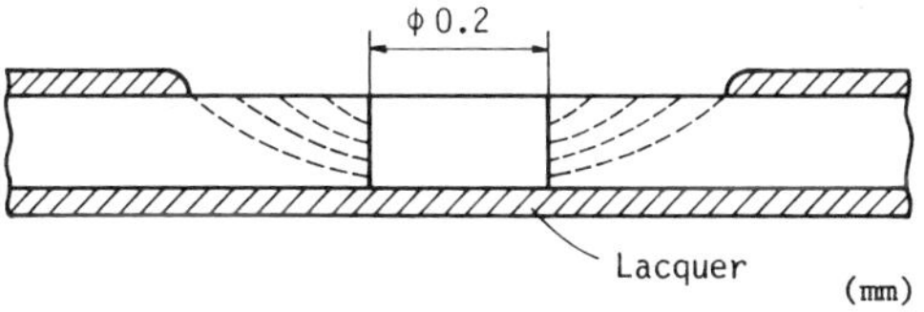

FIG. 11—*Preparation of a foil specimen by electropolishing a sheet specimen.*

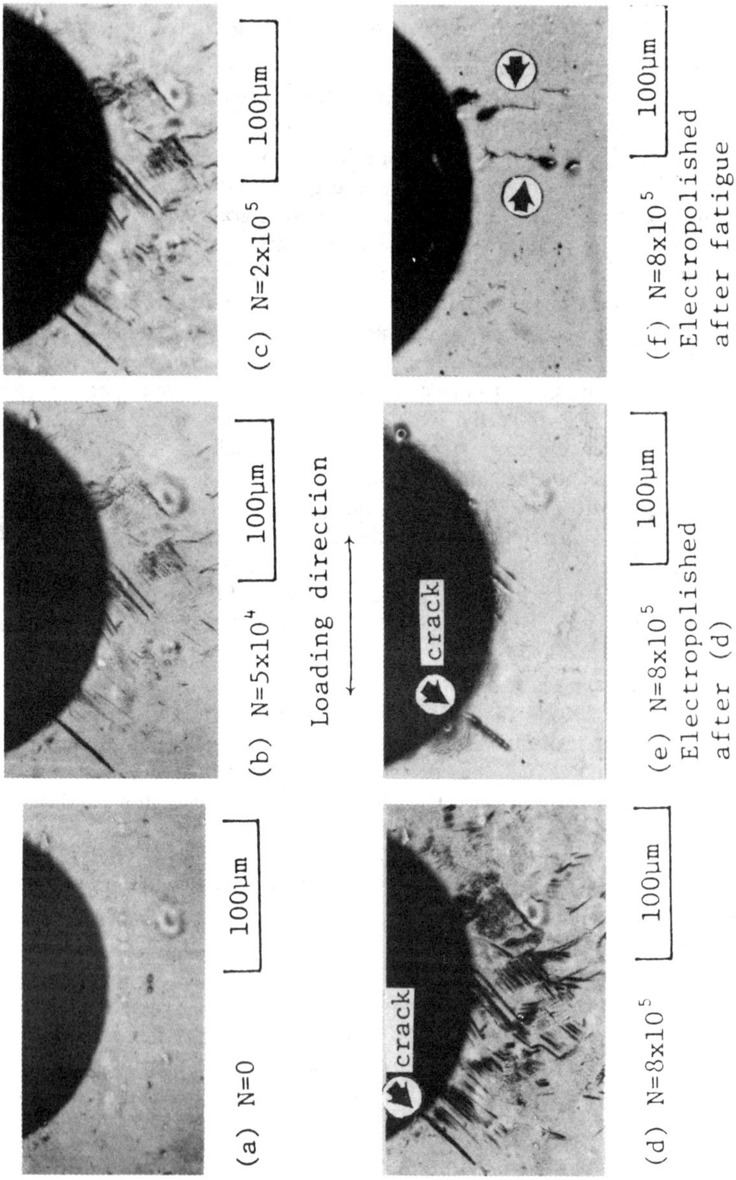

FIG. 12—*Successive observations of fatigue process by an optical microscope.*

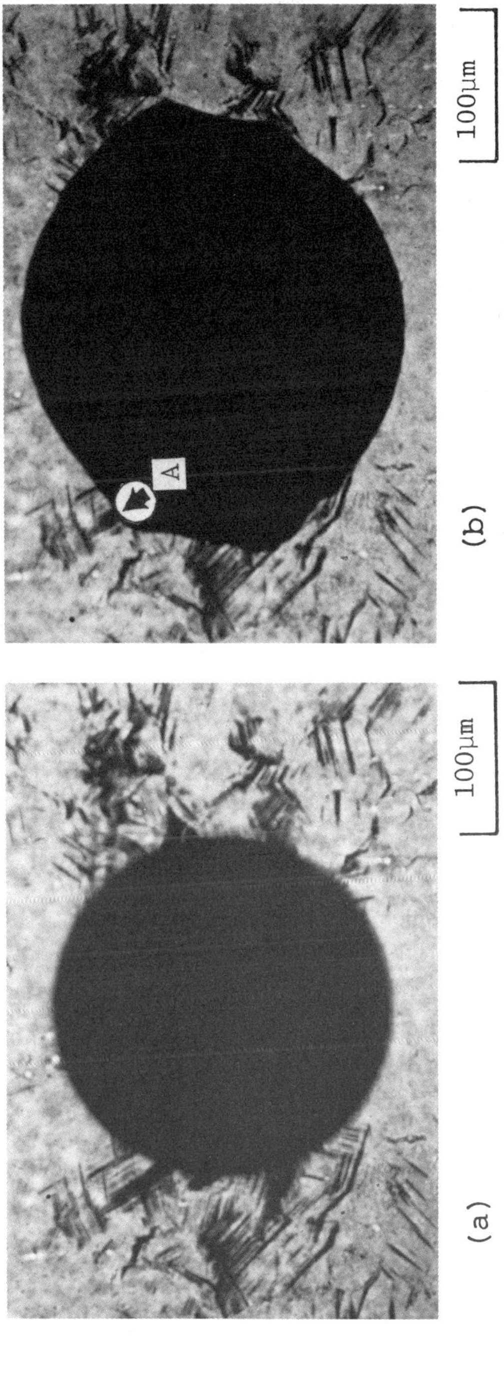

FIG. 13—(a) *Immediately after fatigue test:* N = 2 × 10^5. (b) *Observed side of the foil specimen (same specimen as* (a)).

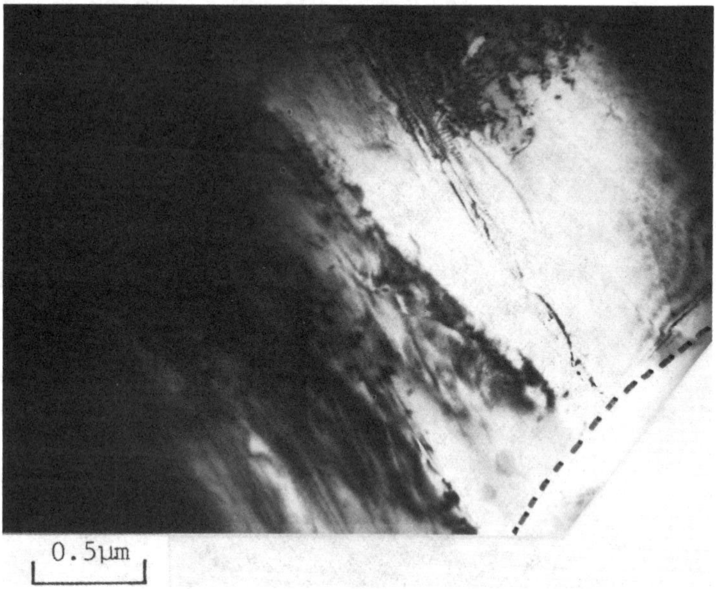

FIG. 14—*Dislocation structure at* A *in Fig. 13*b.

slipbands like Fig. 4*a* correspond to A-A′ in Figs. 5*a* and 5*b*. Extrusions or intrusions may be formed along A-A′ but may not be formed along B-B′.[3]

In the fatigue test of polycrystals the cracks nucleate not only along the PSBs as A-A′ but also along the PSBs as B-B′. Unlike the steps of the PSBs along A-A′, the B-B′ section of PSBs does not form the marked surface roughness due to cyclic slip. This fact implies that the stress concentrations at extrusions and intrusions may not directly affect crack nucleation.

An example of crack nucleation along B-B′ in Fig. 5*b* is shown in Fig. 6 for the high-cycle rotating bending fatigue test of 70/30 brass [*40*]. Nisitani and Yamashita [*41*] observed the same type of crack nucleation in a cyclic compression fatigue test of the same material. Nisitani and Murakami [*42*] observed all types of crack nucleations in rotating bending fatigue of low-carbon steel, that is, crack nucleations along A-A′, B-B′, and along grain boundaries. Thus, it may be concluded that the stress concentrations at extrusions and intrusions may not be the substantial factor in crack nucleation. Accordingly, for the construction of crack nucleation models at PSBs, not only the results of the microscopic observations but also those of the macroscopic observations must be taken into consideration.

Figure 7 shows grain boundary cracks (GB cracks) in low-cycle fatigue of 70/30 brass [*43*]. In the range of low-cycle fatigue of 70/30 brass, all cracks nucleated at grain boundaries. Essman, Goesele, and Mughrabi [*24,38*] proposed a crack nucleation model (PSB-GB cracks) at grain boundaries which was based on the interaction between PSBs and grain boundaries. The commonly observed results are that GB cracks nucleate at the grain boundaries where many PSBs are blocked. In the case of low-cycle fatigue of 70/30 brass the grain boundaries

[3] The line A-A′ in Fig. 5*a* indicates a typical slipband which appears on the surface due to the interaction of slip plane and free surface and is perpendicular to the direction of stress axis. Accordingly, we may expect that there is little slip in the direction of line A-A′. On the contrary, the line B-B′ indicates a slipband which is difficult to observe in static deformation because the slip step at the surface is very small (this is the case where the primary Burgers vector is almost parallel to B-B′). However, after fatigue the slipband like B-B′ becomes visible even when the slip step is very small.

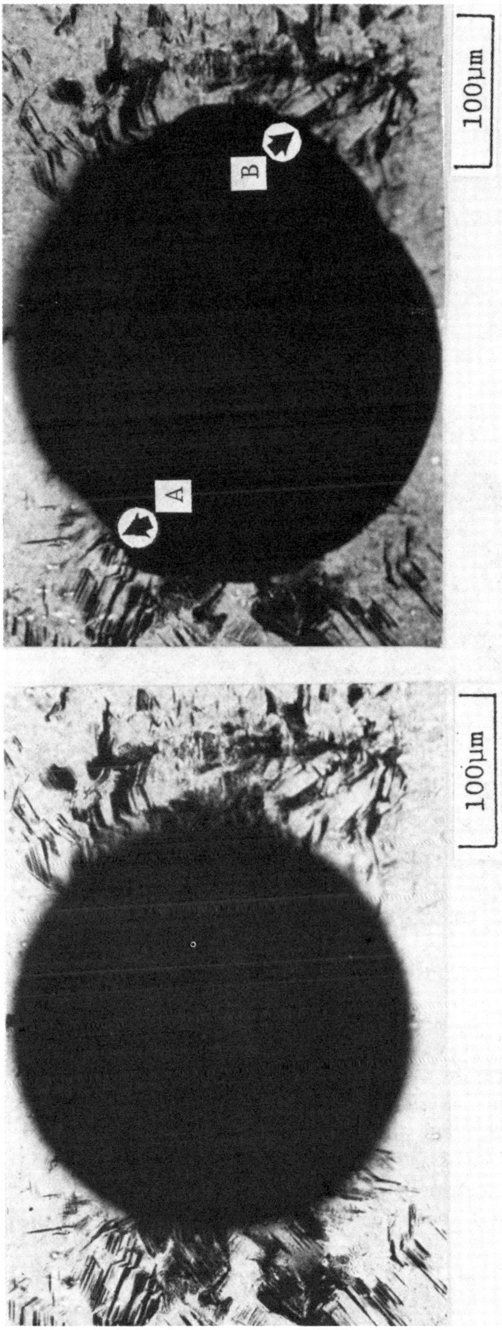

FIG. 15—(a) *Immediately after fatigue test:* N = 4×10^5. (b) *Observed side of the foil specimen (same specimen as* (a)).

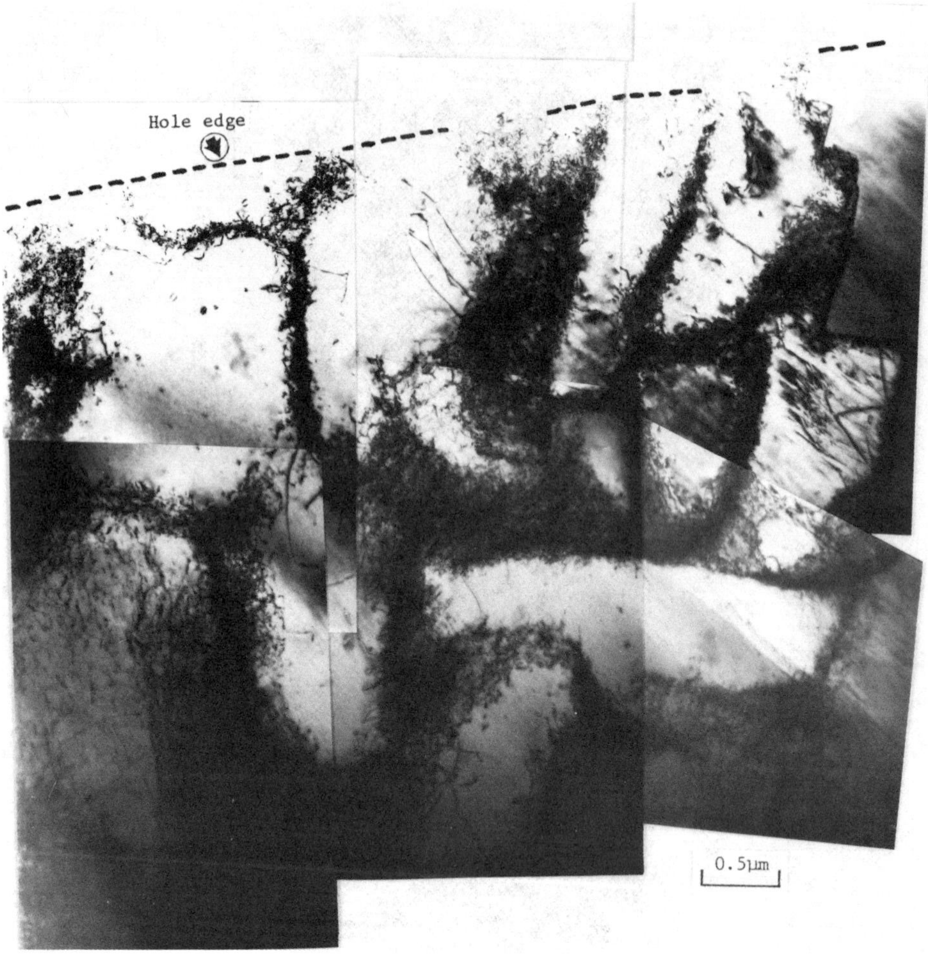

FIG. 16—*Vein structure at* A *in Fig. 15.*

where cracks nucleated were mostly oriented 70 to 90 degs to the axis of the push–pull direction. This indicates that the effect of overall tensile stress must be considered for the criterion of the crack nucleation at grain boundaries. When the plastic strain in a grain of 70/30 brass is sufficiently small and can be accommodated by only one or two PSBs, cracks tend to nucleate along PSBs as shown in Fig. 6. Therefore, the EGM model for PSB-GB cracks would be more realistic if the interactions between many parallel PSBs and a grain boundary were evaluated, though in that case the contribution of the overall tensile stress must be considered.

Figure 8 shows the formation of slipband cracks at the edge of an artificial surface hole of 40 μm diameter in the same specimen as Fig. 7. Because of the existence of a very small artificial hole, the strain concentrates at the PSBs in the small region. As a result, it seems that in the vicinity of the small hole the condition for PSB cracks is predominantly satisfied rather than that of GB cracks. This suggests that the existence of stress concentration and stress gradient may change the evolution of dislocation structures during fatigue and accordingly the crack nucleation mechanism (Fundamental Questions 2 and 5).

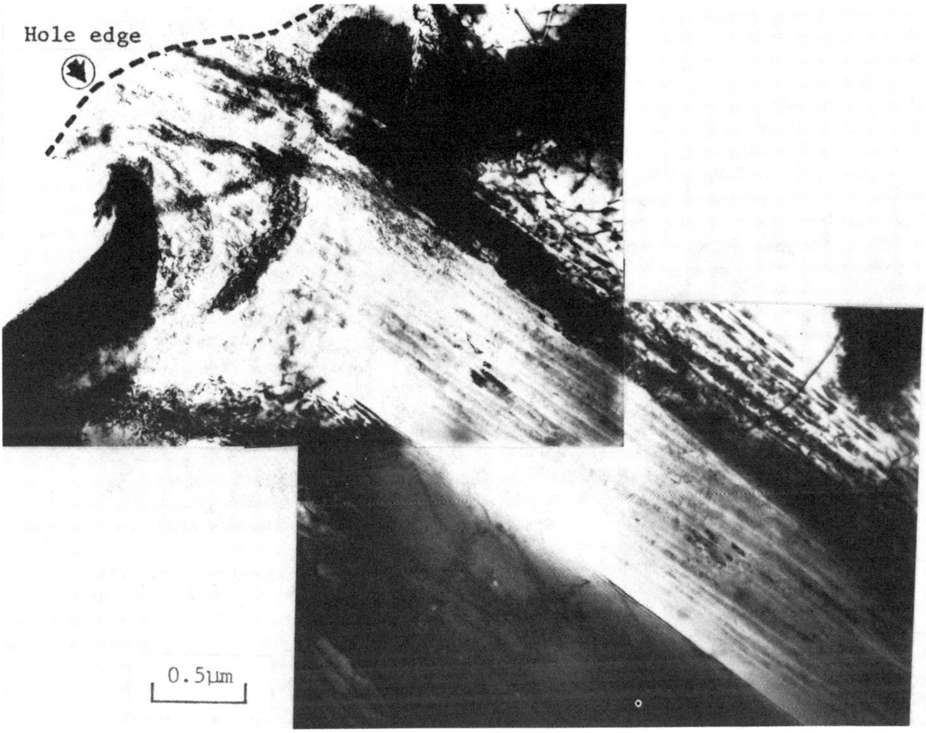

FIG. 17—*PSBs at* A *in Fig. 15 adjacent to vein structure of Fig. 16.*

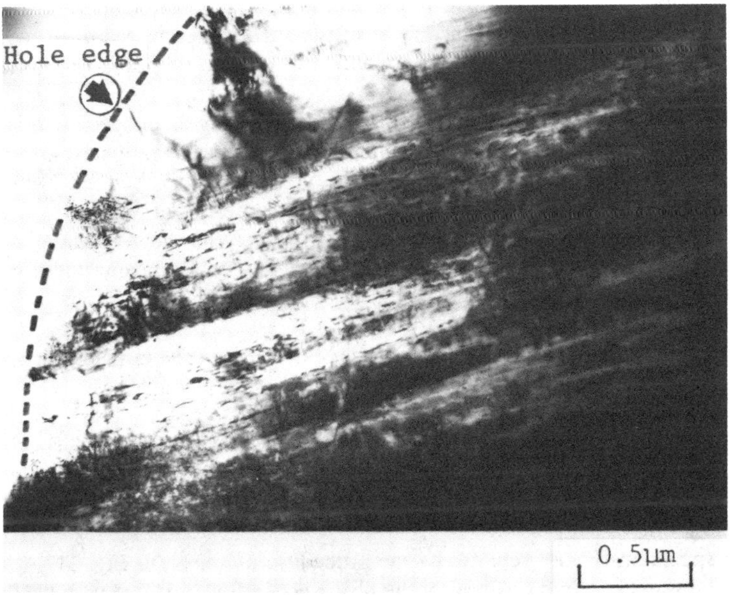

FIG. 18—*PSBs at* B *in Fig. 15.*

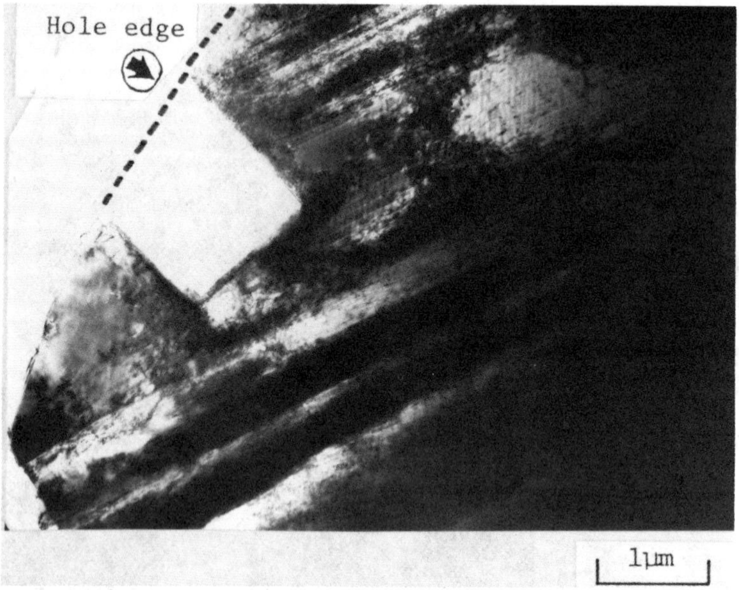

FIG. 19—*PSBs at* B *in Fig. 15.*

Critical Experiments

Material and Test Procedure

The material used in this study was copper polycrystal. The purity of the copper was 99.99%. Figure 9 shows the shape and dimensions of the specimen, which was cut out from a wide copper sheet of 0.1 mm thickness. Sheet specimens were used for convenience of TEM observation. After drilling a small through hole of 200 μm diameter at the center of sheet specimens, 10 μm of the surface layer of the specimens was removed by electropolishing. Then, specimens were annealed at 450°C for 1 h in a vacuum. The grain size after annealing was about 27 μm. These thin sheet specimens were attached by bolts to a steel round bar as shown in Fig. 10. The round bar with a copper sheet specimen on it was put into a rotating bending fatigue testing machine. Although in this fatigue test the round bar is subject to reversed bending, the loading condition of the sheet specimen is approximately tension-compression. The tensile strain in the sheet specimen is determined by the curvature of the round bar subjected to the bending moment. Since the sheet specimen is very thin, it cannot sustain the compressive load, and then it buckles elastically under compressive loading. Accordingly, the sheet specimen is virtually subject to the cyclic tension fatigue test.

Interrupting the fatigue test at definite numbers of cycles N, the fatigue process of a sheet specimen can be monitored with an optical microscope. The fatigue tests were carried out at the nominal tensile stress amplitude $\sigma = 98$ MPa. At this stress level, cracks nucleate along PSBs at the edge of the hole after fatigue of $N \cong 8 \times 10^5$. Many specimens were tested and some of them were detached from the testing machine at definite intermediate numbers of cycles N up to $N = 8 \times 10^5$. All fatigue specimens were thinned to foils for TEM observation.

The foil specimens were prepared by the procedure as shown in Fig. 11. After the one surface of the sheet specimen (we call it the observed side) was coated with lacquer and the

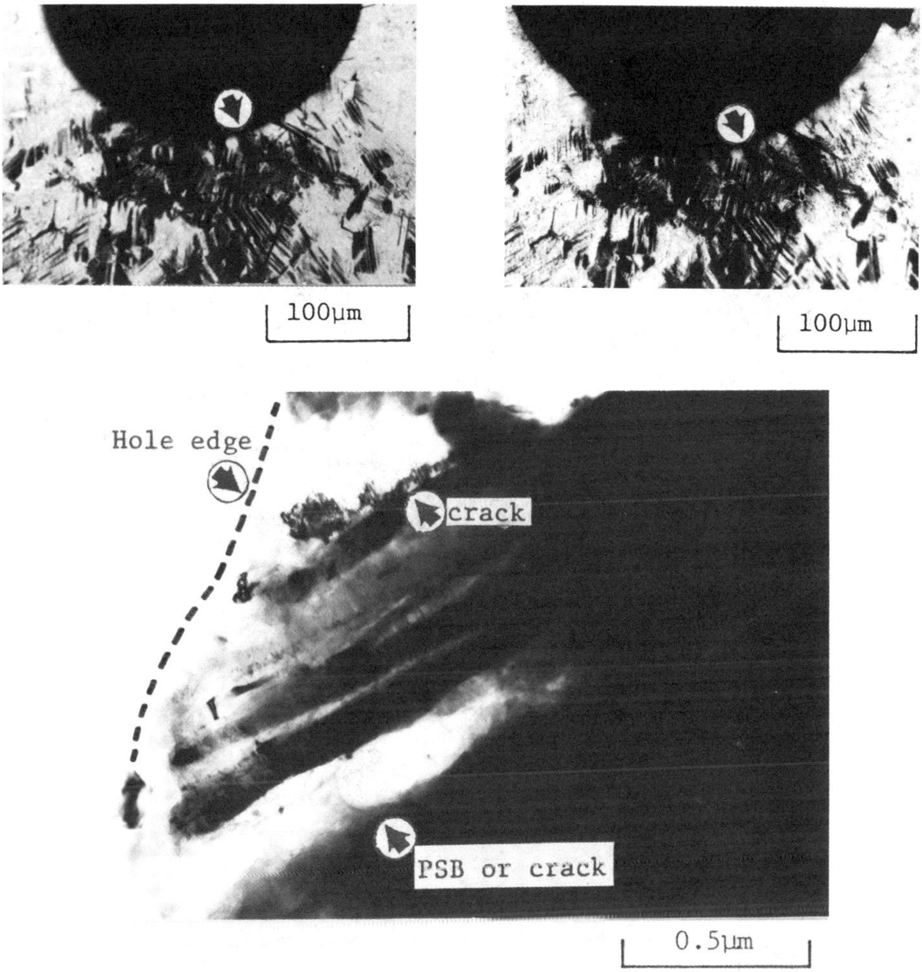

FIG. 20—*Crack nucleations and dislocation structures after fatigue for* N = *8* × *10*^{5}. (a) *Immediately after fatigue test:* N = *8* × *10*^{5}. (b) *Observed side of the foil specimen* (*same specimen as* (a). (c) *TEM observation at the marked point of* (a) *and* (b).

other surface (we call it the polished side) was also coated except for the surrounding of the hole, the specimen was electropolished. By this procedure, the vicinity of the hole edge becomes like a foil and the observed side keeps the fatigued condition unchanged. After the lacquer was removed with acetone, the foil specimen was subjected to observation by TEM at an accelerating voltage of 1000 kV. The TEM used was JEM-1000 at the High Voltage Electron Microscope Laboratory of Kyushu University. This technique enables us to obtain the correlation between surface condition observed with an optical microscope and the dislocation structure of PSBs observed by TEM.

It should be noted that although most previous studies investigated the dislocation structures in the subsurface layer of a specimen, in this study the foil specimens include the

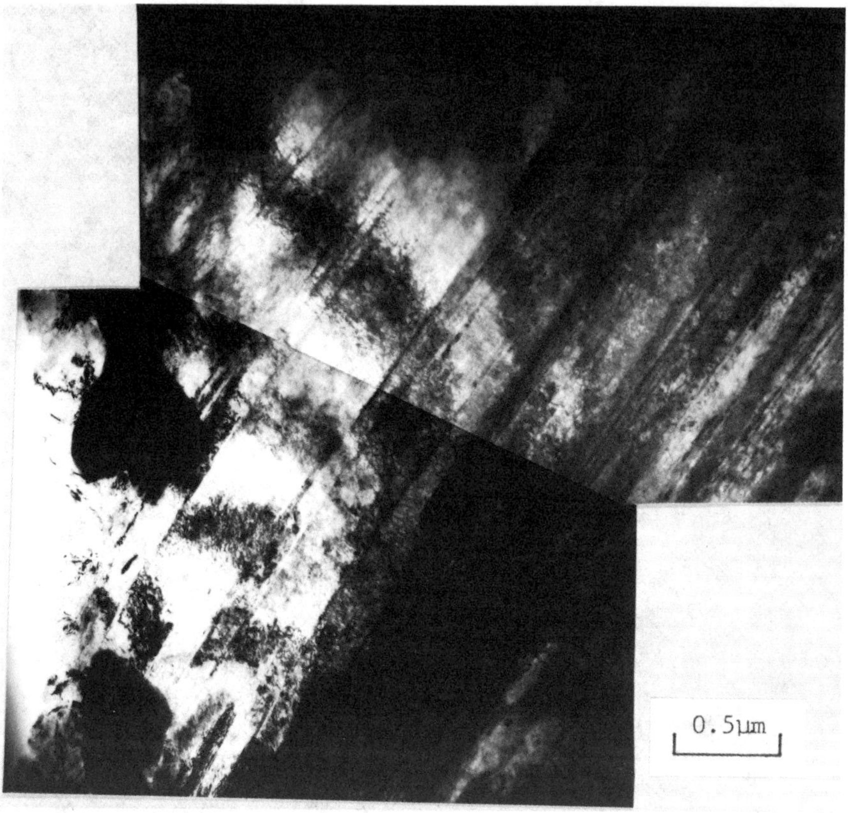

FIG. 21—*Early stage of ladder structure:* N = 4×10^5 *(vein structure is cut with slipbands).*

dislocation structure of the very surface layer; moreover, the location of the dislocation structure is identified.

Results and Discussion

Successive Observation by Optical Microscope

Figure 12 shows the results of successive observations of a sheet specimen during fatigue testing. The fatigue test was interrupted at $N = 5 \times 10^4$, 10^5, 2×10^5, 4×10^5, and 8×10^5 cycles, and the edge of the hole was observed by an optical microscope. The photographs (Fig. 12) indicate that the numbers of slipbands at the edge of the hole increase and the slipbands gradually become dense with increasing numbers of cycles. After the specimen was fatigued for 8×10^5 cycles, cracks nucleated at the edge of the hole (Fig. 12*d*. By electropolishing the specimen of Fig. 12*d* slightly, the cracks at the edge of the hole were revealed more clearly, as seen in Fig. 12*e*. Similar cracks of another specimen are shown in Fig. 12*f*. From Fig. 12*a–e*, it seems that the origin of the cracks and PSBs in Fig. 12*d* and *e* are already beginning to show at 5×10^4 cycles (Fig. 12*b*).

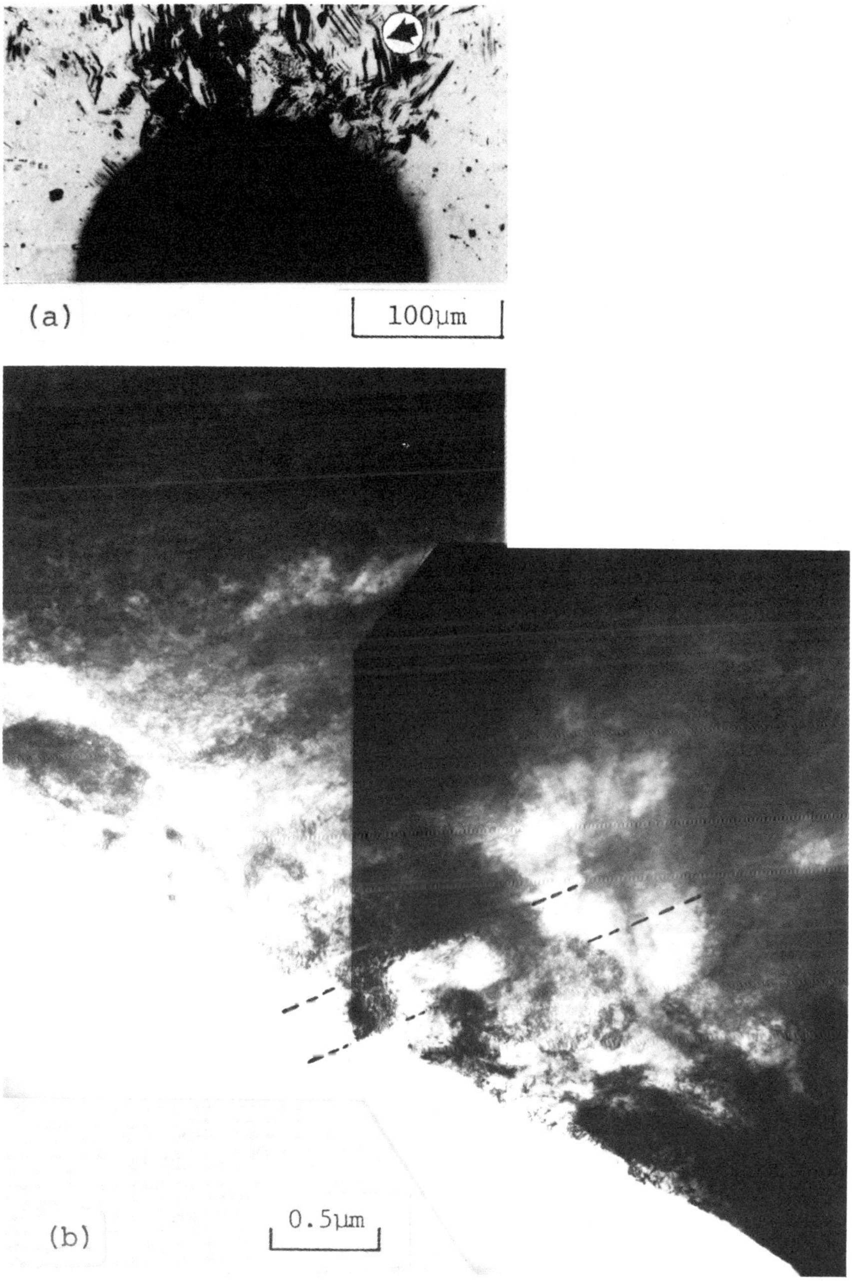

FIG. 22—*Dislocation structures in a grain apart from the edge of a hole:* N = 4×10^5. (a) *After fatigue test and* (b) *dislocation structures at the marked point in* (a).

Observation of Dislocation Structure by TEM

If we could investigate the slipbands continuously by TEM from the initial state ($N = 0$) to crack nucleation ($N = 8 \times 10^5$), the change of dislocation structure might be traced exactly. However, it is impossible to perform this procedure using a single specimen. In this study, as the equivalent procedure for the continuous observation of dislocation structure, many specimens were tested and many foil specimens for TEM observation were cut out of sheet specimens fatigued for 10^5, 2×10^5, 4×10^5, and 8×10^5 cycles.

Figure 13 shows the observations by an optical microscope of the sheet specimen immediately after the fatigue test (Fig. 13*a*) and the foil specimen just before TEM observation (Fig. 13*b*) at $N = 2 \times 10^5$. Comparing Fig. 13*a* and Fig. 13*b*, we can see that Fig. 14 indicates the dislocation structures at the tip region of the long slipbands emerging from the edge of the hole. In Fig. 14, vein structures and slipbands without ladder may be observed, though the photograph is not clear because of the thickness of the foil specimen.

Figure 15 shows the macroscopic observations of the fatigued specimen at $N = 4 \times 10^5$. Figures 16–19 show the examples of its microscopic observation. The locations of TEM observation are indicated with the same marks in Fig. 15 and Figs. 16–19. Although the appearances of the vein structures and the PSBs are indistinguishable in the observation by an optical microscope, it may be understood that the PSBs and the vein structures coexist in one group of slipbands. It should be noted that no ladder structure (rung) was observed in PSBs (see Figs. 17–19).

Figure 20*b* shows the observed side of a foil specimen prepared from a sheet specimen (Fig. 20*a*) fatigued for $N = 8 \times 10^5$ and Fig. 20*c* is the dislocation structure at the same position marked in Fig. 20*a* and *b*. No ladder structure was observed. It may be noted in Fig. 20*c* that cracks nucleated at the edge of the hole are presumably the results of the evolution of PSBs like those shown in Figs. 17–19.

Figure 21 shows the dislocation structure found in a plain specimen (with no artificial hole) fatigued for $N = 4 \times 10^5$. The method of preparation of a foil specimen is the same as for a holed specimen; that is, the one side of a fatigued sheet specimen was coated with lacquer and accordingly the dislocation structure at the very surface was observed. This figure presents a typical early stage of ladder structure, which is being constructed by cutting vein structures with slipbands.

The similar structure, that is, the early stage of ladder structure, was found also in a holed specimen but not at the edge of the hole. Figure 22*b* shows the dislocation structures at the position indicated in Fig. 22*a*. The dislocation structure marked with dotted lines is the early stage of ladder structure. This implies that although no ladder structure was observed at the edge of a hole, there is the possibility that ladder structure may develop in the grains apart from the hole edge.

It seems that there are two types of formation of ladder structure from vein structures. One is the formation of a ladder by cutting the condensed vein structures with slipbands. In this case, both the wall thickness and the wall spacing in a ladder do not change as much in comparison with the original vein structure. Typical examples may be seen in Ref *15*. The other is the formation of a ladder by cutting the uncondensed vein structures with slipbands. In this case, the rungs of a ladder are supposed to be formed from the splitting [*27,28*] or condensation of veins. Figures 21 and 22 may be the examples of the early stage of the latter type. Lukas, Klesnil, and Krejeci [*6*] pointed out the importance of the propagation of PSBs, and Shirai [*28*] showed the PSBs ending within the grain interior. The propagation of a ladder structure within a grain suggests the process of "cutting" or "eating" vein structures by PSBs.

The exact mechanism of the formation of a ladder, however, still seems to be unclear in spite of the simplicity of the structure.

Considering all the results (Figs. 14–22), it may be concluded that although the PSBs with ladder structure can develop in particular grains in a plain specimen and also in particular grains apart from the hole edge of a holed specimen, even at the surface, the PSBs at the edge of a hole are unlikely to form ladder structures. This is presumably because grains at the edge of a hole are more likely to be oriented disadvantageously to the formation of the ladder structure.[4] Concerning this point, the elaborate work by Tabata, Fujita, Hiraoka, and Onishi [27] must be referred to. They found in the fatigue test of copper single crystals that single slip is necessary for the formation of well-developed PSBs (ladder structure) and PSBs were never formed in specimens with [111] and [100] fatigue axes. Therefore, if a grain at the edge of a hole has a chance to be oriented to the direction in which single slip becomes predominant, PSBs with ladder structure may be formed.

Concluding Remarks—Answers to Fundamental Questions

The present study was carried out with special emphasis on the compatibility of the observation by an optical microscope and TEM. Concerning all the results obtained, the authors realize that the *Fundamental Questions* 1–5 posed at the beginning of the paper cannot be answered separately. Therefore, an answer may be directed to not only one particular question but also to other questions. The following numbers refer to those of the *Fundamental Questions* related to the answer.

1, 2, and 5—Two types of PSBs must be considered as the possible sites of crack nucleation. One is the so-called ladder structure, the formation of which is closely related to vein structures. The other is the slipbands which frequently appear at the edge of a hole (the location of stress concentration) and in which no ladder is constructed but cracks nucleate.

3, 4—It seems that PSBs do not show marked differences at the very surface and the interior of a specimen. Although in this study the dislocation structures at the surface were not compared with those at the interior, the early stage of the formation of PSBs was similar to that at the interior shown in many references.

4—Vein structures are closely related to the ladder structure. The wall spacings in PSBs must be dependent on the structural dimensions of veins. The ladder structures may be formed by the process of cutting veins with slipbands.

5—At the locations of stress concentration like the edge of a hole, slipbands are not expected to form a ladder structure. This is presumably because most grains at the edge of a hole are oriented disadvantageously to the formation of ladder structure.

Acknowledgment

The authors wish to thank Mr. E. Tanaka and Dr. Y. Tomokiyo for their generous support and help in observation with the high-voltage TEM of Kyushu University. The authors also wish to express their thanks to Professors T. Imura and H. Saka of Nagaya University and to Dr. H. Shirai of the Electron Technology Center of Nippon Monsan Corp. for several useful discussions and suggestions on this study.

[4] The probability that a certain grain is oriented advantageously to the formation of ladder structure may be the same throughout the specimen. However, the probability that such a grain happens to be at the edge of the hole (especially at the point of stress concentration) is very small.

References

[1] Thompson, N. and Wadsworth, N. J., *Philosophical Magazine*, Vol. 7, 1958, pp. 72–183.
[2] Laufer, E. E. and Roberts, W. N., *Philosophical Magazine*, Vol. 10, 1964, pp. 883–885.
[3] Laufer, E. E. and Roberts, W. N., *Philosophical Magazine*, Vol. 14, 1966, pp. 65–78.
[4] Klesnil, M. and Lukas, P. J., *Journal of the Iron and Steel Institute*, Vol. 203, 1965, pp. 1043–1048.
[5] Lukas, P., Klesnil, M., Krejci, J., and Rys, P., *Physica Status Solidi*, Vol. 15, 1966, pp. 71–82.
[6] Lukas, P., Klesnil, M., and Krejci, J., *Physica Status Solidi*, Vol. 27, 1968, pp. 545–558.
[7] Grosskreutz, J. C., Reimann, W. H., and Wood, W. A., *Acta Metallurgica*, Vol. 14, 1966, pp. 1549–1557.
[8] Feltner, C. E., *Philosophical Magazine*, Vol. 12, 1965, pp. 1229–1248.
[9] Pratt, J. E., *Acta Metallurgica*, Vol. 15, 1967, pp. 319–327.
[10] Feltner, C. E. and Laird, C., *Acta Metallurgica*, Vol. 15, 1967, pp. 1633–1653.
[11] Partridge, P. G., *Acta Metallurgica*, Vol. 13, 1965, pp. 517–525.
[12] Neumann, P., *Zeitschrift für Metalkunde*, Vol. 58, 1967, pp. 780–789.
[13] Kettunen, P. O., *Acta Metallurgica*, Vol. 15, 1967, pp. 1275–1286.
[14] Basinski, S. J., Basinski, Z. S., and Howie, A., *Philosophical Magazine*, Vol. 19, 1969, pp. 899–924.
[15] Woods, P. J., *Philosophical Magazine*, Vol. 28, 1973, pp. 155–191.
[16] Finney, J. M. and Laird, C., *Philosophical Magazine*, Vol. 31, 1975, pp. 339–366.
[17] Antonopoulos, J. G., Brown, L. M., and Winter, A. T., *Philosophical Magazine*, Vol. 34, 1976, pp. 549–563.
[18] Laird, C. in *Fatigue and Microstructure*, 1978 ASM Material Science Seminar, St. Louis, American Society for Metals, Metals Park, OH, 1979, pp. 149–203.
[19] Katagiri, K., Omura, A., Koyanagi, K., Awatani, J., Shiraishi, T., and Kaneshiro, H. in *Fracture 1977*, Proceedings of the Fourth International Conference on Fracture, Waterloo, University of Waterloo Press, ON, Canada, Vol. 2, 1977, pp. 695–702.
[20] Katagiri, K., Omura, A., Koyanagi, K., Awatani, J., Shiraishi, T., and Kaneshiro, H., *Metallurgical Transaction*, Series A, Vol. 8A, 1977, pp. 1769–1773.
[21] Katagiri, K., Awatani, J., Omura, A., Koyanagi, K., and Shiraishi, T. in *Fatigue Mechanisms, ASTM STP 675*, J. T. Fong, Ed., American Society for Testing and Materials, Philadelphia, 1979, pp. 106–128.
[22] Mughrabi, H., Acherman, F., and Herz, K. in *Fatigue Mechanism, ASTM STP 675*, J. T. Fong, Ed., American Society for Testing and Materials, Philadelphia, 1979, pp. 69–105.
[23] Mughrabi, H. in *Continuum Models of Discrete Systems 4*, Proceedings of the Fourth International Conference on Continuum Models of Discrete Systems, Stockholm, 1981, pp. 241–257.
[24] Mughrabi, H., Wang, R., Differt, K., and Essmann, U. in *Fatigue Mechanisms, ASTM STP 811*, J. Lankford, D. L. Davidson, W. L. Morris, and R. P. Wei, Eds., American Society for Testing and Materials, Philadelphia, 1983, pp. 5–45.
[25] Yamamoto, A. and Imura, T. in *Dislocation Modelling of Physical Systems*, Proceedings of the International Conference, Gainesville, FL, 1980, pp. 75–79.
[26] Imura, T. and Yamamoto, A. in *Defects, Fracture and Fatigue*, Proceedings of the Second International Symposium, Mont Gabriel, Canada, 1982, pp. 17–21.
[27] Tabata, T., Fujita, H., Hiraoka, M., and Onishi, K., *Philosophical Magazine*, Series A, Vol. 47, 1983, pp. 841–857.
[28] Shirai, H., Ph.D. Thesis, the Technological Institute, Northwestern University, Evanston, IL, 1984.
[29] Shirai, H. and Weertman, J. R., *Scripta Metallurgica*, Vol. 17, 1983, pp. 1253–1258.
[30] Pohl, K., Mayr, P., and Macherauch, E., *Scripta Metallurgica*, 1980, pp. 1167–1169.
[31] Pohl, K., Mayr, P., and Macherauch, E. in *Proceedings*, International Conference on Low Cycle Fatigue, Stuttgart, West Germany, 1979, pp. 169–193.
[32] Farsyth, P. J. E. and Stubbington, C. A., *Journal of the Institute of Metals*, Vol. 83, 1954–1955, pp. 395–399.
[33] Cottrell, A. H. and Hull, D., *Proceedings of the Royal Society of London*, Series A, Vol. 242, 1957, pp. 211–213.
[34] Mott, N. F., *Acta Metallurgica*, Vol. 6, 1958, pp. 195–197.
[35] Brown, L. M. in *Fatigue 1977* Conference Proceedings, Cambridge, U.K., 1977; also, *Metal Science*, Vol. 11, 1977, pp. 315–320.
[36] Kuhlmann-Wildsdorf, D. and Laird, C., *Materials Science and Engineering*, Vol. 27, 1977, pp. 137–156.

[37] Tanaka, K. and Mura, T., *Journal of Applied Mechanics*, Vol. 103, 1981, pp. 97–103.
[38] Essmann, U., Goesele, U., and Mughrabi, H., *Philosophical Magazine*, Series A, Vol. 44, 1981, pp. 405–428.
[39] Murakami, Y. and Nisitani, H., *Bulletin of the Japan Society of Mechanical Engineers*, Vol. 20, 1977, pp. 271–277.
[40] Murakami, Y., Tazunoki, Y., and Endo, T., *Metallurgical Transactions*, Series A, Vol. 15A, 1984, pp. 2029–2038.
[41] Nisitani, H. and Yamashita, N., *Transactions*, Japan Society of Mechanical Engineers, Vol. 242, 1966, pp. 1456–1461.
[42] Nisitani, H. and Murakami, Y., *Bulletin of the Japan Society of Mechanical Engineers*, Vol. 13, 1970, pp. 325–333.
[43] Murakami, Y., Makabe, C., and Nisitani, H., *Transactions*, Japan Society of Mechanical Engineers, Series A, Vol. 495, 1984, pp. 1828–1837.

Nucleation of Damage in Metal Polycrystals

William J. Baxter[1]

What Are the Kinetics of Slipband Extrusion?

REFERENCE: Baxter, W. J., **"What Are the Kinetics of Slipband Extrusion?,"** *Basic Questions in Fatigue: Volume I, ASTM STP 924,* J. T. Fong and R. J. Fields, Eds., American Society for Testing and Materials, Philadelphia, 1988, pp. 67–80.

ABSTRACT: It is well known that the development of persistent slipbands plays an important role in the initiation of fatigue cracks. This paper describes the application of photoelectron microscopy to measure the kinetics of the early stages of growth of persistent slipbands in 6061-T6 aluminum. The persistent slipbands (PSBs) are shown to consist of a periodic array of approximately semicircular extrusions 1 to 2 μm in diameter. Initially a single extrusion appears. The PSB elongates by the addition of further extrusions. The lengths of individual PSBs increase as the square root of the number of fatigue cycles. A model is proposed based upon an array of cells similar to the dislocation structure which has been observed in the PSBs of other materials. The cyclic strain within these cells is shown to be 10 to 100 times greater than the applied macroscopic strain.

KEY WORDS: persistent slipbands, extrusions, photoelectron microscopy

One of the earliest manifestations of fatigue deformation is the development on the surface of so-called persistent slipbands PSBs within some, but not all, of the grains of the metal. These slipbands are quite distinctive and have long been known to play an important role in, and even become sites for, the initiation of fatigue cracks (see for example Ref *1*). Many models have been proposed to explain this phenomenon in terms of irreversible dislocation motion involving such processes as cross-slip, dislocation intersection, the generation of dislocation dipoles, dislocation annihilation, and the generation of vacancies [*1*].

Perhaps the most striking feature of PSBs, however, is the extent to which thin lamella of material can be extruded [*2*]. Since this extrusion process is unique to fatigue deformation, it is fundamental to the understanding of the early stages of metal fatigue. In many materials, such as copper, aluminum, and iron, it has been shown that the fatigue process also generates so-called ladder and subgrain or cell dislocation structures, particularly in the near-surface regions of the PSBs [*3*]. Thus, the development of extrusions and of these dislocation structures appear to be manifestations of the same process. Indeed the extrusions produced on fatigued crystals of copper have been correlated with the dislocation cell structure [*4*]. A mechanism for the formation of these cells has been described by Kuhlmann-Wilsdorf and Nine [*5*], while more recently Essmann, Mughrabi, and co-workers [*6–8*] have developed a detailed two-stage model of the extrusion process. The first stage is attributed to the annihilation of edge dislocations and vacancy production, while in the second stage the extrusion develops a rough profile by a random process of irreversible slip.

But despite the fundamental importance of PSBs and the effort that has been devoted to analyzing their structure, relatively little attention has been given to the kinetics of their development. This is particularly true from the experimental viewpoint. One reason for this

[1] Senior staff research scientist, Physics Department, General Motors Research Laboratories, Warren, MI 48090-9005.

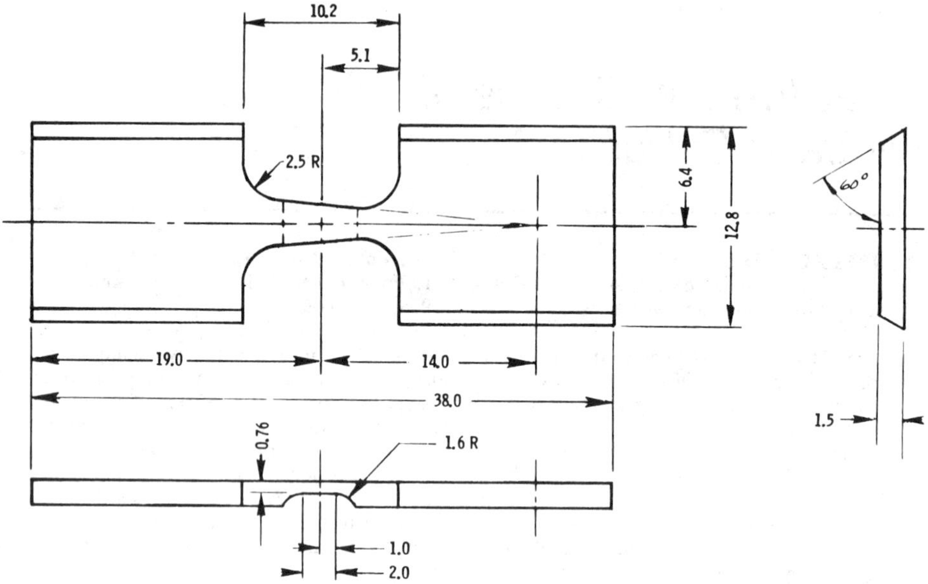

FIG. 1—*Geometry of specimens used in the photoelectron microscope. Dimensions in millimetres.*

is that conventional techniques of surface examination are postmortem in nature, so do not provide sequential information at intervals during a fatigue experiment. Furthermore, the only reported observations are in conflict. Mughrabi et al. [*8*] observed that for copper crystals "The extrusion growth is rapid initially and becomes progressively slower," whereas Lee and Laird [*9*] reached the opposite conclusion in their study of an aluminum alloy.

This paper describes the application of photoelectron microscopy [*10*] to monitor the emergence of PSBs in 6061-T6 aluminum. This technique relies upon the slipband extrusions rupturing a very thin (14 nm) anodic oxide film. In the photoelectron microscope (PEM) the extruded material is imaged as sources of strong electron emission (exoelectrons) [*11*], so that the growth of individual slipbands can be monitored at intervals during a test. A periodic array of small extrusions is shown to develop along the slipband. Initially a single extrusion forms, then the slipband elongates by the addition of more extrusions. This elongation follows a parabolic dependence on the number of fatigue cycles. A simple dislocation model is proposed to explain this behavior in terms of a localized concentration of the cyclic strain within the slipband.

Experimental

Specimens of 6061-T6 aluminum were machined from sheet material 1.5 mm thick, with an average grain size of ~30 μm, and the surface to be examined was mechanically polished. The specimens were degreased with acetone and cleaned by immersion in chromic acid at 75°C for 5 min. After rinsing with water and alcohol, they were anodized in a 3% solution of tartaric acid at a potential of 10 V to form a surface oxide film 14 nm thick.

The goemetry of the specimens for the PEM is shown in Fig. 1. These specimens were fatigued by reverse bending in the specimen chamber of the microscope, to produce surface maximum cyclic strains of $\pm 3.0 \times 10^{-3}$. In the photoemission microscope, electrons emitted from the specimen form a magnified image of the surface on a fluorescent screen [*12*]. As the specimen is fatigued, the development of microcracks in the surface oxide film is identified by the appearance of regions of more intense electron emission (exoelectrons). These appear as white spots or lines on the micrographs shown in the next section.

Results

A typical sequence of photoelectron micrographs, obtained at intervals during fatigue cycling, is shown in Fig. 2. The three long straight lines of emission in these micrographs correspond to fiducial scratches, which identify the location on the surface of the specimen and provide a calibration of magnification. This sequence shows the development of many sources of so-called exoelectron emission, where the 14-nm oxide film has been ruptured by emerging slipbands. A distinctive feature is that in general each slipband appeared initially as a small spot of exoelectron emission and subsequently elongated with continued fatigue cycling. For example, this behavior is clearly illustrated by the slipbands labeled B, and D_3. The slipband at "A" was unusual in that it was first observed as two spots (Fig. 2*a*), which quickly linked together upon further fatigue cycling.

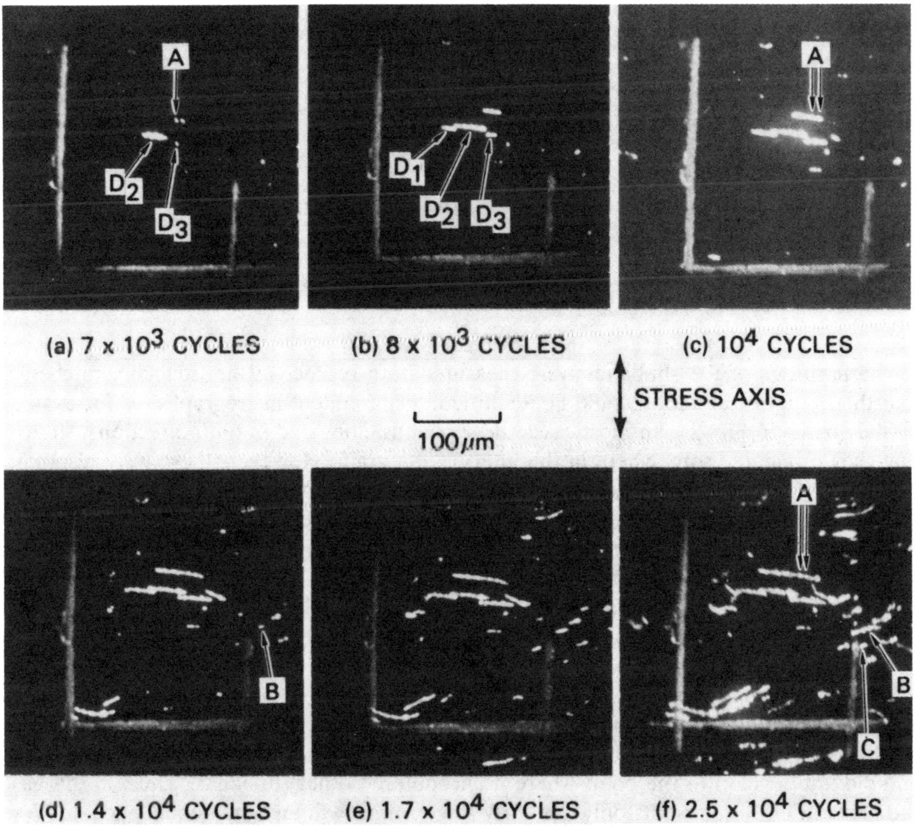

FIG. 2—*Photoelectron micrographs showing development of exoelectron emission from slipbands produced by fatigue cycling of 6061-T6 aluminum.*

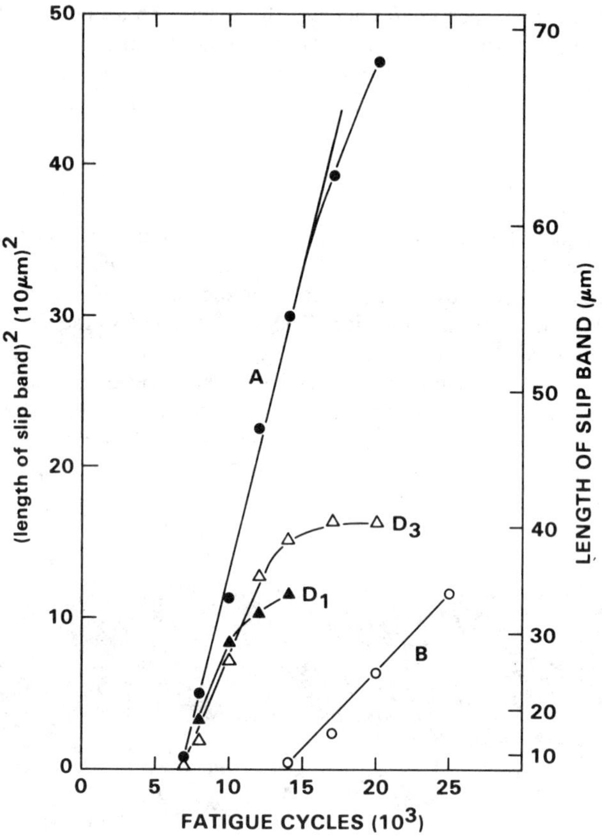

FIG. 3—*Effect of fatigue cycles* (N) *on the square of the length of Slipbands A, B, D_1, and D_3 in Fig. 2.*

The positions of the slipbands were measured with respect to the fiducial scratches, so that their point of initiation could be identified in subsequent micrographs, as for example by the arrows in Fig. 2*f*. In most cases the slipbands elongated in both directions, showing that they originated somewhere in the interior of a grain. However, there were also many situations where the bands elongated primarily in one direction, suggesting that the initiation sites were close to a grain boundary. Slipband "A" provides an example of this behavior as shown by the twin initiation sites indicated by arrows in Fig. 2*c*, *e*, and *f*. The initiation and growth of Slipband D_2 was overlooked; it was already well developed after only 7×10^3 cycles (Fig. 2*a*), and stopped growing after 10^4 cycles (Fig. 2*c*). However, the growth of the parallel slipbands D_1 and D_3 in close proximity to D_2 is well documented.

The elongation of four slipbands is summarized quantitatively in Fig. 3. In the early stages of growth the length of each slipband increased as $(N - N_o)^{1/2}$, where N is the number of fatigue cycles, and N_o is the number of cycles for the initial onset of emission. Slipband B was selected because it did not appear until quite late in the fatigue experiment (Fig. 2*d*) and did not develop to the point where it encountered other slipbands. Thus, in this case, the plot in Fig. 3 represents only the early stages of growth and the dependence on $(N - N_o)^{1/2}$ is maintained throughout.

In the later stages of growth this square-root relationship no longer holds, as illustrated by the more mature slipbands A and D_3. For Slipband A this departure occurred once its length exceeded ~60 μm, a dimension commensurate with the onset of a grain boundary interaction. The cause is more apparent for D_1 and D_3: each encountered either a grain boundary or another slip system at one end, while the other end appears to have been held up by an interaction with the neighboring slipband D_2.

Another noteworthy feature of the photoelectron micrographs in Fig. 2 is that many of the slipbands appear to consist of rows of spots of exoelectron emission, particularly at the end of the fatigue experiment (Fig. 2*f*). Subsequent examination of this specimen in a scanning electron microscope revealed that this was due to a striking periodicity in the structure of the slipband extrusions. This is illustrated by the scanning electron micrograph (SEM) in Fig. 4 which shows the slipbands A, D_1, D_2, and D_3. (Note the left-to-right mirror image relationship between Fig. 2*f* and Fig. 4.) A detailed comparison of this micrograph with the photoelectron micrographs in Fig. 2 provides an identification of the two initiation spots of Slipband A which appeared at 7000 cycles (Fig. 2*a*), as well as the approximate progress of this slipband as indicated by the arrows in Fig. 4.

The dominant feature of the slipbands shown in Fig. 4 is the regular array of individual extrusions, each one extending ~1 to 2 μm along the slipband and having either a semicircular or triangular shape.[2] These extrusions can be seen more clearly at higher magnification in Fig. 5, where each extrusion can be visualized as a tongue of material. These extrusions are the most pronounced on the mature or well-developed slip lines. For example, the micrograph in Fig. 5*a* shows pronounced extrusions on the earliest portion of Slipband A, which first appeared between 7×10^3 and 10^4 cycles; the micrograph in Fig. 5*b* shows the portion which appeared between 1.2×10^4 and 2×10^4 cycles, and the extrusions are smaller toward the (newer) end of the slip line; the other end of the slipband did not propagate until 2×10^4 cycles, and this newest portion does not exhibit pronounced extrusions, those visible in Fig. 5*c* being a part of the earliest portion of the slipband (see Fig. 4).

This difference between "old" and "new" slipbands was confirmed by surveying many others. For example Slipband B (Fig. 2), which did not appear until 1.4×10^4 cycles, had begun to develop some small extrusions (Fig. 6) by 2.5×10^4 cycles (when the test was truncated). But the most recently formed slipbands, for example C in Fig. 2*f*, were scarcely visible in the scanning electron microscope. Nevertheless, C was clearly visible in the more surface-sensitive PEM (Fig. 2*f*). Moreover it appeared as a pair of spots, showing that even at this very early stage of development the slipband already consists of two infant extrusions.

Discussion

These observations, of the rupturing of a surface oxide film during fatigue of 6061-T6 aluminum, have revealed the following (partial) answers to the question of the growth kinetics of persistent slipband extrusions in this alloy.

1. The persistent slipbands in 6061-T6 aluminum appear on the surface as a periodic linear array of thin, approximately semicircular, extrusions ~1 to 2 μm in diameter.

[2] These initial extrusions are judged to be planar, in keeping with the planar geometry of the more-pronounced extrusions which develop after more-prolonged fatigue cycling.

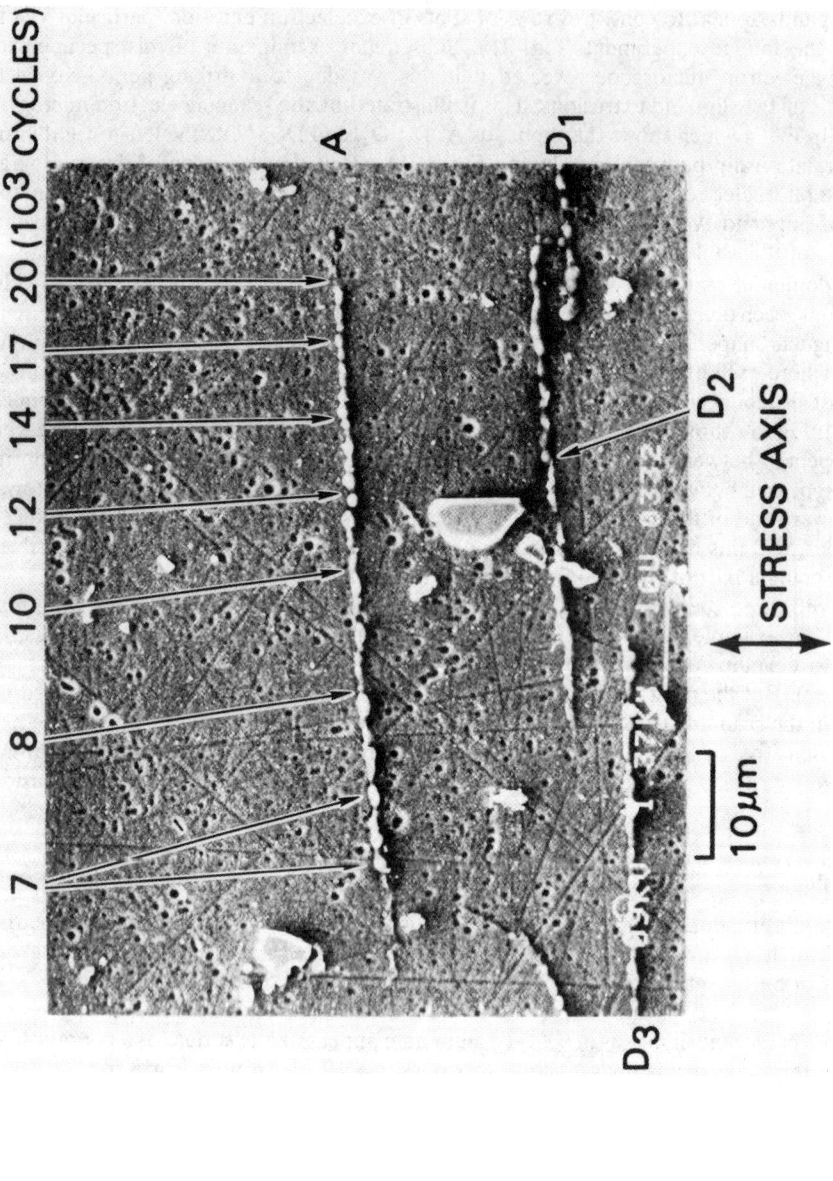

FIG. 4—*Scanning electron micrograph of Slipbands A and D in Fig. 2. Arrows indicate approximate length as a function of fatigue cycles.*

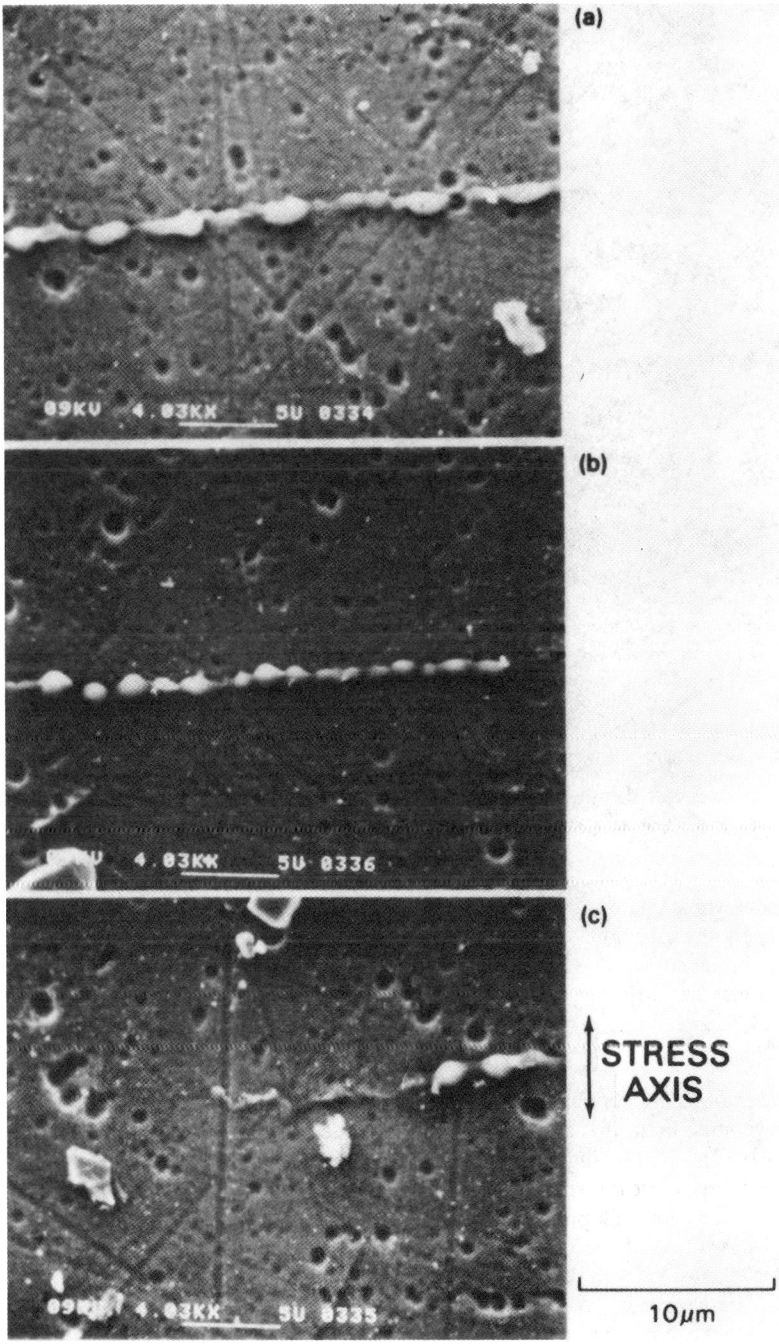

FIG. 5—*Scanning electron micrographs of three portions of Slipband A:* (a) *appeared between 7000 and 10 000 cycles,* (b) *appeared between 12 000 and 20 000 cycles, and* (c) *appeared between 17 000 and 25 000 cycles.*

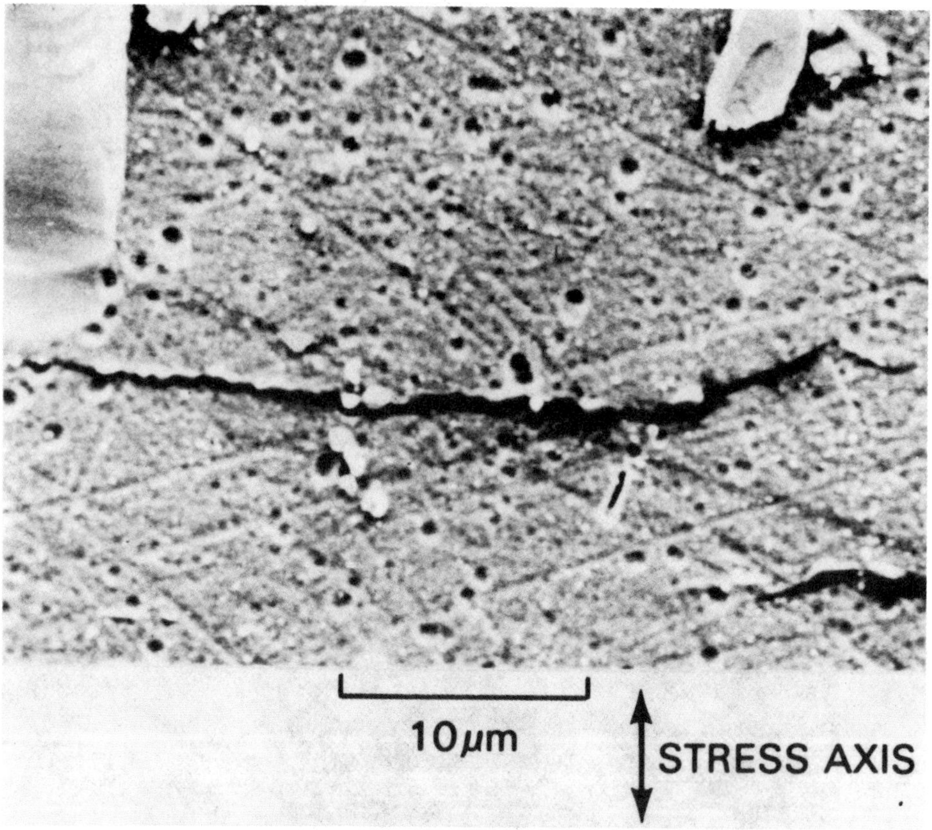

FIG. 6—*Scanning electron micrograph of Slipband B in Fig. 2 (6061-T6 aluminum).*

2. Initially a single extrusion is formed. Then the length of the slipband increases incrementally by the addition of further extrusions, while the initial ones continue to become more pronounced.
3. During the early stages of growth the length of an individual slipband increases as $(N - N_o)^{1/2}$.

This parabolic law for slipband elongation across the surface of a grain is of interest from two viewpoints. First, it is consistent with the vertical extrusion rates reported by Mughrabi et al. [*8*]. (The conflicting data of Lee and Laird [*9*] may have been influenced by the periodic polishing of the specimen during their experiments.) Second, it suggests the possibility of a random walk process, as has been proposed in the literature [*6–8,13*]. However, these observations of the early stages of extrusion, namely, the sequential emergence of individual extrusions with very smooth surfaces, are, in the author's opinion, more suggestive of a systematic process. A random process is more consistent with the irregular nature of the extrusions which develop after more prolonged cycling, in keeping with the model of Mughrabi and co-workers [*6,7*].

The new information revealed in this study immediately raises the more fundamental and often-posed question: "What is the underlying mechanism responsible for this behavior?"

An important clue is provided by the characteristic periodic nature of the slipband extrusions (Fig. 5), which shows that within the slipband there exists an array of barriers to dislocation motion with a spacing of ~1 μm. What are these barriers? This question could be answered experimentally by transmission electron micrographs (TEMs) of the dislocation structure in the near-surface region of the slipbands. Unfortunately, such information for 6061-T6 aluminum is lacking at the present time. However, the 1 to 2-μm tongues of extruded material are not unlike the extrusions of similar magnitude observed by Laufer and Roberts [*4*] on fatigued crystals of copper.

In the Laufer and Roberts study the periodic extrusions were correlated directly with the subgrain or dislocation cell structure of the PSB. These cells consisted of walls of high dislocation density with cell interiors of much lower dislocation density, and a typical cell dimension was 1 to 2 μm. Similar cell structure has also been observed in the near-surface region (that is, within 100 μm) of PSBs of polycrystalline copper and aluminum, single crystals of copper-zinc, and Al-1Mg alloys, as well as in fatigued iron [*3*]. In all cases, the cells are ~1 to 2 μm wide.

Thus, there is circumstantial evidence to suppose that such a cell structure could be present in the surface layers of persistent slipbands in other materials, including 6061-T6 aluminum, and that the cell structure would again correlate with the periodic extrusions observed in this study. (Similar arguments could apply to the ladder type of dislocation structure). In any event, it is certainly true that the periodic nature of the extrusions must originate from a systematic array in the microstructure with a characteristic dimension of ~1 to 2 μm, and that the elements in that array must be able to deform to such an extent that large numbers of dislocations can be emitted from the surface. Thus, whatever their specific nature, it is certainly appropriate to refer to these entities as "cells," and for purposes of discussion we shall consider a dislocation cell structure simply to illustrate our model.

Growth of a Slipband

It is clear that the early stages of formation of a persistent slipband must be considered in terms of the sequential growth of the small individual extrusions. Thus, the results obtained with the PEM provide quantitative information on the kinetics of irreversible[3] dislocation motion, not simply within a single grain, but rather within much smaller (~1 μm diameter) "cells," and the subsequent cooperative action of arrays of "cells" within the slipband.

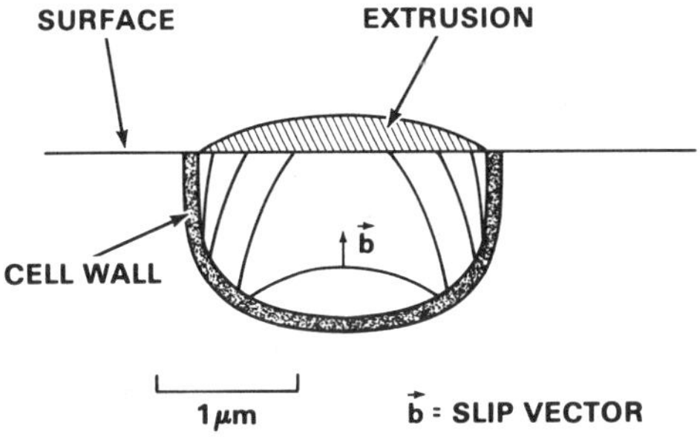

FIG. 7—*Schematic illustration of formation of single extrusion by stress-induced release of edge dislocations from the cell wall.*

[3] Under cyclic stress, dislocations will obviously move back and forth, the surface extrusion resulting from the irreversible component of this process.

An important feature of these PSBs is that they appear initially in the PEM usually in the form of a single spot, occasionally as two spots (feature A in Fig. 2), but never as many spots. Thus there are not several initiatory cells in the slipband, as would be expected to occur if the cell structure was fully developed across the entire grain prior to the appearance of any extrusions. Conversely, the initiation of a single extrusion, followed by the systematic addition of other extrusions, shows that the cell structure is developing either in concert with, or shortly before, the appearance of the single cell extrusions. Thus, the location of the initiatory cell and extrusion is probably determined by an inherent weak spot within the grain, where deformation and the associated buildup of dislocations occur preferentially. After a while the dislocation density in that region increases to the point of instability, so that it is energetically favorable for the dislocations to rearrange into a cell structure, by a mechanism such as that proposed by Kuhlmann-Wilsdorf and Nine [5]. Once the initiatory cell has formed, the sudden ease of dislocation motion within the cell can now result in the rapid growth of an extrusion.

For example, the shear component of an applied stress may bow edge dislocations out from the cell wall, as has been observed by Mughrabi [*14*]. When these dislocations reach the surface, a slip step forms (Fig. 7) as the screw components approach and either pile up against, or become incorporated into, the sidewalls of the cell. In this way, the first extrusion will appear as a single spot in the photoelectron image (Fig. 2).

If after many cycles this initiatory cell has produced a small extrusion, say ~100 nm high, ~300 dislocations will have accumulated irreversibly at the sidewalls of the cell. This is far too many to be accommodated in a pileup, so most will either be incorporated into or effectively penetrate the sidewalls. Thus, we can expect a large increase in dislocation density in the neighborhood outside the cell, and the development of internal stresses adjacent to the cell walls. (Mughrabi [*15*] has shown that the local stress near a cell wall in fatigued copper is double the applied cyclic stress.) These two factors, acting alone or together, may be sufficient to trigger a repetition of the processes of cell formation and extrusion. In this way, the PSB could develop as an array of dislocation cells and a succession of small extrusions of the type shown in Fig. 5.

It should be noted, however, that while the initiation of the extrusions is sequential, all subsequently grow in unison to some extent. This is manifested by the fact that freshly formed slipbands differ from "old" slipbands in appearance. On occasion there are distinct gaps between extrusions (Fig. 5), suggesting that some extrusions were not completely formed.

Strain Concentration in a Slipband

Since $\ell \sim (N - N_o)^{1/2}$, the rate of elongation of the PSB is of the form

$$\frac{d\ell}{dN} \sim \ell^{-1} \tag{1}$$

Thus, the rate at which a new cell and associated extrusion are generated at the tip of a slipband decreases as the slipband elongates and contains more cells. This behavior can be accounted for in terms of a simple model of the concentration of cyclic strain in the slipband.

Let us assume that the cells within the PSB are essentially identical, so that each cell adjacent to the surface accommodates the same amount of cyclic strain, $\Delta\epsilon_c$. The propagation of this PSB requires the creation of a new cell and extrusion at the tip. This additional cell is generated by the emission of dislocations from the leading cell of the PSB. The rate at which this process occurs is assumed to be proportional to the cyclic strain within the cells, that is

$$\frac{d\ell}{dN} = \beta \Delta\epsilon_c \tag{2}$$

where β is a constant.

Consider a grain of diameter D in the surface of the specimen which, as depicted in Fig. 8, has developed a persistent slipband of length ℓ, and thickness t. In a constant-displacement fatigue test, as in these experiments, the total surface cyclic strain ($\Delta\epsilon_T$) accommodated by the grain is constant, and is usually perpendicular to the slipband as viewed in Fig. 8. (See, for example, slipbands in Fig. 2.) In the absence of a PSB this applied strain produces a cyclic change in the diameter of the grain of $D\Delta\epsilon_T$. But when a PSB is present this cyclic strain is no longer uniformly distributed throughout the grain, because the interior of the cells in the PSB are known to be softer than the matrix. For simplicity, the strain distribution across the cell/matrix interface will be ignored, so that the strain can be conveniently divided into independent components. Thus we can write

$$D\Delta\epsilon_T = t\Delta\epsilon_{PSB} + (D - t)\Delta\epsilon_g \tag{3}$$

where $\Delta\epsilon_{PSB}$ is the cyclic strain in the band of material containing the PSB, and $\Delta\epsilon_g$ is the cyclic strain in the matrix. But at this point the PSB does not extend across the entire grain, so

$$\Delta\epsilon_{PSB} = \frac{\ell}{D}\Delta\epsilon_c + \frac{D - \ell}{D}\Delta\epsilon_g \tag{4}$$

where $\Delta\epsilon_c$ is the cyclic strain within the cells of the PSB. (More specifically it is the component of the shear strain parallel to the surface.)

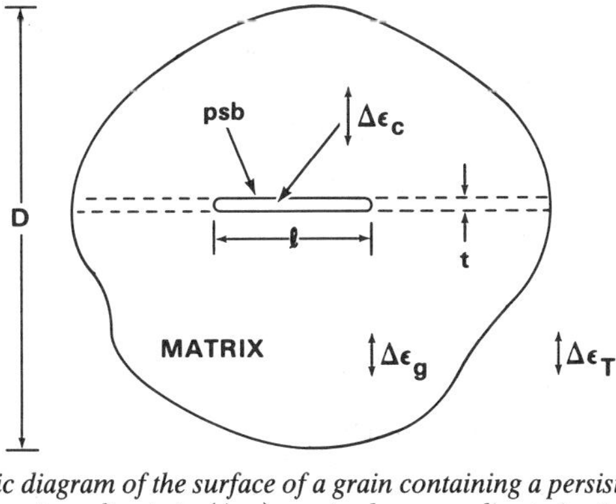

FIG. 8—*Schematic diagram of the surface of a grain containing a persistent slipband (PSB). The applied macroscopic cyclic strain ($\Delta\epsilon_T$) creates larger cyclic strains in the PSB ($\Delta\epsilon_c$) than in the matrix ($\Delta\epsilon_g$).*

Combining Eqs 3 and 4 yields the following expression for the cyclic strain within the cells

$$\Delta\epsilon_c = \frac{D^2}{t\ell}(\Delta\epsilon_T - \Delta\epsilon_g) + \Delta\epsilon_g \quad (5)$$

Substituting in Eq 2

$$\frac{d\ell}{dN} = \frac{\beta D^2}{t\ell}(\Delta\epsilon_T - \Delta\epsilon_g) + \beta\Delta\epsilon_g \quad (6)$$

The first term in this equation is of the same form as the empirical Eq 1. Indeed if the second term can be ignored, integration yields

$$\ell^2 = \frac{2\beta D^2}{t}(\Delta\epsilon_T - \Delta\epsilon_g)(N - N_o) \quad (7)$$

where N_o is the number of cycles required to initiate the first single cell extrusion. This equation is consistent with the results plotted in Fig. 3.

Thus, it may be concluded that the first term in Eq 5 is dominant. This term is simply that portion of the cyclic strain within the cells which is in excess of the cyclic strain experienced by the rest of the grain ($\Delta\epsilon_g$). In other words, it is the process of strain localization within the cells that controls the elongation of the PSB. At this point it is of interest to apply Eq 5 to the case of Slipband A in Figs. 3–5, where $D \sim 70$ μm, $t \sim 0.1$ μm, and $\Delta\epsilon_T = 6 \times 10^{-3}$. For values of $\ell < D$, the geometrical strain concentration factor $D^2/t\ell > 700$, so

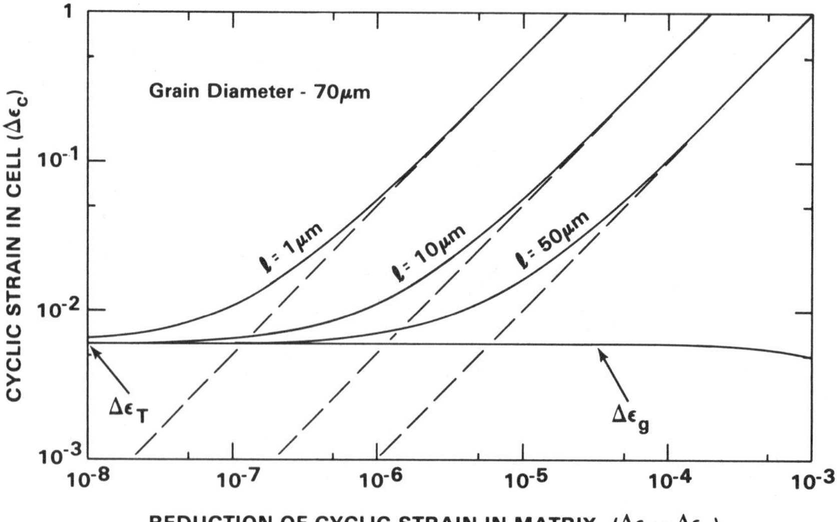

FIG. 9—*Effect of the reduction of cyclic strain in the matrix on the cyclic strain in the cells of the PSB, as calculated from Eq 5 for three values of the length (ℓ) of the PSB in a grain of diameter = 70 μm.*

that a substantial concentration of strain in the cells can be realized by an extremely small reduction of the strain in the matrix of the grain. This is portrayed graphically in Fig. 9, where the calculated value of $\Delta\epsilon_c$ is shown for three values of ℓ. Note that in each case when $\Delta\epsilon_c = 10\Delta\epsilon_T$, the reduction in $\Delta\epsilon_g$ is essentially imperceptible. The dashed lines represent the values of the first term in Eq. 5. Thus it is only the upper portion of the curves, where this component dominates, that the parabolic growth law will hold.

For PSB A the parabolic relationship holds for values of $\ell \leq 50$ μm with a deviation of <10% (Fig. 3). A deviation of this magnitude from the upper linear portions of the curves in Fig. 9 occurs when $\Delta\epsilon_c = 6 \times 10^{-2}$. Thus we conclude that even when $\ell = 50$ μm, the cyclic strain in the cells of PSB A was $\geq 6 \times 10^{-2}$. During the earlier stages of growth, these strains would have been even larger. For example, PSB A was ~10 μm long when it was first observed, so at that point, according to Eq 5, $\Delta\epsilon_c \approx 3 \times 10^{-1}$. Similarly for the first cell, we obtain $\Delta\epsilon_c \sim 1$.

From the curve for $\ell = 50$ μm in Fig. 9 we see that the large values of $\Delta\epsilon_c$ resulted from a reduction in the matrix strain of only $(\Delta\epsilon_T - \Delta\epsilon_g) = 6 \times 10^{-5}$. If we compare Eq 7 with the results for PSB A in Fig. 3, we find that $\beta(\Delta\epsilon_T - \Delta\epsilon_g) = 4.2 \times 10^{-6}$. Therefore, the constant $\beta = 7 \times 10^{-2}$ μm/cycle. This constant can be regarded as a measure of the efficiency with which the cyclic strain in the leading cell of the PSB generates a new cell. Since the cell diameter is ~2 μm, the maximum cyclic distortion would be $2\Delta\epsilon_c$ μm. Thus, only ~3% of $\Delta\epsilon_c$ is consumed in this process.

Summary

The following picture emerges from these measurements and modeling of the growth of persistent slipbands in 6061-T6 aluminum.

1. The PSB consists of an array of approximately semicircular extrusions ~1 to 2 μm in diameter.
2. The PSB elongates by the addition of further extrusions, the total length increasing as $(N - N_o)^{1/2}$.
3. Each extrusion is thought to be associated with a characteristic cell in the microstructure of the PSB.
4. The cyclic strains ($\Delta\epsilon_c$) within the cells of the PSB are very large, encompassing a range 10 to 100 times greater than the applied macroscopic strain.
5. The value of $\Delta\epsilon_c$ decreases as the PSB elongates.
6. The cyclic strain in the leading cell at the tip of a PSB ejects a high density of dislocations into the adjacent material, which in some manner creates a new cell and associated extrusion.

The proposed model of the growth of a PSB is far from complete, but serves to illustrate that many questions remain. Further measurements of the kinetics of the emergence of PSBs can contribute significantly in this regard. Further experiments in the PEM are planned wherein observations at higher magnification and more frequent intervals could reveal more detailed information. In particular, the influence of grain diameter on the rate of elongation predicted by Eq 10 will be investigated. In addition, an examination of other materials, such as pure aluminum and copper, where a dislocation cell structure is known to form would provide an interesting comparison. But to place the model on a firm foundation it is essential to characterize the dislocation structures in the near-surface region of a PSB in 6061-T6 aluminum by transmission electron microscopy.

Acknowledgments

The author is grateful to T. R. McKinney for the photoelectron micrographs in Fig. 2 and to W. Lange for the scanning electron micrographs in Figs. 4, 5, and 10.

References

[*1*] Laird, C. and Duquette, D. J., "Corrosion Fatigue: Chemistry, Mechanics and Microstructure," NACE-2, National Association of Corrosion Engineers, 1971, pp. 88–117.
[*2*] Forsyth, P. J. E., *The Physical Basis of Metal Fatigue,* Blackie and Son, London, 1969.
[*3*] Lukas, P. and Klesnil, M., "Corrosion Fatigue: Chemistry, Mechanics and Microstructure," NACE-2, National Association of Corrosion Engineers, 1971, pp. 118–132.
[*4*] Laufer, E. E. and Roberts, W. N., *Philosophical Magazine,* Vol. 14, 1966, pp. 65–78.
[*5*] Kuhlmann-Wilsdorf, D. and Nine, H. D., *Journal of Applied Physics,* Vol. 38, 1967, pp. 1683–1693.
[*6*] Essmann, U., Gosele, U., and Mughrabi, H., *Philosophical Magazine,* Vol. 44, 1981, pp. 405–426.
[*7*] Essmann, U., *Philosophical Magazine,* Vol. 45, 1982, pp. 171–190.
[*8*] Mughrabi, H., Wang, R., Differt, K., and Essmann, U. in *Fatigue Mechanisms: Advances in Quantitative Measurement of Physical Damage, ASTM STP 811,* American Society for Testing and Materials, Philadelphia, 1983, pp. 5–45.
[*9*] Lee, J. K. and Laird, C., *Philosophical Magazine,* Vol. 47, 1983, pp. 579–597.
[*10*] Baxter, W. J. and Rouze, S. R., *Metallurgical Transactions,* Vol. 7A, 1976, pp. 647–654.
[*11*] Baxter, W. J., *Treatise on Materials Science and Technology,* Vol. 19, *Experimental Methods, Part B,* Academic Press, New York, 1983, pp. 1–66.
[*12*] Baxter, W. J. and Rouze, S. R., *Review of Scientific Instruments,* Vol. 44, 1973, pp. 1628–1629.
[*13*] May, A. N., *Nature,* London, Vol. 185, 1960, p. 303.
[*14*] Mughrabi, H., Ackermann, F., and Herz, K. in *Fatigue Mechanisms, ASTM STP 675,* American Society for Testing and Materials, Philadelphia, 1979, pp. 69–105.
[*15*] Mughrabi, H., *Acta Metallurgica,* Vol. 31, 1983, pp. 1367–1379.

Alan Plumtree[1] *and Linda D. Pawlus*

Substructural Developments During Strain Cycling of Wavy Slip Mode Metals

REFERENCE: Plumtree, A. and Pawlus, L. D., **"Substructural Developments During Strain Cycling of Wavy Slip Mode Metals,"** *Basic Questions in Fatigue: Volume I, ASTM STP 924,* J. T. Fong and R. J. Fields, Eds., American Society for Testing and Materials, Philadelphia, 1988, pp. 81–97.

ABSTRACT: Substructural developments have been monitored during strain-controlled cycling of aluminum at room temperature. This has allowed a relationship between dislocation substructure and stress response to be developed as a function of the number of cycles.

At the onset of saturation, the dislocations were arranged in a loose cellular network for the particular strain range investigated. Continued cycling resulted in a slight but significant decrease in stress response, associated with the formation of a stable structure. With further strain cycling, the stress response and cell size remained relatively constant. The misorientation between the cells also remained constant, although the boundary width decreased due to dynamic recovery. These results indicate that the cyclic stress response is directly related to the dislocation density, and that at saturation the energy absorbed per cycle causes important changes in cell morphology to take place.

Comparison of these results with those of aluminum, copper and iron obtained under both monotonic and cyclic conditions show that the stress-cell size relationship is characteristic for all these wavy slip mode metals.

KEY WORDS: substructural changes, misorientation measurements, strain cycling, dynamic recovery, aluminum

Many investigations have been undertaken to study the formation of substructures during the deformation of metals. Some of these studies have been summarized in recent reviews [*1–5*]. These investigations include research on the substructures formed as a consequence of unidirectional deformation at low [*6–11*] as well as at high temperature [*6,7,10–12*], also during creep [*13–15*] and cyclic deformation [*16–24*]. This study will concentrate on the latter particularly with high strain in mind.

Some work has been undertaken in the area of substructural development during cyclic deformation from the onset of saturation to fracture. Wang and Mughrabi [*25*] showed that microstructural changes occurred in copper monocrystals and polycrystals by the slow increase in slip activity leading to the formation of a cell structure with increasing misorientation. In cyclically deformed α-Fe monocrystals and polycrystals the dislocation cell walls became progressively sharper during saturation, accompanied by increasing misorientations [*17,26,27*].

In the present investigation, the study of substructural development will be taken further. Cell morphology and misorientation changes will be studied in aluminum cyclically strained at room temperature. It is hoped that this work will increase the understanding of the microstructural changes which occur during cyclic deformation.

[1] Professor and former research assistant, respectively, Department of Mechanical Engineering, University of Waterloo, Waterloo, Ontario, Canada. Coauthor Pawlus is now with AMP of Canada, Ltd., Markham, Ontario, Canada.

TABLE 1—*Chemical composition of aluminum studied.*

Elements	Fe	Si	V	Ga	Ti	Cu + Mg + Zn	Al
Weight %	0.12	0.04	0.01	0.02	0.005	<0.01	balance

Experimental Procedure

Polycrystalline aluminum (chemical composition given in Table 1) was received in the form of a 2.54-cm square bar which had been extruded at 400°C from a 17.8-cm-diameter direct chill cast ingot. Due to the inhomogeneity of the grain structure which resulted in a variation in fatigue results during earlier studies (Kayali and Plumtree [*24*]), thermomechanical processing was performed on the as-received material.

The bars were first cut to the approximate length of the fatigue specimens. The gage length was then forged at room temperature to 70% reduction in area. The specimens were subsequently annealed at 600°C for 2 h. This high temperature resulted in a large but uniform grain size of 1.8 mm. Fatigue specimens were machined to 12.7-mm diameter and a gage length of 7.62 mm. To avoid any grip effects, a generous shoulder radius of 101.6 mm was chosen. After machining, the specimens were carefully polished using fine alumina powder.

The specimens were tested in fully reversed strain control at room temperature using an MTS universal testing machine. Total strain ranges of 0.8% to 3.0% were used with corresponding frequencies varying between 0.25 Hz and 0.066 Hz so that the same average strain rate of 4×10^{-3} s^{-1} was used for all testing conditions. The majority of tests were carried out at a frequency of 0.2 Hz and total strain range of 1%. The strain was applied using a sinusoidal waveform and the stress response was continuously recorded on an *x-y* recorder. The maximum stress response was also periodically recorded from a digital voltmeter.

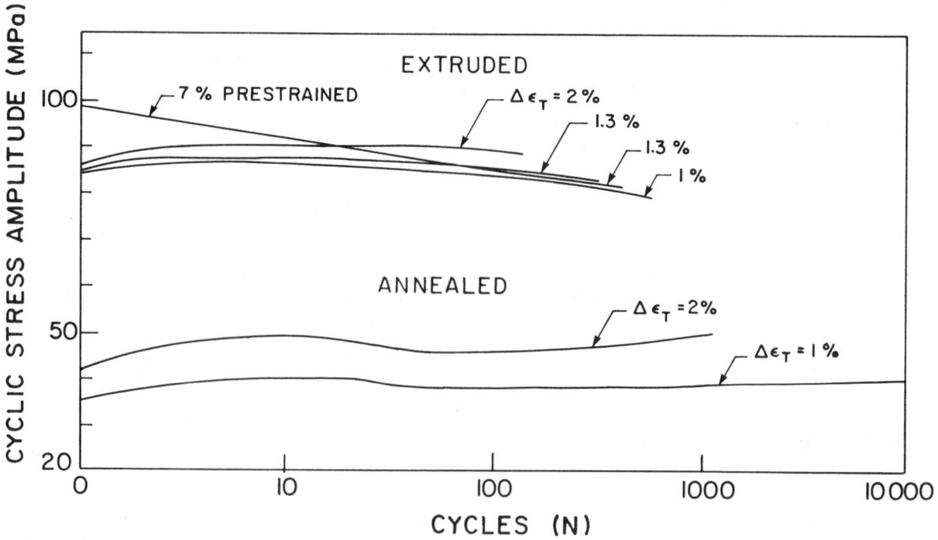

FIG. 1—*Stress response of strain cycled annealed and as-received extruded aluminum sample.*

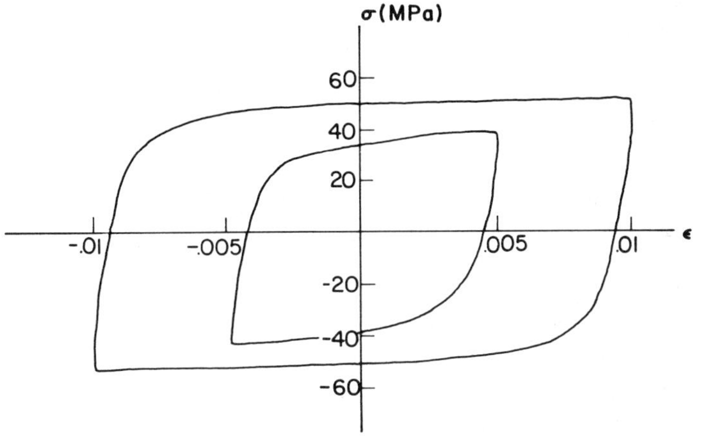

FIG. 2—*Stress-strain hysteresis loops for annealed aluminum cycled at* $\Delta\epsilon_T$ = *1.0% and 2.0%.*

For the substructural investigations each specimen was cycled at a total strain range of 1% for a different number of cycles. This resulted in a collection of specimens representing the condition of the material at various stages of fatigue life which varied from the onset of saturation to failure. The dislocation substructure was examined using transmission electron microscopy. Thin foils were prepared by cutting transverse sections (approximately 0.76 mm thick) perpendicular to the applied strain axis from the gage length. Smaller disks of 3-mm diameter were punched out and thinned by an electrolytic jet polisher. The polishing solution used was 33% nitric acid in methyl alcohol which was kept at 0°C by a water-ice bath. The current was maintained at 250 to 300 mA.

The foils were examined at 100 kV in a Phillips 300 transmission electron microscope (TEM) using a double tilt holder and a cold finger to prevent contamination. The substructural arrangements were noted and photographed. Corresponding Kikuchi patterns were recorded on photographic plate for each cell in the area under investigation. Other regions of the foil were studied in order to obtain a representative distribution of the cell sizes and Kikuchi patterns for each specimen.

Comparative stability of the cell structures at the onset of saturation and at failure was undertaken by annealing the respective thin foils in the TEM using a hot stage. The temperature of the foils was increased at an approximate rate of 10°C/min. and the sequence of changes in each cell structure was recorded on photographic plate.

Results and Discussion

Cyclic Stress-Strain Behavior

The respective stress versus number of cycles plots for the specimens cycled at a total strain range of 1% to 2% are shown in Fig. 1. As expected for an annealed material, all specimens showed an initial hardening stage. During this rapid hardening stage a substructure consisting of dislocation loop patches developed, which impeded dislocation glide, thereby requiring progressively higher applied stresses in order for deformation to continue. The rate of hardening was observed to decrease with increasing number of cycles until the maximum stress response reached steady-state. Once saturation was reached, the stress response remained relatively constant until a large crack developed, after which the stress

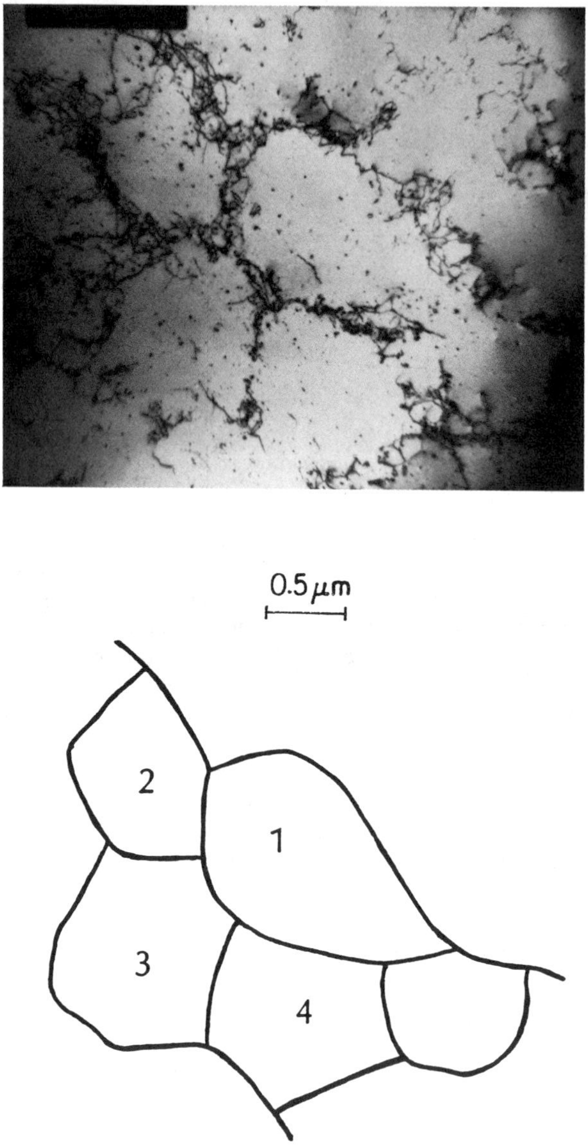

Cells	θ
1/2	2.65°
1/3	3.87°
2/3	3.47°

FIG. 3—*Substructure and misorientation measurements after 30 cycles at* $\Delta\epsilon_T$ = *1.0%.*

decreased. Failure, which was defined as zero stress response, took place after 11 300 cycles at 1% total strain range and 1470 cycles at 2% total strain range.

The hysteresis loops generated during saturation also remained relatively constant. Representative loops are shown in Fig. 2. It will be noted that the amount of plastic strain was quite large as seen by the width of the hysteresis loops at zero stress. The plastic strain range during saturation was 0.875 and 1.875% for the total strain range of 1 and 2%, respectively.

The maximum stress response during saturation increased from 40 MPa to 48 MPa as the applied strain range was increased from 1% to 2%. This was expected.

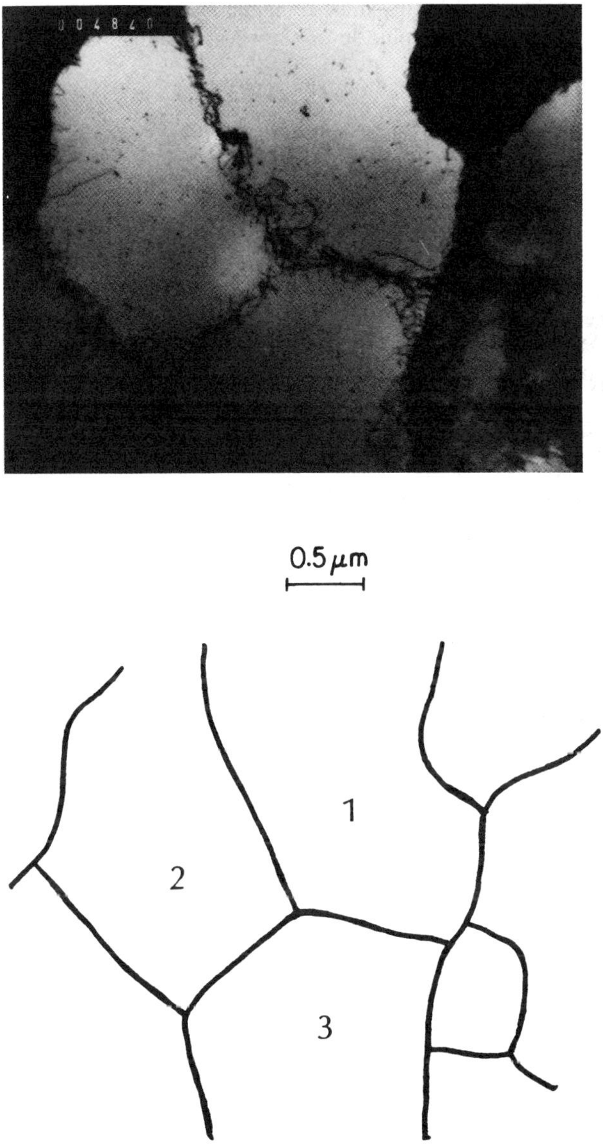

Cells	θ
1/2	1.72°
1/3	2.20°
2/3	2.06°

FIG. 4—*Substructure and misorientation measurements after 5500 cycles at* $\Delta\epsilon_T = 1.0\%$.

During saturation, the maximum stress response has been related to the dislocation density by [*11,24,28*]

$$\tau = \tau_0 + \alpha G b \sqrt{\rho} \tag{1}$$

where

τ = saturation shear stress,
τ_0 = frictional shear stress,
α = a material constant, and
G = shear modulus.

According to this equation, the dislocation density must remain constant in order that the maximum shear stress and its related stress amplitude $\Delta\sigma/2(=\sigma_s)$ remain constant during saturation.

Substructure

The morphology of the cell structure was found to change with increase in the number of cycles. At the onset of saturation (30 cycles for the 1% total strain range), ill-defined

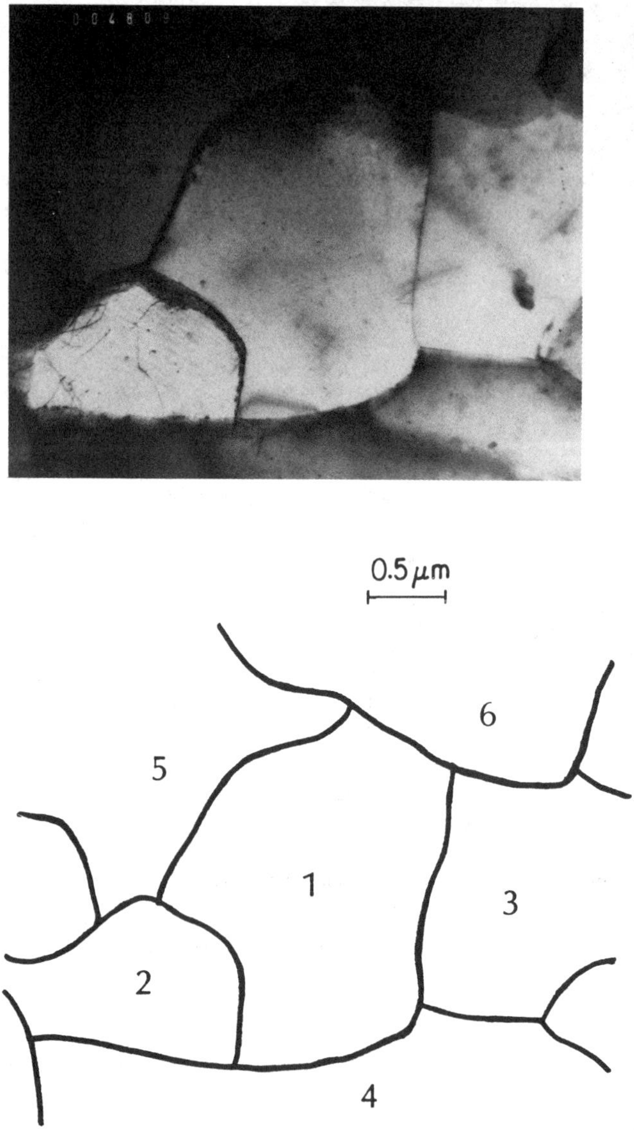

Cells	θ
1/2	2.68°
1/3	3.72°
1/4	2.76°
1/5	2.31°
1/6	3.53°
2/4	2.23°
2/5	1.72°
3/4	3.50°
3/6	2.04°
5/6	3.27°

FIG. 5—*Substructure and misorientation measurements after 11 300 cycles at $\Delta\epsilon_T$ = 1.0%.*

cells were observed as seen in Fig. 3. The walls were thick and uncondensed. There was also a high density of dislocations in the interior of the cells. As the number of cycles increased, the cells became more distinct as seen in Fig. 4, representing 5500 cycles. At failure (11 300 cycles) a well-defined cell structure was observed as illustrated in Fig. 5. The cell walls were narrow and the interior of the cells contained very few dislocations. Several methods for determining the orientation of a crystal using Kikuchi patterns have been reported in the literature [*29–34*]. Here, the procedure described by Kozubowski [*29*] was adopted. This method calculates the beam direction with respect to the crystal lattice by using three Kikuchi lines which enclose the central beam. Misorientation measurements are included in Figs. 3, 4, and 5. After determining the cell misorientation in all the specimens cycled at a total strain range of 1% and for two specimens cycled at 2% strain, it became apparent that no significant changes in misorientation took place during cycling. The results are summarized in Fig. 6.

A series of cell sizes on the same specimens cycled at 1% total strain range was measured by the linear intercept method, and the results are shown in Fig. 7; the error bars represent plus and minus one standard deviation in the measurement. No significant change in cell size was observed with increase in number of cycles. However, the cell size varied inversely with saturation stress amplitude by testing at different strain ranges as demonstrated by Kayali and Plumtree [*24*] (see Fig. 12).

Since the average cell size and the misorientation between the cells remained relatively constant for a given cyclic strain range, there is every indication, then, that the dislocation density also remained constant on attaining saturation. The dislocation density may be expressed by [*20*]

$$\rho = \frac{K}{\lambda h} \tag{2}$$

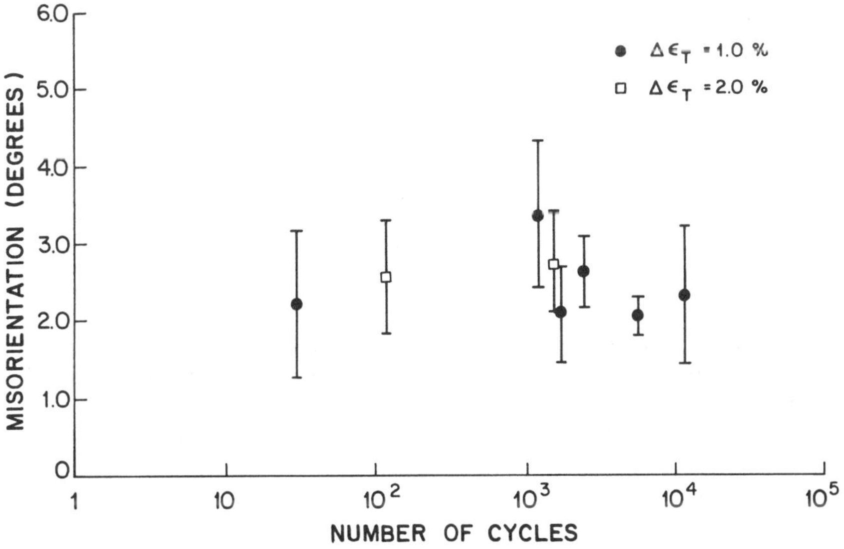

FIG. 6—*Misorientation between cells for* $\Delta\epsilon_T$ = *1.0% and 2.0%. Error bars represent plus and minus one standard deviation.*

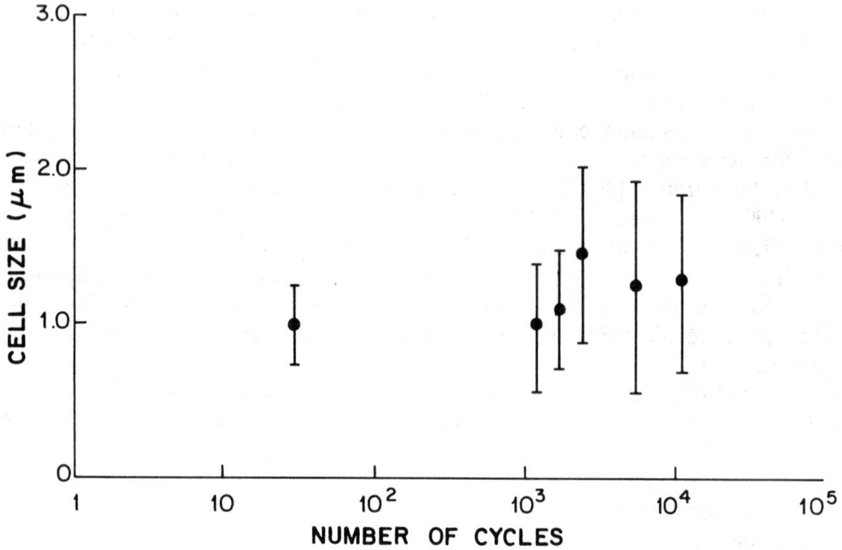

FIG. 7—*Variation of cell size with cycles for* $\Delta\epsilon_T$ = *1.0%. Error bars represent plus and minus one standard deviation.*

where

K = a material constant,
λ = cell size, and
h = spacing of the dislocations in the cell wall.

From the Kikuchi line analysis no significant twist component was observed, hence for a simple tilt boundary

$$h = \frac{b}{\theta} \tag{3}$$

where b is the Burgers vector and θ is the misorientation across the boundary. Substituting Eq 3 into Eq 2 results in

$$\rho = \frac{K\theta}{\lambda b} \tag{4}$$

Therefore, once saturation has been attained and the accompanying cell size, as well as the misorientation between the cells, remain constant for a given strain range, the dislocation density must remain constant.

Mechanism of Cyclic Deformation

During saturation there exists an equilibrium between dislocation generation and annihilation or work-hardening and dynamic recovery. In this stage, the plastic strain should be accommodated [*4*] without any change in dislocation density.

Many theories have been proposed to describe the dislocation behavior during saturation. For any model to be consistent with the present results, it must account for a relatively constant dislocation density, cell size and misorientation, together with change in cell morphology and a large amount of plastic deformation per cycle.

Irreversible bowing of dislocation segments involves the expansion of newly generated loops from the cell walls and subsequent entangling of these dislocations in neighboring walls [*35*]. Dislocations emerge into the free cell volume from pinned segments located within the wall. To obtain the corresponding cyclic flow stress, the length of these sources should be approximately 100 nm, which have been detected [*36*]. Dislocation segments traversing the cell have been seen by Hancock and Grosskreutz [*37*] in specimens which were irradiated under stress to pin the dislocations. It has been proposed that the dislocations originated from Frank-Read sources. However, it is more probable that other sources are activated such as those at cell and grain boundaries. Figure 8 shows boundary sources (for example, A and B) in the strain cycled aluminum that were activated on heating in the electron beam. Slip steps (for example, *S*) tended to indicate that slip took place across the cell with little hindrance to motion.

For a constant dislocation density, the expanded loops must be annihilated each cycle [*36*]. Laird [*38*] suggested that this was not unreasonable because a large density of dislo-

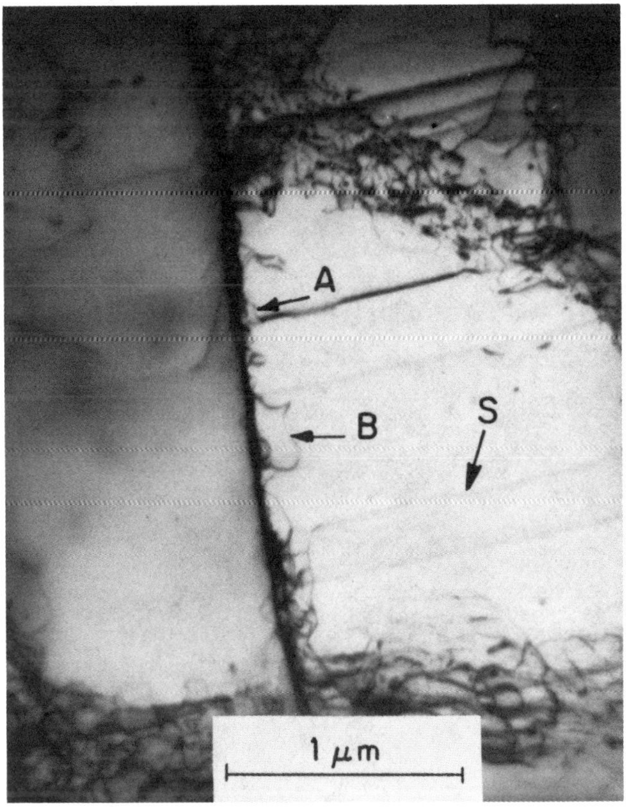

FIG. 8—*Boundary dislocation sources (for example,* A *and* B) *in strain cycled aluminum. Slip steps (for example,* S) *due to heating in electron beam traverse the entire cell.*

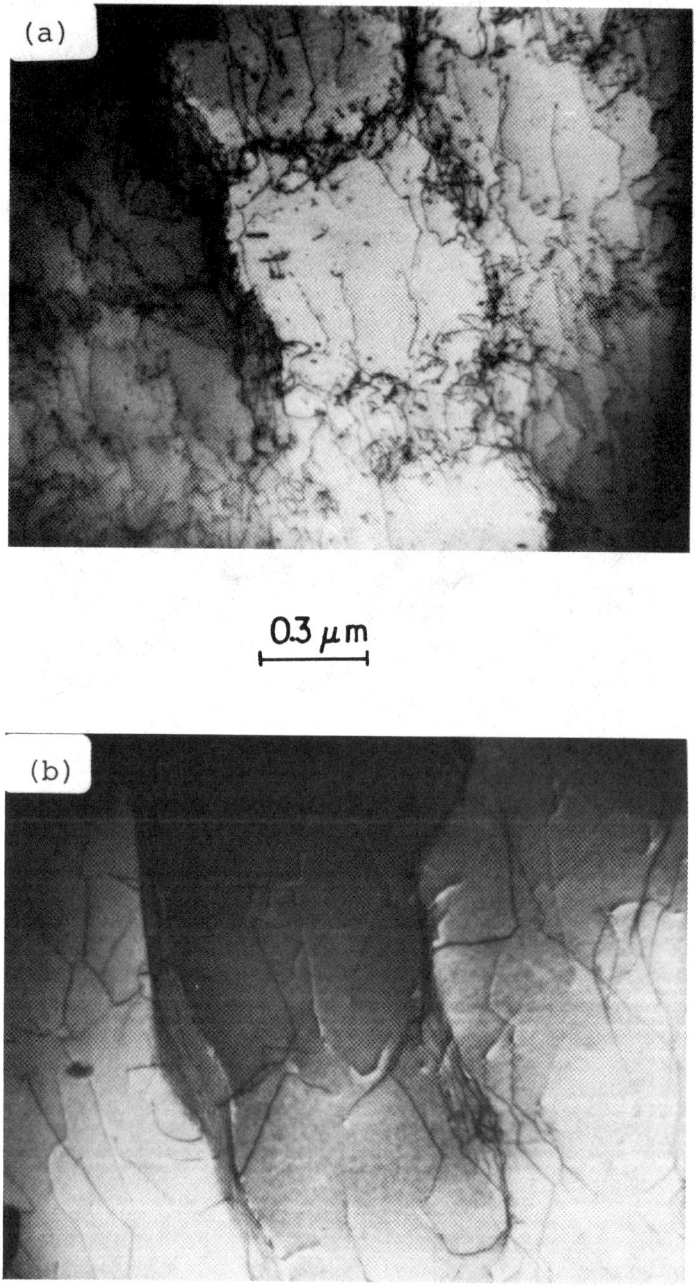

FIG. 9—*Substructure after 30 cycles at* $\Delta\epsilon_T$ = *1.0%* (a) *before heating in TEM,* (b) *after heating at 250°C in TEM.*

cations exist in the cell wall with approximately equal numbers of dislocations of opposite sign. Annihilation occurs by dynamic recovery processes and the change in cell morphology can be explained by dislocation interaction.

This is consistent with a constant dislocation density from the end of one cycle to the end of the next cycle, yet, during each cycle there is an increase in dislocation density as the loops expand and traverse the cells. The dislocation density is subsequently reduced to its equilibrium value as the expanded loops are annihilated in the cell walls. Sasaki and Ochi [*39*] observed a continous hardening and softening process during each half cycle which is consistent with an increase and decrease in dislocation density during each corresponding half cycle. The hardness, and therefore the dislocation density, did not change from the end of one cycle to the end of the next.

Hence, the operative deformation mechanism during saturation in this investigation appears to be that of irreversible bowing of dislocation segments from the cell wall. These dislocation segments are emitted from cell walls, traverse the cell and enter the adjacent walls where they are annihilated by dynamic recovery processes.

Dynamic Recovery During Cyclic Deformation

Thin foils of the specimens cycled for 30 cycles and full life at a total strain range of 1% were annealed in the TEM using a hot stage. The temperature of the specimen cycled for 30 cycles, which correspond to the onset of saturation, was increased at a rate of 10°C/min to 350°C. The dislocation substructure of the same area at room temperature and at 250°C during the anneal is shown in Figs. 9*a* and 9*b*, respectively. The temperature of sections taken from the full life specimen (11 300 cycles) was raised from room temperature to 250°C at the same rate of 10°C/min. The cell structure at room temperature is shown in Fig. 10*a* and the same area at 250°C in Fig. 10*b*.

The cell structure of the full life specimen did not change significantly during the anneal. This was not the case, however, for the specimen cycled to the onset of saturation which showed a considerable amount of dislocation rearrangement and annihilation. Obviously this cell structure was not a stable configuration. By contrast, the well-defined cell structure representative of the full-life specimen having undergone recovery was a more stable arrangement than the loosely defined substructure at the onset of saturation. It is apparent that the process of strain cycling caused the dislocations to be rearranged into a lower energy configuration and the method by which this rearrangement occurs is a dynamic recovery process.

During static recovery treatments, two types of microstructural changes may occur. Type I recovery corresponds to the annihilation and rearrangement of dislocations in the substructure walls and these boundaries become narrow and well defined. This process has been termed polygonization [*40*]. Type II recovery involves subgrain coarsening [*41,42*]. In the present work, since the cell size remained relatively constant, this cyclic recovery process was regarded as being of the Type I class.

During the process of polygonization, incipient tilt boundaries exert a small force on isolated dislocations some distance away which is sufficient for them to glide into the vicinity of the boundary. Each dislocation must then climb to find a niche in the boundary [*43*]. This is the process in static recovery which requires thermal activation and determines the rate of polygonization [*44*]. The driving energy for polygonization is the reduction in strain energy associated with a more stable dislocation configuration.

During static recovery an initial amount of stored energy is released which is associated with Type I recovery. However, the stored energy released during recovery treatments of a fatigued structure is much smaller than the energy released during recovery of a specimen

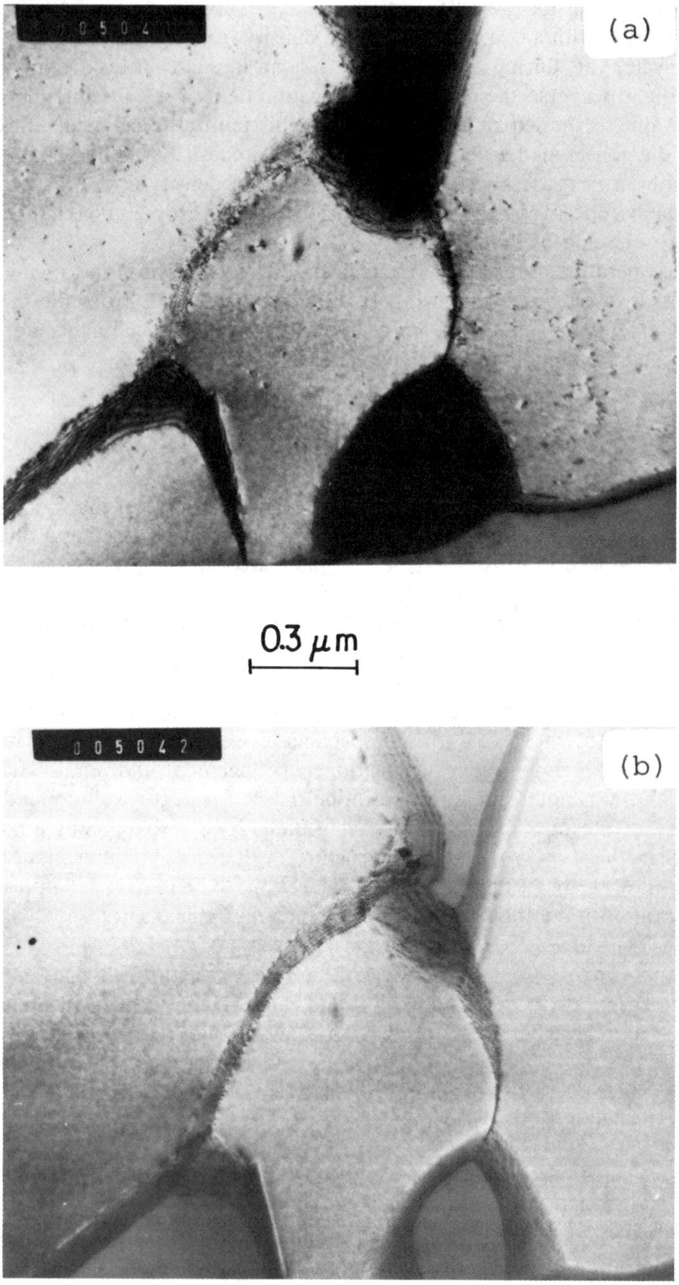

FIG. 10—*Substructure after 11 300 cycles at* $\Delta\epsilon_T$ = *1.0%* (a) *before heating in TEM,* (b) *after heating to 250°C in TEM.*

deformed in tension [*1*]. Therefore, the substructure produced in fatigue is of a lower energy configuration. This was illustrated in the in situ annealing experiment. Albeit the low energy configuration was not developed immediately but only after prolonged cycling.

Dynamic recovery is related to the ability of the dislocations to leave their slip planes and interact with dislocations from other sources. This is exhibited as wavy slip [*45*] and therefore dependent on the stacking fault energy. Low temperature dynamic recovery does not include climb with the result that annihilations are restricted. McQueen [*45*] stated that the mechanisms operative during fatigue involve low temperature dynamic recovery which is enhanced by the back and forth motion of dislocations. Dynamic recovery is clearly demonstrated when fatigue at a lower strain level causes softening of a substructure produced by either cold working or cyclic deformation at a higher strain [*45,46*].

The steady-state regime during cycling involves a constant dislocation density obtained by dynamic equilibrium between dislocation generation and annihilation as discussed previously. The dislocation generation rate is a function of the strain rate and the associated effective stress (local value of net stress acting at a dislocation), yet relatively independent of the accumulated strain. The rate of dislocation annihilation is dependent on the initial dislocation density and on the ease of operation of recovery mechanisms such as climb, cross slip and node unpinning. Therefore, the gradual reduction of strain hardening prior to saturation results from an increase in dislocation density which leads to a continual increase in dislocation annihilation rate until this rate matches that for generation [*47*].

It is well established that dynamic recovery occurs during monotonic creep. Konig and Blum [*48*] compared the substructures produced in aluminum by cycling at room temperature and monotonic creep. They concluded that the development of the dislocation structure during cycling and during monotonic creep was analogous in that the dislocation structure tended to approach steady-state based on dynamic equilibrium between generation and annihilation of dislocations. When considering all these points, one can now deduce that a dynamic recovery process of Type I is the operative mechanism during strain controlled cyclic deformation.

Structural Stability

Since the stable cell structure was more resistant to breakdown on heating, the notion that it may be more resistant to breakdown during cyclic deformation is intriguing. In order to investigate this point the extruded aluminum was strain cycled in its as-received state. In this condition it possessed a well-developed substructure, as shown in Fig. 11. The variation of stress amplitude with number of cycles for the specimens cycled at strain ranges of 1%, 1.3%, and 2% is included in Fig. 1. It will be noticed immediately that the stress responses for the annealed and the as-received extruded conditions cycled at a given strain range do not converge. Although the extruded aluminum immediately displayed a high stress response on cycling, a small amount of hardening occurred. This behaviour indicates that the original substructure was stable and resistant to the changes brought about by the ensuing dynamic recovery process.

The stability of this initial substructure was emphasized further by prestraining an extruded specimen and cycling at a total strain amplitude of 1.3%. Strain softening down to the same stress level as the non-prestrained extruded sample took place. Prestraining obviously produced additional dislocations which were sufficiently mobile to be annihilated by the subsequent dynamic recovery process. However, the 7% prestrain was not sufficient to break down the initial stable microstructure.

FIG. 11—*As-extruded substructure.*

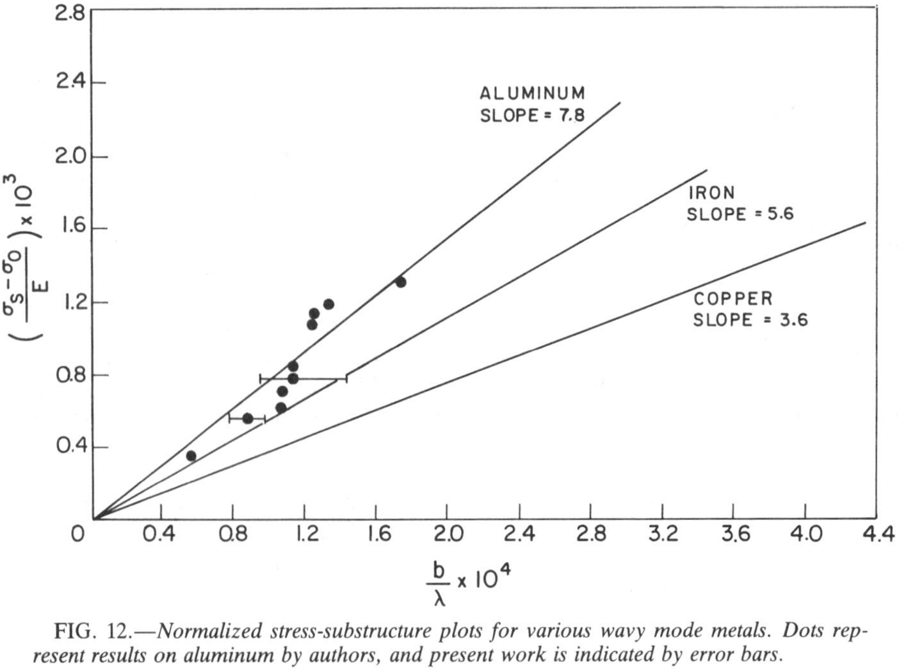

FIG. 12.—*Normalized stress-substructure plots for various wavy mode metals. Dots represent results on aluminum by authors, and present work is indicated by error bars.*

Substructure-Stress Relationship

It has been observed in many investigations dealing with cyclic as well as monotonic deformation of wavy slip mode materials that an inverse relationship exists between the size of the substructure (dislocation cells or sub-grains) and the steady-state flow stress. In order to compare the saturation stress-substructure size relationship for both monotonic (creep and room temperature tensile) and cyclic deformation conditions at various temperatures, data [*6,12,24,49*] for aluminum were collected and presented in Fig. 12. The results of the present investigation are also included. Data for monotonically and cyclically deformed copper [*11,20,35,50*] and α-iron [*18,51,52*] were collected and presented in the same figure. This allowed a comparison to be made of the various saturation stress-substructure size relationships for aluminum, copper and iron, all of which deform by wavy slip. The data are presented as the saturation stress (σ_s) minus the friction stress (σ_0) normalized by the temperature-compensated Young's modulus [that is, $(\sigma_s - \sigma_0)/E$] and plotted against b/λ in order to give a dimensionless plot. The room temperature friction stress (σ_0) values were 3.1 MPa for aluminum, 31 MPa for iron, and 3.9 MPa for copper.

The results of individual investigations involving monotonic and cyclic deformation of aluminum tend to lie on a common straight line [*24*] with a slope of 7.8 as seen in Fig. 12. The results obtained by various investigators on monotonic and cyclic deformation of iron and copper also lie on straight lines [*24*], the slopes of which are 5.6 and 3.6, respectively. Hence the normalized saturation stress-cell size relationship for aluminum, copper, and iron shown in Fig. 12 may be expressed in the following form:

$$\left(\frac{\sigma_s - \sigma_0}{E}\right) = A\,\frac{b}{\lambda} \tag{5}$$

where

E = Young's modulus for the appropriate temperature,
b = Burgers vector,
λ = linear intercept cell size, and
A = an empirical material constant (7.8 for aluminum, 5.6 for iron, and 3.6 for copper) which increases in proportion to the stacking fault energy [*24*].

The free dislocation segment or network link length appears to control the flow stress.

It is clearly seen from Fig. 12 that a unique relationship exists between the saturation (or flow) stress and dislocation cell or subgrain size for the different materials. Apparently, this relationship is not a function of deformation type because all the flow stress-substructure size data obtained by monotonic (tensile or compression or both) and cyclic deformation for a particular metal lie on one common line.

Conclusions

The following conclusions are drawn from this room temperature (0.29 homologous temperature) investigation on aluminum strain cycled at large strain amplitudes:

1. The cell size and misorientation between neighboring cells remained constant from the onset of saturation to failure. This allowed the deduction to be made that the dislocation density remained relatively constant once saturation was attained.

2. The morphology of the substructure continuously changed from an elementary cell structure with thick uncondensed walls and many stray dislocations in the cell interior at

the onset of saturation to a well-defined cell structure with narrow walls and a dislocation free cell interior at full life. This change in morphology is thought to be the result of dynamic recovery processes.

3. During the saturation stage, plastic strain was accommodated during each cycle without any change in the maximum stress response or dislocation density by establishing an equilibrium between dislocation generation and annihilation (or work-hardening and dynamic recovery). The mechanism of deformation taking place during saturation was thought to be irreversible bowing of dislocation segments from the cell wall. The dislocation segments expanded and became entangled and annihilated in neighboring cell walls. This resulted in a relatively constant dislocation density from the end of one cycle to the end of the next.

Acknowledgments

The authors would like to express their thanks to Dr. Paul E. Fortin, Aluminum Company of Canada Ltd., Kingston, ON for supplying the extruded aluminum bars. This work has been supported by the Natural Sciences and Engineering Research Council of Canada through Grant No. A-2770.

References

[*1*] Embury, J. D., *Strengthening Methods in Crystals,* A. Kelly and R. B. Nicholson, eds., Wiley, New York, NY, 1971, pp. 331–402.
[*2*] McElroy, R. J. and Szkopiak, Z. C., *International Metallurgical Review,* 1972, Vol. 17, pp. 175–202.
[*3*] Thompson, A. W., *Metallurgical Transactions,* 1977, Vol. 8A, pp. 833–842.
[*4*] Grosskreutz, J. C. and Mughrabi, M., *Constitutive Equations in Plasticity,* A. S. Argon, ed., MIT Press, Cambridge, MA, 1975, pp. 251–326.
[*5*] Laird, C., *Work Hardening in Tension and Fatigue,* A. W. Thompson, ed., American Institute of Mining, Metallurgical and Petroleum Engineers, New York, NY, 1977, pp. 150–176.
[*6*] McQueen, H. J. and Hockett, J. E., *Metallurgical Transactions,* 1970, Vol. 1, pp. 2997–3004.
[*7*] Michel, D. J., Moteff, J., and Lovell, A. J., *Acta Metallurgica,* 1973, Vol. 21, pp. 1269–1277.
[*8*] Langford, G. and Cohen, M., *Metallurgical Transactions A,* 1975, Vol. 6A, pp. 901–910.
[*9*] Kalish, D. and LeFevre, B. G., *Metallurgical Transactions A,* 1975, Vol. 6A, pp. 1319–1324.
[*10*] Feltner, C. E., *Acta Metallurgica,* 1963, Vol. 11, pp. 817–828.
[*11*] Staker, M. R. and Holt, D. L., *Acta Metallurgica,* 1972, Vol. 20, pp. 569–579.
[*12*] McQueen, H. J., Wong, W. A., and Jonas, J. J., *Canadian Journal of Physics,* 1967, Vol. 45, pp. 1225–1234.
[*13*] Challenger, K. D. and Moteff, J., *Metallurgical Transactions,* 1973, Vol. 4, pp. 749–755.
[*14*] Bendersky, L., Komen, Y., and Rosen, A., *10th Plansee–Seminar, 1981, Reutte, Austria,* 1–5 June, 1981, H. M. Ortner, ed., Vol. 2, pp. 1–13.
[*15*] Sherby, O. D., Klundt, R. H., and Miller, A. K., *Metallurgical Transactions A,* 1977, Vol. 8A, pp. 843–850.
[*16*] Grosskreutz, J. C. and Waldow, P., *Acta Metallurgica,* 1963, Vol. 11, pp. 717–724.
[*17*] Chopra, O. K. and Gowda, C. V. B., *Philosophical Magazine,* 1974, Vol. 30, pp. 583–591.
[*18*] Abdel-Raouf, H., Benham, P. P., and Plumtree, A., *Canadian Metallurgical Quarterly,* 1971, Vol. 10, No. 2, pp. 87–95.
[*19*] Feltner, C. E. and Laird, C., *Transactions TMS-AIME,* 1968, Vol. 242, pp. 1253–1257.
[*20*] Pratt, J. E., *Acta Metallurgica,* 1967, Vol. 15, pp. 319–327.
[*21*] Challenger, K. D. and Moteff, J., *Metallurgical Transactions,* 1972, Vol. 3, pp. 1675–1678.
[*22*] Challenger, K. D. and Moteff, J., *Fatigue at Elevated Temperatures, ASTM STP 520,* American Society for Testing and Materials, Philadelphia, 1973, pp. 68–79.
[*23*] Starker, Jr., E. A. and Lutjering, G., *Fatigue and Microstructure ASM Materials Science Seminar,* Oct. 14–15, 1978, St. Louis, MO, pp. 205–244.
[*24*] Kayali, E. S. and Plumtree, A., *Metallurgical Transactions A,* 1982, Vol. 13A, pp. 1033–1041.
[*25*] Wang, R. and Mughrabi, H., *Materials Science Engineering,* 1984, Vol. 63, pp. 147–163.
[*26*] Mughrabi, H., Kütterer, R., Lubitz, K., and Kronmüller, H., *Physical Status of Solids A,* 1976, Vol. 38, p. 261.

[27] Mughrabi, H., Herz, K., and Stark, X., *International Journal of Fracture,* 1981, Vol. 17, p. 193.
[28] Nabarro, F. R. N., Basinski, Z. S., and Holt, D. L., *Advances in Physics,* 1964, Vol. 13, pp. 193–323.
[29] Kozubowski, J. A., *Physical Status of Solids (a),* 1977, Vol. 43, pp. 535–539.
[30] Otte, H. M., Dash, J., and Schaake, H. F., *Physical Status of Solids,* 1964, Vol. 5, pp. 527–549.
[31] Pumphrey, P. H. and Bowkett, K. M., *Physical Status of Solids (a),* 1970, Vol. 2, pp. 339–346.
[32] von Hemendahl, M., *Physical Status of Solids (a),* 1971, Vol. 5, pp. 137–146.
[33] Ryder, P. L. and Pitsch, W., *Philosophical Magazine,* 1968, Vol. 18, pp. 807–816.
[34] Sheinin, S. S. and Cann, C. D., *Physical Status of Solids,* 1969, Vol. 11, pp. K1–K4.
[35] Feltner, C. E. and Laird, C., *Acta Metallurgica,* 1967, Vol. 15, pp. 1633–1653.
[36] Grosskreutz, J. C., *Physical Status of Solids,* 1971, Vol. 47, pp. 11–39.
[37] Hancock, J. R. and Grosskreutz, J. C., *Acta Metallurgica,* 1969, Vol. 17, pp. 77–97.
[38] Laird, C., *Treatise on Materials Science and Technology,* R. J. Arsenault, ed., Academic Press, New York, 1975, Vol. 6, pp. 101–162.
[39] Sasaki, Shigemi and Ochi, Yasuo, *23rd Japan Congress on Materials Research–Metallic Materials,* March 1980, pp. 27–32.
[40] Byrne, J. G., *Recovery, Recrystallization and Grain Growth,* MacMillan Co., New York, 1965.
[41] Hasegawa, T. and Kocks, U. F., *Acta Metallurgica,* 1979, Vol. 27, pp. 1705–1716.
[42] Hasegawa, T., Yakou, T., and Kocks, U. F., *Acta Metallurgica,* 1982, Vol. 30, pp. 235–243.
[43] Birchenall, C. E., *Physical Metallurgy,* McGraw-Hill, New York, 1959.
[44] Cahn, R. W., *Physical Metallurgy,* R. W. Cahn, ed., North-Holland Publishing Co., Amsterdam, 1965, pp. 925–987.
[45] McQueen, H. J., *Metallurgical Transactions A,* 1977, Vol. 8A, pp. 807–824.
[46] Haasen, P., *Physical Metallurgy,* Janet Mordike, trans., Cambridge University Press, Cambridge, U.K. 1978.
[47] McQueen, H. J. and Jonas, J. J., *Treatise on Materials Science and Technology,* R. J. Arsenault, ed., Academic Press, New York, 1975, Vol. 6, pp. 395–493.
[48] Konig, G. and Blum, W., *Acta Metallurgica,* 1980, Vol. 28, pp. 519–537.
[49] Saga, J., Hayashi, M., and Nishio, Y., *Society of Materials Science, Journal (Zairyo),* 1977, Vol. 26, No. 282, pp. 289–295.
[50] Abdel-Raouf, H. A., Plumtree, A., and Topper, T. H., *Metallurgical Transactions,* 1974, Vol. 5, pp. 267–277.
[51] Keh, A. S. and Weissman, S., *Electron Microscopy and Strength of Crystals,* Gareth Thomas and Jack Washburn, eds., Interscience Publishers, New York, 1962, pp. 231–300.
[52] Abdel-Raouf, H. and Plumtree, A., *Metallurgical Transactions,* 1971, Vol. 2, pp. 1863–1867.

Tsuneshichi Tanaka[1] *and Masahiko Kosugi*[2]

Crystallographic Study of the Fatigue Crack Nucleation Mechanism in Pure Iron

REFERENCE: Tanaka, T. and Kosugi, M., **"Crystallographic Study of the Fatigue Crack Nucleation Mechanism in Pure Iron,"** *Basic Questions in Fatigue: Volume 1, ASTM STP 924,* J. T. Fong and R. J. Fields, Eds., American Society for Testing and Materials, Philadelphia, 1988, pp. 98–119.

ABSTRACT: The fundamental question posed in this study is, What crystallographic conditions are satisfied at fatigue crack nucleation sites in polycrystalline iron? To answer this question it is vital to establish an effective technique to determine the crystallographic orientation of each grain under observation. This problem was solved by developing an etch pit method of grain orientation analysis. Stereographic projection geometry was successfully applied in determining the orientation of each grain with great accuracy from the measured angles between ridge lines of sharp-edged etch pit patterns on the film. The critical experiment, then, is to examine the geometrical relationship between the nucleated fine cracks and the orientations of surrounding grains on fatigued specimen surfaces. The final step is to extract such crystallographic conditions as are commonly satisfied at the sites of crack nucleation.

Electrolytically polished commercial base pure iron plate specimens were tested under completely reversed plane bending stress in the high-cycle fatigue range. Tests were interrupted at about half of the fracture life, and the crystallographic configuration of microcracks with respect to the grain orientation was thoroughly examined by making use of the newly developed technique. Cracks were observed to nucleate only along grain boundaries on the specimen surface. And it was found that there are five conditions satisfied at those cracked sites. These conditions concern the favorable configuration of slip systems, Schmid factor, grain size, and the direction of grain boundaries. Three more conditions were added from information pertaining to the crystallographic structure below the surface gained by observation after removal of the surface layer and of cross sections containing a crack. A brief discussion is also given on some statistical aspects of important parameters.

KEY WORDS: fatigue crack nucleation, pure iron, crystallographic etch pit, stereographic projection, grain orientation, grain boundary, dominating slip system, Schmid factor, orthogonality of slip systems, statistical distribution

Since only a few Stage I cracks emerge on a smooth metal specimen surface under high-cycle fatigue load, the crack nucleation sites must be the most favorable ones satisfying some basic conditions deeply connected with the crystallographic configuration of grains. Without knowing those conditions, it is impossible to establish a reasonable model for analytical treatment of the fatigue crack nucleation mechanism.

As a technique to find the orientations of individual grains in a polycrystalline metal specimen, two methods have been commonly used in the past: the X-ray method and the etch pit method. Kim and Laird adopted the former method and reported on the dependency of crack nucleation sites on the crystallographic configuration using polycrystalline oxygen-free, high-conductivity (OFHC) copper specimens with a mean grain size of 0.3 mm [*1*]. This method is, however, too tedious for determining each grain orientation, and is not

[1] Professor, Ritsumeikan University, Kita-ku, Kyoto, Japan.
[2] Graduate student, Columbia University, New York, NY.

TABLE 1—*Etchants and etching conditions.*

Etchant A			Etchant B		
Reagent		Time	Reagent		Time
			Saturated aqueous solution of		
HCl	0.1 cm^3	20–30 s	$F_eCl_3 \cdot 6H_2O$	10 cm^3	50 s
H_2O_2	1 cm^3	Temperature	H_2O	50 cm^3	Temperature
H_2O	50 cm^3	25°C	HNO_3	5 cm^3	25°C

suitable for our study in which more than 100 grains have to be handled. The etch pit method was, therefore, thoroughly examined and a new analysis method of grain orientation was developed. All the information about grain orientation and the Schmid factor of every slip system can be obtained with great accuracy by this method.

Electrolytically polished pure iron plate specimens were tested under plane bending stress in the high-cycle range with zero mean stress. Every test was stopped at about midpoint in the fracture life, and the specimen surface was carefully checked with a scanning electron microscope (SEM) to locate microcracks, after which etch pits were developed on the surface to determine the orientation of grains near cracks or of all the grains in the test portion.

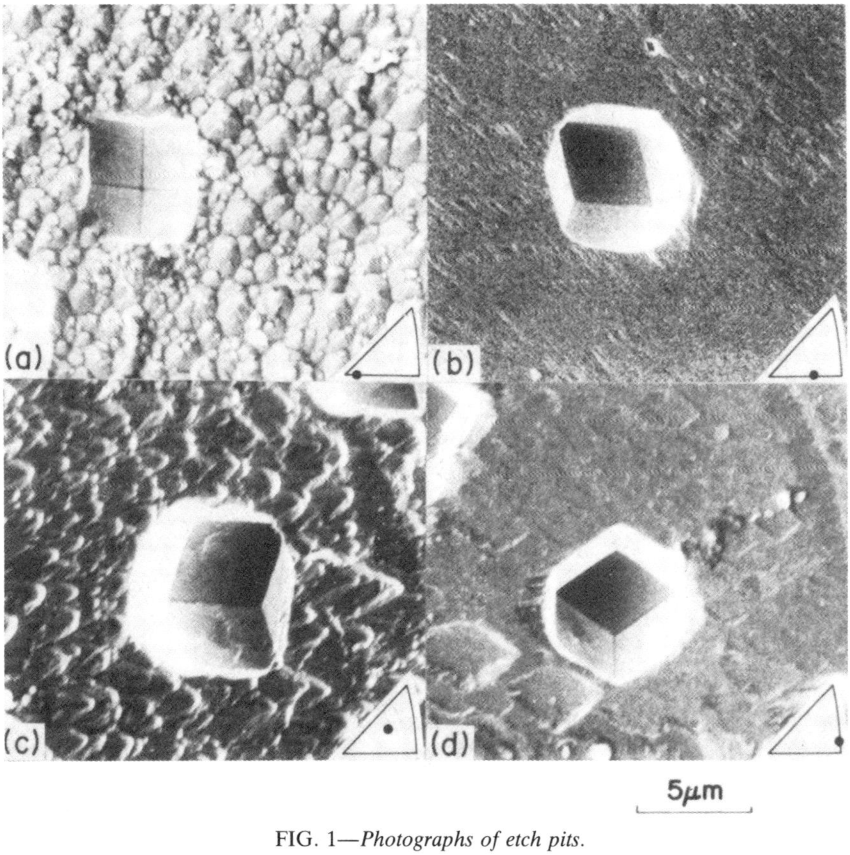

FIG. 1—*Photographs of etch pits.*

Observations and surveys of the crystallographic configuration at crack nucleation sites revealed the existence of several conditions which must be satisfied at those sites. Statistical studies were also made on the distributions of some important parameters included in those conditions.

Grain Orientation Analysis by Etch Pit Method

Experimental techniques to produce crystallographic etch pits on a metal surface were studied intensively by several investigators about two decades ago; we adopted the method developed by Taoka et al. [*2*]. The etchants and etching conditions used in this study are listed in Table 1. These were found to be the best to produce sharp-edged etch pits on surfaces of commercial base pure iron plate specimens, after a number of trials starting from the originally recommended techniques indicated in Ref *2*.

By soaking a specimen first in Etchant A and then in Etchant B under the given conditions, etch pits of different patterns developed in grains of different orientations as shown in Fig. 1. The normal direction of the specimen in each grain is indicated in a standard stereographic projection triangle in each photograph. The ridge lines of an etch pit are clearly observed by SEM owing to its deep focus, and since the areas where etch pits are absent have distinct contrasts depending on individual grain orientations, the grain boundaries can be traced easily.

All the etch pit walls are made of {110} planes and all the ridge lines are in ⟨111⟩ direction. Hence, each etch pit has a shape that is produced by indenting a rhombic-dodecahedron composed only of the {110} planes shown in Fig. 2 onto the specimen surface from an arbitrary direction. Since the etch pit pattern continuously changes its shape with a continuous change of the grain orientation, as illustrated in Fig. 3, one can find the approximate

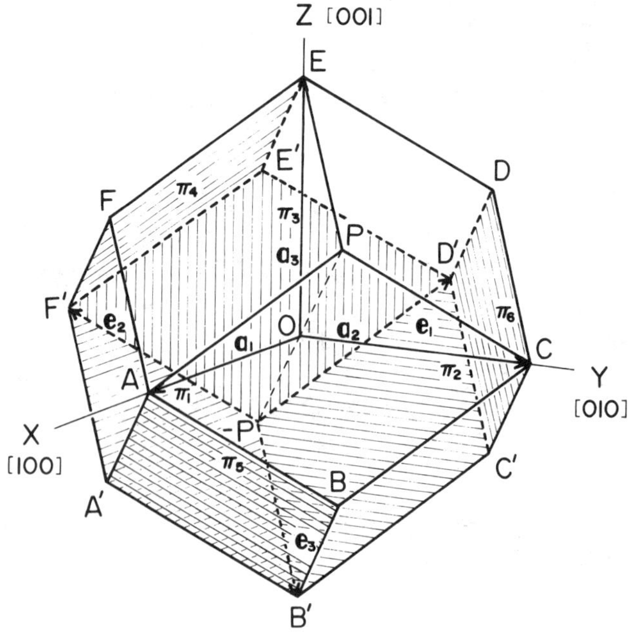

FIG. 2—*Rhombic-dodecahedron.*

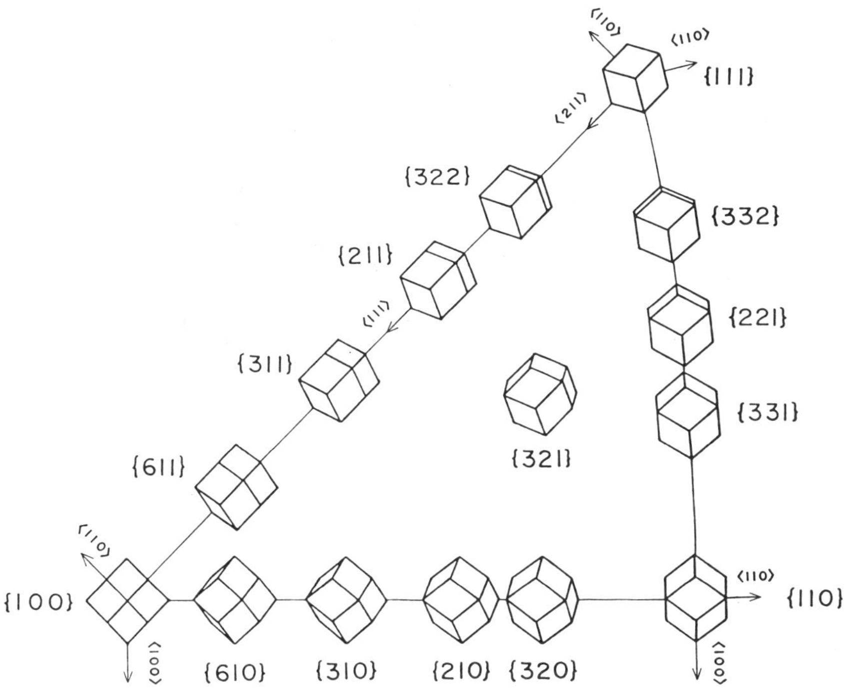

FIG. 3—*Etch pit patterns with different orientations.*

orientation of a given grain by comparing its etch pit photograph with those in Fig. 3. This is the technique commonly used even at present by many investigators [*3,4*]. This method is, however, not accurate enough to determine the geometrical relationship between the crack nucleation sites and the grain orientation. Therefore, a new method was developed for an accurate determination of the grain orientation from etch pit photographs. An outline of this new orientation analysis method is as follows.

In the case of the three ridge lines meeting at a vertex as in Fig. 1*b, c,* and *d,* these lines are all in ⟨111⟩ directions and are equivalent to three ridges emerging from one vertex denoted by $-P$ in the rhombic-dodecahedron depicted in Fig. 2, where three ridges are expressed by vectors $\mathbf{e}_1$, $\mathbf{e}_2$, and $\mathbf{e}_3$, which are here assumed to be unit vectors. In Fig. 2, $\mathbf{a}_1$, $\mathbf{a}_2$, and $\mathbf{a}_3$ are three base unit vectors of the cubic crystal forming a coordinate system *O-XYZ*. $\pi_1 \sim \pi_6$ are all {110} planes composing the rhombic-dodecahedron with the opposite ones parallel to each of $\pi_1 \sim \pi_6$. The angle between each pair of ridge lines is $\theta_0 = \cos^{-1}(-1/3) = 109.5$ deg, and there is a simple relation between **e** and **a** vectors:

$$\begin{bmatrix} \mathbf{e}_1 \\ \mathbf{e}_2 \\ \mathbf{e}_3 \end{bmatrix} = L_0 \begin{bmatrix} \mathbf{a}_1 \\ \mathbf{a}_2 \\ \mathbf{a}_3 \end{bmatrix}, \qquad L_0 = \begin{bmatrix} -1/\sqrt{3} & 1/\sqrt{3} & 1/\sqrt{3} \\ 1/\sqrt{3} & -1/\sqrt{3} & 1/\sqrt{3} \\ 1/\sqrt{3} & 1/\sqrt{3} & -1/\sqrt{3} \end{bmatrix} \tag{1}$$

To form a base system of the specimen, three unit vectors ℓ, **m**, and **n** are introduced, referring to the longitudinal direction (loading direction), the transverse direction, and the direction normal to the specimen surface, respectively. Figure 4 illustrates such a situation in which the point $-P$ (relabeled by O) and three **e** vectors are viewed from an arbitrary

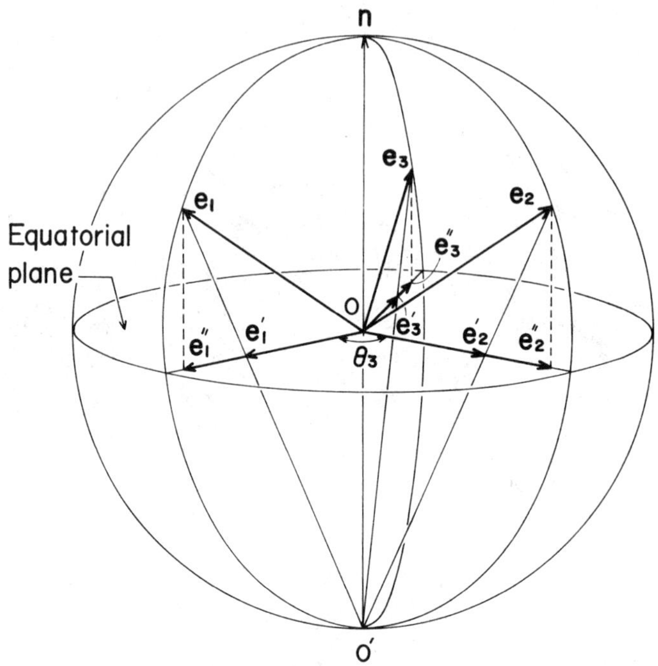

FIG. 4—*Stereographic and orthogonal projections of vectors* $\mathbf{e}_1$, $\mathbf{e}_2$, *and* $\mathbf{e}_3$.

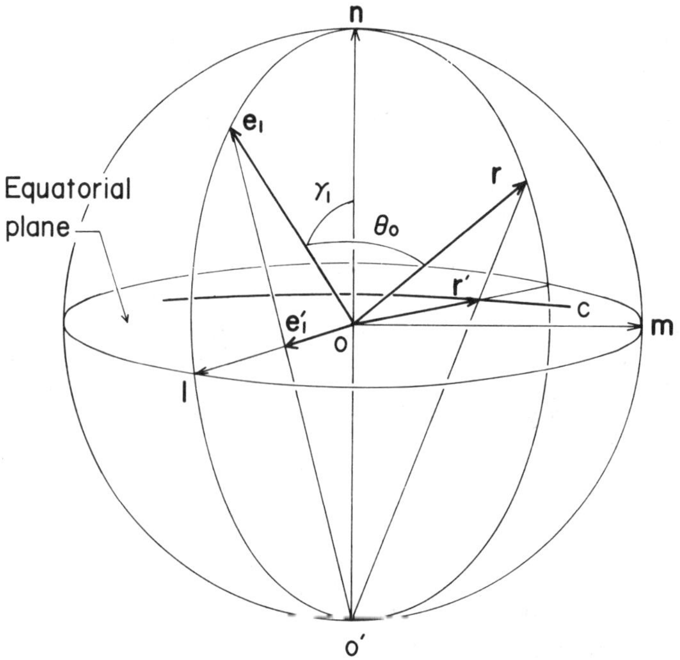

FIG. 5—*Locus of the stereographic projection* $\mathbf{r}'$ *when vector* $\mathbf{r}$ *rotates around a fixed vector* $\mathbf{e}_1$ *forming an angle* θ_0.

direction, which is supposed here as the $-\mathbf{n}$ direction. $\mathbf{e}''_1$, $\mathbf{e}''_2$, and $\mathbf{e}''_3$ are orthogonal projections of $\mathbf{e}_1$, $\mathbf{e}_2$, and $\mathbf{e}_3$ on the equatorial plane (projection plane). The ridge lines seen on the etch pit photograph are in the directions of $\mathbf{e}''$ vectors, and their mutual angles θ_1, θ_2, and θ_3 are directly measured (only θ_3 is shown in Fig. 4). Three vectors $\mathbf{e}'_1$, $\mathbf{e}'_2$, and $\mathbf{e}'_3$ are the stereographic projections of $\mathbf{e}_1$, $\mathbf{e}_2$, and $\mathbf{e}_3$ on the projection plane with respect to the pole O'. Since those $\mathbf{e}'$-vectors are exactly in the same directions as the corresponding $\mathbf{e}''$-vectors, the mutual angles between three $\mathbf{e}'$-vectors are also θ_1, θ_2, and θ_3. This fact enables us to make use of the established geometry of stereographic projection [5]. Derivations of fundamental formulas are provided in the Appendix.

Our tentative problem is to find the direction cosines of three $\mathbf{e}$-vectors with respect to the specimen coordinate system ℓ, $\mathbf{m}$, and $\mathbf{n}$. Now, supposing that one of the ridge line directions on the photograph is the direction of $\mathbf{e}'_1$, we can bring this direction into ℓ-direction by rotating the photograph, as shown in Fig. 5. Then, if it is assumed that a unit vector $\mathbf{r}$ is rotated around a fixed vector $\mathbf{e}_1$ forming the angle θ_0, the tip of vector $\mathbf{r}'$ (the stereographic projection of $\mathbf{r}$) describes a circle C on the projection plane (Eq 11 in the Appendix). Therefore, the tips of $\mathbf{e}'_2$ and $\mathbf{e}'_3$ must be on the circle C, since $\mathbf{e}_2$ and $\mathbf{e}_3$ form the angle θ_0 with $\mathbf{e}_1$, respectively. The radius and the center of this circle depend on an angle γ_1 between $\mathbf{e}_1$ and $\mathbf{n}$.

Circular arcs in Fig. 6 are parts of such circles C corresponding to the respective angles γ_1 for every 10 deg. They will be referred to as "master curves." In addition, the great circle is a unit circle, and each point on the horizontal axis from 0 to the right-hand side indicates the location of the tip of $\mathbf{e}'_1$ for each angle γ_1. Since θ_2 and θ_3 are known, $\mathbf{e}'_2$ and $\mathbf{e}'_3$ are on the straight lines having angles θ_2 and θ_3 with the horizontal line, and unknowns are the locations of the tips of these vectors. These are determined by the trial-and-error method. If the tip of $\mathbf{e}'_1$ is arbitrarily assumed to be located at P'_1 ($\gamma_1 = 60$ deg), those of $\mathbf{e}'_2$ and $\mathbf{e}'_3$ must in turn be located at P'_2 and P'_3 on the circle for $\gamma_1 = 60$ deg. Then, the angle between the corresponding original vectors $\mathbf{e}_2$ and $\mathbf{e}_3$ can easily be found by the geometrical technique of stereographic projection [5]. If the above assumption holds, the angle between

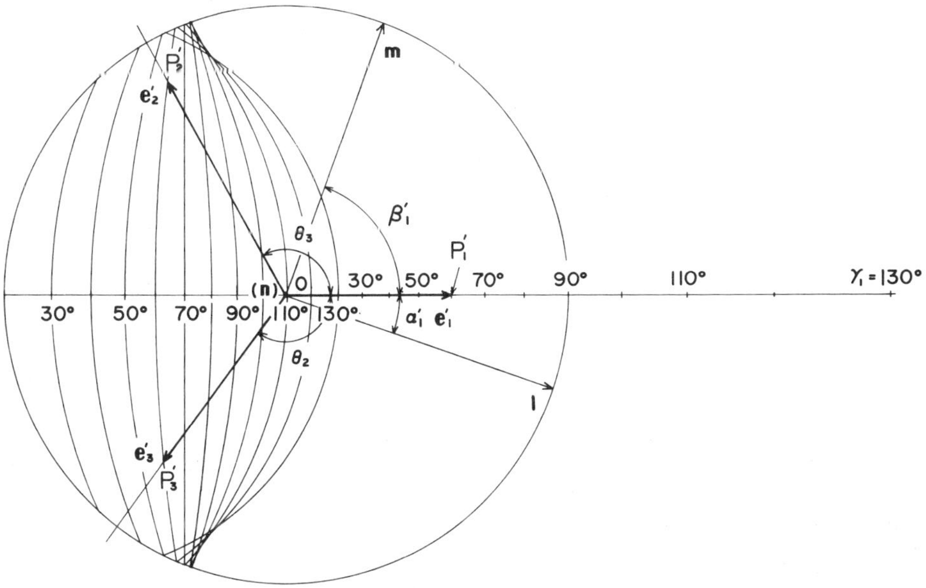

FIG. 6—*Master curves.*

$\mathbf{e}_2$ and $\mathbf{e}_3$ is to be $\theta_0 = 109.5$ deg. If that is not the case, P'_1 is shifted until the angle between $\mathbf{e}_2$ and $\mathbf{e}_3$ becomes equal to θ_0. When the positions of P'_1, P'_2, and P'_3 are thus determined, respective angles γ_1, γ_2, and γ_3 between $\mathbf{e}_1$, $\mathbf{e}_2$, and $\mathbf{e}_3$ and $\mathbf{n}$ are found in the same way by the technique of stereographic projection.

Once three direction cosines $n_i = \cos\gamma_i$ $(i = 1,2,3)$ of $\mathbf{e}$-vectors with respect to $\mathbf{n}$-direction are determined, other direction cosines are found immediately. The three ridge lines on the photograph are now reset with respect to the ℓ- and $\mathbf{m}$-directions, that is, to the longitudinal and transverse directions of the specimen. Then, if $\mathbf{e}'_1$ makes angles α'_1 and β'_1 with ℓ and $\mathbf{m}$ (Fig. 6), direction cosines ℓ_1 and $\mathbf{m}_1$ of $\mathbf{e}_1$ with respect to ℓ and $\mathbf{m}$ are (Eq 8 in the Appendix):

$$\ell_1 = \sqrt{1 - n_1^2}\cos\alpha'_1,\ m_1 = \sqrt{1 - n_1^2}\cos\beta'_1$$

Since $\mathbf{e}_2$ and $\mathbf{e}_3$ can be treated in the same way, we have

$$\ell_i = \sqrt{1 - n_i^2}\cos\alpha'_i,\ m_i = \sqrt{1 - n_i^2}\cos\beta'_i,\ (i = 1,2,3) \tag{2}$$

When all direction cosines of $\mathbf{e}$-vectors are found as above, the transformation equation becomes

$$\begin{bmatrix}\mathbf{e}_1\\ \mathbf{e}_2\\ \mathbf{e}_3\end{bmatrix} = L_1\begin{bmatrix}\ell\\ \mathbf{m}\\ \mathbf{n}\end{bmatrix},\quad L_1 = \begin{bmatrix}l_1 & m_1 & n_1\\ l_2 & m_2 & n_2\\ l_3 & m_3 & n_3\end{bmatrix} \tag{3}$$

This is the solution of our tentative problem. The final solution is a transformation equation between the base vectors $\mathbf{a}_1$, $\mathbf{a}_2$, and $\mathbf{a}_3$ of cubic crystal and the base vectors ℓ, $\mathbf{m}$, and $\mathbf{n}$ of the specimen. This is obtained by combining Eqs 1 and 3:

$$\begin{bmatrix}\mathbf{a}_1\\ \mathbf{a}_2\\ \mathbf{a}_3\end{bmatrix} = L\begin{bmatrix}\ell\\ \mathbf{m}\\ \mathbf{n}\end{bmatrix},\ L = L_0^{-1}L_1 \tag{4}$$

or

$$\begin{bmatrix}\ell\\ \mathbf{m}\\ \mathbf{n}\end{bmatrix} = L^{-1}\begin{bmatrix}\mathbf{a}_1\\ \mathbf{a}_2\\ \mathbf{a}_3\end{bmatrix} \tag{5}$$

The above procedure is for the case in which we have three ridge lines at a vertex as mentioned before. The case where a vertex has four ridge lines (Fig. 1*a*) is dealt with in the same way. Since a vertex with four ridge lines is always adjacent to vertexes with three ridge lines (Fig. 2), and a ridge line from the former vertex has parallel ridge lines from some of the latter ones, the angle relationship between vectors with four ridge lines and those with three ridge lines can be mutually substituted. Consequently, the previous method is equally applicable to the case of four ridge lines.

In the application of the above method, the technique was strictly followed in the first half of the experiment. Later, the "master table" was developed by making the trial-and-error calculations involved in the above technique by computer. In the second half of the

TABLE 2—*Heat treatment and mechanical properties.*

Heat Treatment	Yield Point, σ_s, MPa	Tensile Strength σ_B, MPa	Elongation,[a] δ, %	Vickers Hardness VHN
950°C 3 h, furnace cooling	110.4	209.7	35.0	70.2

[a] $\delta = \Delta l/l$: The cross section of specimens is 1 by 4 mm and the gage length l is 20 mm.

experiment, this table was used to find out the final results immediately from the measured angles on the etch pit photograph.

It is noted that the primary factor controlling accuracy is the measurement of angles θ_1, θ_2, and θ_3. On an etch pit photograph taken by a scanning electron microscope of 5000 magnifications, the measurement error of the angles between ridge lines is in most cases within ±1 deg. The corresponding errors of the direction cosines n_1, n_2, n_3, which vary respectively from zero to unity, are within ±0.05 when the master table is used. The master table was compiled to find out n_1, n_2, n_3 from the measured angles θ_1, θ_2, θ_3 for every 0.25 deg. The corresponding errors of the other direction cosines may be evaluated from Eq 2. In actual measurements, however, the averaging technique must be employed because the measured angles on several etch pit patterns in a grain may vary in some range due to prior deformation of each grain.

Fatigue Test

Pure iron (99.9%) of commercial base was used in the test in the fully annealed state. The heat treatment conditions and mechanical properties are listed in Table 2. The mean grain size was approximately 180 μm.[3] Plate specimens shown in Fig. 7 were prepared with a final finish by electrolytic polishing. A magneto-electric type cantilever plane bending machine was used to test the specimens under reversed bending stress with zero mean stress [6]. The tests were conducted in room atmosphere with the frequency of 60 Hz. The nominal stress amplitude σ at the middle section of the specimen was determined from the bending moment measured by a load cell attached to the fixed end of the specimen. The arm length of the moment was about 20 mm at the point of the middle section.

The obtained *S–N* relation is shown in Fig. 8. Since initial cracks were observed approximately after 1.0×10^5 cycles at 180 MPa and 4.8×10^5 cycles at 140 MPa, tests were interrupted after those stress cycles and the cracks were located by SEM. The specimen surfaces were then treated by the etchants to develop etch pits for the grain orientation analysis.

Observation, Analysis, and Discussion

Before presenting our observation of cracks and analysis, it would be advantageous to give some discussion of the fatigue crack nucleation sites and preferential slip systems, to make clear the standpoint of our experiments.

[3] The mean grain size was determined according to the method of JIS G0552; that is, the number of grains was counted in the area of 25 mm^2 after being magnified 100 times, and the mean grain size d was calculated by $d = \sqrt{a}$ from the average area a of grains included in the above area.

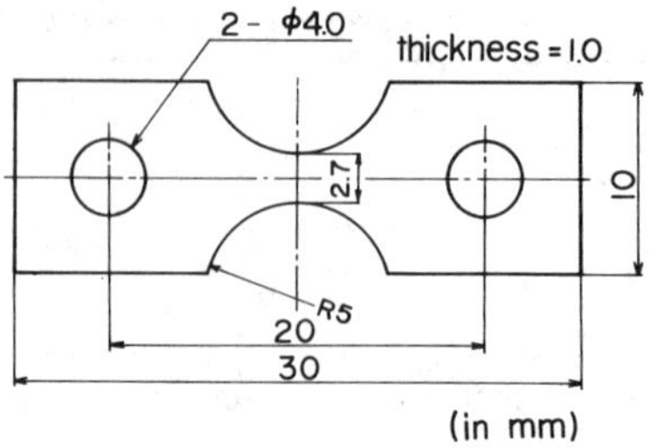

FIG. 7—*Specimen.*

The sites where fatigue cracks nucleate have not yet been conclusively defined for iron and low-carbon steel. Crack nucleation has been reported to occur along persistent slipbands [*7,8*], or it takes place at grain boundaries, at least in the thin surface layers of specimens [*6,9*]. Yoshida et al. carried out rotating bending fatigue tests of low-carbon steel specimens with a shallow notch and observed that the slipband cracks became more pronounced in the specimens work-hardened by surface rolling than in fully annealed ones [*10*]. Kurobe et al. observed the grain boundary cracks in fully annealed low-carbon steel specimens when fatigued in room atmosphere and the slipband cracks when the specimens were tested in vacuum of 8×10^{-5} torr ($\cong 1 \times 10^{-2}$ Pa) [*11*].

Thus, the fatigue crack nucleation site is not inherent in the material but is varied by work-hardening, surface residual stress, surrounding atmosphere, and surface finish conditions. However, in the case of annealed and electrolytically polished specimens, the grain boundaries seem to be preferential crack nucleation sites for iron and low-carbon steel when the observation is made on the specimen surface. This is the case in the authors' previous study [*6*] and also in the present study, as will be shown in the subsequent sections.

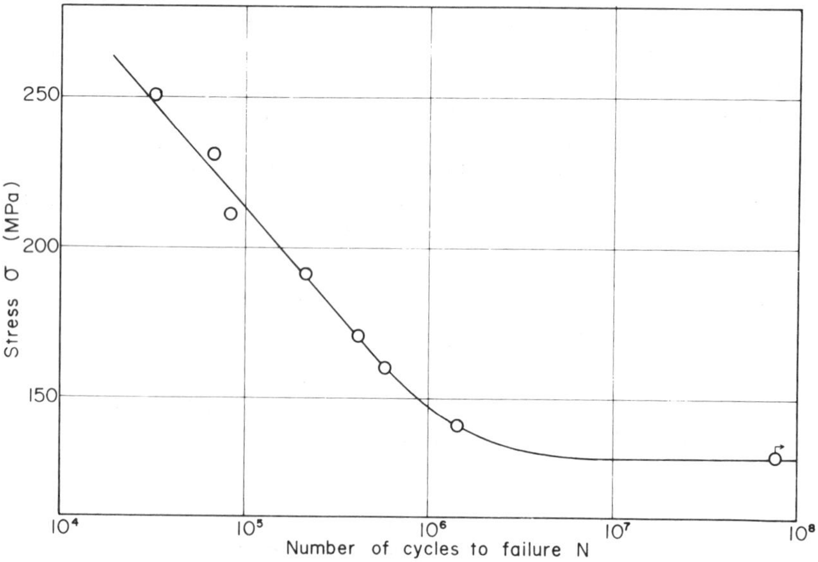

FIG. 8—S–N *diagram.*

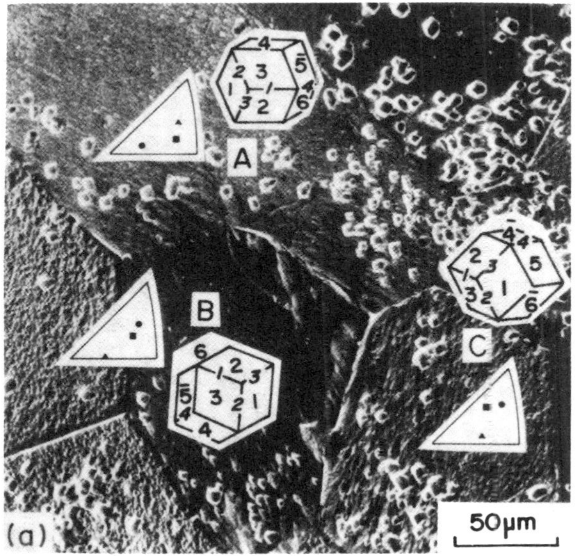

(b)

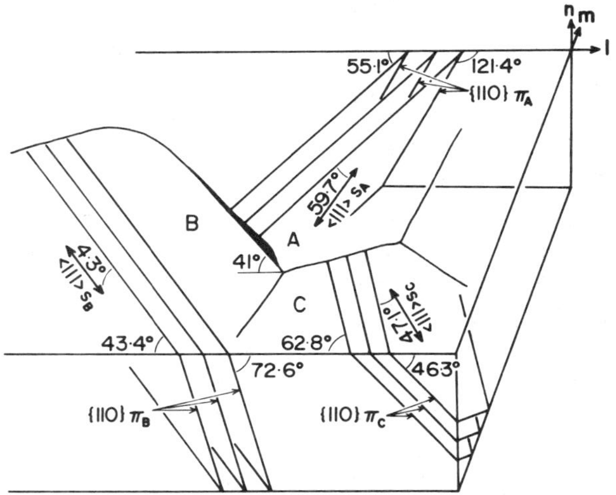

(c)

Grain	s	π	K_I	π	K_{II}	π	K_{III}
A	1	3	0·29	2	0·15	4	0·14
	2	1	0·25	5	0·17	3	0·08
	3	1	0·07	6	0·05	2	0·01
	4	6	0·45	5	0·37	4	0·09
B	1	4	0·49	2	0·31	3	0·19
	2	5	0·31	1	0·23	3	0·09
	3	1	0·48	2	0·31	6	0·18
	4	5	0·30	4	0·22	6	0·08
C	1	3	0·50	4	0·29	2	0·21
	2	5	0·18	3	0·13	1	0·05
	3	6	0·46	2	0·24	1	0·22
	4	5	0·28	6	0·16	4	0·11

FIG. 9—*Microcrack observed on the specimen N44* (σ = *140 MPa*, N = *4.8* $\times$ *10*^{5}). (a) *Photograph of the etch pitted surface.* (b) *Schematic presentation of the configuration of primary slip systems in respective grains.* (c) *Schmid factors* K *of 12 slip systems in grains A, B, and C.*

Regarding the preferential slip systems, it is well known that three slip systems, {110}⟨111⟩, {112}⟨111⟩, and {123}⟨111⟩, are operative in body-centered-cubic (bcc) metals. In the case of fatigue, Wei and Baker observed slips in {110}⟨111⟩ and {112}⟨111⟩ slip systems for pure iron specimens subjected to strain cycles of relatively low magnitudes, but no slips in the {123}⟨111⟩ system [*8*]. However, with respect to the Stage I cracks, Otsuka et al. observed the cracks resting on the {110}⟨111⟩ system for low-carbon steels [*3*]. After that, Asami and Terasawa proposed a simple model of Stage I cracks growing in the ⟨111⟩ direction along {110} planes based on a similar observation for low carbon [*4*]. They also suggested that the shift to another slip system may occur in the later stage following the initiation.

These observations seem to indicate that the above three slip systems are not equally operative under repeated stress in the high-cycle fatigue range, even if these slip systems are in equally favorable orientations, and the {110}⟨111⟩ is the most preferential one in forming Stage I cracks. Consequently, in this study the analysis was made by focusing on the {110}⟨111⟩ slip system, as a first step in the quantitative analysis of the configuration of nucleated cracks and grain orientations.

Observation of Cracks and Surrounding Grains

As mentioned in the previous discussion, microcracks were found only along grain boundaries on the fatigued specimen surfaces. Two examples are shown in Figs. 9 and 10 together with the results of orientation analyses of surrounding grains. Each photograph (*a*) was taken after developing etch pits. The microcracks are seen to have nucleated along the boundary of Grains A and B in both photographs. The orientations of grains A, B, and C determined by the previous method are indicated in each stereographic projection triangle, in which marks ▲, ■, and ● give the directions of ℓ, **m**, and **n** relative to the base vectors of each grain. The etch pit patterns are also given in the respective grains by depicting the corresponding models of a rhombic-dodecahedron with the surface numbers equivalent to $\pi_1 \sim \pi_6$ in Fig. 2. The opposite surfaces are indicated by the numbers with bars. The ridge line numbers of the models correspond to slip directions $s_1 \sim s_4$ of bcc metals all in ⟨111⟩ directions. Their unit vectors are $\mathbf{s}_1 = \mathbf{e}_1$, $\mathbf{s}_2 = \mathbf{e}_2$, $\mathbf{s}_3 = \mathbf{e}_3$, and $\mathbf{s}_4 = \mathbf{e}_1 + \mathbf{e}_2 + \mathbf{e}_3$.

In the case of the {110}⟨111⟩ slip system, since each of four ⟨111⟩ directions has three {110} planes making a pencil, twelve independent slip systems exist; they can be expressed by the combinations of $s_1 \sim s_4$ and $\pi_1 \sim \pi_6$. Figures 9*c* and 10*c* lists the Schmid factor K for these twelve slip systems in the order of $K_{\mathrm{I}} > K_{\mathrm{II}} > K_{\mathrm{III}}$ for each direction of $s_1 \sim s_4$, respectively for grains A, B, and C. The slip systems having the largest Schmid factor are then taken out in each grain, and their slip planes and slip directions are labeled π_A, π_B, π_C, and s_A, s_B, s_C, respectively. These slip systems are defined here as primary slip systems and their geometric relationship with the crack is given in the illustration (*b*) in both figures. The frame is in the directions of the specimen coordinate vectors ℓ, **m**, and **n**, and the load is applied in the ℓ-direction. Some angle relations are also given to help understand the spatial configuration of slip planes and directions.

The illustration (*b*) in Fig. 9 shows that the crack along the grain boundary is almost parallel to the intersection of the plane π_B with the specimen surface, and this is also true in Fig. 10*b*, where the tangential line at point *P* of the slightly curved crack is parallel to the intersection of π_B with the surface. Another interesting point is that at these cracked grain boundaries the primary slip planes in grains A and B intersect approximately at a right angle, and their configuration is such that the slip motion in grain A in s_A-direction on the plane π_A is completely blocked at the boundary by the plane π_B. Such a geometrical relation was also found at other sites where initial cracks were located.

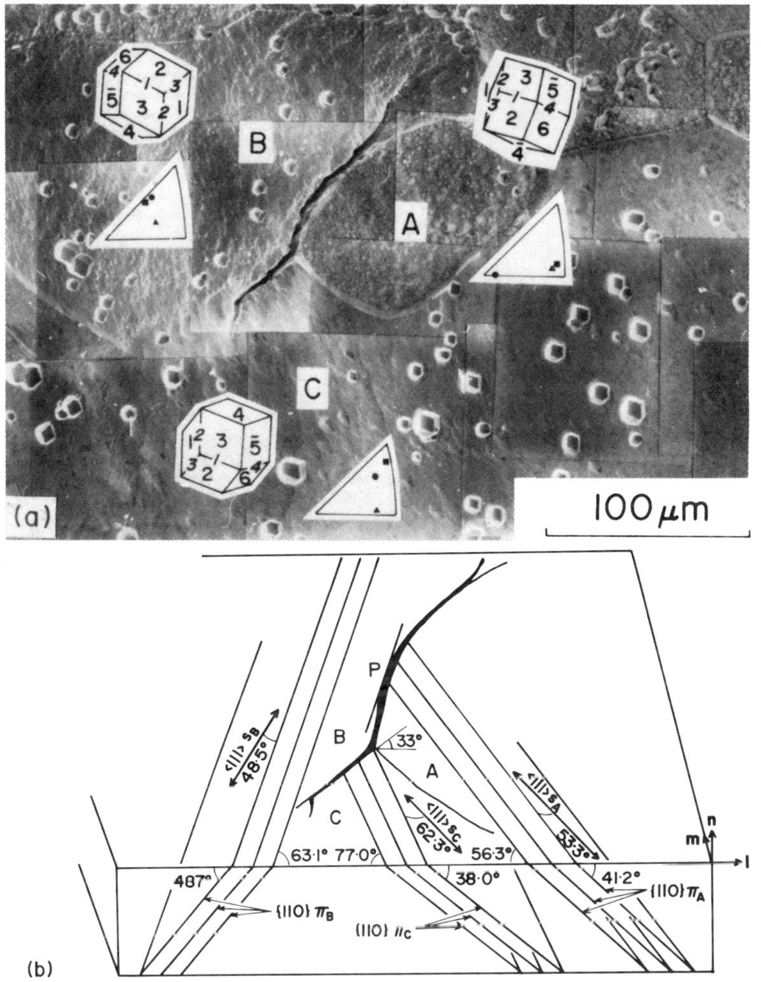

(c)

A	s	π	K_I	π	K_{II}	π	K_{III}
	1	2	0·42	3	0·35	4	0·07
	2	1	0 02	5	0·01	3	0·01
	3	1	0·11	6	0·06	2	0·06
	4	5	0·45	6	0·39	4	0·06

B	s	π	K_I	π	K_{II}	π	K_{III}
	1	4	0·44	2	0·23	3	0·21
	2	5	0·12	1	0·09	3	0·03
	3	1	0·49	6	0·31	2	0·19
	4	5	0·27	4	0·15	6	0·13

C	s	π	K_I	π	K_{II}	π	K_{III}
	1	2	0·49	3	0·26	4	0·23
	2	1	0·22	3	0·15	5	0·08
	3	1	0·23	2	0·15	6	0·08
	4	4	0·49	5	0·26	6	0·23

FIG. 10—*Microcrack observed on the specimen N20* (σ = *180 MPa*, N = *1.0* × *10*^{5}). (a) *Photograph of the etch pitted surface.* (b) *Schematic presentation of the configuration of primary slip systems in respective grains.* (c) *Schmid factors* K *of 12 slip systems in grains A, B, and C.*

Observation of All Grains in the Test Area

Figure 11 shows a result of survey on all grains in the test area of a single specimen (Specimen A), which was subjected to stress cycles of about 5 × 10^5 at 140 MPa. Grains are numbered from 1 to 212, but the total number is 207 because Nos. 6, 58, 61, 82, 192 are absent. After determining the orientation of all of the grains by etch pit method, and this time by making use of the master table mentioned before, Schmid factors were computed

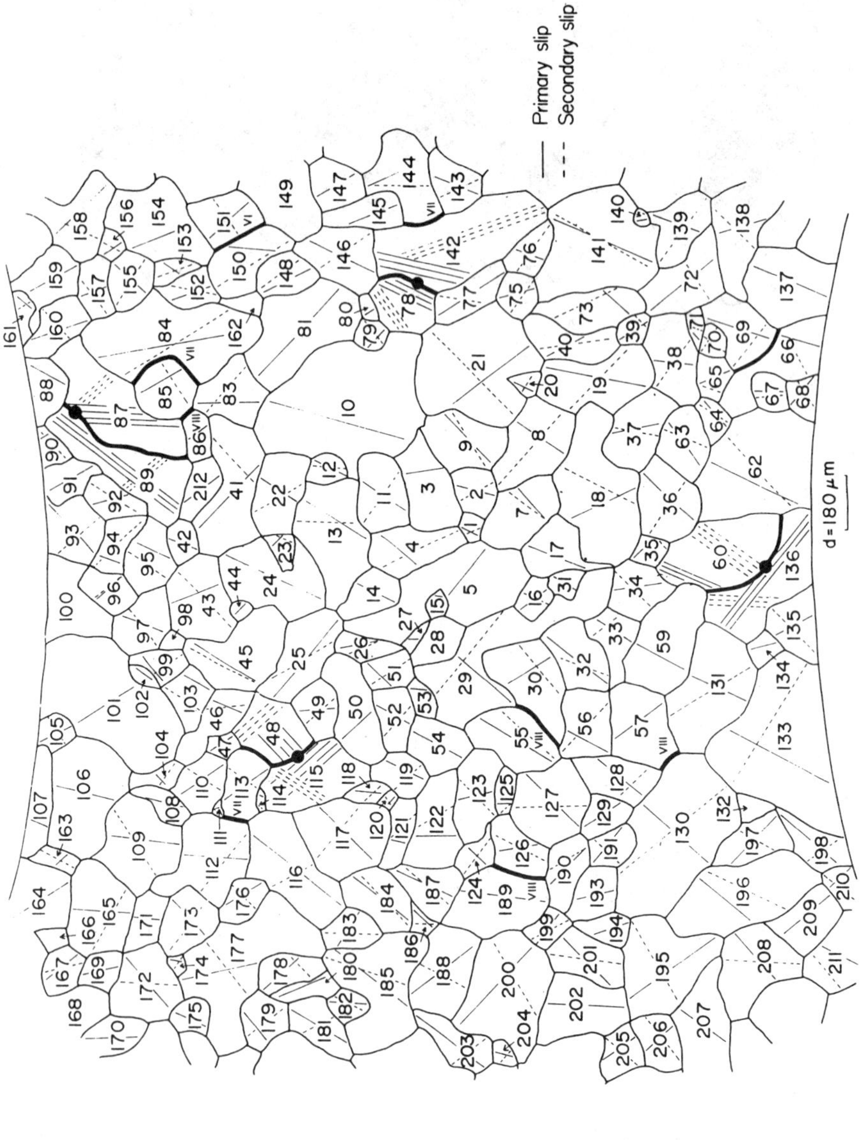

FIG. 11—*All the grains with dominating slip systems in test area on the surface of Specimen A.*

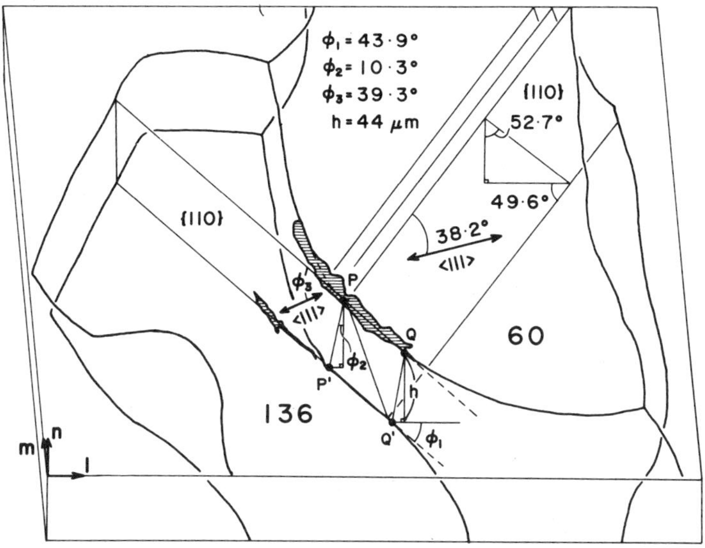

FIG. 12—*Configuration of grain boundaries, dominating slip systems and a microcrack observed at 60–136 grain boundary of Specimen A.*

for twelve slip systems in each grain, and two slip systems having the largest and the second largest Schmid factors were taken out; these are called hereafter dominating slip systems. Intersections of slip planes of such slip systems and the specimen surface are indicated by a fine line (primary slip plane), and a broken line (secondary slip plane) in each grain. These lines increase in number in those grains that neighbor the cracked boundaries (marked by ● in order to identify them easily). On the other hand, only a fine line or no lines are drawn in some grains because such lines are drawn only for dominating slip systems with Schmid factors $K > 0.38$; no trace of slip was visible even by SEM when $K < 0.38$.

Now, by combining the findings in the previous section with the survey of the specimen surface shown in Fig. 11, the following five conditions were derived as a first step that must be satisfied at the crack nucleation sites.

I There must be such dominating slip systems with the Schmid factor $K > 0.38$ in two grains (i, j) adjacent to each other along a boundary that the angle between the slip plane normals $\boldsymbol{\pi}_i$ and $\boldsymbol{\pi}_j$ is larger than 75 deg in acute side angle; that is, $\angle(\boldsymbol{\pi}_i, \boldsymbol{\pi}_j) \geqq 75$ deg.

II In such dominating slip systems, the angle between slip directions $\mathbf{s}_i$ and $\mathbf{s}_j$ must be larger than 65 deg in acute side angle; that is, $\angle(\mathbf{s}_i, \mathbf{s}_j) \geqq 65$ deg.

III Either one of the above dominating slip systems must have an intersection of its slip plane with the surface, being parallel with the grain boundary within 5 deg.

IV The angle between the slip plane normal $\boldsymbol{\pi}_j$ satisfying the Condition III and the slip direction $\mathbf{s}_i$ of the dominating slip system in the opposite grain must be less than 45 deg; that is $\angle(\mathbf{s}_i, \boldsymbol{\pi}_j) \leqq 45$ deg. The former slip system having $\boldsymbol{\pi}_j$ is called hereafter the blocking side and the latter one with $\mathbf{s}_i$ the slipping side.

V Grain sizes[4] d_i, d_j of the grains i, j must be larger than the mean grain size d (=180 μm); that is, $d_i/d \geqq 1.0$ and $d_j/d \geqq 1.0$.

[4] The grain size d_i of each grain was calculated by $d_i = \sqrt{a_i}$ from the measured area a_i of each grain in Fig. 11.

It is noted that, since we have in many cases two slip systems as dominating ones in each of adjacent grains, four combinations of such slip systems are possible at each grain boundary. It is therefore decided that Conditions I–IV preceding are satisfied whenever there is one such combination that satisfies those conditions. The numerical values of angles appearing in these conditions were also decided rather arbitrarily as reference values based on observation results; this point will be discussed later.

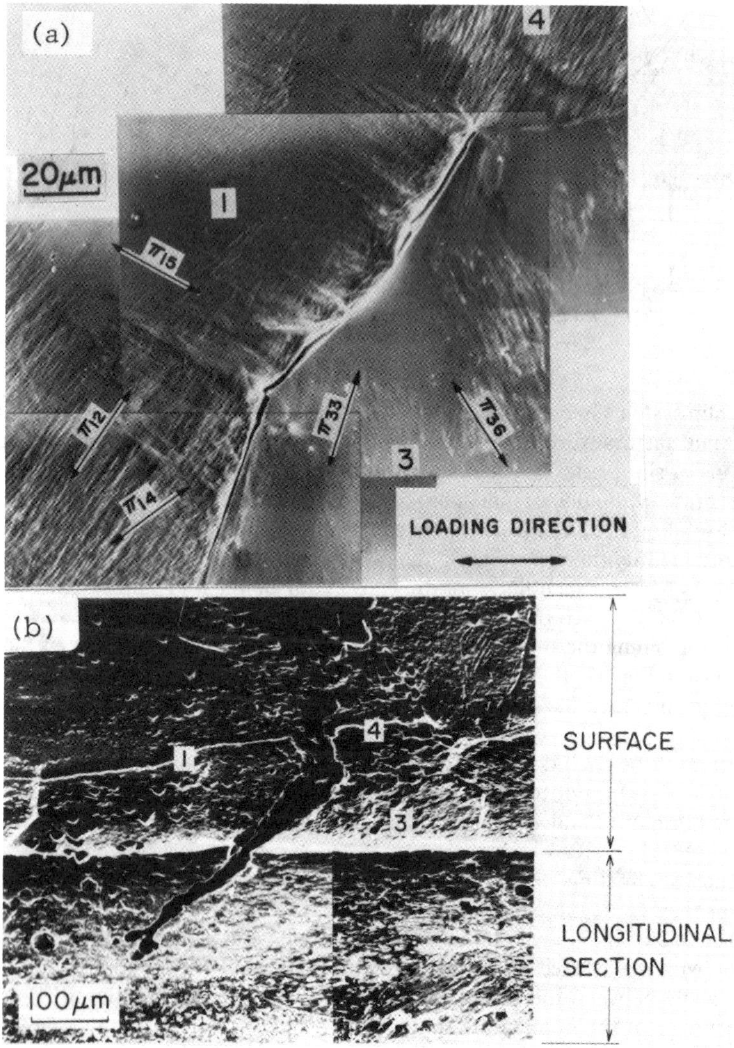

FIG. 13—*Microcrack observed on Specimen B* (σ = *140 MPa,* N = *4.8* × *10*[5]). (a) *Photograph of the fatigued surface.* (b) *Photograph of the longitudinal section containing a crack.* (c) *Schematic presentation of the configuration of the crack, grain boundary, and dominating slip systems.* (d) *Schmid factors* K *of 12 slip systems in grains 1 and 3.* (e) *Orthogonalities between dominating slip systems in grains 1 and 3.*

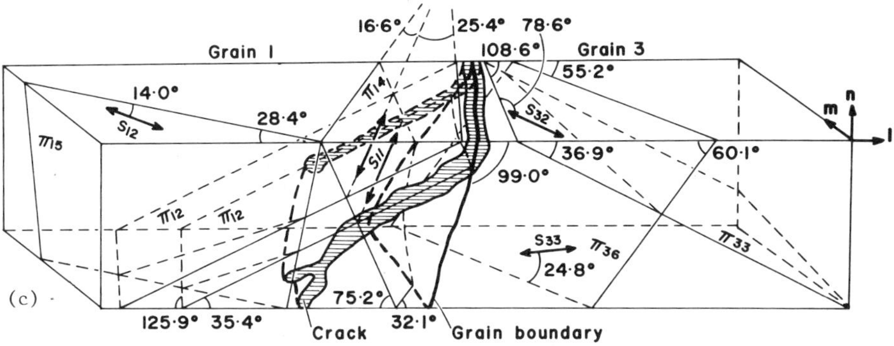

(c)

(d)

GNo.	S_{1i}	π_{1j}	K_I	π_{1j}	K_{II}	π_{1j}	K_{III}	GNo.	S_{3i}	π_{3j}	K_I	π_{3j}	K_{II}	π_{3j}	K_{III}
1	1	2	0·42	4	0·42	3	0·01	3	1	4	0·38	3	0·25	2	0·14
	2	5	0·40	1	0·39	3	0·01		2	3	0·47	1	0·35	5	0·12
	3	2	0·02	1	0·01	6	0·01		3	6	0·49	1	0·29	2	0·20
	4	6	0·03	4	0·02	5	0·01		4	4	0·26	6	0·22	5	0·04

(e)

Slip	Block	Angle (degree)		
		$\angle(\pi_{sj}, \pi_{bj})$	$\angle(S_{si}, S_{bi})$	$\angle(S_{si}, \pi_{bj})$
(S_{32}, π_{33})	(S_{11}, π_{12})	78·5	65·1	25·4
(S_{33}, π_{36})	(S_{11}, π_{14})	84·1	73·5	18·2

FIG. 13—*Continued.*

In Fig. 11, such grain boundaries that satisfy the five conditions are indicated by thick lines, and their total is 14, while microcracks were found at four grain boundaries indicated by the mark ●. This fact suggests that some other conditions must exist for the nucleation of early cracks, and it is supposed that such conditions are related to the information obtained by observations of the structures below the surface. In order to get such information, the surface layer of Specimen A was removed electrolytically by about 44 μm, and the survey was carefully done again. As a result, the following two conditions were added to be satisfied at crack nucleation sites.

VI The grain of the slipping side defined in Condition IV must have a large volume below the surface.

VII There must not be another new grain just below the surface which may disturb some of the given conditions.

In Fig. 11, the thick-lined grain boundaries with symbols VI and VII are those not satisfying the above Conditions VI and VII. Evidence supporting Condition VI had been obtained also for low-carbon steel in the authors' previous study [*6*].

Figure 12 illustrates a geometrical configuration of grain boundaries, dominating slip systems and a microcrack (shadowed parts) that nucleated at the 60–136 grain boundary in Fig. 11. It is observed that two dominating slip systems in grains 60 and 136 make a slip-block pattern at the boundary, and the PQ portion of the crack is parallel to the slip plane in grain 136. It is also observed that the crack originated at the grain boundary on the surface propagated into grain 136 along {110} plane in the ⟨111⟩ slip direction. Thus the grain boundary crack at the specimen surface propagated into a favorable grain below the surface [*6*].

Observation of Cross Section

Figure 13 presents the observation and analysis on another specimen, B. Photograph (*a*) shows a crack that had initiated along the boundary of grains 1 and 3, and propagated into

TABLE 3—*Numerical assessment of 14 grain boundaries as possible fatigue crack nucleation sites.*

No.	Grain Boundary Grain i–Grain j	Schmid Factor K_i–K_j	Angle, deg $\angle(\boldsymbol{\pi}_i, \boldsymbol{\pi}_j)$ (I)	$\angle(\mathbf{s}_i, \mathbf{s}_j)$ (II)	$\angle(\mathbf{s}_i, \boldsymbol{\pi}_j)$ (IV)	Diameter Ratio d_i/d–d_j/d (V)	Increase in Area,[a] % (VI)	Other Grains Below Surface[b] (VII)	V-Shape Crossing[c] (VIII)	Cracking[d]
1	30(1)–55(1)	0.49–0.45	84.6	84.5	5.5	1.4–1.3	1.3	A	O	×
2	48(1)–115(1)	0.50–0.42	87.6	75.8	18.5	1.2–1.4	21.2	A	N	O
3	48(2)–113(2)	0.49–0.44	80.5	80.1	10.3	1.2–1.0	21.2	A	N	
	113(1)–48(1)	0.45–0.50	80.5	84.0	10.3	1.0–1.2	7.7	A	N	×
4	60(2)–136(1)	0.45–0.42	87.3	84.8	30.9	1.6–1.6	48.9	A	N	o
5	66(1)–69(2)	0.45–0.47	84.7	74.4	17.0	1.3–1.2	29.4	A	N	×
6	85(2)–84(1)	0.45–0.48	78.3	68.1	43.9	2.0–1.1	27.7	P	N	×
7	85(2)–86(1)	0.45–0.41	87.2	87.0	3.7	1.1–1.0	27.7	A	O	×
8	87(1)–89(1)	0.50–0.43	79.9	82.7	33.7	1.7–1.5	38.3	A	N	o
9	113(2)–112(1)	0.44–0.45	89.9	89.9	2.6	1.0–1.5	7.7	P	N	×
10	126(1)–189(1)	0.47–0.46	78.5	86.7	25.5	1.0–1.4	50.5	A	O	×
11	130(1)–57(1)	0.42–0.44	74.9	74.4	18.4	2.3–1.3	3.4	A	O	×
12	142(1)–78(1)	0.44–0.44	80.7	80.8	26.3	1.8–1.1	16.4	A	N	o
13	142(1)–144(2)	0.44–0.45	85.2	84.7	8.0	1.8–1.3	16.4	P	N	×
14	151(2)–150(2)	0.45–0.47	78.9	73.5	43.6	1.1–1.2	−31.9	A	N	×

NOTE: All of the above cases satisfy Condition III.
i = slip side.
j = block side.
(1) = primary slip system.
(2) = secondary slip system.
[a] Increase in area a_i of the slip side grain i after removing the surface layer of about 44 μm.
[b] A = absent, P = present at the grain boundary when the surface layer is removed.
[c] O = occurs, N = does not occur.
[d] o = occurs, × = does not occur.

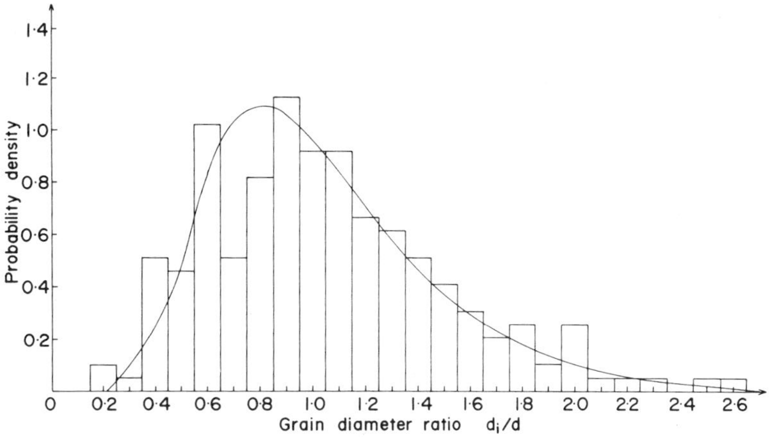

FIG. 14—*Probability density of grain diameter ratio.*

grain 4. One of the longitudinal sections including this crack is shown in photograph (*b*), which was taken after the treatment by etchants. The results of the orientation analysis of grains 1 and 3 are listed in Fig. 13*d* in the same manner as before. From this table, three slip systems (s_{11}, π_{12}), (s_{11}, π_{14}), (s_{12}, π_{15}) in grain 1 and two slip systems (s_{33}, π_{36}), (s_{32}, π_{33}) in grain 3, having higher Schmid factors K in respective grains, are taken out, and their geometrical configuration is depicted in the illustration (*c*) together with the crack and the grain boundary. The intersections of these slip planes π_{ij} with the specimen surface are also indicated in the photograph (*a*). These intersections are in good agreement with the directions of actual slipbands. Angular relationships between two slip systems, respectively, in grains 1 and 3 are examined in table (*e*) for two combinations [(s_{32}, π_{33}), (s_{11}, π_{12})] and [(s_{33}, π_{36}), (s_{11}, π_{14})], in terms of the angles between the slip plane normals and slip directions referring to Conditions I, II, and IV. It is obvious that the latter combination has a closer approach to orthogonality so that the crack is most likely to propagate along the slip plane π_{14}. But the crack actually lies on plane π_{12} and not on plane π_{14} as is seen in the illustration (*c*). Here one can intuitively find that the crack nucleated at the grain boundary cannot proceed along the plane π_{14} which forms a "V shape" with the grain boundary toward the inside. This finding was added as Condition VIII:

VIII The dominating slip plane of the blocking side must not make a "V shape" intersection with grain boundaries.

In Fig. 11, several thick-lined grain boundaries with the symbol VIII indicate that this condition is not satisfied there.

Now, in Fig. 11, only two grain boundaries (66–69, 48–113) are left as sites that satisfy all of the eight conditions but have no crack nucleation. Consequently, there might be still more conditions that must be satisfied at those sites. There is, however, the possibility that cracks may nucleate there if the test is continued. These points will be further studied in the future. Nevertheless, the fact should be emphasized that, in more than 500 grain boundaries, only six places remain as possible crack nucleation sites by imposing the above eight conditions, and at four places out of six such sites, cracks were actually observed.

Table 3 presents a result of a numerical assessment carried out for the 14 grain boundary sites shown by thick lines in Fig. 11. It is assumed that Condition VI holds if the area of

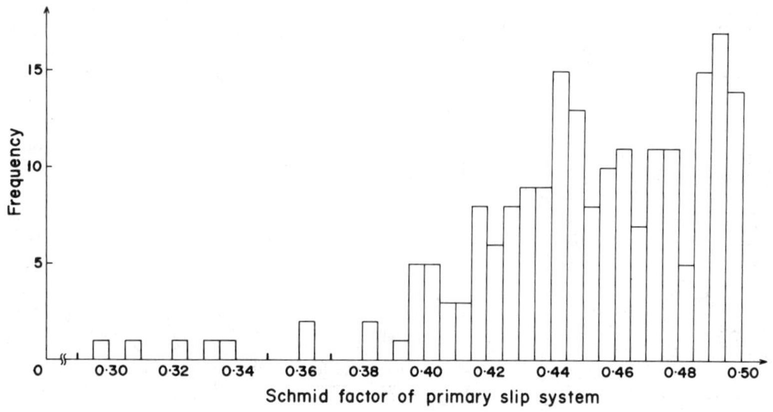

FIG. 15—*Frequency distribution of Schmid factor* K *of primary slip system.*

the slip side grain increases when the surface layer is removed (eighth column). One can understand to what degrees each of the eight conditions is satisfied at respective sites.

Some Statistical Aspects

Although the above discussion was made deterministically, most of the parameters appearing in the discussion have a statistical nature and are essentially random variables. Therefore, strictly speaking, all the conditions should be presented including the probability of occurrences of the involved parameters and related probabilities of crack nucleation. Of

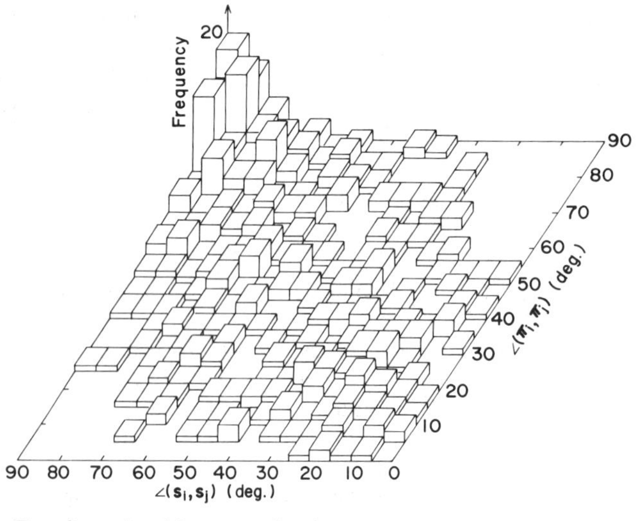

FIG. 16—*Two-dimensional frequency distribution of angles between slip directions and slip plane normals of primary slip systems adjacent to each grain boundary.*

course this is almost impossible at present. But, based on such a statistical viewpoint, those parameter values appearing in the previous conditions were the values chosen to handle the observation results, assuming that the probability of crack nucleation was very small outside the indicated regions of the parameters. Regarding this point, the presentation of the statistical nature of some basic parameters is very helpful to understand an essential feature of the crack nucleation phenomenon.

Figure 14 shows a density distribution of grain diameter ratio d_i/d. The curve gives the density function of a logarithmic normal distribution with the parameters estimated by the experimental values. It is seen that this distribution fits the distribution of grain size very well.

Figure 15 gives a distribution of the largest Schmid factor K in each grain in Fig. 11. A very similar distribution of K had been obtained analytically by one of the present authors for face-centered-cubic (fcc) metals [*12*]. It is noted that the distribution of K is concentrated at the numerically larger side and, therefore, slip motion can easily occur in a number of grains as far as K is concerned.

Finally, Fig. 16 shows a joint distribution of angles $\angle(\mathbf{s}_i, \mathbf{s}_j)$ and $\angle(\boldsymbol{\pi}_i, \boldsymbol{\pi}_j)$ for all pairs of adjacent grains in Fig. 11, where only the slip system with the largest Schmid factor is taken out in each grain. The total number of pairs is 545. Since the distribution is again concentrated at around 90 deg for both angles, one can understand that it is impossible to control the crack nucleation sites only by the orthogonality conditions such as Conditions I, II, and IV. It is noted that the distribution characteristics of the parameters involved in one of the above conditions are important to know the regulation power of that condition for the nucleation of cracks. This poses an interesting problem for our future study.

Conclusion

A new method was developed to determine the grain orientation on the surface of polycrystalline pure iron specimen with great accuracy from the measured ridge line angles of crystallographic etch pit photographs. Especially, a speedy determination of grain orientations became possible by making use of the master table developed in the second half of this study.

Observations and analyses were made on pure iron plate specimens which were subjected to plane bending stress in the high-cycle fatigue range. It was observed that microcracks nucleated only along grain boundaries at the specimen surface, but such microcracks propagated into one of the adjacent grains below the surface along the slip plane and in the slip direction of one of the dominating slip systems of the grain. As favorable crystallographic conditions for crack nucleation, eight conditions were derived that must be satisfied at the crack nucleation sites. These conditions include such factors as geometrical configuration of slip systems in grains adjacent to each other at grain boundary, Schmid factor, grain size, direction of grain boundary, and crystallographic structure below the surface. It is confirmed that these conditions are effective enough to control the favorable sites for crack nucleation at least at the present stage of the study. An explanation is also given on some interesting statistical distribution characteristics of the important parameters appearing in the above conditions.

Acknowledgment

This study was subsidized by the Japanese Ministry of Education, a Grant in Aid for General Research (B) [56460071]. The authors wish to express their thanks for this support.

APPENDIX

In Fig. 17, when a point P (a position vector $\mathbf{r}$) on the unit sphere is mapped to a point P' (a position vector $\mathbf{r}'$) on the equator plane (the projection plane), it is called the stereographic projection. Let $\boldsymbol{\ell}$, $\mathbf{m}$, and $\mathbf{n}$ be the three base vectors along the positive x, y, and z-axes, the vector $\overrightarrow{O'P'}$ is on the vector $\overrightarrow{O'P} = \mathbf{r} + \mathbf{n}$ so that $\overrightarrow{O'P'} = c(\mathbf{r} + \mathbf{n})$. By making use of the fact that the projection of the vector $\overrightarrow{O'P'}$ on the z-axis is equal to unity, the coefficient c can be determined as follows

$$c = 1/(1 + \mathbf{r} \cdot \mathbf{n})$$

Hence

$$\mathbf{r}' = \overrightarrow{O'P'} - \mathbf{n} = (\mathbf{r} + \mathbf{n})/(1 + \mathbf{r} \cdot \mathbf{n}) - \mathbf{n} \tag{6}$$

Let l, m, and n be the direction cosines of the vector $\mathbf{r}$; then

$$\mathbf{r} = l\boldsymbol{\ell} + m\mathbf{m} + n\mathbf{n}, \qquad (l^2 + m^2 + n^2 = 1)$$

Making use of this vector $\mathbf{r}$, Eq 6 is written as

$$\mathbf{r}' = l\boldsymbol{\ell}/(1 + n) + m\mathbf{m}/(1 + n) \tag{7}$$

Now, if α' and β' are the angles formed by the vector $\mathbf{r}'$ on the projection plane and the vectors $\boldsymbol{\ell}$ and $\mathbf{m}$ (Fig. 17)

$$\mathbf{r}' \cdot \boldsymbol{\ell} = r'\cos\alpha' = l/(1 + n)$$

$$\mathbf{r}' \cdot \mathbf{m} = r'\cos\beta' = m/(1 + n)$$

$$r'^2 = (1 - n^2)/(1 + n)^2$$

Hence

$$l = \sqrt{1 - n^2}\cos\alpha',\ m = \sqrt{1 - n^2}\cos\beta' \tag{8}$$

When the vector $\mathbf{r}$ rotates around a fixed vector $\mathbf{e}_1$ $(= l_1\boldsymbol{\ell} + n_1\mathbf{n})$ with a constant angle θ_0, the locus of the point P', that is, the tip of the stereographic projection of $\mathbf{r}$, is as follows: On the stereographic projection plane, let ξ, η be the coordinate variables in the direction of $\boldsymbol{\ell}$ and $\mathbf{m}$; then the coordinates of the point P' are found from Eq 7:

$$\xi = l/(1 + n),\ \eta = m/(1 + n) \tag{9}$$

The inner product is

$$\mathbf{r} \cdot \mathbf{e}_1 = ll_1 + nn_1 = \cos\theta_0,\ (l_1 = \sin\gamma_1,\ n_1 = \cos\gamma_1) \tag{10}$$

By solving Eqs 9 and 10 for l, m and n, and substituting the result into the unit sphere

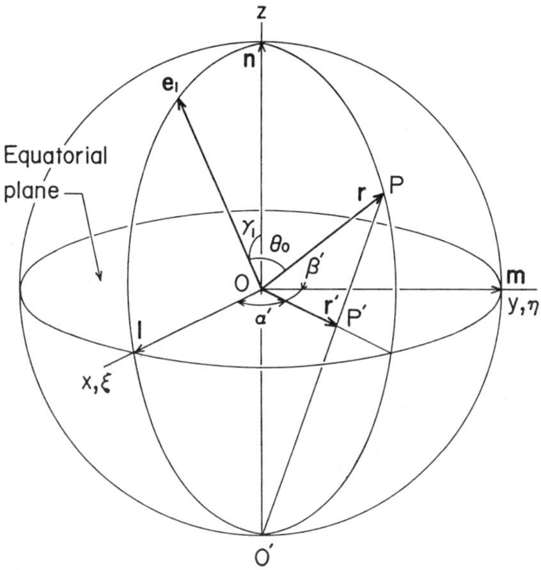

FIG. 17—*Stereographic projection and coordinate systems.*

equation, $l^2 + m^2 + n^2 = 1$, the following equation is derived:

$$\left(\xi - \frac{l_1}{n_1 + \cos\theta_0}\right)^2 + \eta^2 = \left(\frac{\sin\theta_0}{n_1 + \cos\theta_0}\right)^2 \tag{11}$$

Hence the point P' describes a circle, and when $n_1 + \cos\theta_0 = 0$, it becomes a straight line $\xi = \cot\theta_0$.

References

[1] Kim, W. H. and Laird, C., *Acta Metallurgica,* Vol. 26, 1978, p. 789.
[2] Taoka, K. et al., *Journal of the Japan Institute of Metals,* Vol. 30, 1966, p. 820.
[3] Otsuka, A. et al., *Journal of the Society of Materials Science (Japan)*, Vol. 27, 1978, p. 49.
[4] Asami, K. and Terasawa, M., *Journal of the Society of Materials Science (Japan)*, Vol. 30, 1980, p. 803.
[5] Kovacs, T., *Principles of X-Ray Metallurgy,* ILIFFE Books Ltd., London, 1969, p. 52.
[6] Tanaka, T. and Kawase, M. in *Proceedings,* 24th Japan Congress on Material Research, Kyoto, 1981, p. 88.
[7] Nakazawa, H. et al., *Transactions,* Japan Society of Mechanical Engineers, Vol. 40, 1974, p. 9.
[8] Wei, R. P. and Baker, A. J., *Philosophical Magazine,* Vol. 12, 1965, p. 1005.
[9] Sugano, M. and Kojima, K., *Journal of the Society of Materials Science (Japan)*, Vol. 19, 1970, p. 656.
[10] Yoshida, A. et al., *Journal of the Society of Materials Science (Japan)*, Vol. 19, 1970, p. 722.
[11] Kurobe, T. et al., *Journal of the Society of Materials Science (Japan)*, Vol. 26, 1977, p. 355.
[12] Tanaka, T. and Sakai, T., *Bulletin of the Japan Society of Mechanical Engineers,* Vol. 22, 1979, p. 1517.

H. Sehitoglu,[1] *D. F. Socie,*[1] *and D. Worthem*[1]

Crack Growth Studies in Biaxial Fatigue

REFERENCE: Sehitoglu, H., Socie, D. F., and Worthem, D., **"Crack Growth Studies in Biaxial Fatigue,"** *Basic Questions in Fatigue: Volume I, ASTM STP 924,* J. T. Fong and R. J. Fields, Eds., American Society for Testing and Materials, Philadelphia, 1988, pp. 120–135.

ABSTRACT: The influence of strain state on the nucleation and early growth of fatigue cracks was investigated. Thin-walled tubular specimens of Inconel 718 were subjected to combinations of axial and torsional loading in the low-cycle fatigue regime.

The changes in crack growth rates as a function of crack size have been examined at effective strain amplitudes of 0.01 and 0.005. The crack growth rates under pure torsion, combined loading, and axial loading were found to be similar for the same crack size. Crack growth rates were interpreted based on an elastoplastic fracture mechanics characterization. A strain intensity range parameter was defined based on the shear and normal strain calculated on the crack plane. For mixed-mode loading conditions, similar crack growth rates were observed at a given range of strain intensity.

Extensive crack surface rubbing has been observed due to facets formed along the crack. Results indicate that relative changes in the ratio of Mode II/Mode I strain intensity components did not alter the crack growth rates notably at an effective strain amplitude of 0.005. However, at higher strain levels a relative decrease in Mode II/Mode I ratio promoted higher crack growth rates. If the magnitude of Mode I component of loading is high enough to result in crack opening beyond the size of the facets, then Mode II and Mode III components may be fully effective in propagating the crack.

KEY WORDS: Mode I and Mode II fatigue crack growth, Inconel 718, crack closure, small cracks

Many components in engineering applications experience stresses that are multiaxial in nature. The practical importance of Mode II and Mode III crack growth requires a better understanding of the mechanisms of crack nucleation and growth under these conditions. As shown in Fig. 1, when cylindrical members are subjected to combined loading, Mode I, Mode II, and Mode III components of stress (strain) intensity exist along the semi-elliptical (thumbnail) crack front. The crack growth direction shown in Fig. 1 is on the maximum shear strain plane and γ_{max} and ϵ_n denote the maximum shear strain and normal strain amplitudes on the crack plane, respectively. The surface crack length is denoted as $2c$. Note that the Mode II component tends to extend the crack length along the surface while the Mode III component causes the crack to penetrate into the depth of the component. The presence of a cyclic or steady (tensile) normal strain on the crack plane influences the Mode II and Mode III crack growth rates significantly as it provides a Mode I type separation of crack surface, allowing the sliding displacements associated with Modes II and III to occur with minimized abrasion and rubbing. Furthermore, the presence of an axial load may cause a change in the direction of maximum principal and shear plane directions and, depending

[1] Assistant professor, professor, and research assistant, respectively, Department of Mechanical and Industrial Engineering, University of Illinois at Urbana-Champaign, 1206 West Green St., Urbana, IL 61801.

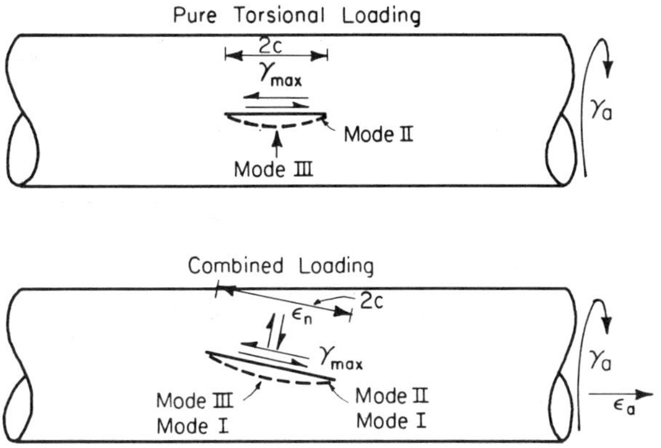

FIG. 1—*Crack growth directions and the normal and shear strain amplitudes on the crack plane for pure torsional and combined loading.*

on the crack orientation, may alter the relative portions of Mode I, Mode II, and Mode III stress (strain) components.

There have been a number of critical investigations on torsional crack growth. Mode III crack growth is produced by a circumferentially notched round bar subjected to pure torsion. Hurd and Irving [*1*] found that Mode III crack growth rates were 10 to 50 times slower than Mode I crack growth rate for the same stress intensity range. Rubbing and abrasion of crack surfaces under crack sliding has been suggested to explain lower growth rates observed. Smeared fracture surface observations provide further support to this effect. Similar observations were made by Ritchie et al. [*2*], and Tschegg et al. [*3*]. Other studies considered a slant crack [*4*] subjected to tensile loading experiencing both Mode I and Mode II components. In this case fatigue cracks tend to deviate to grow under pure Mode I [*5*]. Pook [*6*] found that if Mode I crack formation readily occurs at the tip of a main crack, then the ensuing behavior is controlled by Mode I displacements. Smith [*7*] and Shield and Socie [*8*] observed that in the presence of a compressive strain component on the crack plane, Mode II crack growth is still operative and causes crack advance.

In most of the crack growth experiments reported in the literature on combined loading, the applied stresses have been in the elastic range; linear elastic fracture mechanics (LEFM) representation of crack growth was valid and crack nucleation occurred from a notch or a preexisting crack. Since the notch or the sawcut crack has its own strain field, the interpretation of crack growth behavior in the vicinity of the notch field may require an analysis of elastoplastic strains and stresses. It is known from uniaxial tests [*9*] that notch size has an influence on crack growth rates even when the crack grows beyond the notch plastic zone.

Smooth tubular specimens are considered in this study. The tubular specimen has a uniform stress-strain field in the gage section. Stresses and strains were measured throughout the experiments. Combined axial and torsional loading, axial loading, and pure torsional loading are considered. Early crack growth in the ranges of 0.1 to 1 mm in length has been measured and reported.

Recently, significant attention has been devoted to the early stages of crack growth, and the term "small crack growth behavior" has been coined for the phenomenon of high crack growth rates observed for small cracks. It was found for many materials subjected to Mode I loading that small cracks grow faster than long cracks for a given stress or strain intensity range [*10–11*]. Various hypotheses as to what causes the "small crack effect" have been

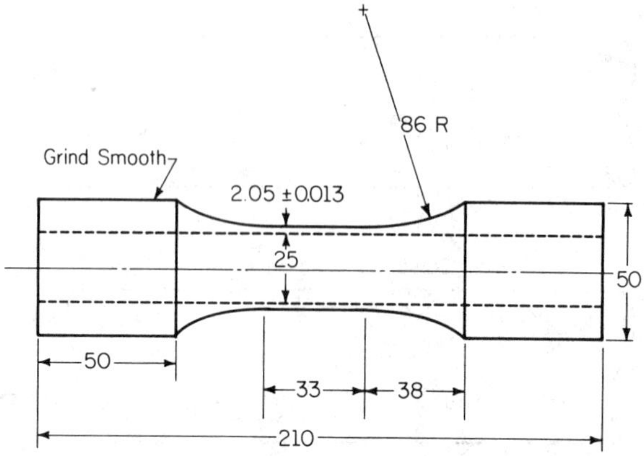

FIG. 2—*Tubular specimen dimensions (0 mm).*

forwarded. Crack closure [*9,12*], plasticity level,[2] grain size, and geometry are some of the many factors considered that may influence early crack growth in a manner different from long crack growth. For mixed-mode loading, the crack path and relative ratios of Mode I and Mode II or Mode I and Mode III components of stress (strain) intensity are additional considerations influencing early crack growth. It would be instructive to study early crack growth under different combinations of Mode I and Mode II cases such that pure Mode I and pure Mode II constitute the extreme cases of crack growth. In this study, interference of crack surfaces under Mode II loading has been identified, and the influence of relative portions of Mode II to Mode I components on crack growth rates is described.

Difficulties arise when fracture mechanics representation of small cracks is attempted. Comparison of short and long crack growth behavior based on a general fracture mechanics parameter appears attractive since this will provide a basis for comparison of crack growth rates for the two cases. The use of elastic-plastic fracture mechanics parameters such as the J-integral, crack tip opening displacement, strain intensity factor, and crack tip plastic zone size has been suggested. The utility of these parameters has been verified for only Mode I loading and only for specific materials and conditions. When Mode II and Mode III components contribute to crack advance, a combined value of these parameters needs to be adopted. There are a number of ways Mode I, Mode II, and Mode III components of crack growth may be incorporated into a unified (total) stress or strain intensity parameter. In this study, a total strain intensity parameter was defined analogous to the summation of energy release rates associated with Mode I, Mode II, and Mode III components of loading. Improvements in the form of fracture mechanics parameters to handle crack growth under combined mode loading are needed. Further emphasis is placed on description and comparisons of crack growth behavior, as a function of crack length, for different strain states.

Experimental Procedure

Tests were conducted on Inconel 718 at room temperature under biaxial loading. The loading cases considered were axial, combined, and pure torsional loading. Defining a shear to axial strain ratio λ as $\Delta\gamma/\Delta\epsilon$, the three cases considered will be referred to as $\lambda = 0$,

[2] Private communication with C. McClung.

TABLE 1—*Experimental test program.*

	Nominal Strain Amplitude		Nominal Stress Amplitude		Mean Stress		Fatigue Life		
Specimen	$\Delta\epsilon/2$	$\Delta\gamma/2$	$\Delta\sigma/2$, MPa	$\Delta\tau/2$, MPa	σ_0, MPa	τ_0, MPa	$N_{0.1}$[a]	$N_{0.1}$[b]	N_f[c]
				Axial $R_\epsilon = 0$					
B-5	0.01	0	1100	0	20	0	240	800	936
B-36	0.005	0	970	0	240	0	2400	7 000	7 029
				Axial $R_\epsilon = -1$					
B-14	0.010	0	1050	0	0	0	200	1 050	1 330
B-12	0.005	0	920	0	0	0	2000	11 000	13 364
				Combined $R_\epsilon = 0$					
B-98	0.007	0.012	740	410	100	+70	225	1 050	1 333
B-17	0.0035	0.0063	610	380	60	100	2000	7 500	9 500
				Combined $R_\epsilon = -1$					
B-35	0.007	0.012	740	430	0	0	200	1 000	1 374
B-26	0.0035	0.0063	630	375	0	0	2000	8 000	12 899
				Torsion $R_\epsilon = 0$					
B-99	0	0.017	0	580	0	10	50	1 000	1 687
B-32	0	0.0085	0	500	0	70	500	4 500	9 526
				Torsion $R_\epsilon = -1$					
B-13	0	0.017	0	570	0	0	100	800	1 674
B-8	0	0.0085	0	510	0	0	2000	7 000	12 942

[a] Cycles to first crack of length 0.1 mm.
[b] Cycles to first crack of length 1.0 mm.
[c] Cycles to failure (10% drop in load).

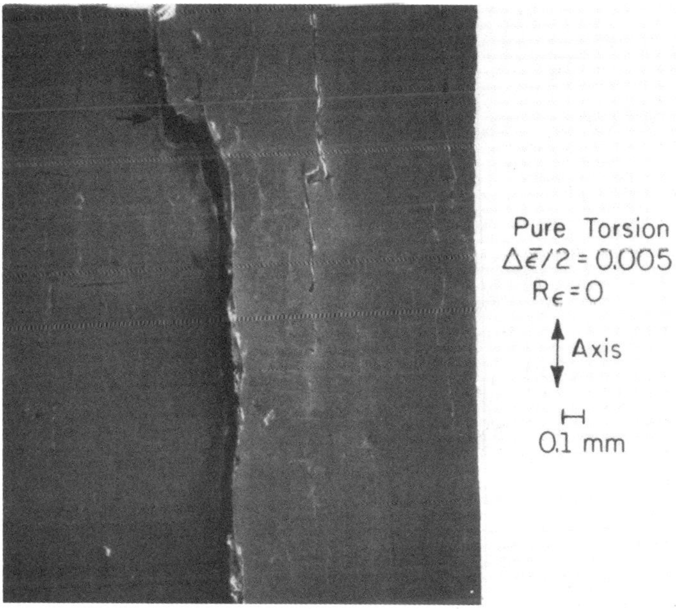

FIG. 3—*Crack growth in the direction of maximum shear strain* ($\Delta\bar{\epsilon}/2$ = *0.005*, R_ϵ = *0 pure torsion*).

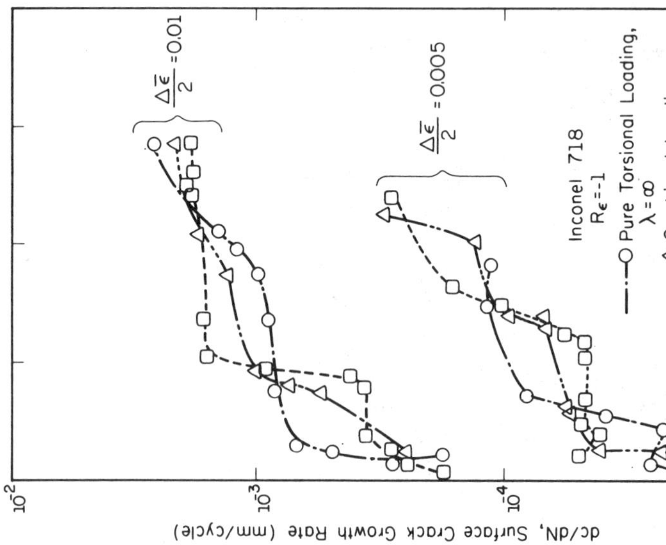

FIG. 4*a*—*Surface crack growth rates as a function of surface crack length* (R_ϵ = −1).

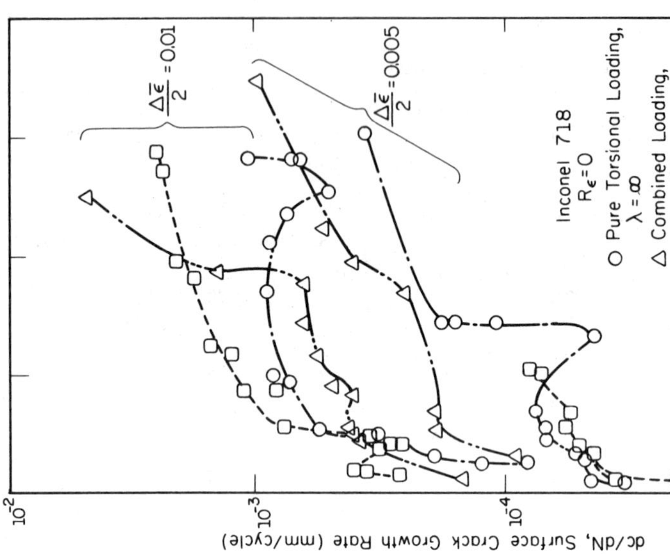

FIG. 4*b*—*Surface crack growth rates as a function of surface crack length* (R_ϵ = 0).

$\sqrt{3}$, and ∞, respectively. The applied strain cycle was either completely reversed $R_\epsilon = -1$ or varied from zero to maximum ($R_\epsilon = 0$).

The effective strains were defined based on the von Mises criterion as

$$\frac{\Delta\bar{\epsilon}}{2} = \bar{\epsilon}_a = \left[\frac{1}{\sqrt{2}}(1 + \nu_{\text{eff}})\right]\left[2\epsilon_a^2(1 + \nu_{\text{eff}})^2 + \frac{3}{2}\gamma_a^2\right]^{1/2} \tag{1}$$

which reduces to the familiar form of

$$\frac{\Delta\bar{\epsilon}}{2} = \bar{\epsilon}_a = (\epsilon_a^2 + \gamma_a^2/3)^{1/2} \text{ for } \nu_{\text{eff}} = 0.5 \tag{2}$$

In the above equations ϵ_a and γ_a denote applied axial and shear strain amplitudes, respectively. This formulation is equivalent to considering elastic and plastic strains separately. The effective Poisson ratio, ν_{eff}, may be defined as

$$\nu_{\text{eff}} = (\nu_e|\epsilon_e| + \nu_p|\epsilon_p|)/(\epsilon_e + \epsilon_p) \tag{3}$$

Tubular specimens were used. Details of the specimen dimensions are shown in Fig. 2. This specimen has the advantage of direct measurement of axial and shear stresses and strains throughout the test. An average shear stress over the thin wall of the cylinder was calculated based on the applied torque. The shear stress at the surface may then be calculated using the shear-stress/shear-strain curve and assuming that shear strains vary linearly over the wall thickness. An extensometer placed inside the specimen allowed axial deflection and

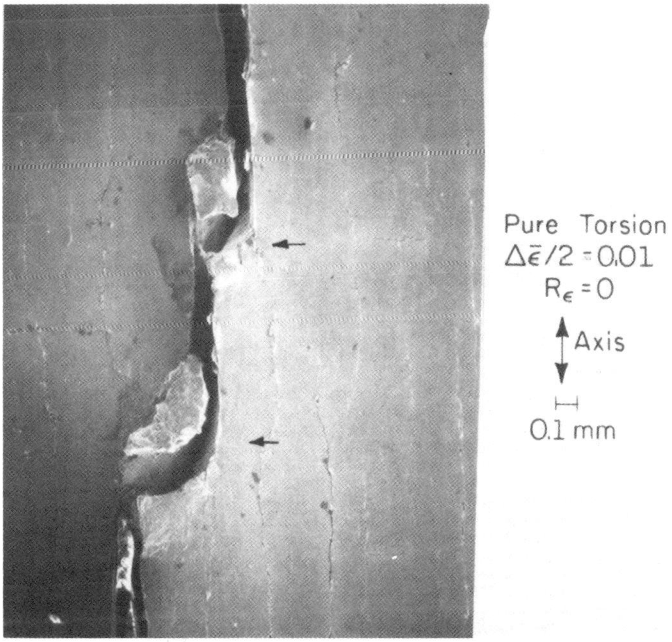

FIG. 5a—*Facets formed along the crack length as indicated by dark arrows.*

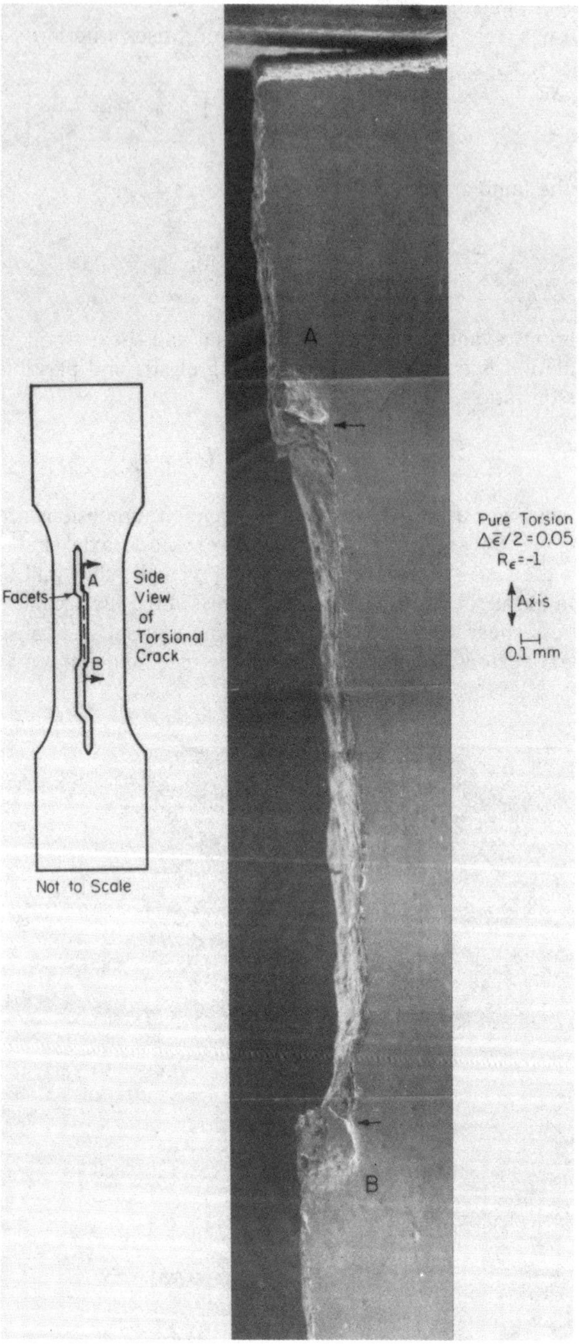

FIG. 5*b*—*Micrographs showing facets for* R_ϵ = −*1 torsional loading case.*

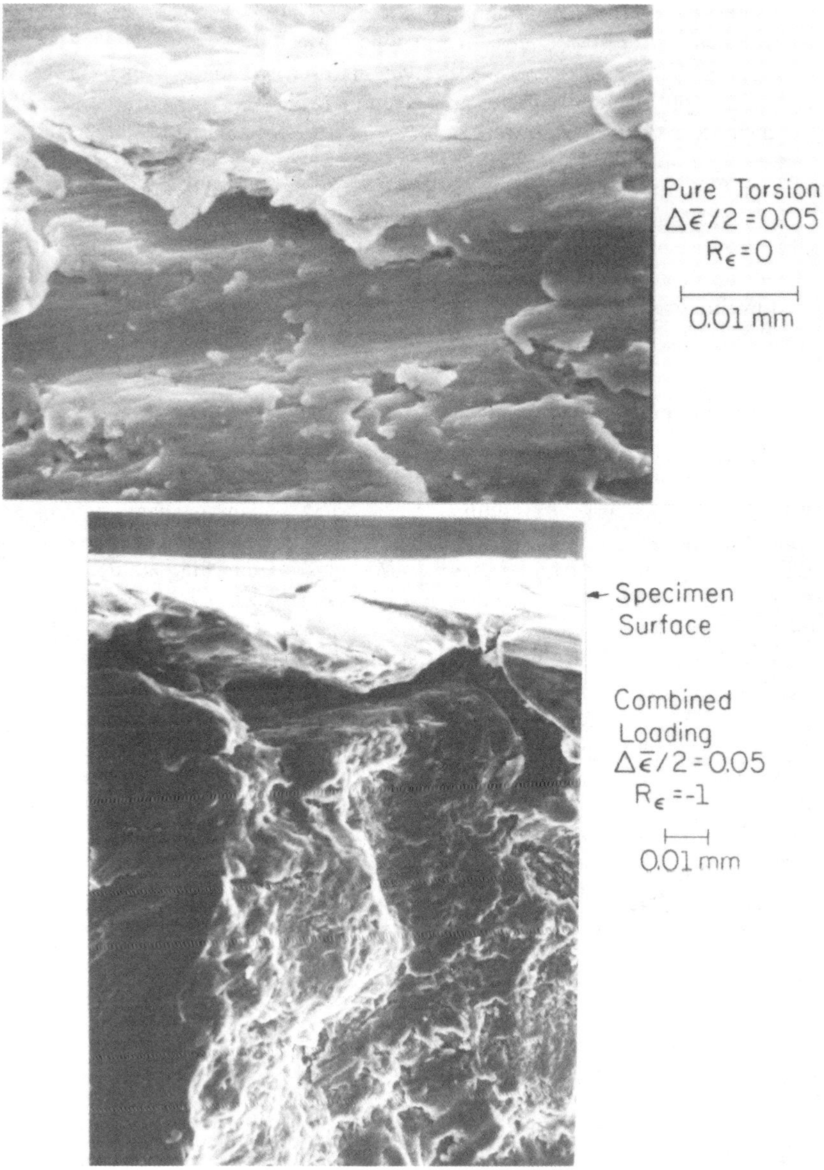

FIG. 5c—*Fracture surfaces under pure torsion* (R_ϵ = *0,* $\Delta\bar{\epsilon}/2$ = *0.005*) *and combined loading* (R_ϵ = *−1,* $\Delta\bar{\epsilon}/2$ = *0.005*) *indicating wear and surface abrasion.*

angle of rotation to be measured. On the outside of the specimen, the strains are 6% greater than the inner surface in pure torsion. The grain size of the material varied from 0.01 to 0.2 mm. An average grain size of 0.03 mm was determined. The specimen thickness was chosen to have at least ten grains in the thickness direction. The monotonic properties of In 718 were: elastic modulus = 209 GPa, 0.2% offset yield strength = 1160 MPa, fracture stress = 1850 MPa, and strain hardening exponent = 0.08. The cyclic strength coefficient was 1530 MPa and cyclic strain hardening exponent was 0.07.

A summary of the fatigue test results is given in Table 1. Fatigue lives to grow a surface crack length, $2c$, of 0.1 and 1 mm are also given in Table 1. It is noted that in most cases, small cracks of less than 0.1 mm were detected as low as 10% of the fatigue life. Shorter fatigue lives were obtained in $R_\epsilon = 0$ tests compared with $R_\epsilon = -1$ tests. This may occur due to the presence of mean tensile stresses in $R_\epsilon = 0$ tests which are tabulated as σ_0 (mean axial stress) and τ_0 (mean shear stress). The crack growth behavior of small and longer cracks under uniaxial loading has been investigated also using center-notched specimens. These results are compared later with crack growth behavior obtained under torsional and combined loading cases. The center-notched specimen used was 38.1 mm wide, 3.175 mm thick, and had an initial notch half width of 1.9 mm.

Experimental Observations

It was observed that cracks nucleated and grew in the direction of maximum shear strain amplitude. The cracks propagated in the direction normal to the principal stress only in the case of axial loading just prior to fracture. Two alternate maximum shear strain planes exist under pure torsion. The dominant cracks grew on the plane along the specimen axis. Crack growth in the direction of maximum shear strain is illustrated in Fig. 3 under pure torsional loading. It is noted in Fig. 3 that multiple small cracks adjacent to the dominant cracks are present. Cracks on the horizontal shear plane are illustrated also in the same figure. Facets along the crack path are shown with a dark arrow. The influence of facets on crack surface rubbing is discussed later. The horizontal cracks decelerated with increasing crack length and were arrested.

Surface crack growth rates dc/dN, are compared for torsional, combined, and axial loading in Figs. 4*a* and 4*b*. It is interesting to note that torsional crack growth rates for the $R_\epsilon = -1$ case (Fig. 4*a*) are similar to the combined and axial loading cases. Fatigue lives tabulated in Table 1 confirm that crack propagation lives for a given effective strain amplitude are similar. For example, fatigue lives corresponding to $\Delta\bar{\epsilon}/2 = 0.01$ and $R_\epsilon = -1$ for the pure torsion, combined, and axial loading cases are 1674, 1374, and 1330 cycles, respectively.

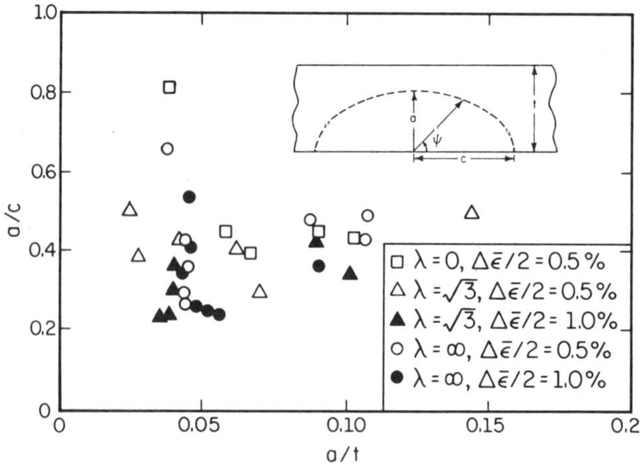

FIG. 6—*Ellipticity ratio of surface cracks and crack geometry of surface cracks.*

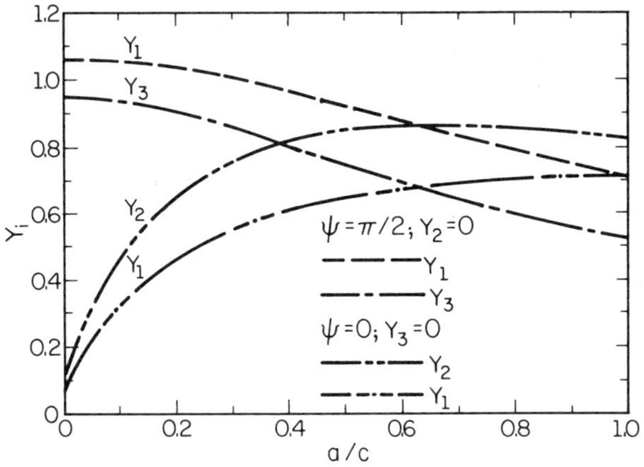

FIG. 7—*Variation of correction factors* Y_1 *and* Y_2 *as a function of aspect ratio* a/c.

The longer fatigue lives observed in pure torsion tests are attributed to the longer crack size at fracture compared with the axial and combined loading cases.

Crack growth for the case of $R_\epsilon = 0$ has been investigated also (Fig. 4*b*). Crack growth rates for the three strain states are again similar. However, a decrease in crack growth rate has been observed for pure torsional loading. The cause and implications of such a decrease in crack growth rate are discussed later. Crack growth rates for axial and combined loading cases appear higher than those observed under pure torsional loading for the $R_\epsilon = -1$ case.

The crack growth rates of secondary cracks ($R_\epsilon = 0$ case) in the axial direction have been observed to decelerate consistently under pure torsion. For the axial and combined loading cases, secondary cracks indicated increasing crack growth rates with increasing crack length. Since crack advance occasionally occurs by crack linking, the crack growth rate of secondary cracks that contribute to major crack growth is important. The influence of secondary cracks on the overall damage will be addressed in a later paper.

The crack growth path under pure torsion loading has been examined. Facets along the crack length have been observed and are indicated in Figs. 3 and 5*a* by dark arrows. Facets are readily formed in the $R_\epsilon = 0$ and $R_\epsilon = -1$ loading cases. In Fig. 5*b*, facets are shown for the $R_\epsilon = -1$ case. Examination of fracture surfaces revealed extensive rubbing and abrasion as shown in Fig. 5*c* both for pure torsional (upper picture) and combined loading (bottom picture). Residual compressive stresses may build along the facets; these stresses need to be overcome before the crack surfaces may slide relative to each other again. The increase in number of facets in the wake of a torsional crack may cause the crack deceleration effect at moderate crack sizes as shown in Fig. 4*b* (pure torsion case).

Analysis

In previous studies on combined mode crack growth, an effective stress intensity has been introduced to characterize crack tip stress-strain fields. When the plastic zone size of the crack is comparable to the crack size, LEFM representation of crack tip stress-strain fields is limited. Then, elastic-plastic fracture mechanics parameters are appropriate to characterize crack growth.

A semi-elliptical surface crack with major axis 2*c* and minor axis 2*a* is shown in Fig. 6. The term *c* indicates half surface crack length; *a* is the crack depth in the thickness direction.

The aspect ratios of these cracks have been determined by disectioning larger number of cracks. The results are summarized on the same figure. The horizontal axis in Fig. 6 is crack depth normalized by the wall thickness. The form of LEFM solutions for semi-elliptical cracks subjected to normal and shear stresses has been derived by Kassir and Shih [*13*]. The state of stress at $\psi = 0$ (Fig. 6) is represented by Mode I and Mode II components of stress intensity. At $\psi = \pi/2$, only Mode I and Mode III components are present. Surface crack growth will be characterized by considering K_1 and K_2 at $\psi = 0$. For cyclic loading

$$\Delta K_1 = Y_1\, 1.2\, \Delta\sigma \sqrt{\pi c} \tag{4}$$

$$\Delta K_2 = Y_2\, 1.12\, \Delta\tau \sqrt{\pi c} \tag{5}$$

where $\Delta\sigma$ and $\Delta\tau$ are normal and shear stress ranges on the crack plane, respectively, and c is half the surface crack length. The correction factors Y_1 and Y_2 depend on the aspect ratio of the semi-elliptical cracks and 1.12 is the surface correction factor. The variation of correction factors Y_1 and Y_2 with the aspect ratio c/a is shown in Fig. 7.

To account for applied stresses and strains in the plastic range, the terms $\Delta\sigma$ and $\Delta\tau$ in the above equations may be replaced by the product of strain ranges $\Delta\epsilon$ and $\Delta\gamma$ and elastic constants E and G, respectively. Then the "stress" intensity expressions are replaced by "strain" intensity terms. It has been shown that the use of strain intensity parameters is similar to the J-integral range for moderately high strain levels [*14*]. The normal and shear

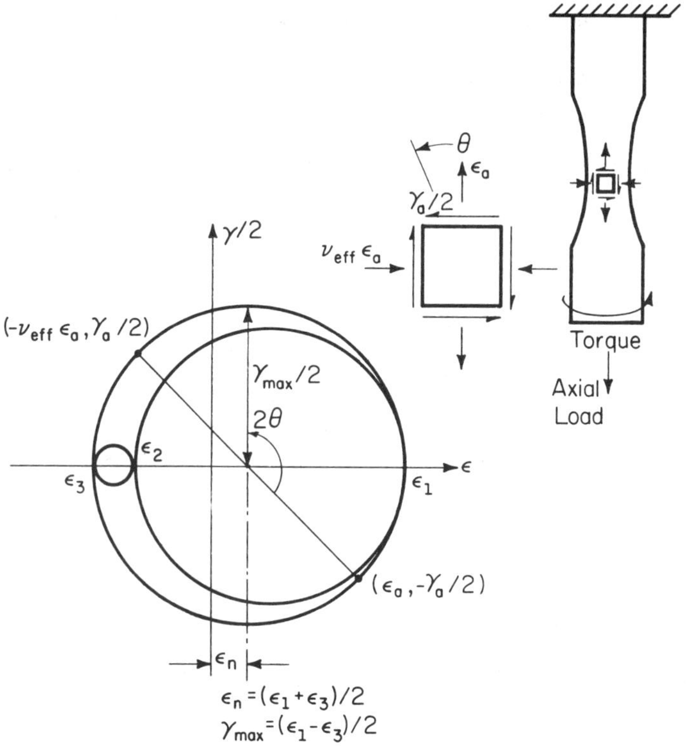

FIG. 8—*Mohr's circle for strain for combined loading.*

TABLE 2—*List of correlation factors and normal and maximum shear strain amplitudes for* $R_\epsilon = 0$ *case.*

$\Delta\bar{\epsilon}/2$	Loading Type	a/c Aspect Ratio	Y_1	Y_2	$\Delta\epsilon_n/2$	$\Delta\gamma_{max}/2$
0.005	torsional	0.4	0.386	0.513	0	0.009
0.005	combined	0.4	0.386	0.513	0.00096	0.0081
0.005	axial	0.4	0.386	0.513	0.0017	0.0069
0.01	torsional	0.3	0.3	0.41	0	0.018
0.01	combined	0.3	0.3	0.41	0.0017	0.0163
0.01	axial	0.3	0.3	0.41	0.003	0.0139

strain ranges on the crack plane may be readily calculated using Mohr's circle (Fig. 8). The case of combined loading under remotely applied ϵ_a (axial strain amplitude) and γ_a (shear strain amplitude) is considered in Fig. 1. The maximum shear strain amplitude, γ_{max}, on the crack plane is identified as $(\epsilon_1 - \epsilon_3)/2$ (see Figs. 8 and 1). The normal strain amplitude, ϵ_n, on the same crack plane is given as $(\epsilon_1 + \epsilon_3)/2$. The plane of maximum shear strain may be measured using Mohr's circle. Alternatively it is given algebraically as

$$|2\theta| = \tan^{-1}[\gamma_a/(1 + \nu_{eff})\epsilon_a]$$

The angle 2θ is illustrated in Fig. 7. It is possible to write the principal strains in algebraic form as follows

$$\epsilon_1 = (1 - \nu_{eff})\epsilon_a/2 + 0.5[\epsilon_a^2(1 + \nu_{eff})^2 + \gamma_a^2]^{1/2} \quad (6)$$

$$\epsilon_2 = -\nu_{eff}\,\epsilon_a \quad (7)$$

$$\epsilon_3 = (1 - \nu_{eff})\epsilon_a/2 - 0.5[\epsilon_a^2(1 + \nu_{eff})^2 + \gamma_a^2]^{1/2} \quad (8)$$

The strain intensity equations may be written as

$$\Delta K_1 = Y_1\, 1.12\, \Delta\epsilon_n\, E\sqrt{\pi c} \quad (9)$$

$$\Delta K_2 = Y_2\, 1.12\, \Delta\gamma_{max}\, G\sqrt{\pi c} \quad (10)$$

where E and G are the elastic modulus and shear modulus, respectively. The values of $\Delta\epsilon_n$, $\Delta\gamma_{max}$, Y_1, and Y_2 used in our analysis for the axial, torsional, and combined loading cases are tabulated in Table 2. The aspect ratio of cracks was taken as 0.4 and 0.3 for the $\Delta\bar{\epsilon}/2 = 0.05$ and $\Delta\bar{\epsilon}/2 = 0.01$ cases, respectively. If the depth of the surface crack is small compared to the specimen thickness (small a/t ratio) no further correction factors are needed in Eqs 9 and 10.

Various forms of combined stress (strain) intensity expressions have been proposed [*15*]. These expressions were based on crack tip displacements for mixed-mode loading [*16*], dislocation model [*17*], and empirical results. The general form of effective stress (strain) intensity range may be defined as

$$\Delta K_{tot} = [(\Delta K_1)^2 + a'(\Delta K_2)^2]^{1/2} \quad (11)$$

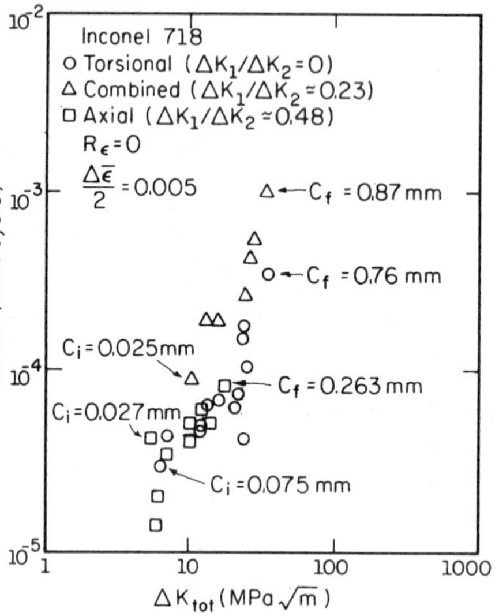

FIG. 9a—*Crack growth rate versus total strain intensity range for the* $R_\epsilon = 0$ *case.*

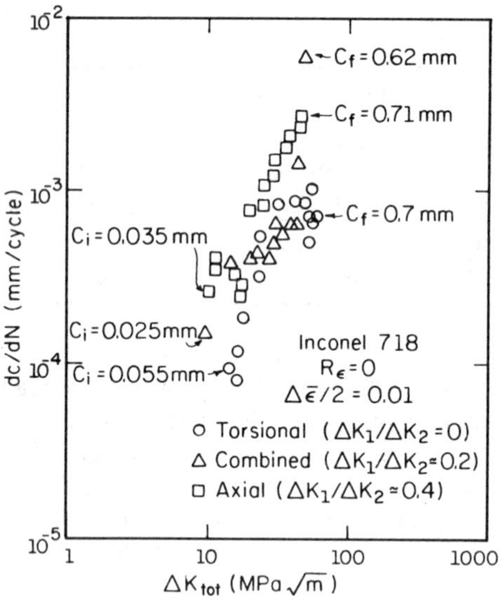

FIG. 9b—*Crack growth rate versus total strain intensity range for the* $R_\epsilon = -1$ *case.*

In some studies product terms of ΔK_1 and ΔK_2 have been considered also. The factor a' has been taken as varying from 1 to 2. Based on Irwin's summation of energy release rates, a may be taken as unity as a first approximation. Considering the estimates made for ΔK_1, ΔK_2 expressions, the choice of a simple representation for ΔK_{tot} appears justifiable.

Crack growth rate behavior as a function of total stress intensity range is given in Fig. 9*a* and Fig. 9*b* for the $R_\epsilon = 0$, $\Delta\bar{\epsilon}/2 = 0.005$, and $\Delta\epsilon/2 = 0.01$ cases, respectively. Sufficient plastic strains are present for both cases; therefore cracks are assumed to be open over the entire strain range. Therefore, total strain range was used in the strain intensity calculations ΔK_1 and ΔK_2. The range of half surface crack lengths measured are indicated in Figs. 9*a* and 9*b*. The terms c_i and c_f indicate the smallest and largest crack size measured in each experiment. Crack lengths measured are in the range 0.025 to 0.90 mm for the $R_\epsilon = 0$ and $R_\epsilon = -1$ cases. In Figs. 9*a* and 9*b*, the stress intensity range ratios $\Delta K_1/\Delta K_2$ for torsional, combined, and axial loading cases are indicated. As shown earlier, crack growth occurred on the maximum shear plane; therefore, the Mode II component of strain intensity range is higher than the Mode I component for the pure torsional, combined, and axial loading cases. The ratios of Mode II to Mode I components of strain intensity range varied from 0 to 0.48 for the $\Delta\bar{\epsilon}/2 = 0.005$ case and from 0 to 0.4 for the $\Delta\bar{\epsilon}/2 = 0.01$ case as indicated in Figs. 9*a* and 9*b*.

An attempt was made to interpret the mixed-mode crack growth data by performing further crack growth tests under pure Mode I loading. Center-notched specimens were used and early crack growth as well as long crack growth was examined for $R_\epsilon = 0$ and $R_\epsilon = -1$ loading cases. The applied stresses and strains were in the elastic range; therefore the tensile portion of the cycle was assumed to be effective in propagating the crack in the $R_\epsilon = -1$ tests. It is noted that the stress and strain intensity factors are identical when the applied strains are in the elastic range. The crack growth rate dc/dN versus stress (strain) intensity range is presented in Fig. 10. If the total stress range was used in the ΔK calculation in the $R_\epsilon = -1$ tests, one would expect the $R_\epsilon = 0$ and $R_\epsilon = -1$ curves in Fig. 10 to be even closer.

Comparison of Figs. 9*a* and 9*b* with Fig. 10 ($R_\epsilon = 0$ case) indicate that the crack growth rates are comparable for the cases examined for a given stress (strain) intensity range. However, localized variations observed in crack growth behavior in the combined-mode loading and pure Mode II loading cases (Figs. 9*a* and 9*b*) are not evident for pure Mode I loading (Fig. 10).

Discussion

Crack growth behavior of Inconel 718 tubular specimens subjected to pure torsional, combined, and axial loading has been studied for effective strain amplitudes of 0.005 and 0.01. It was found that crack growth rates of major cracks under pure torsion are comparable to those under axial and combined loading. Consideration of Mode I and Mode II components of strain intensity indicates that the Mode II component constitutes a significant portion of the total strain intensity range. Further experiments on plate specimens indicated that in the absence of Mode II loading, crack growth rates of small and longer cracks were similar to, if not slower than, those observed under the combined Mode I and Mode II cases. Pure Mode I experiments involved much lower strain ranges than those observed in combined Mode experiments. The crack opening level associated with Mode I loading may decrease with an increase in plasticity level (see footnote 2). Therefore, it may be expected that combined Mode I/Mode II crack growth rates are higher than that of pure Mode I for the cases considered. Previous studies [2] indicate that when a Mode I component is superimposed on the Mode II component, the crack growth rates increase only if Mode I separation minimizes surface abrasion. Higher crack growth rates observed under axial loading ($\Delta K_1/\Delta K_2 = 0.4$, $\Delta\epsilon/2 = 0.01$ case, Fig. 9*a*) may be attributed to the presence of less severe surface abrasion compared with other cases. Examination of fracture surfaces for both pure Mode II and combined Mode I and Mode II indicates that extensive rubbing was present in both cases (Fig. 5*c*). The crack tip opening displacement levels associated with Mode I should exceed the facet size (indicated in Fig. 5*b*) to provide a means for enhanced Mode II crack growth.

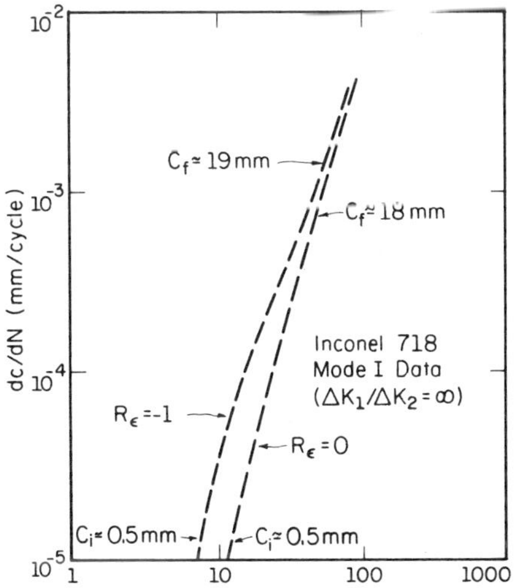

FIG. 10—*Crack growth rate under pure Mode I loading.*

Similar results to those reported here would be obtained if crack growth into the depth of the specimen were considered. In this case, Mode III and Mode I components of strain intensity need to be considered at the location $\psi = 90$ deg in Fig. 6. Crack growth occurred along the specimen surface at a higher rate than crack growth rate into the depth. Therefore, only Mode I and Mode II crack growth analysis has been performed. The crack growth direction into the thickness of the specimen appears to be 45 deg for very small cracks for the axial and combined loading cases. A transition from Stage I to Stage II crack growth occurs rapidly. The use of shear strain at a plane inclined 45 deg to the surface as the driving force for Mode II crack growth may be considered. The use of Mohr's circle of strain with principal strains ϵ_1 and ϵ_2 (see Fig. 8) allows definition of γ and ϵ_n on the 45-deg plane. This modification has not been considered in our analysis for very small cracks.

Further experiments are needed where lower effective strain amplitudes are considered. It is expected that the crack growth direction changes such that the Mode I component is favored over the Mode II component when the crack length reaches a critical size. Then it may be possible to have a larger Mode I component compared to the Mode II component, allowing a large spectrum of combined Mode I and Mode II crack growth cases to be studied.

Conclusions

1. Comparison of crack growth rates for different loading cases is meaningful only when the plasticity levels associated with these cases are similar.
2. Torsional crack growth rates observed are comparable to axial and combined loading crack growth rates based on similar crack size or strain intensity level.
3. Deceleration of crack growth rates under pure torsion has been observed over a narrow range of crack sizes and is attributed to the formation of facets along the crack length.
4. Further experiments that examine crack growth behavior where the Mode I crack driving force exceeds the Mode II component are needed.

Acknowledgments

We would like to acknowledge stimulating discussions held with Dr. J. Fash of the University of Illinois at Urbana-Champaign.

References

[1] Hurd, N. J. and Irving, P. E., "Factors Influencing Propagation of Mode III Fatigue Cracks under Torsional Loading," *Design of Fatigue and Fracture Resistant Structures, ASTM STP 761,* American Society for Testing and Materials, Philadelphia, 1982, pp. 212–233.

[2] Ritchie, R. O., McClintock, F. A., Nayeb-Hashemi, H., and Rifter, M. A., "Mode III Fatigue Crack Propagation in Low Alloy Steel," *Metallurgical Transactions,* Vol. 13A, 1982, pp, 101–110.

[3] Tschegg, E. K., Ritchie, R. O., and McClintock, F. A., "On the Influence of Rubbing Fracture Surfaces on Fatigue Crack Propagation in Mode III," *International Journal of Fatigue,* Vol. 5, No. 1, 1983, pp. 29–35.

[4] Tida, S. and Kobayashi, A. S., "Crack Propagation Rate in 7075-76 Plates Under Cyclic Tensile and Transverse Shear Loadings," *Journal of Basic Engineering, Transactions,* American Society of Mechanical Engineers, Series P, Vol. 91, 1969, pp. 764–769.

[5] Kibler, J. J. and Roberts, R., "The Effect of Biaxial Stresses on Fatigue and Fracture," *Journal of Engineering for Industry, Transactions,* American Society of Mechanical Engineers, 1970, pp. 727–734.

[6] Pook, L. P. and Greenan, A. F., "Fatigue Crack Growth Threshold in Mild Steel Under Combined Loading," *Fracture Mechanics (Eleventh Conference), ASTM STP 677,* American Society for Testing and Materials, Philadelphia, 1979, pp. 23–35.

[7] Smith, R. A., "Mechanics of Mode II Crack Advance" in *Proceedings,* Conference on Micromechanisms of Crack Extension, Cambridge, U.K., 1980.
[8] Socie, D. F. and Shield, T., "Mean Stress Effects in Biaxial Fatigue of Inconel 718," *Journal of Engineering Materials Technology, Transactions,* American Society of Mechanical Engineers, Vol. 106, 1984, pp. 227–232.
[9] Sehitoglu, H., "Crack Opening and Closure in Fatigue," *Engineering Fracture Mechanics,* Vol. 21, No. 2, 1985, pp. 329–339.
[10] Sehitoglu, H., "Fatigue Life Prediction of Notched Members Based on Local Strain and Elastic-Plastic Fracture Mechanics Concepts," *Engineering Fracture Mechanics,* Vol. 18, No. 3, 1983, pp. 609–621.
[11] Dowling, N. E., "Crack Growth During Low-Cycle Fatigue of Smooth Axial Specimens," *Cyclic Stress-Strain and Plastic Deformation Aspects of Fatigue Crack Growth, ASTM STP 637,* American Society for Testing and Materials, Philadelphia, 1977, pp. 97–121.
[12] Sehitoglu, H., "Characterization of Crack Closure," *Fracture Mechanics: Sixteenth Symposium, ASTM STP 868,* American Society for Testing and Materials, Philadelphia, 1985, pp. 361–380.
[13] Kassir, M. K. and Shih, G. C., "Three-Dimensional Stress Distribution Around an Elliptical Crack under Arbitrary Loadings," *Journal of Applied Mechanics,* 1966, pp. 601–610.
[14] Imai, Y. and Matake, T., "Surface Crack Growth During Push-Pull Fatigue of Smooth Specimens," *Journal of the Society of Materials Science, Japan,* Vol. 32, 1983, pp. 1157–1161.
[15] Tanaka, K., "Fatigue Crack Propagation from a Crack Inclined to the Cyclic Tensile Axis," *Engineering Fracture Mechanics,* Vol. 6, 1974, pp. 493–507.
[16] Lardner, R. W., "A Dislocation Model for Fatigue Crack Growth in Metals," *Philosophical Magazine,* Vol. 17, 1968, pp. 71–82.
[17] Weertman, J., "Rate of Growth of Fatigue Cracks Calculated from the Theory of Infinite Dislocations Distributed on a Plane," *International Journal of Fracture,* Vol. 2, 1966, pp. 460–467.

Hironobu Nisitani[1] *and Masahiro Endo*[2]

Unified Treatment of Deep and Shallow Notches in Rotating Bending Fatigue

REFERENCE: Nisitani, H. and Endo, M., **"Unified Treatment of Deep and Shallow Notches in Rotating Bending Fatigue,"** *Basic Questions in Fatigue: Volume I, ASTM STP 924*, J. T. Fong and R. J. Fields, Eds., American Society for Testing and Materials, Philadelphia, 1988, pp. 136–153.

ABSTRACT: As a concept useful for evaluating the strength of notched members, "linear notch mechanics" is proposed. It is explained by this concept that the elastic-plastic behavior near the root of a notch under the condition of small-scale yielding is controlled by the maximum elastic stress σ_{max} and the notch root radius ρ alone, independently of the other notch geometries. In order to confirm the effectiveness of "linear notch mechanics" in fatigue problems, rotating bending fatigue tests are carried out on notched steel specimens having widely varying notch geometries. Moreover, the fatigue notch effects of extremely shallow notches, whose depths are of the order of the grain size, are also discussed.

Summarizing the present studies, a unifying treatment of fatigue notch effects is presented.

KEY WORDS: fatigue (materials), notch effects, fatigue limit, fracture, steels, crack initiation, crack propagation, maximum elastic stress, notch root radius, stress concentration factors, stress-intensity factors, plastic zone sizes, nonpropagating crack, linear fracture mechanics, linear notch mechanics

A schematic representation of notch effects in fatigue is given in Fig. 1. This is obtained typically in a completely reversed fatigue test, such as the rotating bending test, performed on specimens having a notch of various root radii but of constant depth. The fatigue limit based on complete fracture σ_w is determined by the crack initiation limit σ_{w1} for blunt notches and by the crack propagation limit σ_{w2} for sharp notches. σ_{w1} is defined as the maximum nominal stress under which a macrocrack is not formed along the notch root and σ_{w2} is defined as the threshold nominal stress for crack propagation in the range where a nonpropagating macrocrack exists. Isibasi [*1*] has presented evidence concerning the existence of these two fatigue limits σ_{w1} and σ_{w2}, and named the critical point—whether or not a nonpropagating macrocrack exists under σ_w (the intersection of the σ_{w1} and σ_{w2} curves in Fig. 1)—the branch point.

The effect of notches on the fatigue limit is frequently expressed by the strength reduction factor K_f, which is simply the ratio of the fatigue limit of a plain specimen to that of a notched specimen. Several equations have been proposed for relations between the theoretical (elastic) stress concentration factor K_t and the strength reduction factor K_f in the literature [*2*]. These equations are, however, incomplete at least in drawing the distinctions between σ_{w1} and σ_{w2}.

[1] Professor, Faculty of Engineering, Kyushu University, Fukuoka 812, Japan.

[2] Associate Professor, Faculty of Engineering, Fukuoka University, Fukuoka 814-01, Japan.

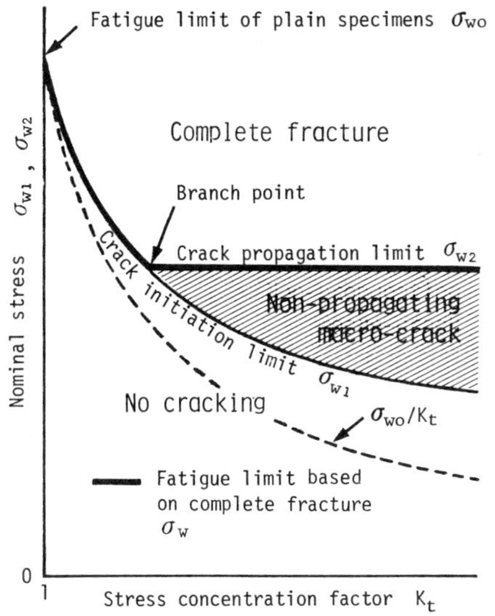

FIG. 1—*Relation between* σ_{w1} *or* σ_{w2} *and* K_t.

Nisitani [3] has provided evidence that the notch root radius at the branch point ρ_0 is constant for a given material,[3] irrespective of the notch depth and the diameter of the minimum section (see Fig. 2) and has explained that the reason why ρ_0 is a material constant is attributed to the fact that the relative stress distribution near a notch root is determined by the notch root radius ρ alone. Based on this fact, the fatigue limit for complete fracture σ_w, in the case of the ordinary notches, can be predicted from a fundamental experiment using a small number of specimens. In relation to the discussion here, Schijve [4] has also pointed out the significance of the combination of K_t and ρ in fatigue of notched specimens.

This prediction method is based on the following three points [3]:

1. $K_t\sigma_{w1}$ is determined mainly by the stress gradient at a notch root χ alone and is independent of the notch depth t or the diameter of the minimum section d [3,5,6].
2. σ_{w2} is dependent on the notch depth t and the diameter d alone and is independent of the notch sharpness (the horizontal line arises as shown in Fig. 1)
3. The notch radius at the branch point ρ_0 is a material constant [3,5,7] (see Fig. 2 [3]).

Although this prediction method is sufficient for practical use, the notch effect in fatigue is not known well:

Q-1 What is the physical background for σ_{w1} *and* σ_{w2}*?*

In order to evaluate the fatigue notch effect unifyingly, the controlling parameters for σ_{w1} and σ_{w2} must be clear. Under σ_{w1}, the material at the root of a notch undergoes the plastic deformation and under σ_{w2}, in addition, it contains a small crack. It is important to discuss whether the elastic quantity is effective even for evaluating such phenomena.

Q-2 Is the concept of effective crack depth (notch depth plus crack length) applicable to the prediction of σ_{w2} *and should* σ_{w2} *be always constant, irrespective of notch sharpness?*

[3] In the extremely shallow notches, ρ_0 is not a material constant [31].

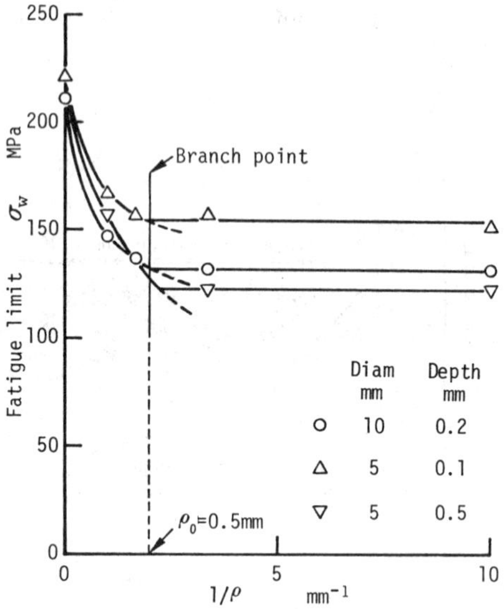

FIG. 2—*Relation between* σ_w *and* $1/\rho$ [3] (*annealed 0.23 carbon steel, rotating bending, circumferential 60-deg V-shaped notch*).

In previous researches, σ_{w2} has usually been treated based on the assumption that σ_{w2} is always constant for a constant notch depth. In such cases, the approach using the concept of effective crack depth is employed. Using this concept, Frost [*8*] and Kobayashi and Nakazawa [*9*] assumed that the equation obtained from the fatigue limits of cracked specimens can be applied in the estimation of σ_{w2} for notched specimens. Smith and Miller [*10*] proposed the equation for giving a relation between σ_{w2} and the threshold stress-intensity factor range for crack propagation, based on this concept.

This concept is empirical; therefore, its reliability must be checked from the physical standpoint.

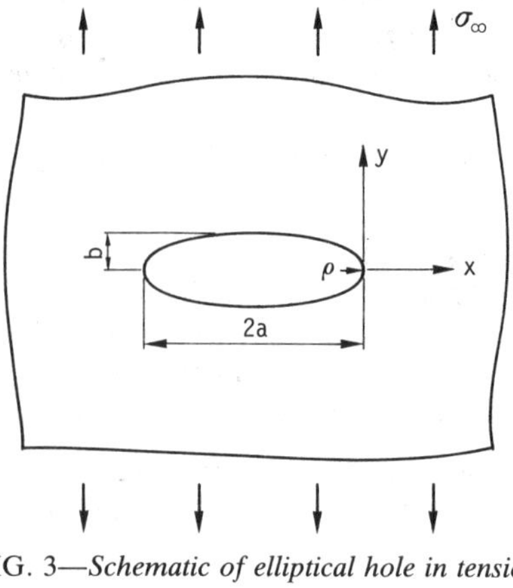

FIG. 3—*Schematic of elliptical hole in tension.*

Q-3 When the notch depth is of the order as the grain, what is the fatigue behavior and what governs the fatigue limit?

It can be easily supposed that the predicting method of notch effects in ordinary notches cannot be applied to such extremely shallow notches. The fatigue behavior of the specimens having an extremely shallow notch is not well known.

It is important to elucidate the notch effects of extremely shallow notches in fatigue, because they are closely related to the effects of surface treatment, surface scratches, or small defects on the fatigue strength.

The principal object of this paper is to answer the above questions. This paper is concerned with the following three items: First, as a concept useful for treating the strength of notched members, "linear notch mechanics" [*11*] is presented and through this concept the physical background for σ_{w1} and σ_{w2} is discussed. Second, the concept is applied to the results obtained in the present tests performed on specimens having widely varying notch geometries and to the results obtained by other investigators; then its effectiveness for fatigue problems is confirmed. Third, and finally, a unifying treatment of effects of notches in fatigue whose depths are from 5 μm to ordinary size is discussed.

The aforementioned questions are to be answered in these discussions.

Concept of "Linear Notch Mechanics" and Its Effectiveness for Fatigue Problems

First, on the basis of a single equation for the stress field of an elliptical hole, the characteristics of the stress fields near the root of a notch and near the tip of a crack are compared with each other. Afterwards, a treatment method for notch effects in fatigue is discussed.

Characteristics of the Stress Fields Generated by a Crack or a Notch

Consider an elliptical hole existing in an infinite plate subjected to tension, as shown in Fig. 3. The stress distribution of the x-axis is given by [*3,11*]

$$\sigma_y(x) = \sigma_\infty \frac{m^4\xi^3 + m^2(m^2 - m - 3)\xi + m + 1}{(m - 1)(m^2 - 1)} \tag{1}$$

where

$$K_t = 1 + 2m, \qquad m = \sqrt{\frac{a}{\rho}}, \qquad \xi = \frac{a + x}{\sqrt{a\rho + 2ax + x^2}}\sqrt{\frac{\rho}{a}} \tag{2}$$

K_t is the stress concentration factor. Based on Eq 1, both characteristics of stress fields due to a crack and a notch are given below.

Stress Field Near the Tip of a Crack—A crack is considered as a notch flattened extremely. Thus, putting $\rho \rightarrow 0$ in Eqs 1 and 2 and considering $x/a << 1$, the stress distribution near the tip of a crack is obtained as follows

$$\sigma_y(x) = \frac{\sigma_\infty\sqrt{\pi a}}{\sqrt{2\pi x}} = \frac{K_I}{\sqrt{2\pi x}} \tag{3}$$

where

$$K_I = \sigma_\infty\sqrt{\pi a} \tag{4}$$

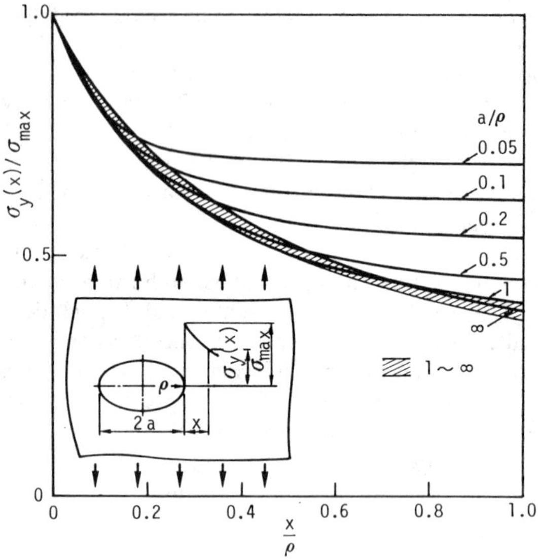

FIG. 4—*Relative stress distribution near the root of a notch.*

K_I is the stress intensity factor for Mode I loading. Equation 3 means that the stress distribution on the x-axis, $\sigma_y(x)$, is determined only by K_I, as is well known.

In crack problems, if the condition of small-scale yielding is satisfied, the load necessary for causing fracture or a given size of plastic zone is controlled by the stress-intensity factor alone, as is well known. This is based on the following two facts [*11*]:

1. When the values of K_I are constant, the stress distributions near the tips of cracks are equal to each other, irrespective of crack size or the other geometrical conditions of cracked members.
2. The additional stress fields due to a given amount of plastic deformation occurring at a given place near the tips of cracks are equal to each other, irrespective of crack size or the other geometrical conditions of cracked members [*12*].

These two facts lead to the following conclusion. If the external loads are adjusted in order that K_I-values are equal in the two members having a crack, the elastic-plastic stress distributions in both members become equal to each other even after the materials near the crack tips undergo slight plastic deformations.

Stress Field Near the Root of a Notch—Letting $x/\rho << 1$ in Eqs 1 and 2, the stress distribution near an elliptical hole is given as [*11*]

$$\sigma_y(x) = \sigma_{max}\{1 + b_1(x/\rho) + b_2(x/\rho)^2 + \cdots\} \tag{5}$$

where

$$\sigma_{max} = K_t\sigma_\infty = (1 + 2m)\sigma_\infty, \quad b_1 = -(3 + 4m)/(1 + 2m),$$

$$b_2 = (15 + 18m)/(2 + 4m) \tag{6}$$

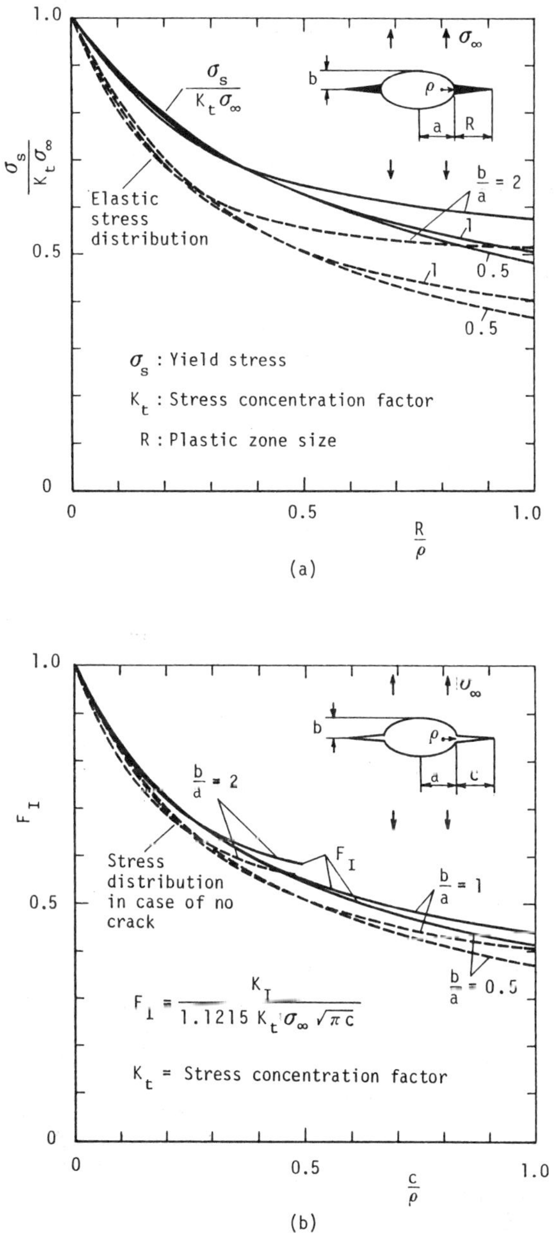

FIG. 5—(a) *Relative plastic zone size* [18]; (b) *nondimensional stress-intensity factor* [22].

TABLE 1—*Chemical composition, weight %.*

C	Si	Mn	P	S	Cu	Ni	Cr
0.45	0.25	0.79	0.01	0.01	0.09	0.03	0.18

In Eq 6 the variations of b_1 and b_2 due to the change in the values of m are small [*11*]. Therefore, it is found from Eq 5 that when x/ρ is small the stress distribution near the notch root is determined by the maximum elastic stress σ_{max} and ρ alone, irrespective of the notch depth a. In Fig. 4, Eq 5 is realized. Figure 4 demonstrates that the stress distribution on the x-axis near an elliptical hole is determined by σ_{max} and ρ alone, as long as the notch is not extremely shallow. Schijve [*4,13*] has shown that not only $\sigma_y(x)$ but also the whole stress field near the notch root is determined uniquely by σ_{max} and ρ alone.

Although the above discussion is limited to the elliptical hole as an example of notches, this significant situation holds similarly in the hyperboloidal notches subjected to tension [*3*] or bending [*3*] and in the round bars having a circumferential notch subjected to tension [*14*] or bending [*15,16*].

A situation analogous to crack problems exists in notch problems; that is, if the condition of small-scale yielding is satisfied, the load necessary for causing fracture or a given size of plastic zone is controlled by σ_{max} and ρ alone [*17*]. This is based on the following two facts [*11*], similar to the case of crack problems:

1. In the case where ρ is constant, when the values of σ_{max} are constant, the stress distributions near the roots of notches are equal to each other, irrespective of notch depth or the other geometrical conditions of notched members.
2. In the case where ρ is constant, the additional stress fields due to a given amount of plastic deformation occurring at a given place near the roots of notches are equal to each other, irrespective of notch depth or the other geometrical conditions of notched members [*11,12*].

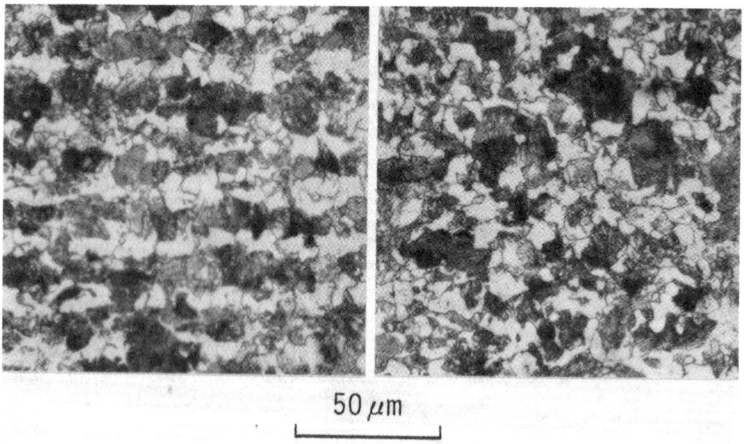

FIG. 6—*Metallographical structure:* (left) *longitudinal section;* (right) *transverse section.*

TABLE 2—*Mechanical properties.*

Lower Yield Point, MPa	Ultimate Tensile Strength, MPa	True Fracture Stress, MPa	Reduction of Area, %
364	632	1158	45.8

Due to the foregoing two facts, if the external loads are adjusted in order that σ_{max} values are equal in the two members having a notch with the same size of ρ, the elastic-plastic stress distributions in both members become equal to each other even after the materials near the notch roots undergo slight plastic deformations.

"Linear Notch Mechanics"

The similarity of stress distribution after a small plastic deformation is assured by the aforementioned two facts in both cases of crack and notch. As the scale of the severity of stress field, therefore, we should employ σ_{max} and ρ for the notch problems as well as K_I for the crack problems.

Here we shall call the concept that treats the strength of cracked members using stress-intensity factors, "linear crack mechanics," and the concept that treats the strength of notched members using σ_{max} and ρ, "linear notch mechanics." "Linear crack mechanics" of course means linear fracture mechanics. "Linear notch mechanics" is considered as an extension of "linear crack mechanics." Both concepts are based on the similarity of relative stress distributions and the equivalence of response against small plastic deformations, and stand on a common physical basis.

Effectiveness of "Linear Notch Mechanics" for Fatigue Problems

In order to evaluate the fatigue strength of notched members, it is necessary to clarify the controlling factor of σ_{w1} and σ_{w2} on the physical basis.

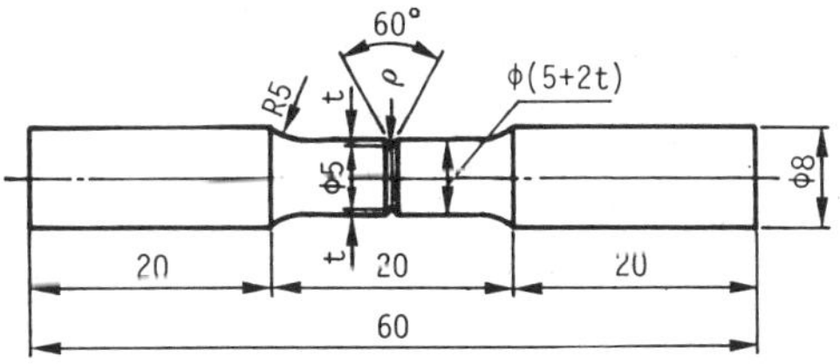

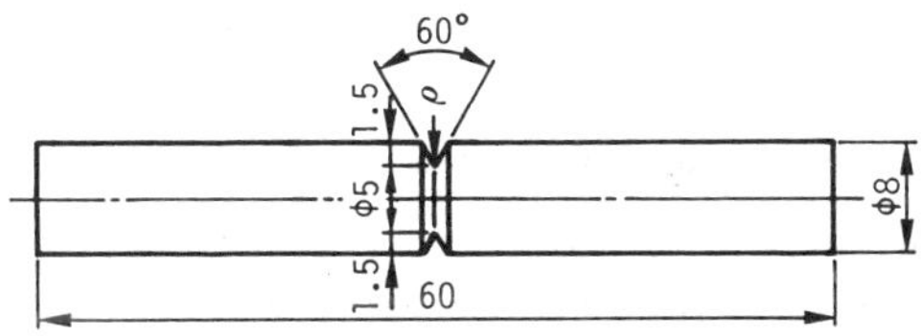

FIG. 7—*Dimensions of notched specimens:* (top) t = *0.1 mm and 0.5 mm,* (bottom) t = *1.5 mm.*

TABLE 3—*Relation between notch geometry and σ_{w1} or σ_{w2}.*

t, mm	ρ, mm	K_t	σ_{w1}, MPa	σ_{w2}, MPa	$K_t\sigma_{w1}$, MPa	$K_t\sigma_{w2}$, MPa
			PLAIN SPECIMEN			
—	20	1.04	280	—	291	—
			NOTCHED SPECIMEN			
0.005	0.05	1.67	245	275	409	459
0.005	0.02	2.06	240	275	494	567
0.005	0.01	2.52	215	280	542	706
0.01	0.05	1.95	215	245	419	478
0.01	0.02	2.52	205	250	517	630
0.01	0.01	3.19	180	245	574	782
0.1	0.6	1.58	210	—	332	—
0.1	0.3	1.89	190	—	359	—
0.1	0.1	2.72	...	170	...	462
0.1	0.05	3.54	...	180	...	637
0.1	0.02	5.21	...	180	...	938
0.5	0.6	1.86	180	—	335	—
0.5	0.3	2.39	150	—	359	—
0.5	0.1	3.80	100	140	380	532
0.5	0.05	5.19	80	140	415	727
0.5	0.02	7.94	60	145	476	1151
0.5	0.01	11.0	55	145	605	1595
1.5	0.6	1.91	175	—	334	—
1.5	0.3	2.52	140	—	353	—
1.5	0.1	4.09	95	125	389	511
1.5	0.05	5.66	75	125	425	708
1.5	0.02	8.78	60	125	527	1098
1.5	0.01	12.3	...	130	...	1599

Figure 5*a* shows the relation between plastic zone size and stress at infinity in an infinite plate with an elliptical hole under tension [*18*]. The calculation is based on the model of Refs *19* and *20*. The ordinate is the ratio of the yield stress σ_s to the maximum elastic stress σ_{max} $(=K_t\sigma_\infty)$ calculated by neglecting plastic defomation, $\sigma_s/(K_t\sigma_\infty)$, and the abscissa is the nondimensional plastic zone size R. The dotted lines in this figure indicate the elastic stress distributions (see Fig. 4). Figure 5a demonstrates that when ρ is constant the same values of σ_{max} result in the same plastic zone size. A similar trend is also presented in Ref *21*.

Similarly, the stress-intensity factors of cracks of the same length emanating from the roots of elliptical notches are determined by the maximum elastic stress σ_{max} calculated by neglecting the existence of the crack and ρ alone, irrespective of the notch depth [*13,22*], as shown in Fig. 5*b* [*22*].

Considering the above examples, it can be expected that the crack initiation and nonpropagation phenomena at notch roots can be evaluated by "linear notch mechanics"; that is, we can evaluate σ_{w1} and σ_{w2} by using two parameters of σ_{max} and ρ alone.

Material Used, Dimensions of Specimens, and Test Procedures

The materials used are rolled cylindrical bars of 0.45 carbon steel. The chemical composition is given in Table 1. The specimens were turned after annealing for one hour at 845°C. The metallographic structures of the annealed material are shown in Fig. 6. The mean linear intercepts [*23*] of ferrite and pearlite were 6 and 9 μm, respectively. The

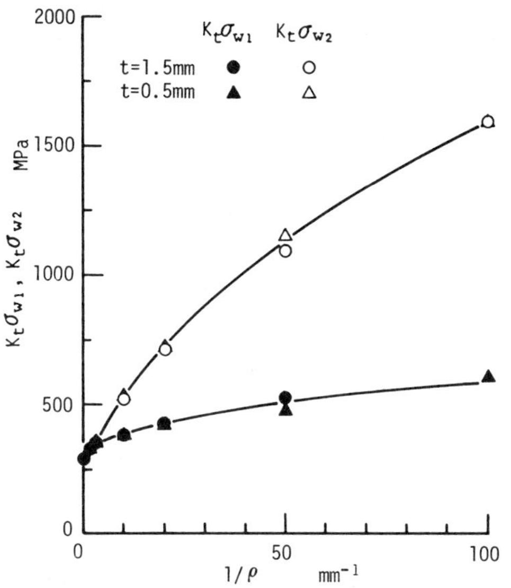

FIG. 8—*Relation between* $K_t\sigma_{w1}$ *or* $K_t\sigma_{w2}$ *and* $1/\rho$ *for ordinary notched specimens.*

mechanical properties are given in Table 2, and the dimensions of notched specimens in Fig. 7. The notches are of the circumferential 60-deg V-shaped type. The notch root radius ρ was widely varied from 0.01 to 0.6 mm. The notch depths t were taken from 0.005 to 1.5 mm. Before testing, all specimens were electropolished after annealing *in vacuo* for one hour at 600°C. The electropolishing solution used was 2000 cm^3 phosphoric acid containing 66 g gelatin and 66 g oxalic acid.

The equipment used in the fatigue tests was a rotating bending fatigue testing machine of uniform bending moment type with a capacity of 15 Nm operating at 3000 rpm.

The crack initiation limit σ_{w1} is defined as the stress required to develop microcracks after 10^7 cycles whose size is approximately the same size as that of a nonpropagating microcrack

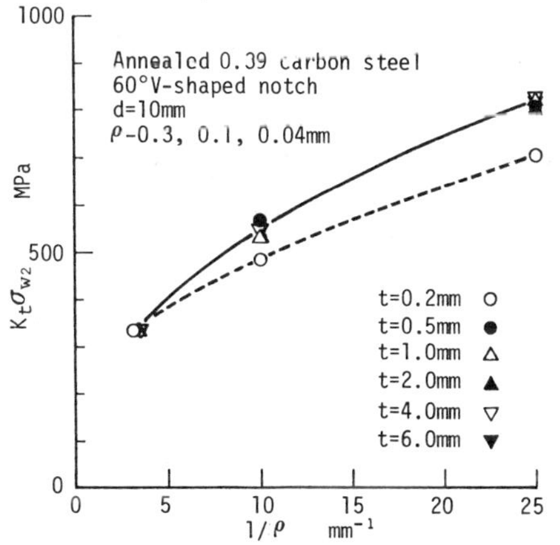

FIG. 9—*Relation between* $K_t\sigma_{w2}$ *and* $1/\rho$ [28].

TABLE 4—*Relation between* $K_t\sigma_{w2}$ *and* ρ *for annealed 0.36 carbon steel specimens having a circumferential 60-deg V-shaped notch of constant root radius* (*Kobayashi and Nakazawa* [9]).

Diameter of Minimum Section, d, mm	Notch Root Radius, ρ, mm	Notch Depth, t, mm	K_t[a]	σ_{w2}, MPa	$K_t\sigma_{w2}$, MPa
13.0	0.2	1.00	4.20	87	365
13.0	0.2	0.70	3.89	93	362
13.0	0.2	0.50	3.58	102	365
13.0	0.2	0.30	3.13	116	363
13.0	0.2	0.15	2.56	136	348
13.0	0.2	0.10	2.29	151	346

[a] Accurate value calculated by the body force method [*16,25*].

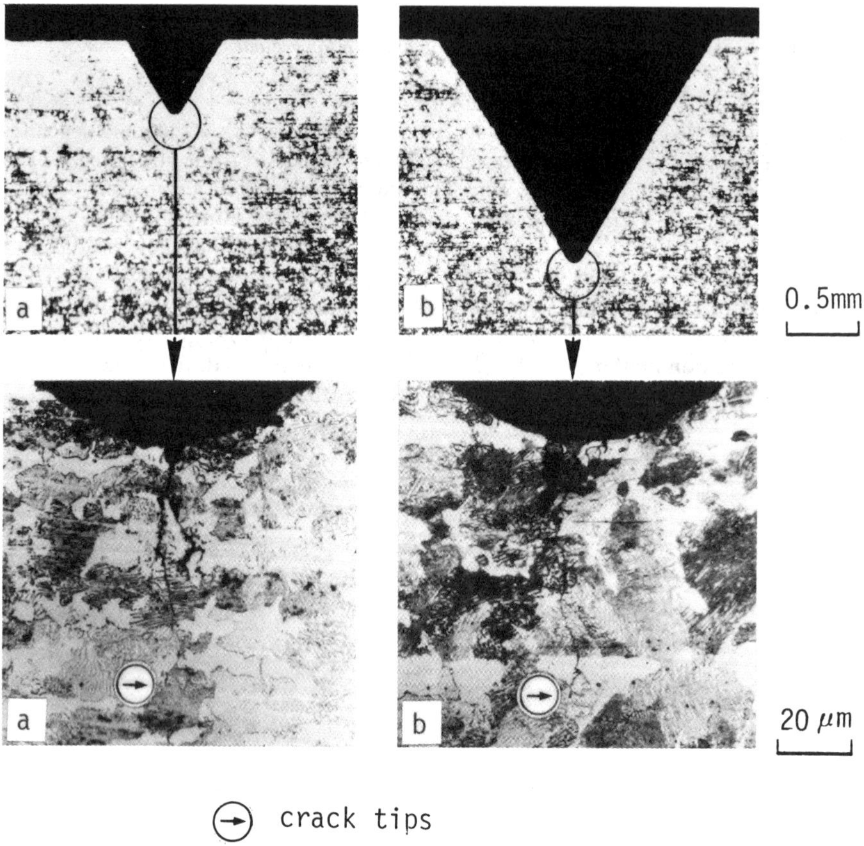

FIG. 10—*States of notch roots observed after* 10^7 *cycles of* σ_{w2}: (a) $\sigma_{w2} = 140$ *MPa,* $K_t\sigma_{w2} = 727$ *MPa,* $\rho = 0.05$ *mm,* $t = 0.5$ *mm,* $N = 10^7$. (b) $\sigma_{w2} = 125$ *MPa,* $K_t\sigma_{w2} = 708$ *MPa,* $\rho = 0.05$ *mm,* $t = 1.5$ *mm,* $N = 10^7$.

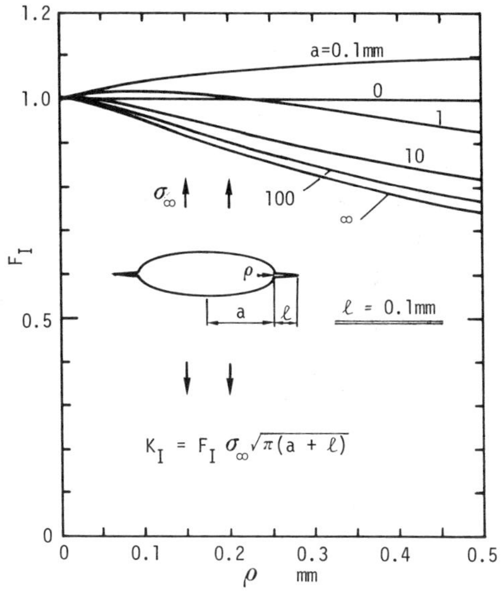

FIG. 11—*Change of* K_I *values against notch sharpness* [29].

observed at the fatigue limit of a plain specimen [*24*]. The definition of the crack propagation limit σ_{w2} was given in the introduction. The step of stress level for deciding σ_{w1} and σ_{w2} was 5 MPa. The fatigue limit σ_w taken in this paper means the maximum nominal stress under which a specimen endures more than 10^7 cycles.

Results and Discussion

Table 3 gives the experimental results. The results for 0.005- and 0.01-mm-deep notches are already reported in Ref *24*. In this paper, the accurate values of stress concentration factor K_t [*16,25,26*] are employed. The values of K_t for the notches whose $t \geqq 0.1$ mm were calculated recently by one of authors [*16,25*] by means of the body force method [*27*]. As the values of K_t for $t \leqq 0.01$ mm, the analytical solutions presented by Isida [*26*] for the semi-infinite plates with a circular-arc notch under tension are used.

Notch Effects of Ordinary Notches in Fatigue

It is inferred from the earlier discussion that both σ_{w1} and σ_{w2} are determined by σ_{max} and ρ alone. In Fig. 8, the results of notched specimens having a notch of ordinary depth (t = 0.5 and 1.5 mm) are arranged by use of σ_{max} and ρ. The scale of the ordinate is the maximum elastic stress $K_t\sigma_{w1}$ or $K_t\sigma_{w2}$ repeated at the notch root under σ_{w1} or σ_{w2}. The abscissa is the reciprocal of the notch root radius $1/\rho$. It appears from this figure that both σ_{w1} and σ_{w2} are determined by σ_{max} and ρ alone, independently of the other notch geometries. The root radii at the branch point ρ_0 become the same for such deep notches and are approximately 0.25 mm for the present material (see footnote 2).

It has been shown in the previous result [*3,5,6*] that $K_t\sigma_{w1}$ is determined by the stress gradient at the notch root χ alone. Since χ is nearly proportional to $1/\rho$, these previous results mean that $K_t\sigma_{w1}$ is determined mainly by ρ alone.

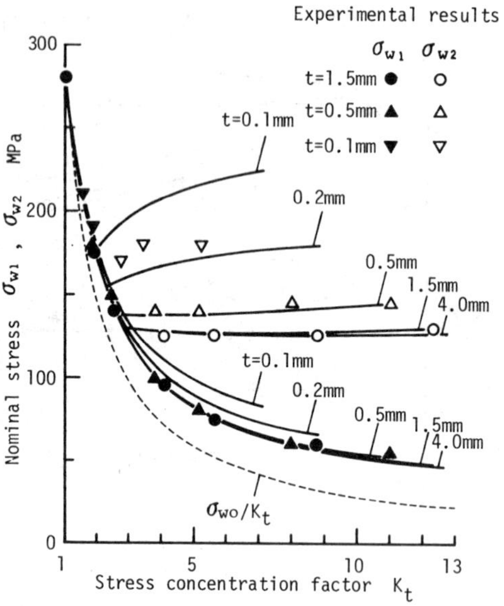

FIG. 12—*Relation between nominal stress and* K_t *estimated from Fig. 8.*

The fact that $K_t\sigma_{w2}$ is determined by ρ alone can also be recognized in the other previous data [*9,28*]. Figure 9 shows the previous results obtained in the rotating bending fatigue tests performed on the annealed 0.39 carbon steel specimens [*28*]. $K_t\sigma_{w2}$ in Fig. 9 is determined mainly by ρ alone, irrespective of the notch depth, similar to the case of the present results. Table 4 gives Kobayashi and Nakazawa's results obtained using the annealed 0.36 carbon steel specimens having a circumferential 60-deg V-shaped notch of depth varying from 0.10 to 1.00 mm under the condition of constant notch root radius (ρ = 0.2 mm). It is found from this table that $K_t\sigma_{w2}$ settles in the same value for the constant root radius except for the case where $t < 0.2$ mm. In Fig. 9 and Table 4, the accurate K_t-values [*16,25*] were used.

Figure 10 shows the longitudinal sections of two notches, whose root radii are the same and depths are different, observed after 10^7 cycles of σ_{w2}. This figure suggests that the same values of $\sigma_{max} = K_t\sigma_{w2}$ and the same values of ρ in two notches result in the same fatigue behavior; in other words, the whole fatigue process from crack initiation to nonpropagation is governed by σ_{max} and ρ alone.

From the above discussion, σ_{w1} and σ_{w2} can be evaluated by using σ_{max} and ρ alone. Once a single curve of the relation between $\sigma_{max} = K_t\sigma_w$ and $1/\rho$ is obtained in a fundamental experiment, we can estimate the value of σ_w for an arbitrary ordinary notched specimen from this curve using K_t and ρ. This estimation method is superior to the previous method [*3*] from the standpoint of clarity in physical meaning and simplicity.

Reliability of the Concept of Effective Crack Depth

The reliability of this concept in evaluating σ_{w2} can be examined by comparing the severity due to a crack emanating from a notch with the severity due to the corresponding effective crack. From this point of view, the changes in K_I-values of elliptical holes with a crack are shown in Fig. 11 [*29*]. When the crack depth is smaller than the notch radius, the concept of effective crack depth is not available. Since the length of nonpropagating cracks observed

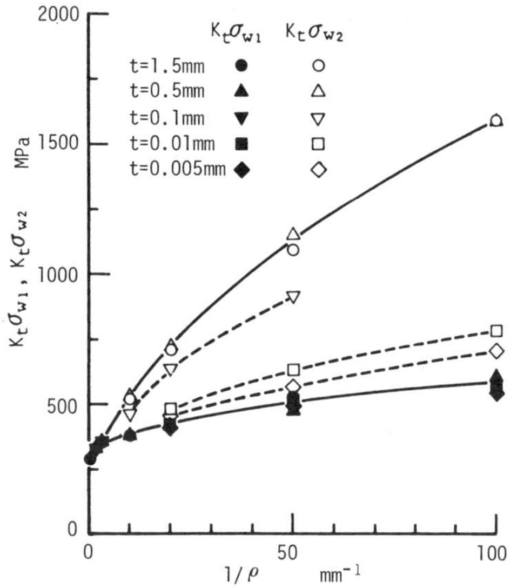

FIG. 13—*Relation between* $K_t\sigma_{w1}$ *or* $K_t\sigma_{w2}$ *and* $1/\rho$.

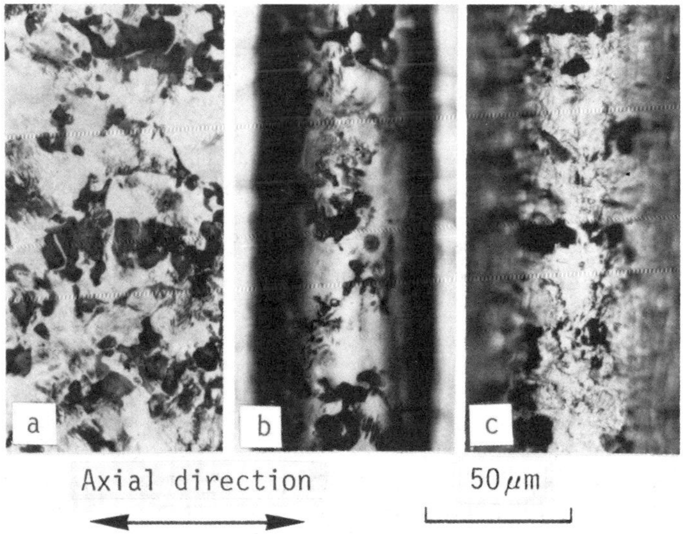

FIG. 14—*Surface states observed after* 10^7 *cycles under fatigue limit* σ_{w0} *(plain specimen) or* σ_{w1} *(notched specimens)*: (a) *plain specimen,* σ_{w0} = *280 MPa.* (b) t = *0.01 mm,* ρ = *0.05 mm,* σ_{w1} = *215 MPa,* $K_t\sigma_{w1}$ = *419 MPa.* (c) t = *1.5 mm,* ρ = *0.05 mm,* σ_{w1} = *75 MPa,* $K_t\sigma_{w1}$ = *425 MPa.*

under σ_{w2} is not always larger than the notch radius [*9*], it is difficult to explain the constancy of σ_{w2} consistently with this concept. In the previous researches, strictly speaking, there are several results where σ_{w2} is not constant.

Figure 12 shows the theoretical relations between σ_{w1} or σ_{w2} and K_t, which are estimated from Fig. 8 based on the concept of "linear notch mechanics." In some cases σ_{w2} is constant but in the others σ_{w2} is not always constant.

The above discussion suggests that the prediction based on the concept of effective crack depth does not always provide the correct value of σ_{w2}.

Notch Effects of Shallow or Extremely Shallow Notches in Fatigue

When the notch is sufficiently small, the fatigue strength of the specimen is somewhat influenced by the statistical factor [*24*]. In this paper, however, the discussion is limited to the mechanical factor. According as the notch depth becomes small, the extent of relative stress distribution determined by the notch root radius ρ alone becomes small (see Fig. 4). This fact is significant in the problems of the fatigue notch effects of extremely shallow notches.

Plotting the results of σ_{w1} and σ_{w2} for the shallow notched specimens whose $t \leqq 0.1$ mm in Fig. 8, we obtain Fig. 13.

As seen from this figure, $K_t\sigma_{w1}$ is controlled by ρ alone for the notches whose depths are from 5 μm to ordinary size, while $K_t\sigma_{w2}$ is not controlled by ρ alone and depends on the notch depth in case of the extremely shallow notches. This difference is attributed to the difference in the size of region related to σ_{w1} and the one related to σ_{w2}.

For the present material, the region related to σ_{w1} is the thin surface layer of 5 to 10 μm thickness [*24*]. The relative stress distribution in such a thin surface layer is almost independent of notch depth, if the notch depth is larger than 5 μm. Therefore, the same relation between $K_t\sigma_{w1}$ and $1/\rho$ is obtained from the deeply notched specimens and from the extremely shallow notched specimens.

Figure 14 shows the surface states observed after 10^7 cycles under σ_{w0} and σ_{w1} (σ_{w0}: fatigue limit of plain specimen). The slipped regions are blackish. Figures 14*b* and 14*c* suggest that under the condition of constant ρ the phenomenon at σ_{w1} is determined by σ_{max} ($= K_t\sigma_{w1}$) alone, irrespective of the notch depth (t = 0.01 and 1.5 mm).

On the other hand, under σ_{w2}, an initiated crack ceases to propagate after a small amount of propagation. Considering this fact, it can be said that the region related to σ_{w2} is larger than the one related to σ_{w1}. As shown in Fig. 4, when the notch is sufficiently shallow the relative stress distributions near the notch root are not determined by σ_{max} and ρ alone and become higher for smaller t. Consequently, even when ρ is constant, $K_t\sigma_{w2}$ is dependent on the notch depth t and becomes smaller for smaller t. The same trend is recognized in Fig. 9 and Table 4.

As shown in Fig. 13, the $K_t\sigma_{w1}$ versus $1/\rho$ curve is located below the $K_t\sigma_{w2}$ versus $1/\rho$ curve for the 5-μm-deep notch, which is the lower limiting curve for fracture in the present experiment. Physically, $K_t\sigma_{w2}$ versus $1/\rho$ curves are never located below $K_t\sigma_{w1}$ versus $1/\rho$ curves.

When the notch depth is of the same order as the grain size, σ_{w2} is determined mainly by the notch depth alone. Figure 15 shows the states of notch roots under σ_{w2} in the case when the notch depths are 5 and 10 μm [*24*]. The depth of nonpropagating cracks is greater than the notch depth. In such cases, the influence of notch sharpness is so small that σ_{w2} is determined by t alone, irrespective of notch sharpness [*24*].

If the notch depth becomes smaller and smaller, the notch no longer affects the fatigue limit. It is known that the critical size of notches not affecting the fatigue limit is closely

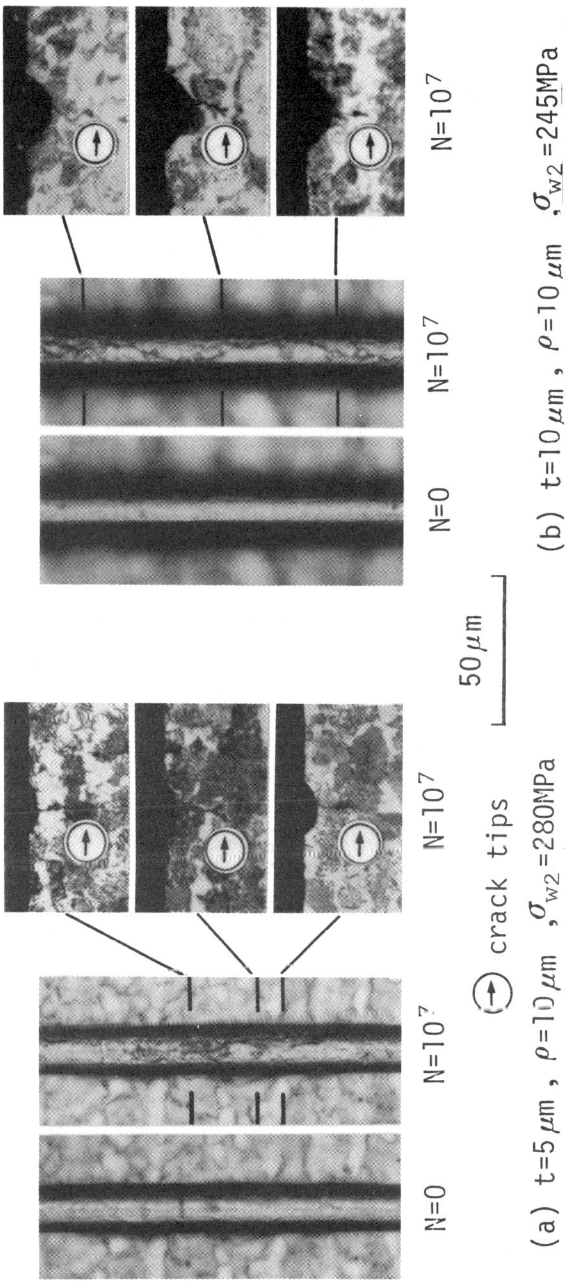

FIG. 15—*Surface states at roots of extremely shallow notches and their longitudinal sections under fatigue limit.*

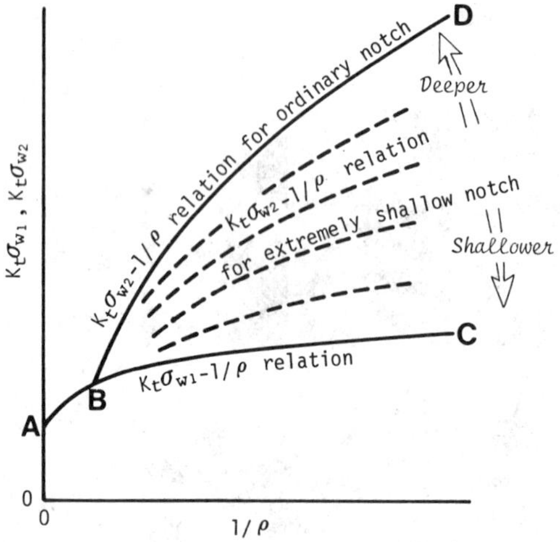

FIG. 16—*Schematic figure of* $K_t\sigma_{wl}$ *versus* $1/\rho$ *relation and* $K_t\sigma_{w2}$ *versus* $1/\rho$ *relation.*

related to the maximum size of nonpropagating microcracks observed at the fatigue limit of plain specimens σ_{w0} [*24,30*]. When the notch depth is 5 μm, σ_{w2} is nearly equal to σ_{w0}, as shown in Table 4.

Conclusions—A Unifying Treatment of Notch Effects in Fatigue

According to the concept of "linear notch mechanics" presented in this paper, the effects of notches under the condition of small-scale yielding can be evaluated by the maximum elastic stress at a notch root σ_{max} and the notch root radius ρ alone. In order to confirm the usefulness of this concept in fatigue notch problems, rotating bending fatigue tests were carried out on annealed 0.45% carbon steel specimens having a notch of various depths and sharpnesses. The physical basis of "linear notch mechanics" is fundamentally the same as that of linear fracture mechanics, which is well known as a concept useful for evaluating the fatigue strength of various kinds of materials containing cracks. Therefore, a result obtained in this paper is strongly expected to be also applicable to other materials. Summarizing the present study, a unifying treatment of notch effects in fatigue is described as follows:

The $K_t\sigma_{w1}$ versus $1/\rho$ relation and the $K_t\sigma_{w2}$ versus $1/\rho$ relation are shown in schematic form in Fig. 16, where σ_{w1} and σ_{w2} are the limiting nominal stresses for crack initiation and crack propagation, respectively, and K_t is the elastic stress concentration factor. In this figure, the points A and B correspond to the fatigue limit of a plain specimen and the branch point, respectively. Every result of fatigue notch effects never fails to correspond to some point on the curve A-B or in the region bounded by the curves B-C and B-D. The arrangement shown in Fig. 16 stands on a clearer physical basis compared with Fig. 1 and is fit for unifying treatment of the fatigue notch effects.

In case of the ordinary notch, the fatigue limit based on fracture σ_w is estimated from the curve A-B (when $\rho > \rho_0$) or B-D (when $\rho < \rho_0$). Once the curve A-B-D is obtained from a fundamental experiment using a small number of specimens, we can predict σ_w using K_t and ρ alone.

In the case of the extremely shallow notch, the $K_t\sigma_{w2}$ versus $1/\rho$ curve deviates from the curve B-D and is located lower as the notch becomes shallower, but is never located below the curve B-C. Then, using the curve B-C, we can do a conservative evaluation of the fatigue limit for fracture σ_w for the extremely shallow notch. It should be noted that this B-C curve can be obtained also from a fundamental experiment using ordinary notched specimens.

References

[1] Isibasi, T., *Prevention of Fatigue and Fracture of Metals*, Yokendo, Tokyo, 1954.
[2] Moore, H. F. in *Proceedings*, American Society for Testing and Materials, Philadelphia, Vol. 45, 1945, p. 507; Peterson, R. E., *Stress Concentration Design Factors*, Wiley, New York, 1953, p. 8; and Heywood, R. B., *Engineering*, Vol. 179, 1955, p. 146.
[3] Nisitani, H., *Bulletin of the Japan Society of Mechanical Engineers*, Vol. 11, No. 48, 1968, pp. 947–957.
[4] Schijve, J., *Fatigue of Engineering Materials and Structures*, Vol. 3, No. 4, 1980, pp. 339–356.
[5] Nisitani, H. and Takao, K., *Engineering Fracture Mechanics*, Vol. 6, 1974, pp. 253–260.
[6] Siebel, E. and Stieler, M., *Zeitschrift-VDI*, Vol. 97, No. 5, 1955, pp. 121–126.
[7] Nisitani, H., *Mechanics of Fatigue-AMD*, Vol. 47, T. Mura, Ed., American Society of Mechanical Engineers, 1981, pp. 151–166.
[8] Frost, N. E. in *Proceedings*, Institution of Mechanical Engineers, Vol. 173, No. 35, 1959, pp. 811–827.
[9] Kobayashi, H. and Nakazawa, H., *Transactions*, Japan Society of Mechanical Engineers, Vol. 35, No. 277, 1969, pp. 1856–1863.
[10] Smith, R. A. and Miller, K. J., *International Journal of Mechanical Science*, Vol. 20, 1978, pp. 201–206.
[11] Nisitani, H., *Transactions*, Japan Society of Mechanical Engineers, Vol. 49, No. 47, 1983, pp. 1353–1359.
[12] Nisitani, H., Chen, D. H., and Isida, M., *Transactions*, Japan Society of Mechanical Engineers, Vol. 50, No. 451, 1984, pp. 341–350.
[13] Schijve, J., Report of Delft University of Technology, LR-330, 1981, p. 1.
[14] Murakami, Y., Noda, N., and Nisitani, H., *Transactions*, Japan Society of Mechanical Engineers, Vol. 47, No. 423, 1981, pp. 1194–1205.
[15] Murakami, Y., Noda, N., and Nisitani, H., *Transactions*, Japan Society of Mechanical Engineers, Vol. 48, No. 427, 1982, pp. 312–320.
[16] Murakami, Y., Noda, N., and Nisitani, H., *Transactions*, Japan Society of Mechanical Engineers, Vol. 48, No. 430, 1982, pp. 800–809.
[17] Nisitani, H., *Science of Machine*, Vol. 27, No. 1, 1975, pp. 15–20.
[18] Nisitani, H. in *Proceedings*, Third International Conference on Fracture, Munich, Germany, 1973, p. I-513.
[19] Dugdale, D. S., *Journal of the Mechanics and Physics of Solids*, Vol. 8, 1960, pp. 100–104.
[20] Bilby, B. A., Cottrell, F. R. S., and Swinden, K. H., *Proceedings of the Royal Society of London*, Series A, *Mathematical and Physical Sciences*, London, Vol. 272, 1963, pp. 304–314.
[21] Dowling, N. E. in *Fracture Mechanics, ASTM STP 677*, American Society for Testing and Materials, Philadelphia, 1979, pp. 247–273.
[22] Nisitani, H. and Isida, M., *Transactions*, Japan Society of Mechanical Engineers, Vol. 39, No. 317, 1973, pp. 7–14.
[23] Weinberg, F., *Tools and Techniques in Physical Metallurgy*, Marcel Dekker, New York, Vol. 1, 1970.
[24] Nisitani, H. and Endo, M., *Engineering Fracture Mechanics*, Vol. 21, No. 1, 1985, pp. 215–227.
[25] Nisitani, H. and Noda, N., *Transactions*, Japan Society of Mechanical Engineers, Vol. 51, No. 461, 1985, pp. 54–62.
[26] Isida, M., Data from the Japan Society of the Promotion of Science, Committee 129, 1964.
[27] Nisitani, H. in *Mechanics of Fracture*, Vol. 5, G. C. Sih, Ed., Nordhoff, Leyden, The Netherlands, 1978, pp. 1–68.
[28] Nisitani, H. and Kawano, K. in *Proceedings*, 13th Japan Congress on Materials Research—Metallic Materials, 1968, pp. 49–51.
[29] Nisitani, H. in *Proceedings*, Joint Conference on Experimental Mechanics, Hawaii, 1982, pp. 150–155.
[30] Murakami, Y. and Endo, T., *International Journal of Fatigue*, Vol. 2, 1980, pp. 23–30.
[31] Nisitani, H. in *Proceedings*, International Conference on the Mechanical Behavior of Materials, Vol. 2, Society of Materials Science, Japan, 1972, pp. 312–322.

Fatigue of Steels—Mechanical Loading

Norman A. Fleck[1]

Influence of Stress State on Crack Growth Retardation

REFERENCE: Fleck, N. A., **"Influence of Stress State on Crack Growth Retardation,"** *Basic Questions in Fatigue: Volume I, ASTM STP 924,* J. T. Fong and R. J. Fields, Eds., American Society for Testing and Materials, Philadelphia, 1988, pp. 157–183.

ABSTRACT: Several mechanisms have been proposed in the literature to account for crack growth retardation following a single peak overload.

In order to determine the dominant mechanism, overload tests were performed on thick and thin specimens made from BS4360 50B structural steel. The baseline stress intensity range was 25 MPa $\sqrt{\text{m}}$ and the load ratio (= minimum load/maximum load of fatigue cycle) was 0.05, while the overload was of stress intensity range 50 MPa $\sqrt{\text{m}}$.

It was observed that the crack growth and closure responses were different at the surface and in the bulk of the thick specimen; no such variations in behavior occurred along the crack front for the thin specimen. For both thicknesses of material, the crack growth rate predicted by measurements of the crack opening load was in agreement with the observed crack growth rate, except for the period when crack growth rates were recovering from the slowest transient growth rate to the post-overload stabilized value. This discrepancy was due to the phenomenon of discontinuous closure—the crack first closed at a location far from its tip. Fracture surface profiles showed that the crack path deviated by only a small amount after application of the overload; hence the retarded growth cannot be due to crack branching.

It is concluded from these tests and from a critical examination of the literature that, at high baseline ΔK levels, retardation is due to plasticity-induced crack closure. At low baseline ΔK levels approaching the threshold value, retardation may be due to plasticity-induced crack closure or to irregularities of the crack front.

KEY WORDS: crack propagation, crack closure, fatigue (materials), steel, variable-amplitude loading

Since Elber first discovered the crack closure phenomenon in the late 1960's [*1,2*], there has been much controversy over the ability of crack closure to account for crack growth rates under variable-amplitude loading, and in particular crack growth retardation following a single peak overload. We are left with the fundamental question.

What is the cause of crack growth retardation following an overload?

In order to answer this question for the case of above-threshold growth in a low-strength structural steel, the crack growth rate and closure responses following a single peak overload were investigated for thick and thin compact tension specimens made from BS4360 50B structural steel.

[1] Lecturer, Cambridge University Engineering Department, Trumpington St., Cambridge, CB2 1PZ, United Kingdom.

A Critical Review of Previous Work

Phenomenology of Crack Growth Retardation After an Overload

It is well known that a single peak overload slows down or even arrests a long fatigue crack; see, for example, Refs *2–5*. The crack growth transient following an overload depends upon the stress state at the crack tip [*3,4*].

Consider first the plane stress case, Fig. 1*a*. Assume the loading consists of a constant baseline stress intensity range, ΔK_b, with a single peak overload. On application of the overload, the crack first accelerates [*3,6,7*]. Then the crack slows down to a minimum growth rate over a crack growth increment of about one-quarter the forward plastic zone size due to the overload [*8,9*]; this phenomenon is termed "delayed retardation" [*9*]. Finally, the crack accelerates to a stabilized value which is approximately equal to the pre-overload growth rate, as the crack advances to near the boundary of the overload plastic zone [*9,10*].

Now consider the plane strain case. Bernard et al. [*3,4*], in a careful and detailed study of crack growth retardation, found that the response to a single peak overload varies along the crack front in thick specimens. At the surface of a thick specimen, conditions are close to plane stress and the retardation behavior is similar to that of a thin specimen. Behavior is modified, however, at the center of thick specimens where plane strain conditions prevail. There, retardation occurs for fewer cycles and over a smaller crack advance increment than at the surface of the thick specimens or along the whole crack front of the thin specimens. Also, the crack growth rate at the center of thick specimens attains the minimum value almost immediately after application of the overload, Fig. 1*b* [*3,4,11–14*]. The work of Bernard et al. [*3,4*] clearly shows the futility of taking only surface measurements when we wish to elucidate the details of crack growth retardation in plane strain.

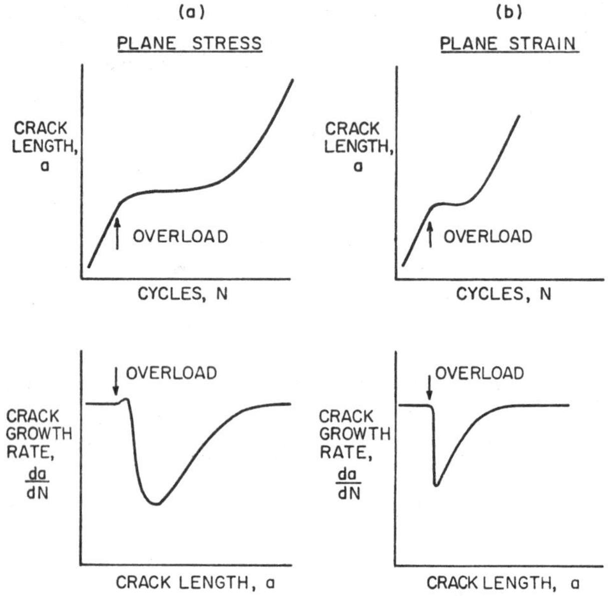

FIG. 1—*Typical retardation responses after a single peak overload. Tests performed using constant* ΔK *before and after overload.*

Mechanisms Causing Retarded Growth

Several mechanisms have been proposed to account for crack growth delay following a single peak overload:

1. Crack tip blunting [*15*].
2. Crack tip strain hardening [*16*].
3. Irregular crack front [*12,17,18*].
4. Residual compressive stresses ahead of crack tip [*19*].
5. Plasticity-induced crack closure [*1,2*].

It is now shown from a critical examination of the literature that the *dominant* cause of retardation is either an irregular crack front induced by the overload or plasticity-induced crack closure. The other mechanisms are secondary for most structural metals.

Crack Tip Blunting—It has been argued that the overload cycle blunts the crack and turns it into a notch: crack growth retardation is due to the finite number of cycles required to reinitiate and propagate the crack from this notch [*15*]. While this mechanism can account for retarded growth in plastics such as polycarbonate [*20*], it fails in structural metals, for the following reasons.

1. The crack tip blunting model suggests that retardation is immediate following an overload. In reality, delayed retardation occurs for thin specimens [*3,4,8,9,14*].

2. Taylor and Knott [*21*] have found that crack tip blunting need not cause retardation. Specifically, they investigated the reinitiation of fatigue cracks from blunted cracks in A 533B pressure vessel steel. Four-point bend specimens of width 20 mm were precracked and overloaded to give a variety of crack tip root radii. The specimens were then stress-relieved and the fatigue loading recommenced using a baseline stress intensity range, ΔK_b, of 5 MPa $\sqrt{m}$ and a load ratio, R, ($= K_{min}/K_{max}$) of 0.3. The loading was close to threshold, since the threshold stress intensity range was about 4.2 MPa $\sqrt{m}$, for $R = 0.3$. Taylor and Knott found that the crack growth rate following stress-relief was at least the value associated with constant-amplitude loading, provided the blunted crack tip root radius was less than about 15 μm. The overload stress intensity corresponding to such a root radius is about 80 MPa $\sqrt{m}$ [*22*]. Similar tests were performed using a higher baseline stress intensity range, ΔK_b, of 25 MPa $\sqrt{m}$ and a load ratio, R, of 0.3. Again, retarded growth was not observed unless the blunted crack tip root radius exceeded about 45 μm, corresponding to an overload stress intensity of about 140 MPa $\sqrt{m}$. It is concluded from this work that blunting does not lead to retardation at both high and low baseline ΔK_b levels, unless the overload is very large.

3. It is deduced from the theoretical work of Cameron and Smith [*23*] that crack tip blunting leads to a much smaller retarded crack growth increment than that found experimentally after a single peak overload. Cameron and Smith show that the crack growth increment over which a crack is retarded, Δa_n, when it grows from a notch of length D and root radius ρ, is given by

$$\Delta a_n = 0.21 \sqrt{D\rho} \qquad (1)$$

Equation 1 may be used to deduce the crack growth increment over which blunting can retard growth rates, under plane stress conditions.

The notch length, D, is identified with the crack length at overload, a_{OL}; and the notch root radius, ρ, corresponds to the crack tip radius produced by the overload. Hence, by Rice [*22*]

$$\rho \approx 0.5(K_{OL}^2/\sigma_y E) \tag{2}$$

where

K_{OL} = overload stress intensity,
σ_y = yield stress, and
E = Young's modulus.

Also, for simplicity, put $K_{OL} = \sigma_{OL}\sqrt{\pi a_{OL}}$ where σ_{OL} is the nominal applied stress at overload. The overload forward plastic zone size, $(2r_p)_{OL}$ is approximately given by

$$(2r_p)_{OL} = \frac{1}{\pi}(K_{OL}^2/\sigma_y^2) \tag{3}$$

Hence, by Eqs 1–3

$$\Delta a_n \approx 0.26\sqrt{\sigma_y^3/E\sigma_{OL}^2}\cdot(2r_p)_{OL} \tag{4}$$

Typically, $\sigma_{OL} \approx \sigma_y/2$ and $E \approx 10^3\,\sigma_y$ for overload tests on structural metals; thus equation (4) reduces to

$$\Delta a_n \approx 0.02(2r_p)_{OL} \tag{5}$$

The above analysis shows that the crack growth increment over which blunting may retard growth, Δa_n, is only 2% of the overload plastic zone size, $(2r_p)_{OL}$, created by the overload. Since retardation ensues for an increment of at least $(2r_p)_{OL}$, [*9*], it is inferred that blunting can have only a very minor influence on crack growth following a single peak overload when plane stress conditions apply. Although the numerical values of Eqs 1–4 are slightly different for the case of plane strain deformations, the same conclusion applies.

Crack Tip Strain Hardening—An overload leads to severe strain hardening of material at the crack tip. Can such strain hardening lead to retarded growth?

The following evidence is cited to show that strain hardening has also a second order influence on the crack growth response.

1. The strain hardening model erroneously predicts immediate retardation following a single peak overload.
2. Jones [*16*], Schijve [*24*], and Legris et al. [*25*] have examined the influence of prestrain on Ti-6Al-4V titanium alloy, 2024-T3 aluminum alloy, and SAE 1010 steel, respectively. In all cases, cracks grew faster through the prestrained, strain-hardened material than through the as-received material. It seems that strain hardening leads to increased growth rates by a reduction in the material's ductility [*24*], rather than leading to crack growth retardation.
3. The application of an underload immediately after a single peak overload reduces or eliminates any subsequent retardation [*26*], in conflict with the strain-hardening model.

Irregular Crack Front—The stress intensity factor is a function not only of the applied stress and the crack length, but also of the crack geometry: The stress intensity factor is reduced by crack tip branching, microcracking ahead of the crack tip, or by an increase in crack front length associated with increased fracture surface roughness [*12,27*]. Several workers have suggested that these irregularities in crack front contribute in a dominant or secondary manner to the retarded growth after a single peak overload.

Schijve [*17*] argues that overload cycles lead to a 45-deg slant mode of growth in thin aluminum alloys. This slant growth is incompatible with the normal 90-deg mode of growth associated with the small-amplitude load cycles, and contributes to the retardation associated with the small-amplitude loading.

Powell et al. [*18*] have noted that overloads lead to an increase in the fracture surface roughness and to a slant mode of failure in high-purity aluminum-zinc-magnesium (Al-Zn-Mg) alloy, but not in En 58B stainless steel. They argue that the associated increases in crack front length are secondary causes of retardation in the aluminum alloy, and that the primary cause of retardation in both materials is plasticity-induced crack closure. Powell et al. supported these arguments by stress-relieving some of the specimens after the overload: retardation was eliminated in the stainless steel and greatly reduced in the aluminum alloy.

Bucci et al. [*28*] have found that periodic overloads in low-purity aluminum alloys of the 7000-series cause separation of the matrix-intermetallic interfaces near the overload locations, the subsequent development of branch cracks, and also retarded crack growth. High purity forms of these alloys gave rise to much less crack tip branching and to smaller retardations. Bucci et al. concluded that the retarded growth is due to crack tip branching. More recently, Fleck [*29*] has applied periodic overloads to BS4360 50B structural steel and observed retardations of similar magnitude to those reported by Bucci et al. [*28*]. No significant secondary cracks were found, however, in the steel.

Schijve [*17*], Powell et al. [*18*], and Fleck [*29*] investigated crack growth rates far above threshold, while Bucci et al. [*28*] examined crack growth down to threshold rates. Suresh [*12*] argues that crack branching may be one of the causes of retardation when growth rates are near threshold. Vecchio et al. [*30*] support the arguments of Suresh by finding that crack branching and microcracking accompany crack growth retardation in ACO50, ACO62, and 2024-T3 aluminum alloys and ASTM A514F high strength steel, tested at ΔK levels close to threshold.

It is not clear how much crack branching and microcracking contribute to crack growth retardation in thick and thin specimens, following a single peak overload. It is known that the stress intensity K at the tip of a branched crack in a linear elastic material is less than for a straight crack by a factor of about $\sqrt{2}$ [*12,30–32*]. However, Lankford and Davidson [*7*] have found that overload-induced crack branching is associated with accelerated growth in thin specimens made from 2024-T4, 7075-T6, and 6061-T6 aluminum alloys. They observed that initial crack extension after the overload occurs at an accelerated rate along heavily deformed shear bands induced by the overload. Typically, these shear bands are at an angle of 45 to 70 deg to the main cracking plane. Crack growth retardation occurs only after the crack has realigned itself normal to the loading axis.

Clearly, microstructural effects may outweigh any decrease in K associated with branching, and give rise to a net acceleration of growth rate. While Lankford and Davidson [*7*] considered only plane stress behavior, it is plausible that under plane strain conditions also, microstructural effects outweigh the decrease in K associated with crack branching. Much further work is required to resolve these issues.

Residual Compressive Stress Ahead of the Crack Tip—On application of a single peak overload, compressive residual stresses are generated ahead of the crack tip. It is argued [*19*] that these compressive stresses retard crack growth; this assumption forms the basis of the Wheeler [33] and Willenborg et al. [34] retardation models. Several objections may be raised to this proposed mechanism:

1. The largest compressive residual stresses are present immediately ahead of the crack tip following the overload. Therefore, immediate retardation is predicted; in reality, delayed retardation occurs in thin specimens.
2. Crack growth rates remain retarded even when the crack has propagated far beyond the overload reversed plastic zone [12,14]. Such behavior cannot be explained on the basis of overload-induced compressive residual stresses ahead of the crack tip.
3. Mean stress relaxation occurs in the reversed plastic zone ahead of an advancing crack tip. Therefore, the "process zone" at the crack experiences fully reversed loading, regardless of the sign or magnitude of any residual stress field.

Błazewicz [35] has conducted a critical experiment where he made ball impressions on uncracked 2024-T3 aluminum alloy sheet specimens. A compressive residual stress field was thus generated between the impressions. There was a slight delay in crack growth rate when a crack was grown through the zone between ball impressions, but a much larger delay when the crack had grown beyond the impressions. It appears that the deformation associated with the ball impressions led to plasticity-induced crack closure behind the crack tip and thence to retardation. Błazewicz's experiment provides strong evidence that retardation is not due to compressive residual stresses ahead of the crack tip.

Plasticity-Induced Crack Closure—Elber [2] has argued that an overload induces large tensile plastic deformations ahead of the crack tip. Crack advance through this stretched region results in a wake of "extra" material on the crack flanks; this leads to interference of the crack surfaces when the loading is still tensile. The stress intensity range for which the crack is open, ΔK_{eff}, is then much less than the nominal value of ΔK. Elber postulated that this phenomenon of plasticity-induced crack closure forms the basis for an understanding of crack growth under constant- and variable-amplitude loading. There remains much controversy over the relevance of plasticity-induced crack closure to constant- and to variable-amplitude fatigue cracking in thick and thin specimens. For example, several investigators [6,14,36,37] have found that plasticity-induced crack closure is able to account quantitatively for crack growth retardations following a single peak overload in thick and thin specimens, while others [5,12,13,38] have observed that plasticity-induced crack closure fails to account for the retarded growth. Much of the controversy stems from the use of insensitive equipment to measure crack closure [39,40]. Also, the different responses to an overload at the surface and in the bulk of thick specimens [3,4,14] have pointed to a need to separate the surface and bulk closure responses. The difficulty of obtaining an unambiguous measure of the crack opening load and crack length at the center of a specimen has led to further confusion.

Summary

Crack growth retardation following a single peak overload is due either to the development of an irregular crack front such as crack tip branching, or to plasticity-induced crack closure. The dominance of one of these two mechanisms over the other may be universal, or may depend upon the baseline ΔK_b level and the stress state.

The objective of the present study is to determine the dominant mechanism causing retardation in thick and thin specimens made from BS4360 50B structural steel, at ΔK_b levels far above threshold.

Experimental

The crack growth rate and closure responses following a single peak overload were investigated in great detail for a thick and thin compact tension (CT) specimen made from BS4360 50B structural steel. The steel was of percentage composition by weight 0.14C-1.27Mn-0.41Si-0.017P-0.004S-0.073Al-Fe. The microstructure consisted of alternate layers of ferrite and pearlite; the ferrite grains were equiaxed and had a mean size of 10 μm. Mechanical properties in the roll direction were: yield stress 352 MPa, tensile strength 519 MPa, and elongation to failure 36% on a gage length of 25.4 mm. Specimens were 50 mm wide and of thickness 3 mm or 24 mm. They were machined from a single parent plate of thickness 25 mm, such that the specimen loading axis lay in the roll direction. Prior to testing, all specimens were stress-relieved for 1 h at 650°C in a vacuum furnace followed by a slow furnace cool.

The test conditions were identical for both thicknesses of specimens: The baseline stress intensity range, ΔK_b, was kept constant at 25 MPa$\sqrt{m}$ by manually shedding the applied loads after crack growth increments of 0.25 mm. The load ratio, R, $(= K_{min}/K_{max})$ was 0.05. A single peak overload, K_{OL}, of 51 MPA$\sqrt{m}$ was applied after the crack had grown at least 5 mm from the machine starter notch of length 15 mm; thus, an overload ratio, $(K_{OL} - K_{min})/(K_{max} - K_{min})$, of 2 was employed.

Crack Closure Measurements

Crack closure measurements were taken periodically throughout each fatigue test. The bulk closure responses of both thick and thin specimens were monitored using crack mouth displacement and backface strain gages. In addition, a novel pushrod displacement gage was used on the thick test pieces, Fig. 2. This gage has been used successfully to monitor the crack closure response of through cracks under constant-amplitude loading [*40,41*], and part-through cracks after a single peak overload [*14*]. In order to operate the pushrod gage, the test was interrupted immediately prior to overload and two parallel holes were drilled to a location 1 mm behind the crack front, as shown in Fig. 2. The end of one of the holes lay just above the cracking plane, while the end of the other hole lay just below the cracking plane. The relative displacement of the ends of the two holes was monitored by the pushrod gage, upon recommencement of the fatigue test. New holes were drilled and the pushrod gage was relocated after crack growth increments of 3 mm. It was verified that the pushrod gage did not influence the crack growth and closure behavior by repeating the test without use of this gage.

The crack closure responses at the surface of both thick and thin test-pieces were monitored using an Elber gage and a twin-cantilever displacement gage, Fig. 3. These gages give an accurate measure of the crack opening load, K_{op}, at the surface of a specimen, provided they are placed less than 2.5 mm behind the crack tip [*40*].

The fraction of the load range for which the crack is open, U, was defined in terms of the crack opening stress intensity, K_{op}, for all types of gages. Hence

$$U = \frac{K_{max} - K_{op}}{K_{max} - K_{min}} \quad \text{for } K_{op} > K_{min}$$
$$U = 1 \quad \text{for } K_{op} \leq K_{min} \tag{6}$$

The crack opening stress intensity, K_{op}, was determined from the point at which a load-displacement $(P - \delta)$ trace became linear on loading; see Fig. 4. Discrimination of this

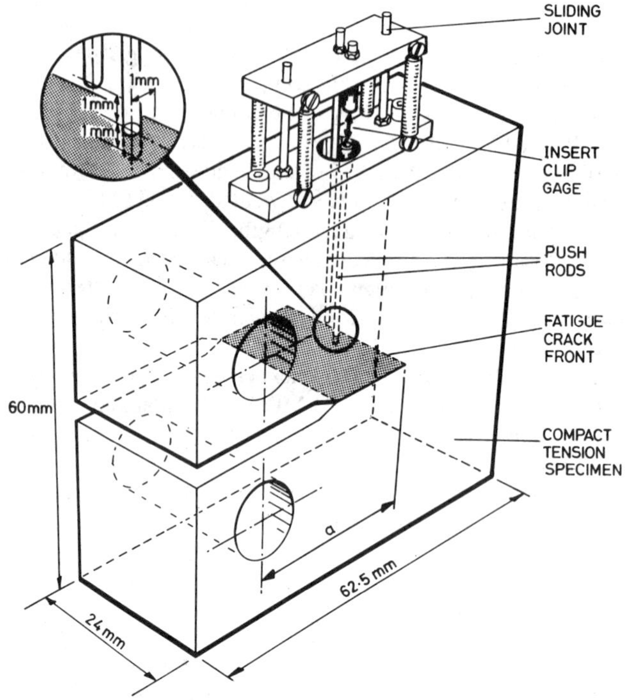

FIG. 2—*Pushrod closure gage.*

point was improved by offsetting the displacement δ by a fraction α of the load *P*, such that the offset displacement ξ is given by

$$\xi = \delta - \alpha \cdot P \tag{7}$$

By adjusting α to equal the compliance of the cracked specimen, ξ is equal to zero when the crack is open, Fig. 4. This offset arrangement greatly improves the sensitivity of closure measurements; it was first employed by Kikukawa et al. [*42*].

Third-order low-pass filters of cutoff frequency 1 Hz were also employed to aid discrimination of the crack opening load [*39*], by reducing electrical noise. The fatigue test frequency of 5 Hz was therefore reduced to 0.05 Hz when taking closure measurements, in order to avoid signal distortion. This change in frequency has no influence on the crack closure response [*39*].

Crack Growth Measurements and Replication Technique

Crack growth at the center of the thick and thin specimens was measured using the dc potential-drop technique [*40*], while surface growth was followed with a traveling microscope.

Plastic replicas were taken of the crack flanks at the surface of the thin specimen during application of the overload cycle and also after subsequent crack growth. These replicas were developed into positive replicas by evaporation with lead [*39,43*]. The mechanisms of crack opening and crack growth could thus be elucidated.

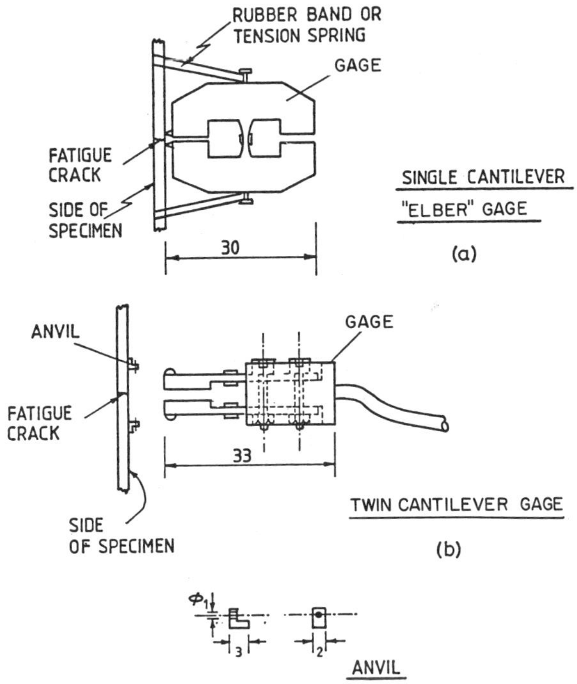

FIG. 3—*Crack tip compliance instrumentation for monitoring the crack closure response at the surface of a specimen. In each diagram it is assumed that crack growth is into the plane of the page. All dimensions are in mm.*

Results

Crack Growth Transient

The crack growth rate responses of both thick and thin specimens are given in Fig. 5. A comparison of crack lengths measured by the potential-drop method and by the traveling microscope showed that the crack growth rate did not vary along the crack front of the thin specimen: the entire crack front was in a state of plane stress. In contrast, the crack growth response differed at the surface and in the bulk of the thick specimen, Fig. 5. Crack growth retardation was more immediate but slightly less severe at the center of the thick specimen than at the surface, in agreement with the findings of Bernard et al. [*3,4*]. This is consistent with the fact that plane strain conditions prevailed near the center of these specimens, while near the surface the crack front experienced a stress state closer to plane stress. Delayed retardation is evident in all cases, Fig. 5. After the crack grew a distance, Δa^*, from the overload location, the growth rate stabilized to an almost constant value, which was slightly less than the pre-overload growth rate, Fig. 5.

The overload forward plastic zone size, $(2r_p)_{OL}$, was easily measured at the surface of the specimen, since Lüders bands of yield were formed. The measured and calculated values for $(2r_p)_{OL}$ are given in Table 1, together with the overload-affected crack growth increment, Δa^*. The following deductions may be made from Table 1:

- The calculated and measured values of the overload plastic zone size are in good agreement for the thin test-pieces. The measured plastic zone size at the surface of the

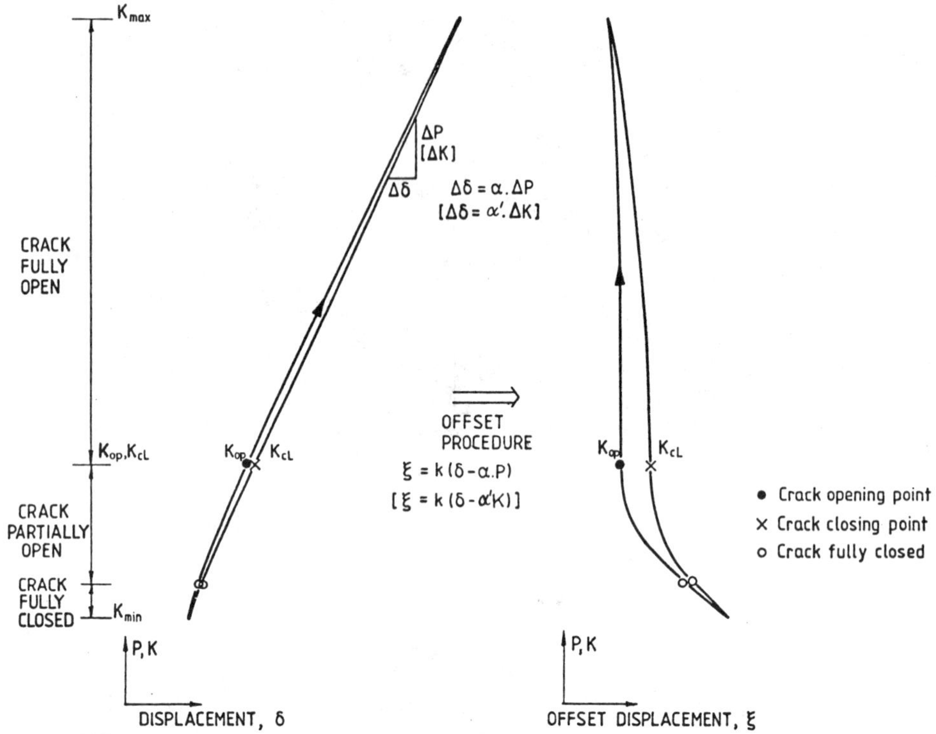

FIG. 4—*Determination of the crack opening stress intensity,* K_{op}*, from the specimen's compliance. Typical response, showing hysteresis due to crack tip plasticity. The crack closing stress intensity,* K_{cl}*,* $\approx K_{op}$.

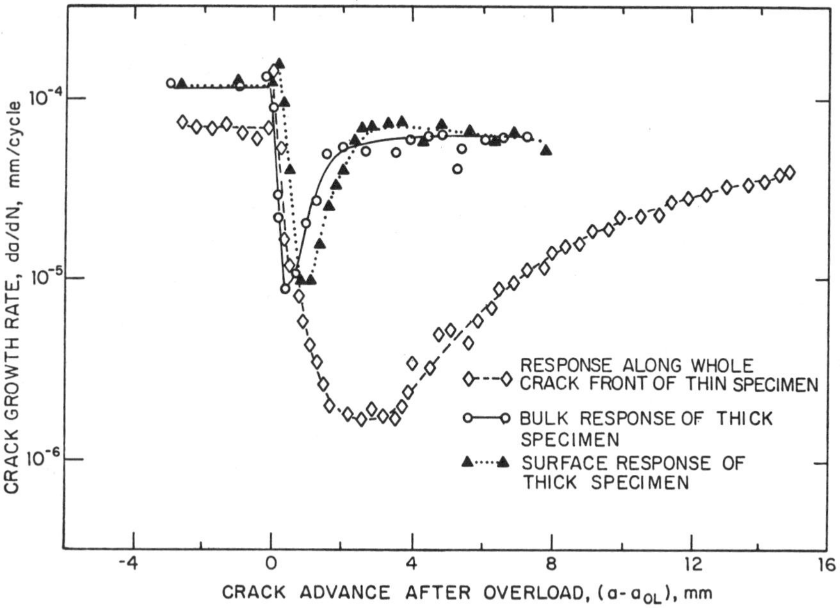

FIG. 5—*Comparison of crack growth rates at the surface of the thick specimen, at the center of the thick specimen, and along the whole crack front of the thin specimen.*

TABLE 1—*Comparison of overload-affected crack growth increment, Δa^*, with measured and calculated overload plastic zone sizes* (K_{OL} = *51 MPa$\sqrt{m}$*).

Specimen and Location	Measured Overload-Affected Crack Growth Increment, Δa^*, mm	Measured Overload Plastic Zone Size, mm	Calculated Overload Plastic Zone Size, (Plane stress)[a] mm	Calculated Overload Plastic Zone Size, (Plane strain)[b] mm
Whole crack front of thin specimen	>15	6.5	6.7	...
Surface of thick specimen	2.8	3.4	6.7	2.2
Bulk of thick specimen	2.4	...	...	2.2

[a]Plane stress plastic zone size $= \frac{1}{\pi}(K_{OL}^2/\sigma_y^2)$.

[b]Plane strain plastic zone size $= \frac{1}{3\pi}(K_{OL}^2/\sigma_y^2)$.

thick specimen is about half the size of the measured plastic zone for the thin specimen; Fleck [*40*] has observed this same ratio of forward plastic zone sizes for thick and thin specimens of BS4360 50B steel, under constant-amplitude loading. Plainly, the bulk, plane strain regions of the thick specimen influence the superficial size of the plastic zone. It appears that the plastic zone size at the surface of the thick specimen is intermediate between the plane stress and plane strain values, see Table 1.

- The overload-affected crack growth increment, Δa^*, for the thin specimen is more than twice the corresponding measured or calculated overload plastic zone size. In contrast, Δa^* at the surface of the thick test-piece is comparable with the measured, superficial plastic zone size. Also, Δa^* at the center of the thick specimen is comparable with the calculated, plane strain plastic zone size.

Fractography

Examination of the fracture surfaces in the scanning electron microscope revealed that crack advance was always by the formation of poorly defined striations, Figs. 6 and 7. Stage I near-threshold growth did not occur: crack growth rates exceeded 10^{-6} mm/cycle with no evidence of faceted growth.

Crack Closure Transient

Consider first the thick specimens. The pushrod gage, crack mouth gage, and backface strain gage all showed the same closure response at the center of the thick test-piece to within ±5%. Hence the crack mouth gage and backface strain gage indicate correctly the bulk closure response; the laborious technique of using the pushrod gage is not needed for future tests. The twin cantilever displacement and Elber gage both displayed the same closure response at the surface of the thick test-piece. This response differed slightly from the bulk behavior; see Fig. 8.

Now consider the thin specimen. The crack mouth displacement, backface strain, and crack tip closure gages all indicated the same U-values to within ±5%; surface and bulk

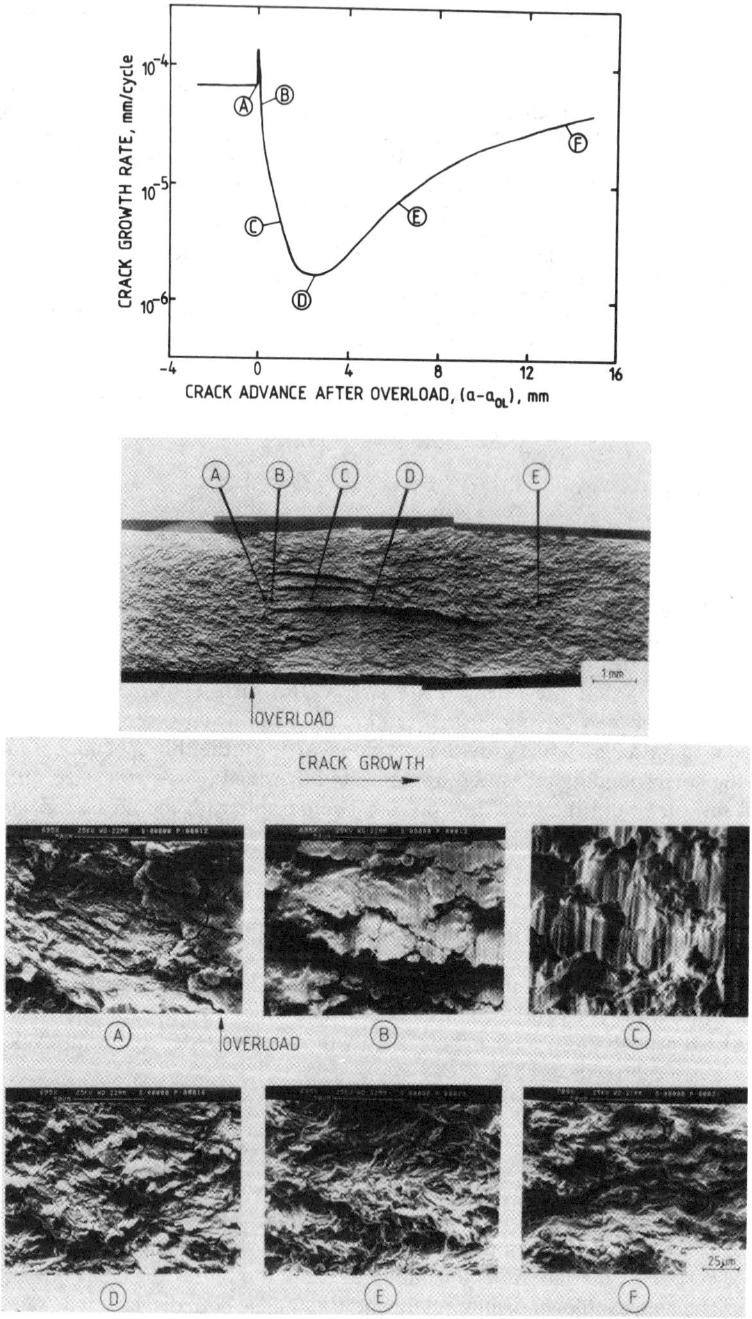

FIG. 6—*Effect of single peak overload on fracture surface morphology, thin specimen.*

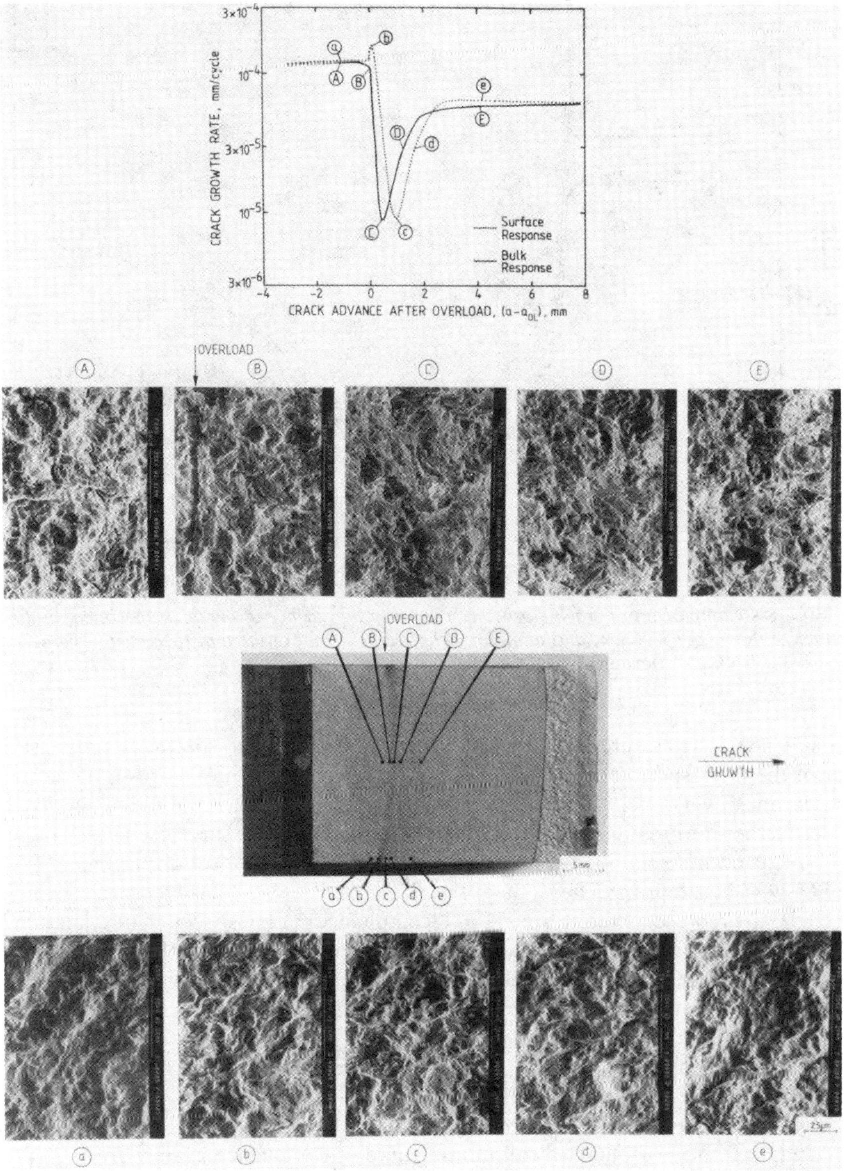

FIG. 7—*Effect of single peak overload on fracture surface morphology, thick specimen.*

behaviors were identical. Therefore, only a single curve is given in Fig. 8 to show the closure response along the whole crack front of the thin specimen.

In each case, the shape of the closure transient is similar to that of the crack growth rate transient; compare Figs. 5 and 8. On application of the overload, the crack is stretched fully open and the U-value increases discontinuously to 1.0. With subsequent crack growth the closure value, U, drops steeply to a minimum level and then slowly returns to the pre-

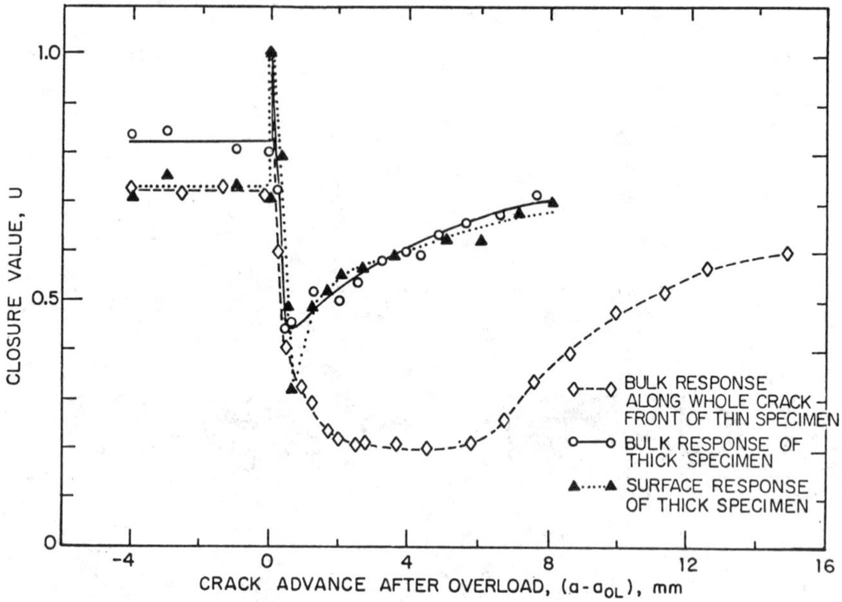

FIG. 8—*Comparison of crack closure responses at the surface of the thick specimen, at the center of the thick specimen, and along the whole crack front of the thin specimen. Averaged responses from the relevant closure gages.* U = $\Delta K_{eff}/\Delta K$.

overload value. The closure transient is most severe for the thin specimen and least at the center of the thick specimen.

The distance over which the crack is closed at minimum load of the fatigue cycle, $\Delta a'_{min}$, provides further information on the mechanics of crack closure. This closed crack increment may be deduced from the load-displacement ($P - \delta$) or load-offset displacement ($P - \xi$) trace [*37,39,40*], as follows: The slope of the $P - \delta$ or $P - \xi$ trace is measured at minimum load; this slope represents the stiffness (= 1/compliance) of the specimen at minimum load. The length over which the crack is open at minimum load, a'_{min}, is then deduced from a standard calibration curve of crack length against specimen compliance [*39*]. The closed crack increment at minimum load, $\Delta a'_{min}$, equals the measured crack length, a, at maximum load minus the inferred open crack length at minimum load, a'_{min}. Good agreement has been found between $\Delta a'_{min}$ deduced from compliance measurements and the closed crack increment measured from plastic replicas [*39*]. In the present tests, replicas confirmed that the crack was shut everywhere between a'_{min} and a, for a load P_{min}. Hence the procedure of deducing a'_{min} from a standard calibration curve of compliance against crack length was valid.

The effect of a single peak overload on $\Delta a'_{min}$ and a'_{min} is shown in Fig. 9 for the bulk response of the thick and thin specimens.

Consider first the thin specimen, Fig. 9*a*. Prior to the overload, the crack closes back to the notch at minimum load of the fatigue cycle, and $\Delta a'_{min}$ increases linearly with crack length. On application of the overload, the crack flanks are pulled sufficiently apart such that the closed crack increment, $\Delta a'_{min}$, becomes zero ($U = 1$). Thereafter, the crack closes back to near the overload location at minimum load of the fatigue cycle; $\Delta a'_{min}$ again increases linearly with crack advance. Visual observations of the crack flanks with a traveling microscope confirm these findings.

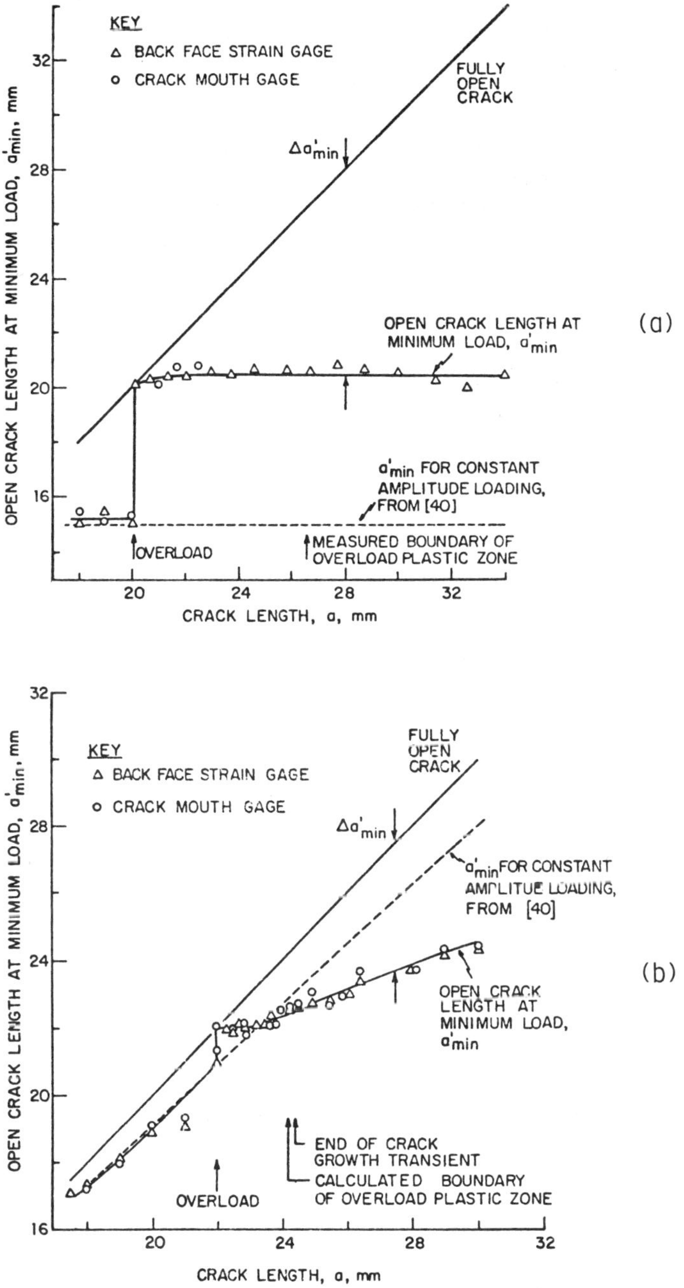

FIG. 9—*Effect of overload upon the closed crack increment at minimum load,* $\Delta a'_{min}$, *and the open crack length at minimum load,* a'_{min}. (a) *Bulk response of thin specimen;* (b) *bulk response of thick specimen.*

The bulk response of the thick specimen was somewhat different; see Fig. 9*b*. Prior to application of the overload, the effective crack length lags behind the crack length by a constant value, $\Delta a'_{min}$, of about 1 mm. On application of the overload, the crack flanks are again pulled apart and $\Delta a'_{min}$ is reduced discontinuously to zero. During the subsequent retarded growth, the effective crack length at minimum load, a'_{min}, remains close to the overload location, Fig. 9*b*. When the growth rate has reestablished itself to a constant value (slightly less than the pre-overload rate Fig. 5), a'_{min} remains smaller than the constant-amplitude value, but increases slowly with crack advance. It is likely that the differences in response shown by the thick and thin specimens in Fig. 9 are due to greater plastic constraint and a smaller residual plastic wake behind the crack tip in the thick specimens.

Brown and Weertman [*44*] have also investigated the effect of a single peak overload on the closed crack increment at minimum load, $\Delta a'_{min}$. They considered center-cracked panels made from 7050 aluminum alloy and deduced $\Delta a'_{min}$ from a $P - \delta$ trace, where δ was the crack opening displacement on the specimen centerline. Their specimens were sufficiently thick for plane strain conditions to prevail. In conflict with the results shown in Fig. 7*b*, Brown and Weertman found that $\Delta a'_{min}$ reached a minimum value when the crack had almost traversed the overload plastic zone. No explanation was given for their results. Recent work [*39,40*] has shown that the crack mouth gage with no offset procedure or filter arrangement is insufficiently sensitive for taking such closure measurements.[2] Therefore, the discrepancy may be due to differences in instrumentation rather than to differences in material response.

Discussion

Ability of Crack Closure to Account for Transient Growth Rate

The crack growth transient may be predicted from the measured closure response, using the modified Paris law [*40*]

$$\frac{da}{dN} = 5.91 \times 10^{-9} (\Delta K_{eff})^{3.24} \text{ mm/cycle} \tag{8}$$

where $\Delta K_{eff} = U \cdot \Delta K$ and ΔK_{eff} is in units of MPa $\sqrt{m}$. The regression given by Eq 8 was calculated from the constant-amplitude crack growth response of 3-mm-thick and 24-mm-thick specimens made from BS4360 50B steel; crack growth rates were in the range 10^{-6} to 10^{-4} mm/cycle and the load ratio was in the range 0.05 to 0.5 [*40*].

These predictions are compared with the observed crack growth responses in Fig. 10. In each case, the crack growth rates predicted by measurements of the crack opening load are in agreement with the observed crack growth rates, except for the period when crack growth rates are increasing from the slowest transient growth rate to the post-overload stabilized value. During this period, growth rates are faster than predicted by closure readings. Recent work, using plastic replicas of the crack flank opening response at the surface of thin through-cracked specimens [*40,45*] and at the surface of thick part-through cracked specimens [*14*], has shown that the anomalous response is due to *discontinuous closure*—the crack first closes at a location far from the crack tip; see Fig. 11.

[2] When the closed crack increment $\Delta a'_{min}$ is small, there is only a small change in the specimen compliance and in the slope of the load versus crack mouth opening displacement (CMOD) trace. The large hysteresis of a load versus CMOD trace then masks the crack opening load [*39,40*]. The backface strain gage shows less hysteresis and is therefore a more reliable closure gage.

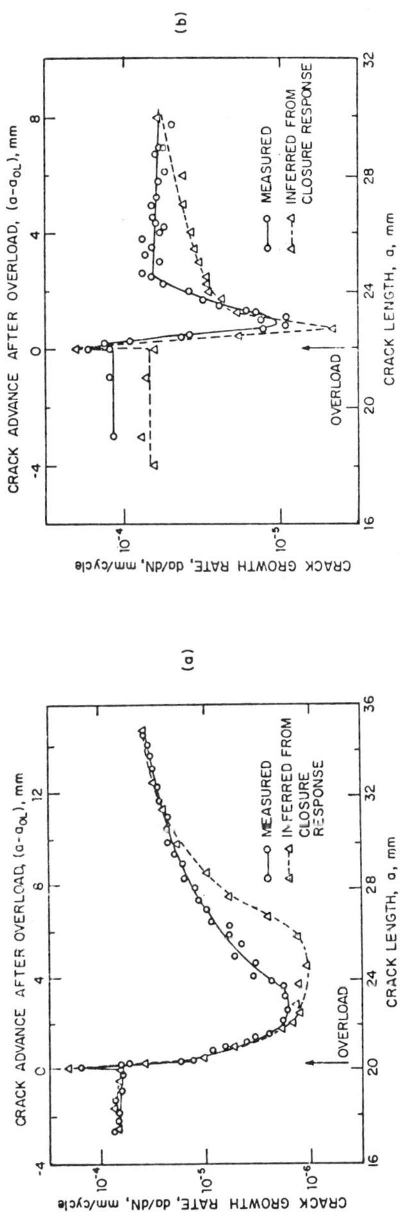

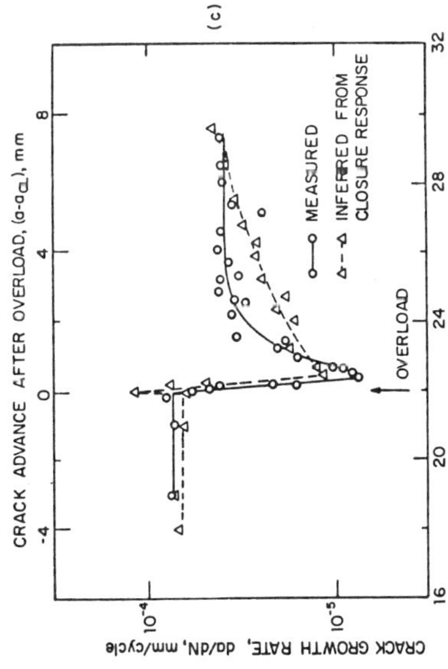

FIG. 10—*Comparison of measured crack growth response and predicted response from closure measurements.* (a) *Thin specimen;* (b) *surface response of thick specimen;* (c) *bulk response of thick specimen.*

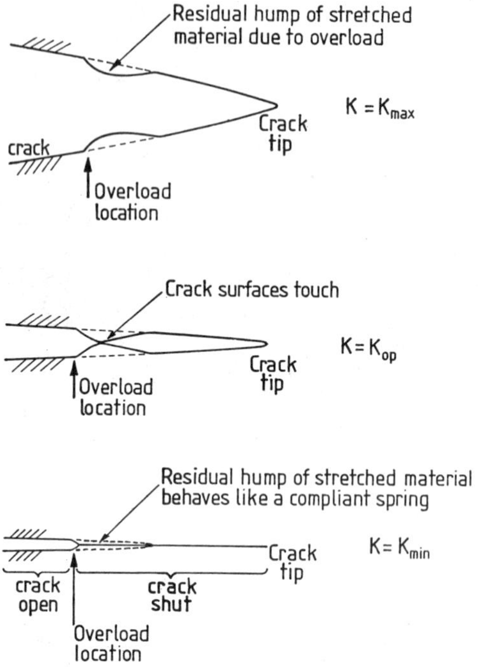

FIG. 11—*The phenomenon of discontinuous closure.*

It appears that a residual hump of stretched material at the overload location leads to the phenomenon of discontinuous closure. This material acts like a compliant spring and allows cyclic crack tip displacements to occur at loads below the crack opening load, Fig. 11. Thus the stress intensity range experienced by the crack tip is greater than suggested by closure measurements, and crack growth rates exceed the predicted values, Fig. 10. The anomaly between predicted and observed growth rates is thus explained.

This discrepancy between measured and predicted growth rates is apparent in much previous work [*5,14,46–48*] and has caused many investigators to conclude that crack closure is unable to account for crack growth retardation following a single peak overload. It is clear from the present work that crack closure is the dominant cause of retardation, once discontinuous closure is properly taken into account.

Discontinuous closure appears to be significant in low-strength steels which strain-harden appreciably rather than in high-strength aluminum alloys of low work-hardening capacity. For example, Robin et al. [*5*] found that the closure value *U* dropped to a steady low value after an overload was applied to E36 steel specimens (σ_y = 380 MPa). They concluded that crack closure was unable to account for the crack growth transient caused by the overload. Paris and Herman [*37*], however, found that crack closure could fully account for the crack growth transient in 2024-T351 aluminum alloy. It is likely that in steels the residual hump of stretched material work-hardens and thereby maintains its size; in the case of aluminum alloys, the residual hump is reduced in size by compressive yield as the crack grows through the overload plastic zone.

In the preceding discussion it has been assumed that retardation is due to plasticity-induced crack closure rather than crack branching or any other retardation mechanism. Evidence is now presented to substantiate this claim.

Evidence for Plasticity-Induced Crack Closure

1. The existence of a residual hump of stretched material at the overload location would provide strong evidence that retardation is due to plasticity-induced crack closure. Such residual deformations have been observed by measuring the overlap between the mating fracture surfaces after termination of the fatigue tests on BS4360 50B steel. The procedure was as follows.

The crack surfaces of each specimen were separated at the end of the fatigue tests and a roughness traverse made of corresponding locations on each face of the crack. The traverse was taken in the crack growth direction, over a crack growth increment corresponding to the retarded growth. The resulting traces from the two fatigue fracture surfaces were superimposed and the overlap determined, Fig. 12. This overlap equals the residual displacement caused by the overload. Results are presented in Fig. 12.

It appears that the overload induces a residual displacement, $2s$, of up to 25 μm in the thin specimen and 15 μm in the thick test-piece. The residual displacements are greatest near the overload location, as expected. Considerable scatter is associated with these measurements since the fracture surface roughness is of the order of 3-μm centerline-average (CLA) [*40*].

Figure 9 provides confirmation that large residual plastic displacements are generated at the overload location in both thick and thin specimens: after application of the overload, the crack closes back to near the overload location, even when the crack has propagated beyond the overload plastic zone. This residual hump of material also accounts for the observation that the closure value U does not attain the pre-overload value until the crack has propagated well beyond the overload plastic zone, for both thick and thin specimens; compare Table 1 and Fig. 5.

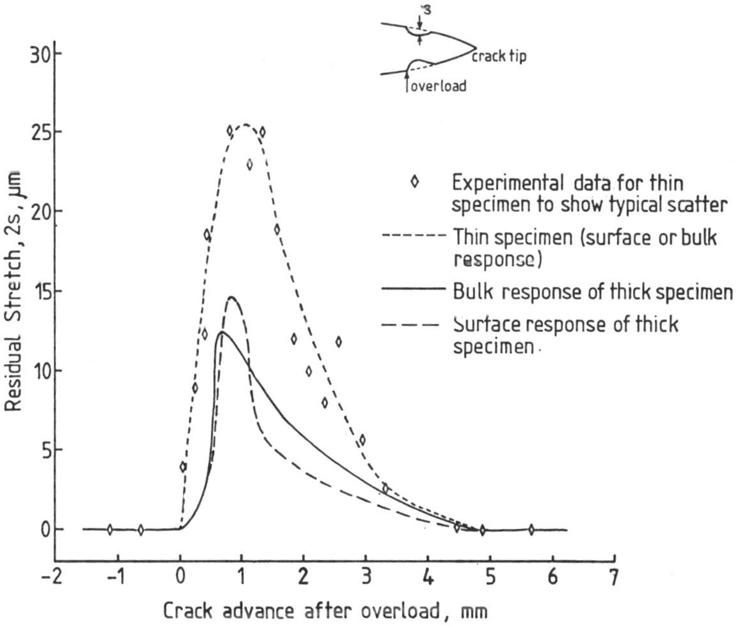

FIG. 12—*Residual displacements left in the wake of the fatigue crack, due to a single peak overload. Measurements taken using a Ferranti Surfcom with a probe tip radius of 1 μm.*

Nowack et al. [*48*] have also observed residual displacements of magnitude 10 μm following a step decrease in loading. They investigated thin specimens made from 2024-T3 aluminum alloy, and monitored the crack flank profile using a traveling microscope.

Indirect evidence for the existence of a residual hump of stretched material due to overloads has been given by Trebules et al. [*49*]. They used thin compact tension specimens made from 2024-T3 aluminum alloy. A fatigue crack was grown at constant ΔK during which a high load sequence of 100 overloads was applied. The overload ratio was 1.5. Immediately after application of the overloads the crack slowed down to a minimum rate and then increased to a stabilized post-overload value. This stabilized growth rate was 25% less than the pre-overload rate. The fatigue test was then interrupted and a sawcut was made to just behind the crack tip. On resumption of the test, the crack growth rate increased discontinuously to the pre-overload value. Trebules et al. correctly concluded that the step increase in growth rates was due to removal by the saw cut of the residual hump of stretched material associated with the overloads.

2. The relation between crack growth rate, da/dN, and effective stress intensity range, ΔK_{eff}, is preserved except for the crack growth increment where discontinuous closure is significant. If retardation were due to a mechanism such as an irregular crack front, the constant-amplitude da/dN versus ΔK_{eff} relation would be violated: the local ΔK_{eff} following the overload would be less than the measured value of ΔK_{eff} and crack growth rates would thus be less than predicted by the constant-amplitude da/dN versus ΔK_{eff} relation.

3. There is fractographic evidence to suggest that retardation is plasticity-induced. Examination of the fatigue fracture surfaces of the thick and thin specimens revealed the presence of oxide debris and abrasion immediately ahead of the overload locations, Figs. 6 and 7. This damage to the fracture surface is much greater than that shown at the same growth rates under constant-amplitude loading. It is well established that plasticity-induced crack closure occurs under constant-amplitude conditions yet gives rise to no damage of the fatigue fracture surfaces [*1,9,12,40*]. How can plasticity-induced crack closure lead to such damage after a single peak overload?

Allison [*50*] has measured the nominal compressive stresses across the flanks of closed cracks in 1045 steel, using the X-ray method. Measurements were conducted on the surface of thin compact tension specimens at zero load. He applied a single peak overload, grew the crack partway into the overload plastic zone, and then measured the compressive residual stresses across the crack flanks. Allison found that the stress reached a maximum of about two-thirds the compressive yield stress at a point immediately ahead of the overload location. The compressive residual stress distribution was also determined for the constant-amplitude case, prior to application of the overload; then the stresses reached a maximum of only one-third the compressive yield stress behind the crack tip. Similar observations have also been made by Taira and Tanaka [*51*].

It seems that overloads induce a much larger compressive stress field than that associated with constant-amplitude loading. Therefore, it is not surprising that an overload can induce severe abrasion and oxidation on the fatigue fracture surfaces, Figs. 6 and 7.

Crack branching and similar deflections in crack path lead to the development of significant Mode II stress intensities at the crack tip, [*12,31*]. Thus, it is likely that Mode II wear scars would be apparent on the fracture surface if retardation were due to branching [*52,53*]. No Mode II wear scars or associated debris were found in the overload-affected region of the fracture surface, for both thick and thin specimens. It is concluded that the fracture surface damage immediately ahead of the overload location is consistent with the retardation mechanism of plasticity-induced crack closure.

After application of the overload to the thin specimen, when crack growth rates were decreasing to the minimum value, some Mode III sliding occurred behind the crack tip; this was evidenced by the formation of Mode III wear scars, Fig. 6. Such sliding behavior was associated with a slant mode of growth in the overload plastic zone. Overload tests on other thin specimens [*45*] gave rise to no Mode III behavior or slant growth: it seems that this anomalous behavior was due to slight-misalignment of the specimen in the test machine. Since similar tests gave rise to no Mode III wear scars but to retardations of the same magnitude as that shown in Fig. 5 for the thin BS4360 50B steel specimen [*45*], it is concluded that the irregular crack front and roughness-induced closure associated with the Mode III behavior did not contribute significantly to the retarded growth in this test.

4. Suresh [*12,31,54*] argues that near-threshold closure mechanisms, such as oxide-induced and roughness-induced closures, operate when the post-overload growth rate is near threshold. This is fully consistent with the prior observations of Hertzberg and Mills [*55*], who found that the character of the fatigue fracture surface (and thus the associated crack advance mechanisms and closure mechanisms) is the same after an overload as in a constant-amplitude test, when the comparison is made at similar growth rates. The operation of near-threshold closure mechanisms and the associated increase in K_{op} and decrease in growth rates are *consequences* and *not* the cause of retarded growth following an overload.

It is now argued that near-threshold closure mechanisms do not significantly increase the size of the overload-affected crack growth increment corresponding to retarded growth. In a conventional threshold test where loads are shed to threshold and then increased again, no retardation or hysteresis loop is shown on the da/dN versus ΔK plot. A hysteresis loop would be expected if near-threshold growth were maintained once it had been triggered. The results of the present study suggest that retardation is prolonged to beyond the overload cyclic plastic zone by the process of plasticity-induced discontinuous closure.

It seems that oxide-induced and roughness-induced crack closures can decrease ΔK_{eff} when growth rates are near threshold in a constant-amplitude or overload test. However, these forms of crack closure are neither the cause of retardation, nor are they responsible for prolonging retarded growth to beyond the overload cyclic plastic zone.

Evidence For and Against the Crack Branching Mechanism of Retardation

It is now shown from the results of the present study and from the literature that, for both thick and thin specimens, crack branching is not the dominant cause of retardation at high-baseline, ΔK_b, levels. At low ΔK_b levels, retardation may be due to crack branching or to plasticity-induced crack closure.

Overload Tests at High Levels of Baseline ΔK—1. It is deduced from the shape of the crack path in the BS4360 50B steel specimens that retarded growth is not due to crack branching.

In order to determine the crack path following overload, fracture surface profiles were taken in the direction of crack advance using a Ferranti Surfcom. For each of the specimens, only one of the two mating fracture surfaces was examined. No deviation in crack path was found at the center of the thick BS4360 50B specimen, Fig. 13, while small deviations were found at the surface of the thick specimen and for the whole crack front of the thin specimen.

These measurements show that crack branching does not lead to retarded growth in the bulk of the thick specimen, since no deviation in crack path was found. At the surface of the thick specimen and along the whole crack front of the thin specimen, the crack advanced by about 0.5 mm beyond the overload location and slowed down by an order of magnitude before the crack deviated from its path; see Figs. 5 and 13. Therefore, crack branching does not cause retarded growth for plane stress conditions as well as for plane strain conditions.

What causes the crack to change its direction of growth after an overload is applied to

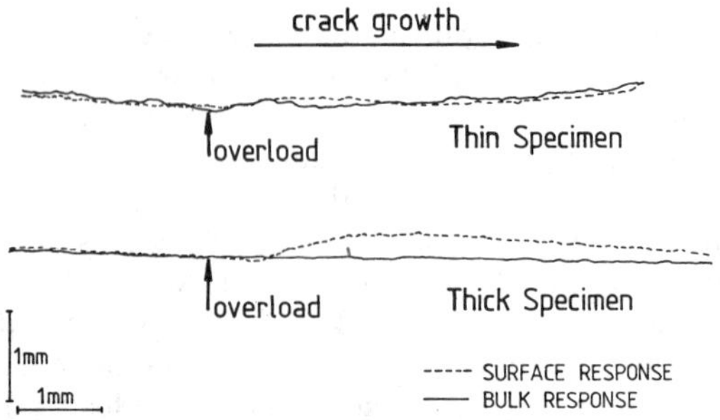

FIG. 13—*Deviation in crack path following a single peak overload, for thin and thick specimens.*

the thin specimen? To answer this question, plastic replicas were taken of the crack upon application of the overload and also after some subsequent crack growth; see Fig. 14.

On application of the overload, intense shear bands developed at the crack tip, Fig. 14*a,b*. Crack advance occurred at about twice the preoverload rate along one of these shear bands, such that the crack grew in a self-similar direction, Fig. 14*b,c*. Significant branching of the crack did not occur until the crack had propagated by about 0.5 mm into the overload plastic zone. A careful examination of the crack path in relation to the overload plastic zone in the thin specimen, Fig. 15, revealed that the crack attempted to exit the overload zone as quickly as possible and then grow through elastic material between the Lüder's bands of yield. The overload plastic zone was asymmetric with respect to the plane of the crack; it seems that the deviations in crack path after overload were a result of the crack growing away from the more intense side of the plastic zone. Lankford and Davidson [*10*] have similarly observed asymmetric overload plastic zones in 6061-T6 aluminum alloy, and that the fatigue crack deviates away from the most intense deformation in order to exit the plastic zone via the nearest elastic-plastic boundary.

2. Cracks may retard by several orders of magnitude or even arrest when the baseline ΔK is far above threshold; see for example, Refs *3* and *4*. Crack branching is unable to account for such large decreases in growth rate, since the drop in local ΔK due to branching is only by a factor of about $\sqrt{2}$ [*12,30–32*].

Thus, irregularities of the crack front are unable to explain quantitatively retarded growth following overloads in tests at high baseline ΔK levels.

3. There is evidence that deflected cracks grow at about the same rate as "straight" cracks. For instance, before application of the overload to the thin BS4360 50B steel specimen, an increase of test frequency from 1 to 10 Hz caused the crack to abruptly change direction by about the same amount as that later induced by the overload. Yet, no change in crack growth rate was observed.

Lankford and Davidson [*10*] have monitored the crack propagation response of cracks in 6061-T6 aluminum alloy after application of a single peak overload. They found that the crack growth rate quickly achieved the pre-overload rate once the crack had crossed the boundary of the overload plastic zone, even though the crack was still highly deviated in path.

Overload Tests at Low Levels of Baseline ΔK—At low ΔK_b levels, growth rates are highly sensitive to ΔK and significant retardations or arrest will occur with any small decrease in crack driving force. Further work is required before it is known whether the decrease in driving force associated with an overload is due to crack branching, plasticity-induced crack closure, or to some other mechanism.

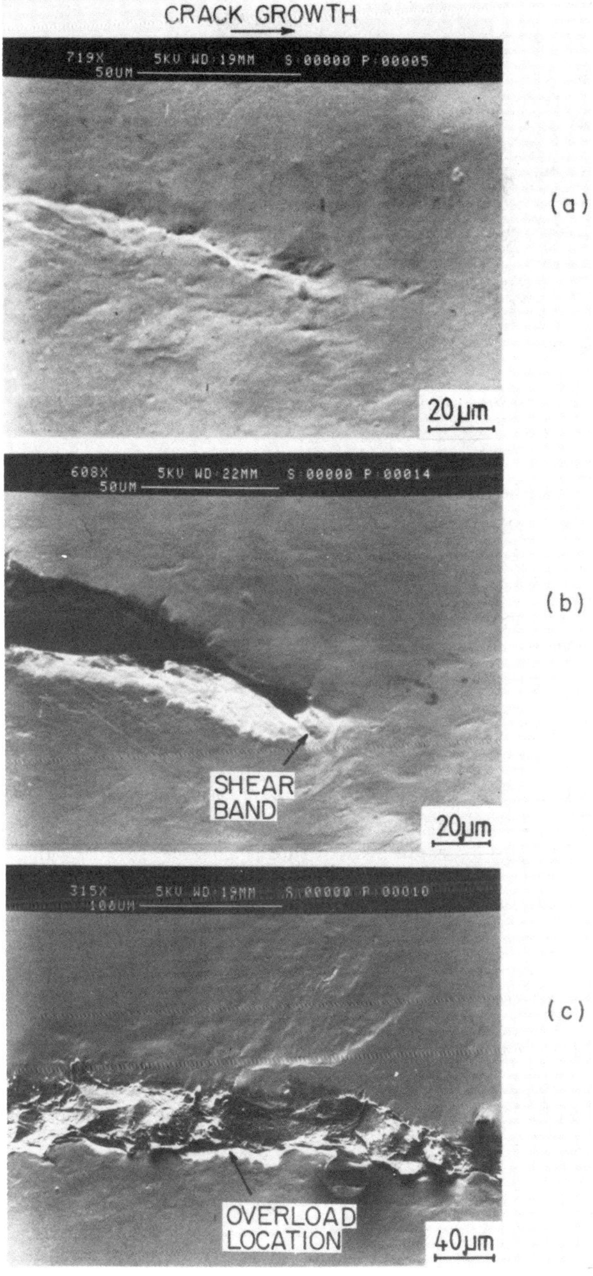

FIG. 14—*Replicas of the crack flanks at surface of thin specimen.* (a) *Crack tip immediately prior to application of overload,* $K = K_{min}$; (b) *crack tip upon application of overload,* $K = K_{OL}$; (c) *crack flanks at overload location. Crack has advanced 5 mm beyond overload location,* $K = K_{min}$.

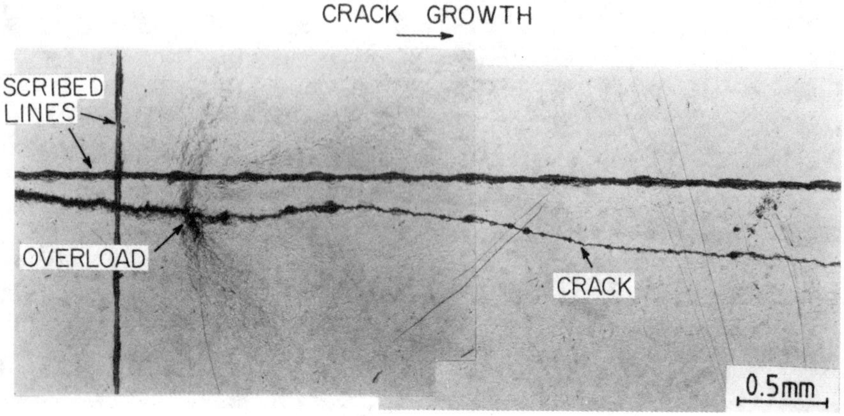

FIG. 15—*Optical view of the crack path after overload in relation to the overload plastic zone, thin specimen.*

Crack branching and similar irregularities of the crack front can reduce ΔK by a factor of about $\sqrt{2}$ [*12,30–32*], as discussed previously. Thus, provided the baseline ΔK_b following an overload is less than $\sqrt{2} \cdot \Delta K_{th}$ where ΔK_{th} is the nominal threshold stress intensity range, crack branching has the capacity to account for large retardations and crack arrest. Suresh [*12*] and de Castro and Parks [*13*] report such crack arrest immediately after an overload in ASTM 542 low-alloy steels, when tests are performed at low ΔK levels.

Conclusions

Tests on BS4360 50B steel specimens and a critical examination of the literature have led to the following conclusions on the physics and phenomenology of crack growth retardation following a single peak overload; these conclusions relate specifically to high levels of baseline loading such that the pre-overload growth rate is far above threshold.

1. Retarded growth, both in thick specimens suffering plane strain conditions, and in thin specimens suffering plane stress conditions, is due primarily to plasticity-induced crack closure.
2. The crack growth and closure transients at the surface of a thick specimen differ from the bulk, plane strain response. No such differences in behavior occur when the specimen is sufficiently thin for plane stress conditions to exist along the whole crack front.
3. A residual hump of stretched material is left on the fatigue fracture surfaces by an overload. This residual material leads to the phenomenon of discontinuous closure.

It is not yet clear whether retardation associated with baseline ΔK loading close to the threshold value, ΔK_{th}, is due to plasticity-induced crack closure or to irregularities of the crack front.

Acknowledgments

The author wishes to thank Professors J. W. Hutchinson and S. Suresh and Dr. D. L. Davidson for helpful discussions, and Drs. D. Taylor and J. F. Knott for details of their experiments on blunted cracks.

References

[1] Elber, W., "Fatigue Crack Closure Under Cyclic Tension," *Engineering Fracture Mechanics,* Vol. 2, 1970, pp. 37–45.
[2] Elber, W., "The Significance of Fatigue Crack Closure" in *Damage Tolerance in Aircraft Structures, ASTM STP 486,* American Society for Testing and Materials, Philadelphia, 1971, pp. 230–242.
[3] Bernard, P. J., Lindley, T. C., and Richards, C. E., "The Effect of Single Overloads on Fatigue-Crack Propagation in Steels," *Metal Science,* Aug./Sept. 1977, pp. 390–398.
[4] Bernard, P. J., Lindley, T. C., and Richards, C. E., "Mechanisms of Overload Retardation During Fatigue Crack Propagation," *Fatigue Crack Growth Under Spectrum Loads, ASTM STP 595,* American Society for Testing and Materials, Philadelphia, 1976, pp. 78–97.
[5] Robin, C., Louah, M., and Pluvinage, G., "Influence of an Overload on the Fatigue Crack Growth in Steels," *Fatigue of Engineering Materials and Structures,* Vol. 6, No. 1, 1983, pp. 1–13.
[6] Tokaji, K., Ando, Z., and Kojima, T., "Fatigue Crack Retardation of low Carbon Steel in Salt Water," *Journal of Engineering Materials and Technology, Transactions,* American Society of Mechanical Engineers, Vol. 106, Jan. 1984, pp. 38–42.
[7] Lankford, J. and Davidson, D. L., "The Effect of Overloads upon Fatigue Crack Tip Opening Displacement and Crack Tip Opening/Closing Loads in Aluminium Alloys," *Advances in Fracture Research, Proceedings,* Fifth International Conference on Fracture, Cannes, France, D. François et al., Eds., Pergamon Press, Oxford, U.K., Vol. 2, 1981, pp. 899–906.
[8] Schijve, J., "Fatigue Crack Propagation in Light Alloy Sheet Material and Structures," Report MP195, National Luchtvaart Laboratorium, Amsterdam, Aug. 1960.
[9] von Euw, E. F. J., Hertzberg, R. W., and Roberts, R., "Delay Effects in Fatigue Crack Propagation," *Stress Analysis and Growth of Cracks, Proceedings of the 1971 National Symposium on Fracture Mechanics, Part I, ASTM STP 513,* American Society for Testing and Materials, Philadelphia, 1972, pp. 230–259.
[10] Lankford, J., Jr., and Davidson, D. L., "Fatigue Crack Tip Plasticity Associated with Overloads and Subsequent Cycling," *Journal of Engineering Materials and Technology, Transactions,* American Society of Mechanical Engineers, Jan. 1976, pp. 17–23.
[11] Bathias, C. and Vancon, M., "Mechanisms of Overload Effect on Fatigue Crack Propagation in Aluminium Alloys," *Engineering Fracture Mechanics,* Vol. 10, 1978, pp. 409–424.
[12] Suresh, S., "Micromechanisms of Fatigue Crack Growth Retardation Following Overloads," *Engineering Fracture Mechanics,* Vol. 18, No. 3, 1983, pp. 577–593.
[13] de Castro, J. T. P. and Parks, D. M., "Decrease in Closure and Delay of Fatigue Crack Growth in Plane Strain," *Scripta Metallurgica,* Vol. 16, 1982, pp. 1443–1446.
[14] Fleck, N. A., Smith, I. F. C., and Smith, R. A., "Closure Behavior of Surface Cracks," *Fatigue of Engineering Materials and Structures,* Vol. 6, No. 3, 1983, pp. 225–239.
[15] Christensen, R. H., *Metal Fatigue,* McGraw-Hill, New York, 1959.
[16] Jones, R. E., "Fatigue Crack Growth Retardation After Single-Cycle Peak Overload in Ti-6Al-4V Titanium Alloy," *Engineering Fracture Mechanics,* Vol. 5, 1973, pp. 585–604.
[17] Schijve, J., "Fatigue Damage Accumulation and Incompatible Crack Front Orientation," *Engineering Fracture Mechanics,* Vol. 6, 1974, pp. 245–252.
[18] Powell, P. M., Forsyth, P. J. E., and Williams, T. R. G., "The Effect of Overloading on Fatigue Crack Propagation in Two Aluminium Alloys and an Austenitic Stainless Steel," *Advances in Fracture Research, Proceedings,* Fifth International Conference on Fracture, Cannes, France, D. François et al., Eds., Vol. 4, Pergamon Press, Oxford, U.K., 1981, pp. 1781–1788.
[19] Schijve, J. and Broek, D., *Aircraft Engineering,* Vol. 34, 1962, p. 314.
[20] Banasiak, D. H., Grandt, A. F., Jr., and Montulli, L. T., "Fatigue Crack Retardation in Polycarbonate, "*Journal of Applied Polymer Science,* Vol. 21, 1977, pp. 1297–1309.
[21] Taylor, D. and Knott, J. F., private communication, results reported in a preliminary form in the Ph.D. thesis of Braid, J. E. M., "Fatigue Crack Propagation in Residual Stress Fields," Department of Metallurgy and Materials Science, Cambridge University, U.K., Aug. 1982.
[22] Rice, J. R., "Mechanics of Crack Tip Deformation and Extension by Fatigue," *Fatigue Crack Propagation, ASTM STP 415,* American Society for Testing and Materials, Philadelphia, 1967, pp. 247–311.
[23] Cameron, A. D. and Smith, R. A., "Upper and Lower Bounds for the Lengths of Non-Propagating Cracks," *International Journal of Fatigue,* Jan. 1981, pp. 9–15.
[24] Schijve, J., "The Effect of Pre-Strain on Fatigue Crack Growth and Crack Closure," *Engineering Fracture Mechanics,* Vol. 8, 1976, pp. 575–581.
[25] Legris, L., El Haddad, M. H., and Topper, T. H., "The Effect of Cold Rolling on the Fatigue Properties of a SAE 1010 Steel," *Materials, Experimentation and Design in Fatigue, Proceedings,*

Fatigue '81, Society of Environmental Engineers Conference, 24–27 March, 1981, F. Sherratt and J. B. Sturgeon, Eds., Westbury House, 1981.
[26] Stephens, R. I., Chen, D. K., and Hom, B. W., "Fatigue Crack Growth with Negative Stress Ratio Following Single Overloads in 2024-T3 and 7075-T6 Aluminium Alloys," *Fatigue Crack Growth Under Spectrum Loads, ASTM STP 595,* American Society for Testing and Materials, Philadelphia, 1976, pp. 27–40.
[27] Schijve, J., "The Effect of an Irregular Crack Front on Fatigue Crack Growth," *Engineering Fracture Mechanics,* Vol. 14, 1981, pp. 467–475.
[28] Bucci, R. J., Thakker, A. B., Sanders, T. H., Sawtell, R. R., and Staley, J. T., "Ranking 7XXX Aluminium Alloy Fatigue Crack Growth Resistance Under Constant Amplitude and Spectrum Loading," *Effect of Load Spectrum Variables on Fatigue Crack Initiation and Propagation, ASTM STP 714,* American Society for Testing and Materials, Philadelphia, 1980, pp. 41–78.
[29] Fleck, N. A., "Fatigue Crack Growth Due to Periodic Underloads and Periodic Overloads," *Acta Metallurgica,* Vol. 33, 1985, pp. 1339–1354.
[30] Vecchio, R. S., Hertzberg, R. W., and Jaccard, R., "On the Overload-Induced Fatigue Crack Propagation Behavior in Aluminium and Steel Alloys," to be published in *Fatigue of Engineering Materials and Structures.*
[31] Suresh, S., "Crack Deflection: Implications for the Growth of Long and Short Fatigue Cracks," *Metallurgical Transactions A,* Vol. 14A, Nov. 1983, pp. 2375–2385.
[32] Congleton, J., "Stress Corrosion Cracking and Hydrogen Embrittlement of Iron Base Alloys," National Association of Corrosion Engineers, 1978, p. 473.
[33] Wheeler, O. E., "Spectrum Loading and Crack Growth," *Journal of Basic Engineering, Transactions,* American Society of Mechanical Engineers, Vol. 4, March 1972, pp. 181–186.
[34] Willenborg, J., Engle, R. M., and Wood, H. A., "A Crack Growth Retardation Model Using an Effective Stress Concept," Technical Memorandum 71-1-FBR, Air Force Flight Dynamics Laboratory, Wright Patterson Air Force Base, OH, Jan. 1971.
[35] Błazewicz, W., Results reported in Schijve, J., "Lecture II—Fatigue Cracks, Plasticity Effects and Crack Closure," *Engineering Fracture Mechanics,* Vol. 11, 1979, pp. 182–196
[36] Tanaka, K., Matsuoka, S., Schmidt, V., and Kuna, M., "Influence of Specimen Geometry on Delayed Retardation Phenomena of Fatigue Crack Growth in HT80 Steel and A5083 Aluminium Alloy," *Advances in Fracture Research, Proceedings,* Fifth International Conference on Fracture, Cannes, France, D. François et al., Eds., Pergamon Press, Oxford, U.K., Vol. 4, 1981, pp. 1789–1798.
[37] Paris, P. C. and Herman, L., "Twenty Years of Reflection on Questions Involving Fatigue Crack Growth," *Proceedings,* First International Symposium on Fatigue Thresholds, Stockholm, J. Bäcklund et al., Eds., EMAS Publications, Warley, U.K., Vol. 1, 1982, pp. 3–32.
[38] Chanani, G. R. and Mays, B. J., "Observation of Crack-Closure Behavior after Single Overload Cycles in 7075-T6 Single-Edge-Notched Specimens," *Engineering Fracture Mechanics,* Vol. 9, 1977, pp. 65–73.
[39] Fleck, N. A., "The Use of Compliance and Electrical Resistance Techniques to Characterise Fatigue Crack Closure," Report CUED/C/MATS/TR.89, Cambridge University Engineering Department, U.K., Jan. 1982.
[40] Fleck, N. A., "An Investigation of Fatigue Crack Closure," Report CUED/C/MATS TR.104 (Ph.D. Thesis), Cambridge University Engineering Department, U.K., May 1984.
[41] Fleck, N. A. and Smith, R. A., "Crack Closure—Is It Just a Surface Phenomenon?" *International Journal of Fatigue,* July 1982, pp. 157–160; correction in *International Journal of Fatigue,* Oct. 1982, p. 243.
[42] Kikukawa, M., Jono, M., and Tanaka, K., "Fatigue Crack Closure Behaviour at Low Stress Intensity Level," *Proceedings,* Second International Conference on Mechanical Behavior of Materials, Boston, 1976, pp. 254–277.
[43] Brown, C. W., "The Initiation and Growth of Fatigue Cracks in Titanium Alloys," Ph.D. Thesis, Department of Metallurgy and Materials Science, Cambridge University, U.K., 1980.
[44] Brown, R. D., and Weertman, J., "Effects of Tensile Overloads on Crack Closure and Crack Propagation Rates in 7050 Aluminium, *Engineering Fracture Mechanics,* Vol. 10, 1978, pp. 867–878.
[45] Shin, C. S. and Fleck, N. A., "Overload Retardation in Structural Steel," *Fatigue and Fracture of Engineering Materials and Structures*, Vol. 9, 1987, pp. 379–393.
[46] Gan, D. and Weertman, J., "Crack Closure and Crack Propagation Rates in 7050 Aluminium," *Engineering Fracture Mechanics,* Vol. 15, No. 1-2, 1981, pp. 87–106.
[47] Gan, D. and Weertman, J., "Fatigue Crack Closure after Overload," *Engineering Fracture Mechanics,* Vol. 18, No. 1, 1983, pp. 155–160.

[48] Nowack, H., Trautmann, K. H., Schulte, K., and Lütjering, G., "Sequence Effects on Fatigue Crack Propagation; Mechanical and Microstructural Contributions," *Fracture Mechanics, ASTM STP 677,* American Society for Testing and Materials, Philadelphia, 1979, pp. 36–53.
[49] Trebules, V. W., Jr., Roberts, R., and Hertzberg, R. W., "Effect of Multiple Overloads on Fatigue Crack Propagation in 2024-T3 Aluminium Alloy," *Progress in Flaw Growth and Fracture Toughness Testing, ASTM STP 536,* American Society for Testing and Materials, Philadelphia, 1973, pp. 115–146.
[50] Allison, J. E., "Measurement of Crack-Tip Stress Distributions by X-Ray Diffraction," *Fracture Mechanics, ASTM STP 677,* American Society for Testing and Materials, Philadelphia, 1979, pp. 550–562.
[51] Taira, S. and Tanaka, K., "Local Residual Stress Near Fatigue Crack Tip," *Transactions,* Iron and Steel Institute of Japan, Vol. 19, 1979, pp. 411–418.
[52] Smith, M. C. and Smith, R. A., "The Formation of Spherical Wear Debris in Mode II Fatigue Cracks," *Wear*, Vol. 76, 1982, p. 105.
[53] Lankford, J. and Davidson, D. L., "Wear Debris Due to Mode II Opening of Mode I Fatigue Cracks in an Aluminium Alloy," *Metallurgical Transactions A,* Vol. 14A, June 1983, pp. 1227–1230.
[54] Suresh, S. and Vasudevan, A. K., "Application of Fatigue Threshold Concepts to Variable Amplitude Crack Propagation," *Proceedings,* International Symposium on Fatigue Crack Growth Threshold Concepts, Philadelphia, Oct. 3–5, 1983, D. L. Davidson and S. Suresh, Eds., The Metallurgical Society of the American Institute of Mining, Metallurgical and Petroleum Engineers, 1983, pp. 361–378.
[55] Hertzberg, R. W. and Mills, W. J., "Character of Fatigue Fracture Surface Micromorphology in the Ultra-Low Growth Rate Regime," *Fractography—Microscopic Cracking Processes, ASTM STP 600,* American Society for Testing and Materials, Philadelphia, 1976, pp. 220–234.
[56] Dugdale, D. S., "Yielding of Steel Sheets Containing Slits," *Journal of the Mechanics and Physics of Solids,* Vol. 8, 1960, pp. 100–104.

Takeshi Kunio,[1] Kunihiro Yamada,[1] and Min Gun Kim[1]

Critical Behavior of Nonpropagating Crack in Steel

REFERENCE: Kunio, T., Yamada, K., and Kim, M. G., **"Critical Behavior of Nonpropagating Crack in Steel,"** *Basic Questions in Fatigue: Volume I, ASTM STP 924,* J. T. Fong and R. J. Fields, Eds., American Society for Testing and Materials, Philadelphia, 1988, pp. 184–193.

ABSTRACT: A study has been made on the critical condition for the nonpropagation of a crack in Stage II growth at the endurance limit of pearlitic steel with special emphasis on the closure behavior of the crack.

It is found from fatigue experiments and measurements of crack opening displacement that a tip of the nonpropagating crack (NPC) is closed at the stress level of the endurance limit. It is also found that the tip of the NPC opens again under the original endurance limit when the NPC is vacuum-annealed. Therefore, it turns out that the crack which has propagated in Stage II growth stops its propagation by the closure effect resulting mainly from the local residual compressive stress at the crack tip.

Further experimental evidence implies that not only the crack closure due to residual compressive stress (plasticity induced), but also the closure associated with the oxidation during the reduced crack propagation, is responsible for the existence of NPC at the endurance limit.

KEY WORDS: fatigue, microcrack, pearlitic steel, endurance limit, nonpropagating crack, crack closure, residual compressive stress, oxidation, threshold condition

There remain some unknown properties concerning the threshold behavior of short cracks [*1,2*] at a stress level of the endurance limit. This might be due to the fact that there is no appropriate method to evaluate the threshold behavior of short cracks in contrast with that of large cracks, and the fact that the microstructure-sensitive properties of propagation of such short cracks [*3*] make it difficult to characterize the relationship between microstructural parameters and mechanistic parameters.

Recently the authors reported that for plain carbon steel the level of the endurance limit of smooth specimens can be determined by the critical stress for the onset of growth of the nonpropagating crack (NPC), having a relatively large crack length of 200 to 300 μm in Stage II growth, rather than a length of single grain size order [*4,5*].

There still remain some uncertainties on the threshold condition for the cracks, that is, the nonpropagation of this relatively large-size crack subjected to constant stress amplitude.

The question arises: What is the principal mechanism responsible for a nonpropagating crack?

In this paper, a study has been made of the experimental approach to explain the above question, with special emphasis on the crack-tip closure behavior [*6–9*] of NPC in smooth specimens of carbon steel.

[1] Professor, associate professor, and graduate student, respectively, Department of Mechanical Engineering, Keio University, 3-14-1 Hiyoshi, Kohoku-ku, Yokohama 223, Japan.

TABLE 1—*Chemical composition of material, weight %.*

C	Si	Mn	P	S	Cu	Ni	Cr
0.84	0.22	0.45	0.014	0.027	0.03	0.02	0.10

Material and Experimental Procedures

The material employed was pearlitic steel of 0.84% carbon with chemical composition as given in Table 1. This was annealed at 1200°C for 2 h to give a decarburized ferrite domain in a surface microstructure providing a soft spot for crack nucleation. A micrograph of the material is shown in Fig. 1. The material was machined into solid hourglass-shape specimens having a 9-mm diameter and a 20-mm radius of curvature at the gage length so as to make microscopic observation easier. The stress-concentration factor of this geometry is approximately 1.06, so these specimens may be regarded as smooth test pieces. Every specimen was then vacuum-annealed at 640°C for 1 h and electropolished approximately 10 μm to eliminate residual stress and strain due to machining. Rotating bending fatigue tests were carried out on these specimens with a frequency of 48 Hz.

The mechanical properties and microstructural parameters of the specimen are given in Table 2.

Crack opening displacement was measured by the change of the distance of two micro-Vickers indentation marks spanning the crack [*10*]. The distance *d* was obtained from the projection (×100) of photographic negatives taken at ×400 magnification (Fig. 2). A measurement of crack length was made along the direction perpendicular to the specimen axis by use of an optical microscope.

Results and Discussion

Crack Opening or Closure Behavior or Both Associated with Stress Release Annealing

An endurance limit was obtained as σ_{w0} = 210 MPa for the present material under rotating bending fatigue tests.

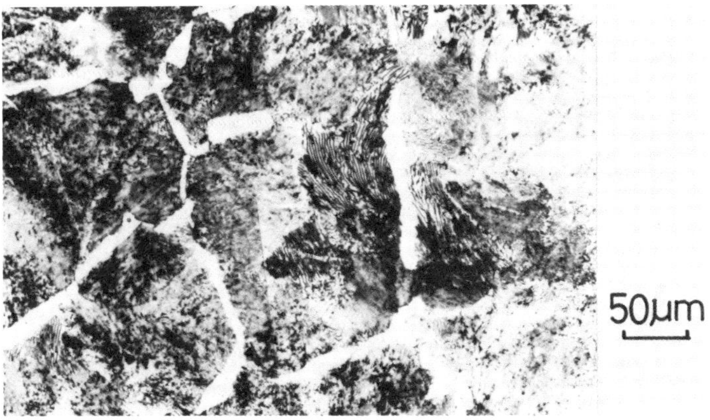

FIG. 1—*Typical feature of surface microstructure.*

TABLE 2—*Mechanical properties and microstructural parameters.*

Volume fraction of ferrite at surface, %	6
Total decarburized depth, mm	0.95
Pearlite interlamellar spacing, μm	0.54
Micro-Vickers hardness of ferrite, 20 g	150
Micro-Vickers hardness of pearlite, 50 g	257
0.2% proof stress, MPa : σ_y	277
Ultimate tensile strength, MPa	714
Endurance limit, MPa : σ_{w0}	210

Since the validity of the concept of linear fracture mechanics is not assured in evaluating the threshold behavior of surface microcracks [*1,11*], the length of the critical NPC was estimated in the following manner.

A number of precracked specimens having a wide range of crack lengths were prepared by fatigue at a stress level 10% above the endurance limit and were restressed in fatigue at the original endurance limit, σ_{w0} = 210 MPa, in order to evaluate the maximum allowable crack length which can survive after 10^7 stress cycles. It has already been mentioned in previous papers concerning medium-carbon steel [*4,5*] that the maximum allowable crack length obtained from the above procedure coincides with the critical NPC length. From this we have concluded in our experiment that the influence of precracking stress level, that is, 10% above the endurance limit, can be negligible in an evaluation of NPC length. Results obtained are shown in Fig. 3. The maximum allowable crack length was found to be about 340 μm from this figure and this length can be regarded as the critical NPC at the endurance limit. It is noted that this length involves a few tens of micrometres of propagation from the precrack. Nonpropagation of this crack has been confirmed by an application of an additional 10^7 cycles of the same stress amplitude.

For the purpose of examining the critical NPC behavior, the specimens having such an NPC were vacuum-annealed at 640°C for 1 h and the surface was slightly electropolished to eliminate slip markings ahead of the NPC. After this annealing and electropolishing, the

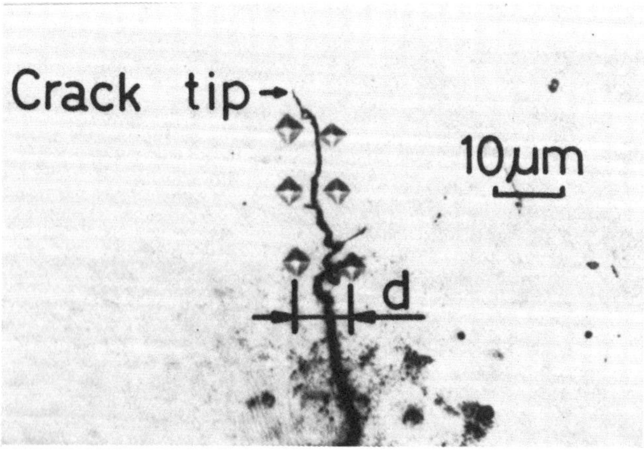

FIG. 2—*Micro-Vickers indentation marks which were used for the measurements of crack opening displacement.*

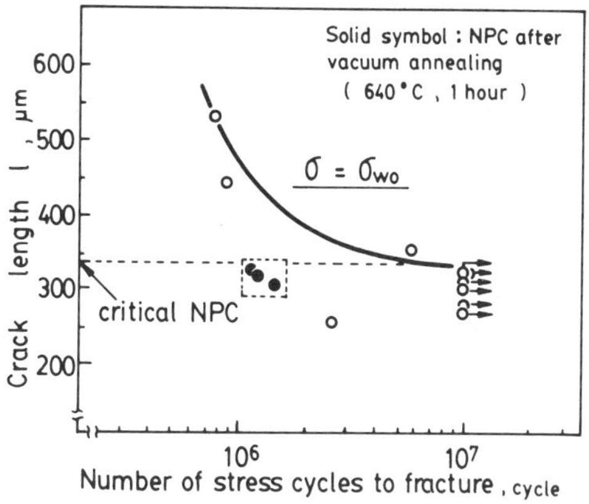

FIG. 3—*Evaluation of the maximum allowable crack length (that is, the critical NPC) at the endurance limit employing precracked specimens prepared at 10% above the original endurance limit.*

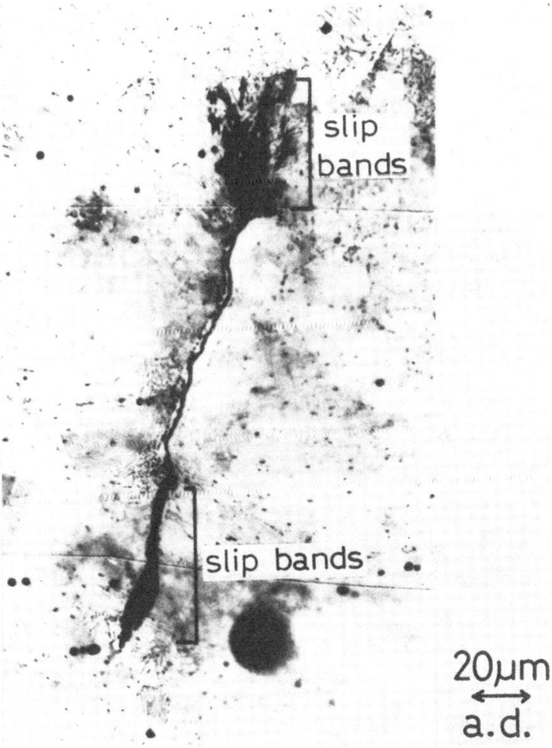

FIG. 4—*Formation of new slipbands appeared at the crack tip of vacuum-annealed specimen:* $\sigma = \sigma_{w0}$, N = *5* $\times$ *10*^{5}, *length = 90* μ*m.*

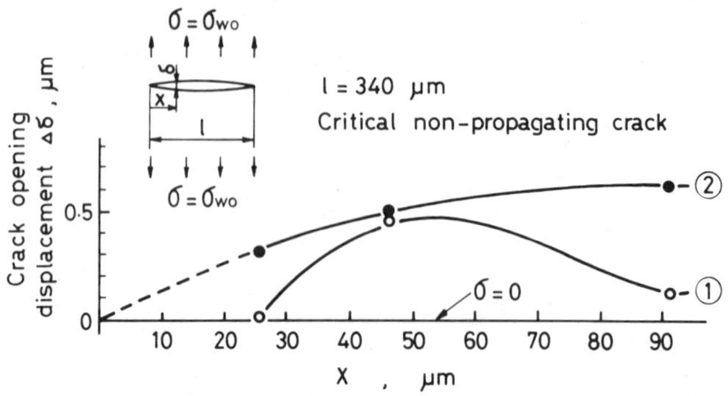

FIG. 5—*Crack opening displacement in the vicinity of the NPC before* (curve ①), *and after vacuum-annealing treatment* (curve ②).

specimens were again fatigue tested at a stress level equal to the original endurance limit. The crack, which had been one of the NPC's, began its propagation accompanying a formation of new slipbands at the crack tip and led to a final fracture of the specimen after approximately 10^6 stress cycles. These results are also denoted in Fig. 3 by three solid symbols. In addition, it should be noted that the onset of new propagation from the NPC occurs only after the vacuum-annealing treatment irrespective of the length of NPC's.

Figure 4 shows an example of formation of new slipbands observed at the annealed crack of length 90 μm after fatigue stressing of 5×10^5 at the stress level of the endurance limit.

On the other hand, the specimen which had the NPC of 138 μm was electropolished to eliminate slipband markings and was fatigue tested again under the stress level of endurance limit to observe whether the NPC began to propagate or not. After 10^7 stress cycles, no slipbands have been nucleated at the crack tip and no propagation of crack has been observed; that is, the crack remained as a nonpropagating one.

These evidences indicate that the vacuum-annealing treatment on the specimen having NPC's turns the NPC into a propagating crack and suggest that this treatment changes the tip of the crack from closed to open.

Measurements of Crack Opening Displacement

The change of crack opening displacement of the particular NPC of 340 μm was compared before and after annealing treatment on this crack under static bending stress equivalent to the endurance limit. The results are given in Fig. 5. The curve marked ① in the figure

TABLE 3—*Micro-Vickers hardnesses of pearlitic structure before and after vacuum-annealing.*

50 g, 100 points	
Loading and Treatment	Hv
① Before loading	257
② After loading at $\sigma = \sigma_{w0}$ for 10^6	261
③ Vacuum annealing of ②	237
④ After loading at $\sigma = 1.1\sigma_y$ for 10^4	255

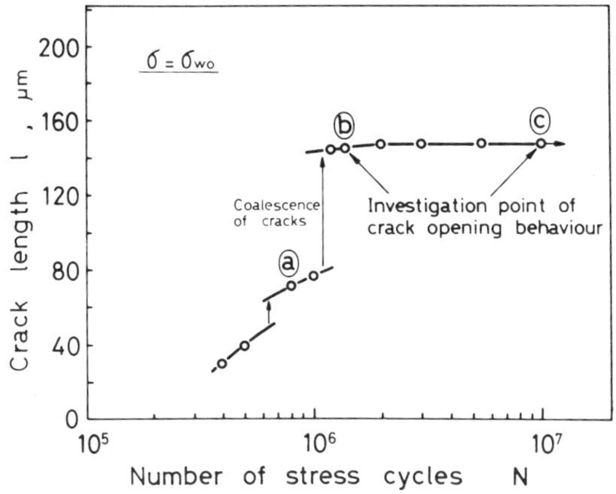

FIG. 6—*Typical behavior of crack propagation on the surface of smooth specimen under constant stress amplitude.*

indicates the results of NPC, that is, before annealing, while the curve marked ② indicates those after annealing. It is found from this figure that the tip of this NPC, which was closed after $N = 10^7$ of stress cycles, opens even under the stress level of endurance limit when the NPC is vacuum-annealed.

In addition, micro-Vickers hardness measurements as shown in Table 3 revealed that there are no appreciable differences in the pearlitic structure of a fatigued specimen before and after the annealing treatment. This implies that the simple effects of strain hardening are negligible for the existence of NPC mentioned above.

Therefore, the question raised at the beginning of the paper can be answered in the following manner:

The crack, once propagated under a constant stress amplitude, stops due to the closure at the crack tip, which resulted mainly from the local residual compressive stress associated with the localized plasticity introduced during cyclic loading.

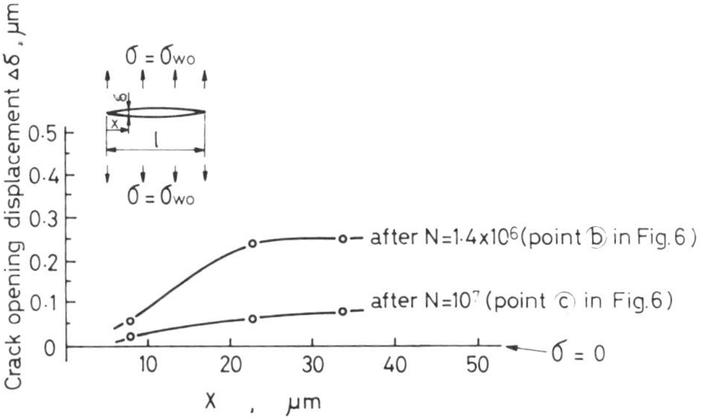

FIG. 7—*Change in crack opening displacement during fatigue loading.*

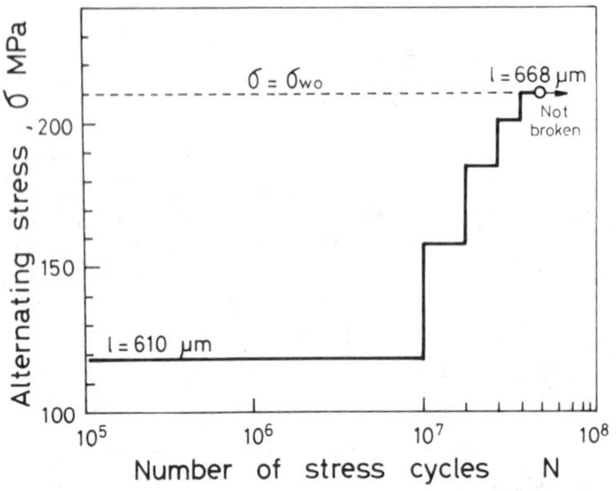

FIG. 8—*Example of stepwise loading sequence on the fatigue precracked specimen.*

Crack Tip Closure Associated with the Reduced Rate of Crack Propagation Near Threshold

Figure 6 illustrates typical behavior of a propagation curve of the crack which finally became an NPC under the constant stress amplitude of endurance limit. Measurements of crack-tip opening displacement were made on the NPC at a point marked ⓒ and also on the crack at the onset of growth into the pearlitic structure, the point marked ⓑ. Results are given in Fig. 7. It is found from these measurements that the crack tip is closed at point ⓒ but is still open at point ⓑ. Since the crack length at points ⓑ and ⓒ are 146 μm ($N = 1.4 \times 10^6$) and 148 μm ($N = 10^7$), respectively, the transition from crack opening to closure must have occurred during a small increment of crack length such as 2 μm with a consumption of the order of 10^6 cycles of stress in this experiment. This implies that the crack closure behavior can be related to the considerably reduced rate of crack propagation (e.g. $10^{-8} \sim 10^{-9}$ mm/cycle) in the pearlitic structure during the stress cycles between 1×10^6 and 10×10^6.

In order to examine whether or not the reduced rate of propagation is related to crack closure, an experimental consideration is offered in the following manner.

A loading sequence as shown in Fig. 8 indicates a stepwise incremental loading with each step of $N = 10^7$ cycles. This sequence was quite effective to realize either the reduced rate of crack propagation or no propagation of the crack. It is found from this experiment that

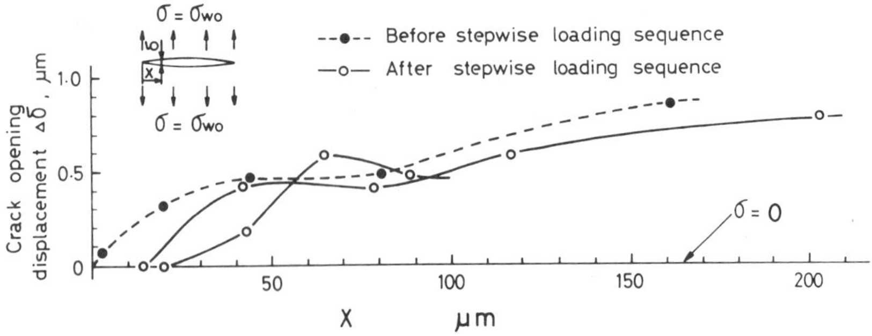

FIG. 9—*Crack opening displacement of the precracked specimen before and after the stepwise incremental loadings.*

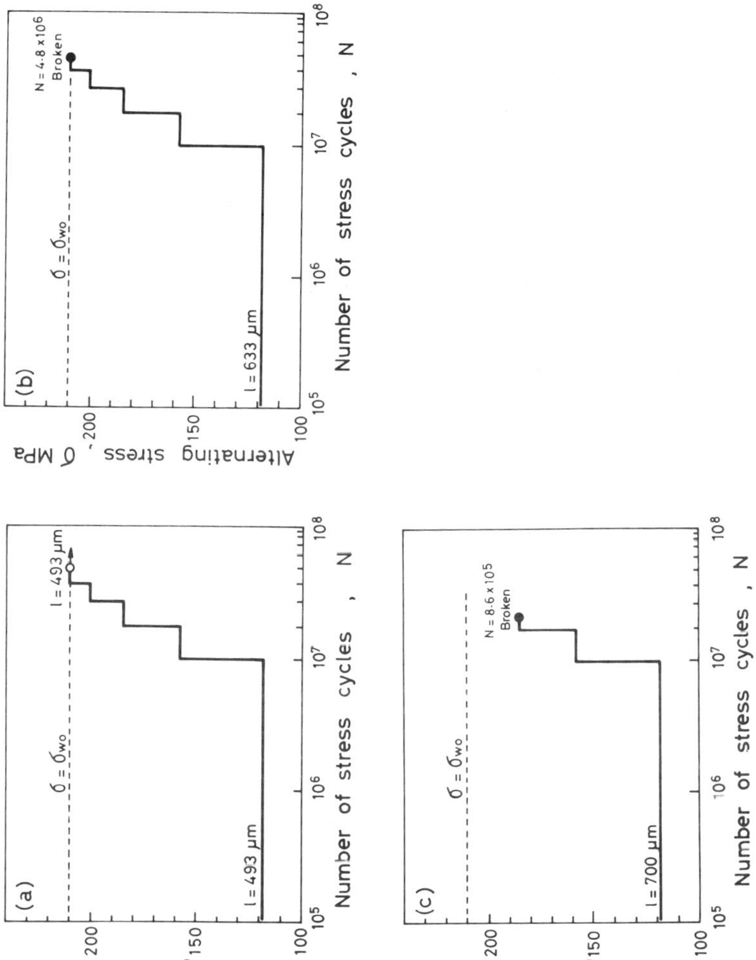

FIG. 10—*Stepwise loading sequence on the precracked specimens with different crack lengths.*

the crack of length 610 μm became a nonpropagating one with a small increment of crack length after applying the above loading sequence. Since this crack is approximately twice as long as the critical NPC (340 μm), it would not become NPC under the stress level of the endurance limit unless the above loading history is applied.

The crack opening displacement was measured for this particular crack after a final loading sequence as shown in Fig. 9. Two solid lines with open symbols in the figure represent the results obtained from both ends of the crack. Thus, the tips of this crack were found closed. The broken line with solid symbols in the figure represents the crack opening displacement of the same type of precrack of 650 μm before the stepwise loading sequence. The crack tip opens in this case.

It is confirmed from this result that the tip of the crack which had been initially opened became closed after the above particular stepwise loading sequence.

Effect of Oxidation on the Closure

In order to examine the effect of stepwise incremental loading on the nonpropagation of the crack, an additional three specimens with crack lengths 493, 633, and 700 μm, respectively, were fatigue tested with the same loading sequences as for the 610-μm crack. Results are given in Fig. 10. It is found from these figures that the stepwise loading sequence is quite effective for the nonpropagation of cracks shorter than approximately 600 μm.

Since this stepwise loading sequence involves five steps of 10^7 loading sequences, which corresponds to approximately 300 h of fatigue loading, this period would be enough to introduce an oxidation effect due to ambient air on the crack propagation [7].

Therefore, in order to examine the effect of oxidation on crack propagation, the same loading sequence was applied to the crack of 357 μm which was prevented from exposure to ambient air by use of plastic replication films of 0.06 mm thickness. Results are shown in Fig. 11. This figure shows that even a smaller crack than 600 μm could not become nonpropagating crack under the above experimental condition.

This may imply that oxidation at the crack tip during the stepwise loading sequence is associated with the critical condition for the crack closure, that is, the existence of NPC.

Consequently, it may be concluded that not only the plasticity-induced closure, but also the closure associated with oxidation [7] during the reduced crack propagation, is responsible for the existence of NPC at the endurance limit.

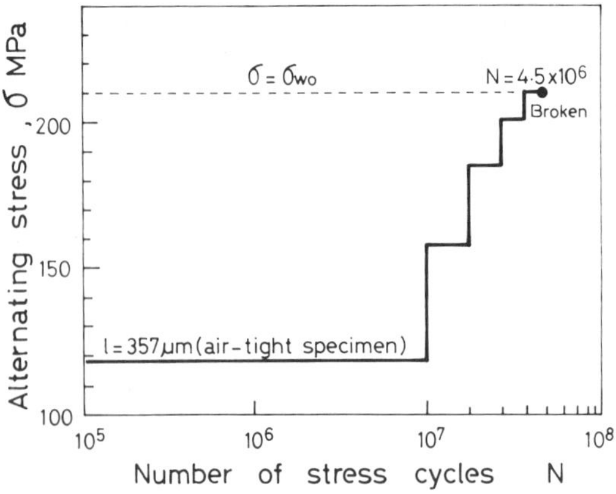

FIG. 11—*Stepwise loading sequence of precracked specimen under airtight conditions.*

Conclusion

A study has been made on the threshold condition for a microcrack at the endurance limit of pearlitic steel with special emphasis on the closure behavior of the critical nonpropagating crack. Results obtained are summarized as follows.

1. The crack, once propagated in Stage II growth under a constant stress amplitude, stops due to the suppression of opening at the crack tip, that is, the crack closure.

2. It may be concluded that not only a plasticity-induced closure, but also the closure associated with oxidation during reduced crack propagation near the threshold, is responsible for nonpropagation of the microcrack.

Acknowledgments

We wish to thank Professor Masao Shimizu for his discussions and encouragement of the present work.

References

[*1*] Kitagawa, H., Takahashi, S., Suh, C. M., and Miyashita, S., "Quantitative Analysis of Fatigue Process—Microcracks and Slip Lines Under Cyclic Strains," *Fatigue Mechanisms,* Proceedings of an ASTM-NBS-NSF symposium, Kansas City, Mo., May 1978, J. T. Fong, Ed., *ASTM STP 675*, American Society for Testing and Materials, Philadelphia, 1979, pp. 420-449.
[*2*] Hudak, S. J., *Journal of Engineering Materials and Technology,* Vol. 103, 1981, p. 26.
[*3*] Kunio, T. et al. in *Proceedings*, International Conference on Fatigue Thresholds, Stockholm, Vol. 1, 1981, p. 409.
[*4*] Kunio, T. and Yamada, K., "Microstructural Aspects of the Threshold Condition for Nonpropagating Fatigue Cracks in Martensitic-Ferritic Structures," *Fatigue Mechanisms,* Proceedings of an ASTM-NBS-NSF symposium, Kansas City, Mo., May 1978, J. T. Fong, Ed., *ASTM STP 675*, American Society for Testing and Materials, Philadelphia, 1979, pp. 342-370.
[*5*] Kunio, T. et al. in *Proceedings*, Conference on Fatigue, The University of Birmingham, Birmingham, U.K., 1984, p. 817.
[*6*] Elber, W. in *Significance of Fatigue Crack Closure, ASTM STP 486*, American Society for Testing and Materials, Philadelphia, 1970, p. 230.
[*7*] Surresh, S. et al., *Metallurgical Transactions*, Vol. 12A, 1981, p. 1435.
[*8*] Minakawa, K. and McEvily, A. T., *Scripta Metallurgica*, Vol. 15, 1981, p. 633.
[*9*] Kobayashi, H., *Bulletin of the Japan Institute of Metals*, Vol. 21, No. 5, 1982, p. 329.
[*10*] Kim, M. G. et al., *Transactions*, Japan Society of Mechanical Engineers, Vol. 51, No. 461, 1985, p. 10.
[*11*] McClintock, F. A. and Irwin, G. R. in *Symposium on Fracture Toughness and Its Application, ASTM STP 381*, American Society for Testing Materials, Philadelphia, 1965, p. 84.

E. R. de los Rios,[1] *M. W. Brown,*[1] *K. J. Miller,*[1] *and H. X. Pei*[1]

Fatigue Damage Accumulation During Cycles of Nonproportional Straining

REFERENCE: de los Rios, E. R., Brown, M. W., Miller, K. J., and Pei, H. X., **"Fatigue Damage Accumulation During Cycles of Nonproportional Straining,"** *Basic Questions in Fatigue: Volume I, ASTM STP 924,* J. T. Fong and R. J. Fields, Eds., American Society for Testing and Materials, Philadelphia, 1988, pp. 194–213.

ABSTRACT: Nonproportional biaxial fatigue tests were performed using cruciform-shaped specimens of AISI 316 stainless steel at room temperature, in order to study the effect of nonproportional straining on crack growth rate. While the fluctuating stress applied perpendicular to the crack was common to all the tests, including the uniaxial test, the stress parallel to the crack was applied for only half the former period and only on either the raising or the downward part of the cycle.

The crack growth rate of the specimen subjected to uniaxial loading was more than five times faster than the growth rate of the cracks in the biaxial specimen, which is contrary to the reduced endurance observed for nonproportional loading of 1% chromium-molybdenum-vanadium steel in tension-torsion.

Transmission electron microscopy observations of regions next to the fracture surface showed a very different dislocation substructure relating to uniaxial and biaxial loading. The degree of strain hardening in both conditions was estimated by employing dislocation models suitable for each case. If a "slipping off" mechanism for crack propagation is adopted, then the difference in crack growth rate experienced by the biaxial specimens can be explained by considering the magnitude of the strength of the slipband in each case.

KEY WORDS: biaxial stresses, cyclic loads, crack growth rate, fatigue (materials), plastic deformation, dislocation structures, fractography

Fatigue life data collected in the laboratory or under service loading situations are conventionally collated and interpreted in terms of component lifetime. Even in variable-amplitude loading, the various cycle counting techniques that have been postulated use the successive peak values or turning points of stress or strain in order to assess the fatigue damage accrued. Although this is universally accepted as the correct way to categorize fatigue loading cycles (probably for historical reasons), this method of analysis makes an implicit assumption that the damage caused by each loading cycle or each reversal is uniquely described by the peak strain values.

However, the mechanics of fatigue crack extension, while clearly being heavily dependent on stress or strain amplitude, may also depend on the strain path taken between successive peak values. Indeed this will probably be the case of ductile materials where crack extension is a gradual process, building up throughout the cycle, as opposed to an instantaneous jump in crack length on achieving the peak strain value. This distinction between steady growth and a sudden growth step may be academic in most laboratory fatigue tests conducted under

[1] University of Sheffield, Department of Mechanical Engineering, Mappin St., Sheffield, S1 3JD, U.K. Co-author Pei is a visitor from the Academia Sinica in Shenyang, China.

uniaxial loading—for example, center-cracked panel, compact tension, single-edge notch bend, push-pull, and rotating bend type tests—since the stress/strain loading path is fully defined for the single loading mechanism by the stress amplitude and mean stress level. For multiaxial loading situations, however, nonproportional loading paths can be readily devised that do not concur with a proportional path, although they show the same amplitude and mean stress levels. Despite the need for more general multiaxial data, nonproportional loading is very important since it simulates more closely many practical cases. Very limited experimental results are available in the literature [*1–5*], though the topic has been reviewed recently [*3,6*].

The presentation of experimental data has proved to be a problem, particularly the selection of appropriate parameters which would best correlate the results [*2*]. Some workers used bulk cyclic deformation parameters; for example, Taira et al. [*7*] integrated the octahedral shear strain through a loading cycle, while Dietmann and Issler [*4*] considered all the stresses on the octahedral plane both alternating and mean. Zamrik and Frishmuth [*5*] found that a parameter they called "total strain," defined by

$$\epsilon_t = (\epsilon_1^2 + \epsilon_2^2 + \epsilon_3^2)^{1/2} \tag{1}$$

gave the minimum amount of scatter. Note that ϵ_t is also equivalent to the octahedral shear strain for proportional cycling, for Poisson's ratio equal to 0.5. Garud [*8*], likewise, proposed the plastic work per cycle, and subsequently a modified version of it defined for tension-torsion loading by

$$W_p^* = \oint [\sigma \, d\epsilon_p + 0.5\tau \, d\gamma_p] \tag{2}$$

where σ/ϵ and τ/γ are the axial and torsional stress/strain, respectively. This was shown to improve the correlation of experimental data. However, in almost all the cases above, only one material was examined.

All the work cited previously has one drawback; it does not consider the crack. Fatigue is a fracture process involving the initiation and propagation of cracks; therefore the selection of the critical parameter should best be based on the physical processes of deformation and fatigue crack growth. Following this line of thought the authors have previously proposed that the two fundamental strains to consider in multiaxial fatigue, which govern the direction of crack growth and crack growth rate, are the maximum shear strain range and the normal strain on the plane of maximum shear [*9,10*]. This seems to agree very well with two established observations. First, the initiation of cracks is intimately related to the formation of persistent slipbands which are clearly observed in high-cycle fatigue when the stress level is sufficient to initiate slip only along the direction of maximum shear [*11*]; and second, even in Stage II crack growth when cracks grow on a plane normal to the maximum stress [*12*], the direction and size of the plastic regions developed ahead of the crack depend on the shear and normal strains on the maximum shear plane.

The exhaustion of ductility in a metal is related (in a tension test, for example) to the fracture stress. Similarly, in fatigue crack growth the fracturing of the material at the tip of the crack depends on its ductility and consequently on the normal stress on the plane of maximum shear.

This approach was shown to be applicable to out-of-phase cyclic loading, so that prediction not only of fatigue life, but also of the planes on which cracks will form, could be made [*3*].

Tension-Torsion Test Results

Full details of this work are given elsewhere [*13*]; here, only the results which are relevant to the present work are reported, being derived from tension-torsion tests on tubular specimens of a 1% chromium-molybdenum-vanadium (1Cr-Mo-V) steel at room temperature to determine low-cycle fatigue endurance.

Four types of strain cycle were used; see Fig. 1. All waveforms of the axial and torsional strain cycle were sinusoidal. For cycles of Type I, both torsional and axial sine waves had the same period while for cycles of Type II(*a*), (*b*), and (*c*) the period of the intermittent axial sine wave was one-tenth that of the torsional cycle. The phase difference is represented by ϕ for Type I cycles and by h for Type II cycles. The strain ratio, $\lambda = \gamma_a/\epsilon_a$, where γ_a and ϵ_a and the sine wave amplitudes for torsional and axial strain, was varied from 0 to infinity for Type I tests and kept constant at 2.14 for Type II tests. The values of λ and the phase function h were restricted in the Type II tests so that the peak torsional strain remained the undisputed point of reversal for the full load cycle, the axial strains affecting only the straining path.

When the experimental results were plotted in terms of the peak normal strain ϵ_{np} on the plane of maximum shear, all data being at 3% maximum shear strain range as in Ref *3*, the data from Type II cycles (solid symbols) consistently fell above the Type I data (open symbols) (see Fig. 2). This indicates an effect of straining path on fatigue life, when the straining path is changed from Type I to Type II. This is because for $\lambda = 2.14$, Type I tests with the phase angle $\phi = 90$ deg have identical strain ranges to the Type II(*a*) tests, for $-0.7 < h < 0.7$ [*13*], but the fatigue endurance for the Type II(*a*) nonproportional loading tests is double that observed for Type I in Fig. 2.

A further observation made by Jordan et al. [*13*] was that Type II(*a*), (*b*), and (*c*) cycles gave almost identical endurance, which implied that only one of the axial strain peaks had affected the behavior of fatigue cracks for Type II(*a*) and (*c*). The only consistent conclusion drawn was that fatigue damage had probably occurred only during the tension-going part of the cycle, that is, when the cracks were being opened, and that the straining path during

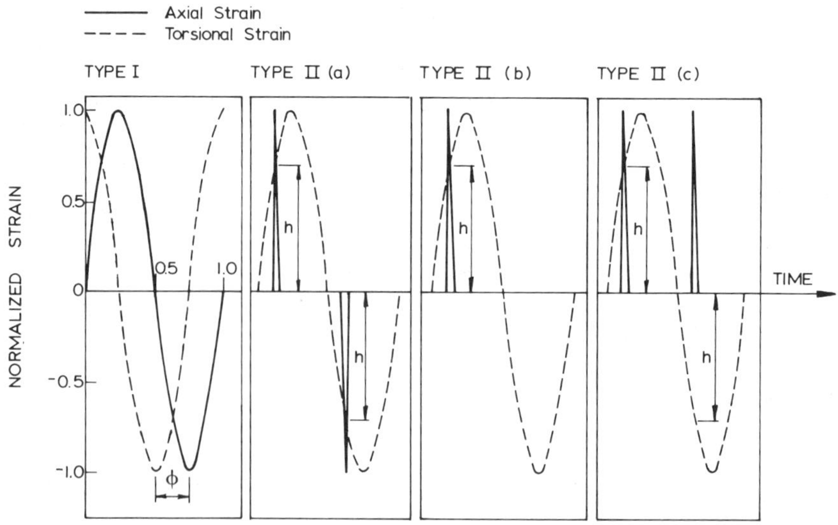

FIG. 1—*Definition of cyclic types, phase angle* ϕ*, and phase parameter* h *for tension-torsion tests (note for* $\phi = 90$ *deg,* h = *0).*

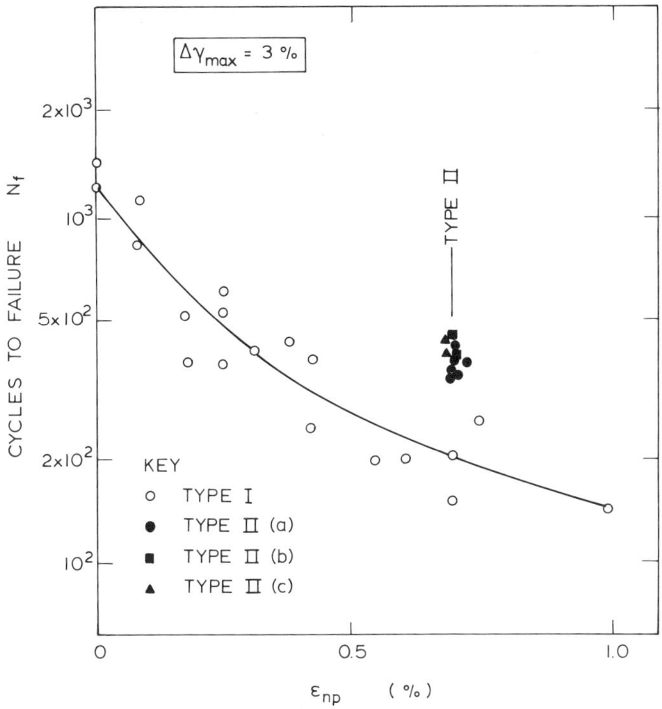

FIG. 2—*Fatigue life as a function of peak normal strain,* ϵ_{np}, *for* $\Delta\gamma_{max}$ = 3%.

this half of the cycle alone governed the rate of crack growth. It should be emphasized, however, that these are endurance test results, and low-cycle fatigue behavior reflects the propagation of small cracks. Therefore, for clarification of straining path effects, critical experiments need to be performed in which crack growth rate is measured directly. These are discussed in the following section.

To summarize the results on the bainitic 1Cr-Mo-V steel, the conclusions drawn were:

1. Fatigue endurance depends on the strain path followed in a cycle in addition to the multiaxial cyclic strain amplitudes, so that life cannot be predicted from the strain amplitude alone.
2. In low-cycle fatigue with reversed plasticity, the mean value of strain has no effect on life.
3. The variation of the normal strain on the maximum shear plane during the unloading half-cycle has no effect on endurance. Fatigue damage accrues during the crack opening half-cycle of shear strain.
4. The cracks developed were Stage I shear mode cracks, forming on the plane experiencing the greatest range of shear deformation.

Crack Propagation Studies

The observed strain path dependence has been investigated further using cruciform-shaped biaxially stressed specimens, measuring the growth of fatigue cracks directly and performing a thorough metallographic examination of the test specimens. The stress normal to the Mode

I cracks followed a simple cycle, identical in all tests, whereas the stress parallel to the crack varied but always returned to zero when the stress normal to the crack reached a turning point. The waveforms used were chosen to emphasize those parts of the cycle in which fatigue damage was believed to occur [*13*], Fig. 3, without altering either the applied stress amplitude or biaxial stress state at the critical turning points of the stress normal to the cracks. If the accumulation of fatigue damage occurs primarily during the tension part of the cycle, then the Type III cycle should exhibit faster crack growth than Type II, because the *x*-axis stress is applied during the tension-going portion of the cycle only. However, Types I and II in Fig. 3 should produce identical crack growth rates, as both have identical tension-going loading paths.

Experimental Details and Test Results

The material used in the present investigation was an AISI 316 stainless steel, Heat No. 8092297, from the Oak Ridge National Laboratory. The grain structure of the material was mixed (coarse grains intermingled with fine grains) and elongated in the direction of rolling, Fig. 4. The chemical composition, values of the mechanical properties, and characterization of the microstructure are given in Tables 1, 2, and 3, respectively.

Fatigue tests were performed in a servo-hydraulic biaxial test facility. Tensile or com-

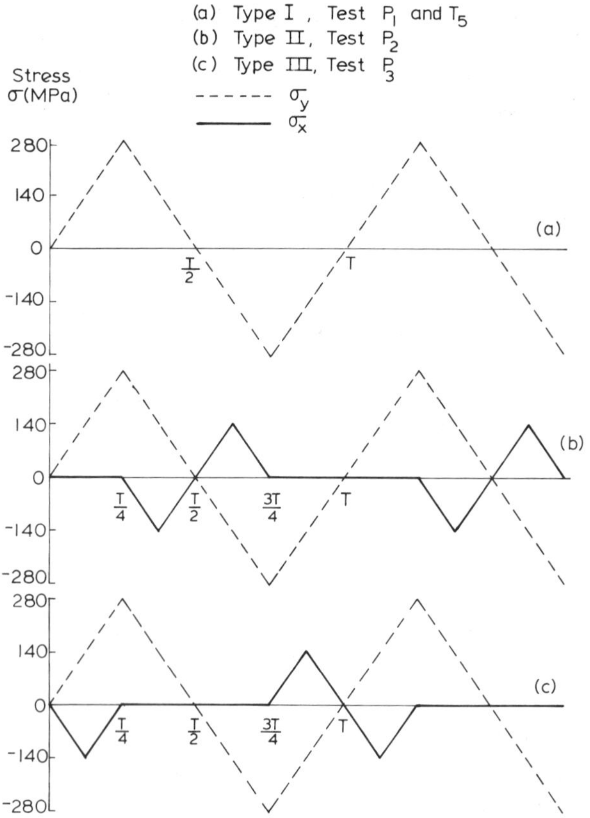

FIG. 3—*Stress cycles for cruciform specimens; frequency 0.1 Hz.*

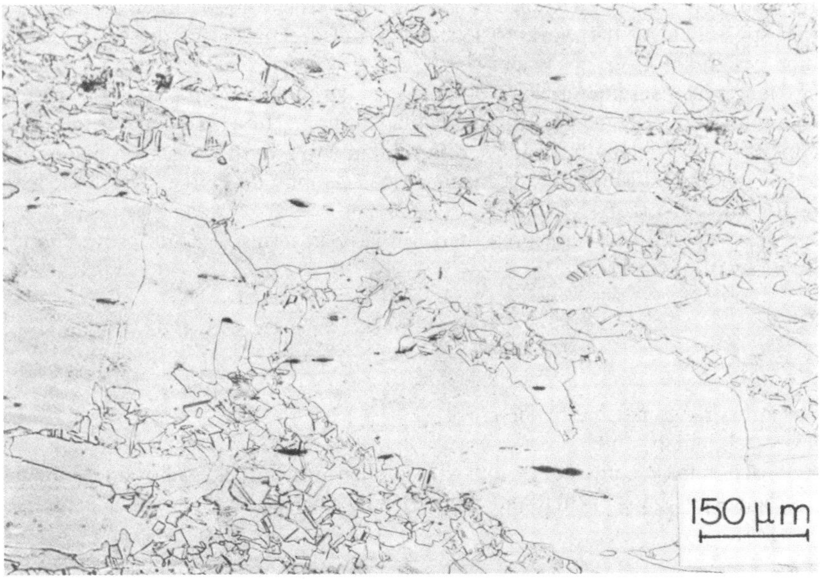

FIG. 4—*Material microstructure (rolling direction horizontal).*

TABLE 1—*Material chemical composition, weight %.*

C	Si	Mn	Cr	Ni	Mo	S	P
0.06	0.62	1.88	17.30	13.40	2.34	0.020	0.023

TABLE 2—*Room temperature tensile properties in rolling direction.*

0.2% Proof Stress MPa	Ultimate Tensile Strength, MPa	Elongation, %	Reduction of Area, %
393	609	57	75

TABLE 3—*Microstructure characterization.*

Mean Linear Intercept, $\bar{\ell}$, µm	Mean Diameter, $\bar{d} = 1.68\,\bar{\ell}$, µm	ASTM No.	Range of Grain Size, $\ell_{max} - \ell_{min}$, µm	$\frac{\ell_{max}}{\bar{\ell}}$	Aspect Ratio
31	52	6.6	92	3.66	1.75

pressive loads may be applied to each pair of arms of a cruciform specimen (Fig. 5) developing a biaxial stress field in the working section. A full description of the testing system and specimen design are given elsewhere [*14,15*].

Crack length measurements were obtained by the potential-drop method at several intervals during the test. Crack growth rates were determined by a least-squares fit of a parabola to groups of five crack length readings, differentiated to give the extension per cycle according to the ASTM Test Method for Constant-Load-Amplitude Fatigue Crack Growth Rates Above 10^{-8} m/Cycle (E 647-83).

The crack growth results are presented in Fig. 6 in terms of crack growth rate plotted against ΔK, with [*14*]

$$\Delta K = \Delta\sigma\sqrt{\pi a \sec(\pi a/w^*)} \tag{3}$$

with

$\Delta\sigma$ = stress range normal to the crack,
a = crack length, and
w^* = 101.9 mm = equivalent width of specimen, which makes allowances for the finite height of the specimen and the stiffening due to the fillet radius at its periphery [*14*].

For the stress range used in these tests, $\Delta\sigma$ = 560 MPa, linear elastic fracture mechanics cannot strictly be employed, but the use of ΔK in Fig. 6 provides an expedient method of comparing crack growth rates at a given crack length for each of the four tests conducted, since the finite-width correction in Eq 3 can be employed.

Specimens subjected to uniaxial loading, P_1 and T_5, showed a faster crack growth rate than biaxial specimens P_2 and P_3 by a factor between 5 and 10. Figure 6 shows also the crack

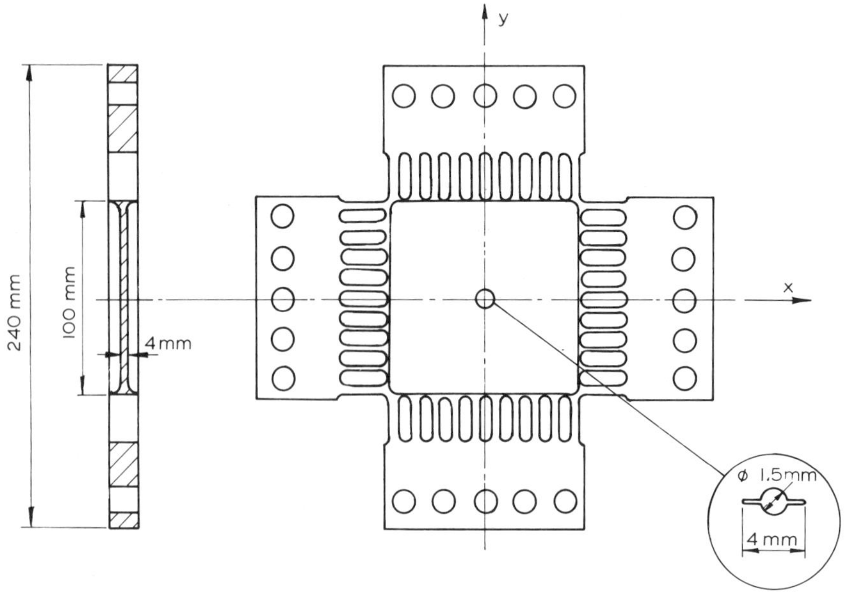

FIG. 5—*Specimen geometry and initial central crack.*

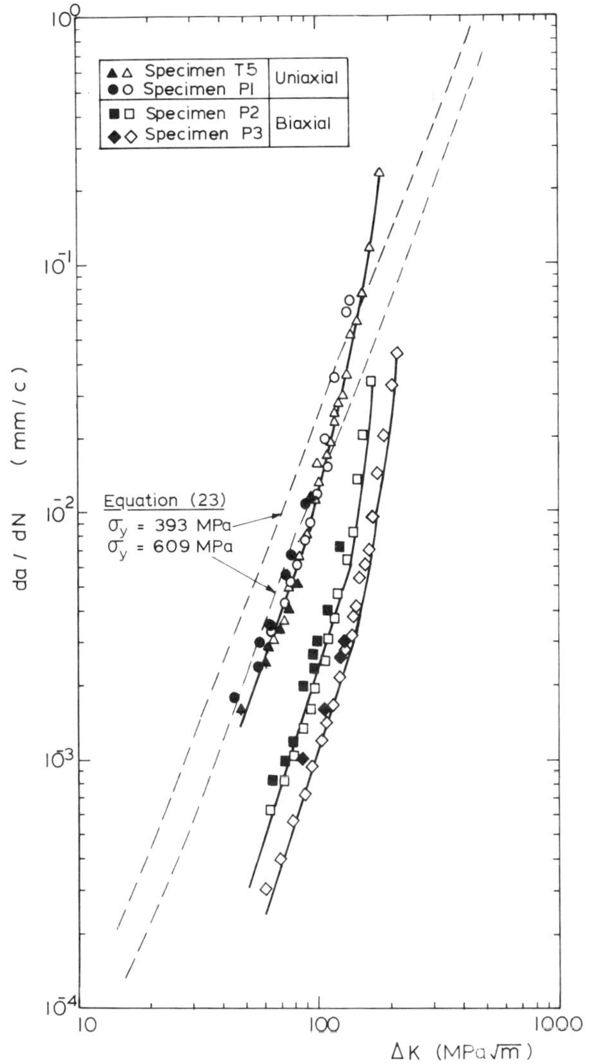

FIG. 6 *Crack growth rate as a function of* ΔK, *from potential-drop measurements (open symbols), and striation spacings (closed symbols).*

growth rates determined by striation spacing counting (closed symbols). Details of striation spacing determinations are given in the following subsection.

These results indicate a significant effect of straining path on crack growth rate; the application of a stress parallel to the specimen axis reduced the crack growth rate by more than five times. Also, they indicate apparently conflicting behavior to that reported for the 1Cr-Mo-V steel. In that instance the application of an axial load to a specimen subjected to torsion loading resulted in a decrease in low-cycle fatigue life of more than three times [*13*], implying increased crack propagation rates.

In order to explain this apparent discrepancy, a metal structure study was carried out to investigate the deformation behaviour of the AISI 316 stainless steel uniaxially and biaxially stressed specimens within the enclave of the plastic zone. Rolling direction of the plate was shown to be unimportant by Tests P_1 and T_5, where the crack plane was normal to and parallel with the rolling direction, respectively (see Fig. 6).

Microscopic Examination

Fracture surfaces were first examined in the scanning electron microscope (SEM). All the fracture surfaces showed a crystallographic type of fracture by the notch and a striation fracture thereafter, Fig. 7. The extent of crystallographic fracture in Specimens P_2 and P_3 was about double that in Specimens P_1 and T_5; in other words, striation fracture occurs earlier in Specimens P_1 and T_5, after approximately a grain diameter.

Crack growth rates can be estimated while observing the fracture surfaces in the SEM, by determining the striation spacing. The crack growth rate is equal to the striation spacing assuming that there is a one-to-one relationship between number of striations and number of cycles. Figure 6 shows the crack growth rates determined by this method, where they are compared with corresponding rates obtained using the potential-drop technique. The close agreement found between the two methods reiterates the validity of the experimental results, highlighting again the difference in propagation behavior between Specimens P_1 and T_5 on the one hand and Specimens P_2 and P_3 on the other.

Besides the differences in the extent of the crystallographic region on the fracture surfaces and the striation spacing for a given crack length, the fatigue fracture surfaces of Specimens P_2 and P_3 also showed that striation-free regions intermingled with the striation regions, Fig. 7*c*. These striation-free regions are at an angle to the plane of the striations, with the result that the overall plane of the fatigue fracture becomes inclined to the plane of the notch and initial fatigue fracture to produce a slant mode fracture. Even though these striation-free regions and the resultant inclination of the crack plane may play some part in the large difference in crack growth rates observed between Specimens P_1/T_5 and P_2/P_3, they do not seem to be the dominant feature because even in the initial part where the crack is on the same plane as the notch, the differences in crack growth rate are already significant.

If the type of cracking is the same (that is, striations), but crack growth rate is markedly different for a given ΔK, the reason may rest on the events occurring within the plastic zone. For example, changing the loading conditions may change the material response in terms of its hardening behavior. Obviously this is very difficult to measure during the test, but it can be studied after the test by looking at the dislocation substructure next to the fracture surface (within the plastic zone).

To accomplish this study, thin foils for transmission electron microscopy (TEM) were prepared from a region not more than 100 μm below the fracture surface and along the line of fracture.

Significant differences were observed in the dislocation substructures of Specimens P_1 and T_5 on the one hand, and of Specimens P_2 and P_3 on the other. The former showed that the dominant form of deformation was by plane slip; the dislocations are seen to remain in the slip planes in all the foils investigated. Figure 8 shows the dislocation structure of two foils of Specimen P_1—one taken at 1.5 mm and the other at 14.5 mm from the notch, where the half-width of the initial slot was 2 mm and the depth 0.18 mm. While showing the same kind of slip, the foil farther from the notch shows also a high dislocation density and the operation of a second slip system. The effect of crack length on the plastic zone is, therefore, one of intensification of strain rather than changing of the slip character.

The dislocation substructure shown by biaxial Specimens P_2 and P_3 is of a different kind; it is a cell type, Fig. 9, which is developed when multiple slip prevails. Slip in multiple systems leads to the interaction and rearrangement of dislocations into cell boundaries.

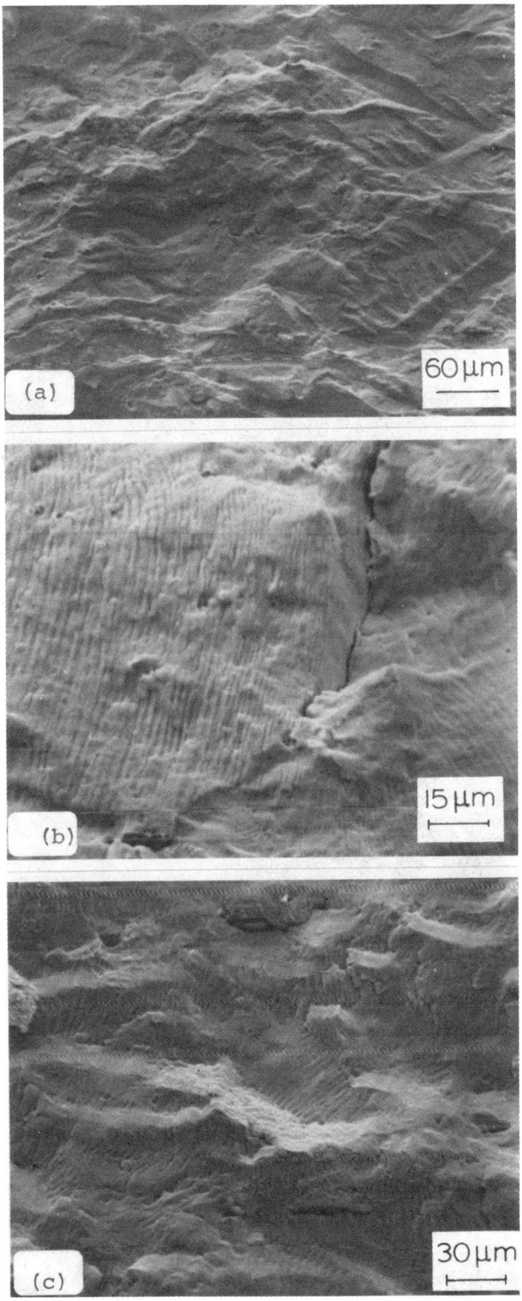

FIG. 7—*SEM photographs of the fatigue fracture surface:* (a) *crystallographic fracture by the notch,* (b) *striation fracture at a later stage, and* (c) *striation-free regions in Specimens P_2 and P_3.*

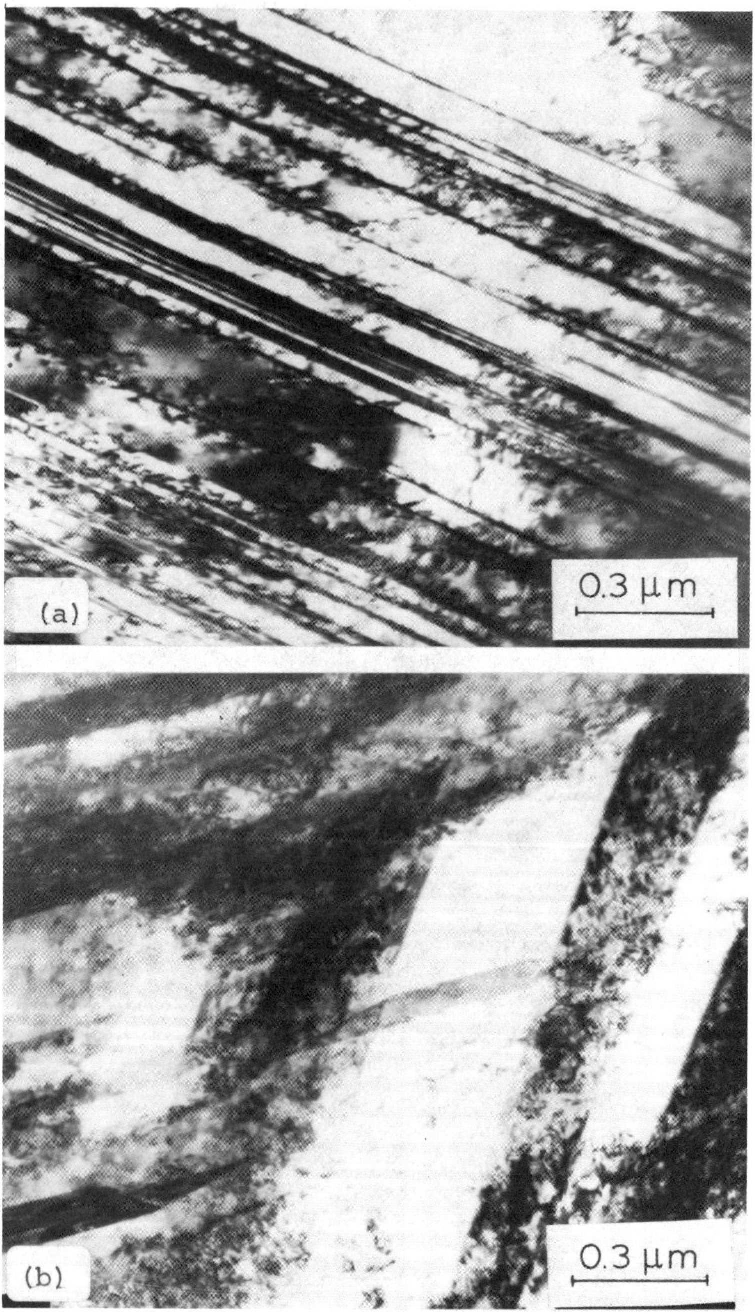

FIG. 8—*Typical dislocation substructure of Specimens P_1 and T_5:* (a) *1.5 mm from the notch tip and* (b) *14.5 mm from the notch tip.*

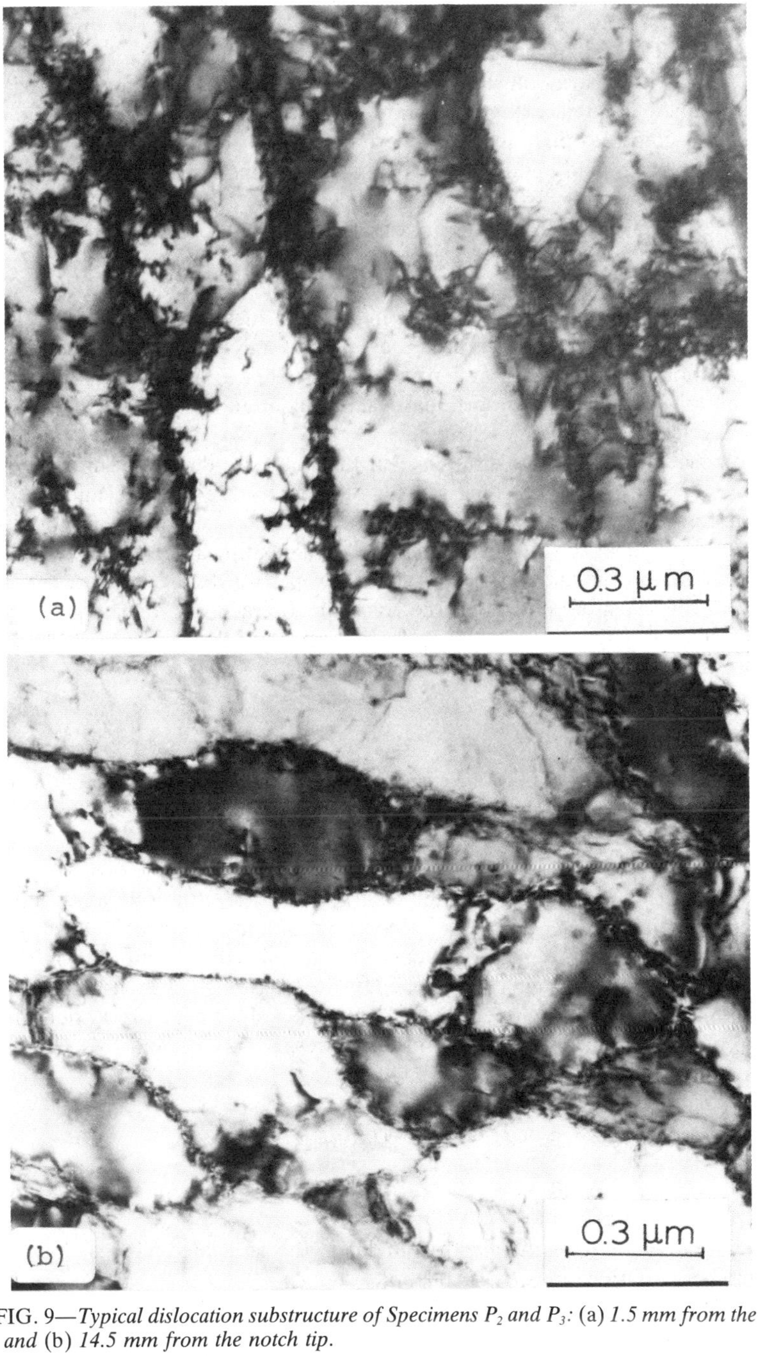

FIG. 9—*Typical dislocation substructure of Specimens* P_2 *and* P_3*:* (a) *1.5 mm from the notch tip and* (b) *14.5 mm from the notch tip.*

Again, as in the previous case, Fig. 9 shows the substructure prevailing at distances 1.5 and 14.5 mm from the notch of Specimen P_2. The effect of crack length on the substructure within the plastic zone is again one of intensification of strain, reflected in this instance in the refinement of the cell structure.

This difference in dislocation substructure observed for the two cases (uniaxial, biaxial) may be explained to some extent by considering the size and shape of the plastic zones developed in each case. Miller and Kfouri [*16*] used an elastic-plastic finite-element analysis to study the crack tip field under biaxial loading conditions for three cases: uniaxial, shear, and equibiaxial. Their results for the first two cases, of relevance here, showed that the orientation of the plastic ears (plastic zone) was different for monotonic loading, being an inclination to the crack plane of 65 deg for the uniaxial case as compared to 50 deg for the shear case. For the shear case, the length of the plastic zone radiating from the crack tip was more than eight times larger for a similar level of maximum principal stress. This difference in the distribution of strain at the crack tip may be a primary factor responsible for the operation of multiple slip systems in the biaxial case and for the formation of cells, because the loading was nonproportional, varying continuously between the uniaxial and shear stress states during the cycle (Fig. 3.)

It is proposed now that this difference in the plasticity developed in the plastic zone accounts for the difference in crack growth rate experienced by specimens tested under uniaxial and biaxial loading. Many fatigue crack growth mechanisms proposed in the literature are based on some sort of slip ahead of the crack tip. Crack growth rate is therefore related to characteristic parameters such as crack opening displacement, which are representations of the magnitude of slip at the crack tip. If crack growth rate is directly related to this slip ahead of the crack tip, then factors affecting the type and magnitude of slip should also influence crack growth rate.

Hardening Mechanisms

As illustrated previously, there are two dominant dislocation structures in the present tests: plane slip (Specimens P_1 and T_5) and intragranular dislocation cells (Specimens P_2 and P_3). The magnitude of hardening in both cases will now be estimated.

Plane Slip—The dislocations generated at the sources in each cycle move along the slip planes until they are stopped at the grain boundaries, and pileups of dislocations are consequently formed. The hardening developed due to this configuration of dislocations has been studied before [*17,18*] and is determined by the long-range stress from the accumulated pileups acting in a newly developed slipband. With a wall of edge dislocations at each end of the slip plane, there is a fairly uniform back-stress across a grain. In shear stress terms this gives approximately

$$\tau = \psi n \mu b / \pi \ell$$

where

τ = stress on the slipband,
ψ = relief factor affecting the pileups and due to the associated pileups in the next grain,
n = number of dislocations in the pileup,
μ = rigidity modulus,
b = Burgers vector, and
ℓ = separation between slipbands.

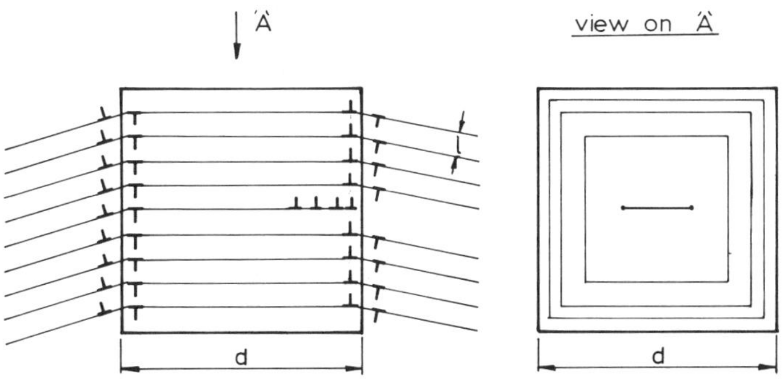

FIG. 10—*Schematic representation of plane slip and pileup configuration.*

In this model it is considered that each dislocation loop expands to the diameter of the grain d (more accurately to the diameter of the center of gravity of the pileups), Fig. 10.

With n dislocations in each slipband, the total length of dislocation lines in each slipband is $n4d$. If the slipband spacing is ℓ and five slip systems are assumed to operate, then the dislocation density ρ is

$$\rho = \frac{5n \cdot 4d}{d^2\ell} = \frac{20n}{d\ell} \tag{5}$$

It follows that

$$\tau = \psi\mu b\rho d/20\pi \tag{6}$$

This expression shows that the hardening in the plastic zone is dependent on the length of the slipband, which is approximately equal to the grain diameter d, and on the dislocation density ρ, which depends on grain size also.

Dislocation Cells—As a consequence of multiple slip, the dislocations manage to arrange themselves on walls or cell boundaries, so that the long-range stress from the grain boundary pileups is generally relieved and the main resistance to dislocation motion is from the short-range intersects between the glide dislocations in the slipband and the dislocations in the cell boundaries [*19*]. To take an extreme case, for example, where the dislocations are distributed to form cell walls with a single-layer square network of dislocations, Fig. 11, and considering a prismatic volume of base d_c^2 and unit length, the total length of dislocation lines within this volume is $2d_c/k$, where k is the spacing between adjacent dislocations in the cell wall, and therefore

$$\rho = \frac{2d_c}{d_c^2\, k} = \frac{2}{d_c k} \tag{7}$$

The shear stress required to push a single dislocation through a wall is [*19*]

$$\tau_0 \cong \frac{\mu b}{2k} \cong \frac{\mu b \rho d_c}{4} \tag{8}$$

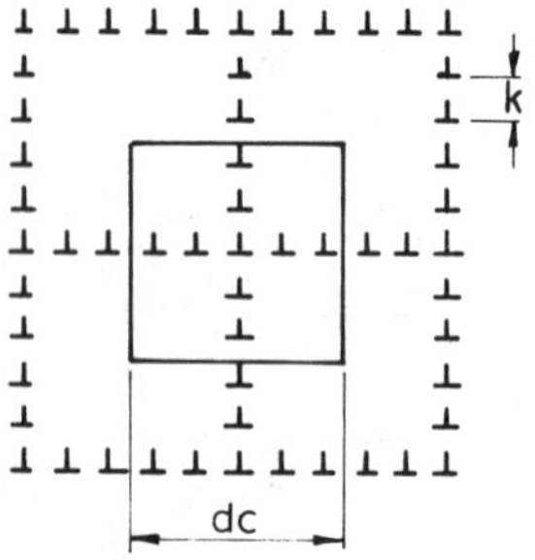

FIG. 11—*Dislocation arrangements to form cells (schematic).*

If n dislocations pile up across a clean cell under a shear stress τ, the stress on the leading dislocation is $n\tau$. The number of dislocations in the pileup is determined from the shear strain γ, where

$$\tau = \mu\gamma \tag{9}$$

or

$$\tau = \mu(nb/d_c) \tag{10}$$

Therefore for penetration of the walls

$$n\tau = \tau_0 \tag{11}$$

or

$$\tau^2 \frac{d_c}{\mu b} = \frac{\mu b \rho d_c}{4} \tag{12}$$

Hence

$$\tau = \frac{1}{2} \mu b \rho^{1/2} \tag{13}$$

The hardening is in this case proportional to the square root of the dislocation density.

Crack Growth Rate

The hardening discussed above gives the magnitude of the stress required in the plastic zone to operate a slipband. The detailed mechanism of crack extension and how different substructures in the plastic zone play such a dominant part are discussed next.

Many of the mechanisms of crack growth proposed in the literature involve the operation of slip or shear bands from the tip of the crack [*20–25*]; crack advance per cycle is the result of crack sharpening and blunting [*20*], "shear plane decohesion" or "sliding off" on the plane of maximum shear stress [*21*], or shear on alternate planes [*25*].

Assuming that crack propagation occurs by a slip decohesion mechanism with at least two active slip planes, Pelloux [*26*] argued that the crack plane should be the bisecting plane.

For face-centered cubic (fcc) metals in which the slip planes are the {111}, the crack plane would be either a (110) or a (100) plane. Bowles and Broek [27] showed this to be the case for 7075-T6 aluminum, cracking being a slipping-off process along the {111} planes. Etch pit studies showed the crack plane to be either the (110)(100) or the (111) shear plane.

Regardless of the precise mechanism of crack extension, all these theories have a common factor, namely, that slip precedes crack growth and, consequently, crack growth rate should be a function of the magnitude of slip in front of the crack. This magnitude of slip is determined by the strength of the slipband, so in turn crack growth rate should be proportional to it.

The average back stress τ acting over a slipband of length r in a single grain is

$$\tau = \frac{\mu n b}{r} \tag{14}$$

When the plastic deformation extends for more than one grain, the stress fields of the pileups of dislocations in one grain interact with those of the neighboring grains. This interaction results in the partial cancellation of the stress fields of some pileups. This effect will depend on the ratio of the slip distance to the plastic zone size r/r_p; consequently

$$\tau = \frac{r}{r_p}\left(\frac{\mu n b}{r}\right) \tag{15}$$

By taking the cyclic plastic zone size [28] for plane strain as

$$r_p = \frac{1}{24\pi}\left(\frac{\Delta K}{\sigma_y}\right)^2 \tag{16}$$

Eq 15 gives the strength of the slipband, nb, as

$$nb = \frac{\Delta K^2 \tau}{24\pi \mu \sigma_y^2} \tag{17}$$

If crack growth rate is proportional to the strength of the slipband, then

$$\frac{da}{dN} = \frac{f\,\Delta K^2 \tau}{24\pi \mu \sigma_y^2} \tag{18}$$

Crack growth rate is in this case proportional to the stress acting on the shear band within the plastic zone (τ). This is because crack extension per cycle is equal to the shear decohesion on the shear band, that is, the displacement along the shear band developed ahead of the crack to accommodate the crack opening. Obviously the displacement (nb) will depend on the local stress-strain relationship, and in particular on the shear stress acting on the shear band which will be given by the strain hardening opposing dislocation movement. Previously it was shown that the dislocation substructure within the plastic zone of the uniaxial and the biaxial specimens was very different, and the degree of hardening was calculated for both cases.

In the uniaxial case, the dislocations are constrained from moving out of their slip planes because of the low stacking fault energy for this material, and classical pileups form. The hardening can be estimated with Eq 6: When $\psi = 0.1$, $\mu = 76$ GPa, $d = 0.050$ mm, $b =$

0.25 nm and, as calculated from the micrographs, $\rho = 3.5 \times 10^8$ mm^{-2}; then $\tau = 529$ MPa.

In the biaxial case several slip systems are activated and a rearrangement of dislocations in cell boundaries occurs. The high stresses from the dislocation pileups are relaxed to such an extent that hardening is given by Eq 13. Taking the same values for μ and b as above but with $\rho = 1.2 \times 10^8$ mm^{-2}, this gives $\tau = 104$ MPa. In both cases, ρ was determined for a crack length of 3.5 mm, at 100 μm above the crack tip plane.

The ratio of the hardenings is approximately 5, which is of the same order of magnitude as the ratio between the experimentally determined crack growth rates (5.5 between Specimens P_1 and P_2 and 10 between Specimens P_1 and P_3).

The microscopic observations gave no indication that a different crack growth rate for Specimens P_2 and P_3 should be anticipated. Both have the same kind of substructure, and the dislocation densities appear to be of the same order. The observed difference has to be assigned either to experimental scatter or to reasons beyond the scope of the present investigation, in view of the limited number of tests completed so far.

Discussion

The viability of Eq 18 as a model for crack propagation may be assessed from its predictive capability. Since the crack tip deformation has been modeled by a pair of slipbands emanating from the crack tip with plane strain conditions prevailing, a suitable formula for plastic zone size is that derived by Rice [*29*]

$$r_{ps} = \frac{\pi}{64} \sin^2\theta(1 + \cos\theta)K^2/\tau_y^2 \tag{19}$$

where r_{ps} is the length of the shear bands at inclinations $\pm\theta$ to the crack plane. The stress τ_y is the shear stress sustained within the slipband, taken by Rice to be $(\sigma_y/\sqrt{3})$. For the maximum value of r_{ps}, $\theta = 70.53$ deg, and for cyclic loading one may replace K by ΔK and σ_y by $2\sigma_y$ to obtain the cyclic plastic zone size

$$r_p = 0.0436(\Delta K/\sigma_y)^2 \tag{20}$$

which compares favorably with Eq 16 but provides a more accurate solution for an ideal plastic material.

Since the crack extension per cycle is equated to the shear decohesion at the tip of the slipband, we expect $f = 1$ in Eq 18 if the cyclic plastic zone formula is correct. Thus

$$\begin{aligned} \frac{da}{dN} &= nb \\ &= (\tau/\mu)r_p \end{aligned} \tag{21}$$

The value of τ in this equation should best be taken at the grain boundary adjacent to the crack tip, being the location of the first dislocation pileup in the macroscopic shear band. To allow for the influence of strain hardening within the plastic zone, one may assume that a Hutchinson-Rice-Rosengren singularity [*30,31*] operates within the plastic enclave, so that along the slipband at $\theta = 70.5$ deg, the shear stress distribution can be represented by [*32*]

$$\tau/\tau_y = (r_p/r)^{n/(n+1)} \tag{22}$$

for a strain hardening exponent n, where r is the distance along the shear band. Taking the mean linear intercept $\bar{\ell}$ as a representative average value for distance from the crack tip to the grain boundary, Eqs 20, 21, and 22 give

$$\frac{da}{dN} = \left(\frac{\sigma_y}{\sqrt{3}\mu}\right) (\bar{\ell})^{-n/(1+n)} \left(\frac{\sqrt{0.0436}}{\sigma_y} \Delta K\right)^{2(1+2n)/(1+n)} \tag{23}$$

This equation has been plotted in Fig. 6, using $\bar{\ell} = 31$ μm, $\sigma_y = 393$, and $\mu = 76$ GPa with $n = 0.30$, the strain hardening exponent for this heat of stainless steel for both monotonic and cyclic loading, where

$$\sigma = k(\epsilon)^n \tag{24}$$

A similar power-law dependence on plastic zone size has been observed in biaxial stress proportional loading tests [*33*]. Reasonable agreement for the slope of the first portion of the experimental crack growth curve is observed, with Eq 23 overestimating the propagation rate by a factor of 2. The discrepancy in growth rates is due primarily to use of Eq 20, which fails to incorporate the effect of strain hardening, thus overestimating the plastic zone size. This may be allowed for by noting that the cyclic flow stress at the crack tip is more closely approximated by tensile strength [*33*], and therefore a second line is plotted in Fig. 6 for $\sigma_y = 609$ MPa. A good correlation for the uniaxial test is now observed.

Equation 22 predicts a stress τ of 384 MPa at 100 μm from the crack tip for $\sigma = 280$ MPa and a crack length of 3.5 mm. Alternatively, a stress τ of 486 MPa is obtained with $\sigma_y = 609$ MPa. This is comparable to the stress of 529 MPa measured for Tests P_1 and T_5 under uniaxial stress. However, the much lower stress (104 MPa) and corresponding crack propagation rate observed for nonproportional loading imply that the crack tip stress field solutions from elastoplastic fracture mechanics for monotonic loading may not be able to describe the conditions in Tests P_2 and P_3, where the incremental nature of plastic flow theory dominates within the crack tip plastic enclave.

Crack Propagation Summary

To summarize the foregoing discussion it is apparent that nonproportional loading in austenitic steel causes a reduction in crack propagation rate, compared to an apparent increased growth rate in bainitic steel. The following differences have been observed:

1. Mode I growth was studied in the stainless steel, compared to Mode II/III shear cracks in tension-torsion.
2. A change in cyclic deformation response and the local flow stress at the crack tip has been detected for the low stacking fault energy stainless steel.
3. For both steels, crack propagation rate depends on the strain path adopted in the cycle. In the case of stainless steel, however, the change in deformation behavior must be added to strain path effects, and is probably the dominant influence.

Conclusions

1. AISI 316 stainless steel has shown a load path dependence of fatigue crack growth different to that observed in 1Cr-Mo-V steel.

2. For AISI 316 stainless steel the uniaxial test gives a higher crack growth rate than the biaxial nonproportional loading test. This effect is attributed to the operation, within the plastic zone, of a different hardening mechanism.

3. The "slipping off" mechanism for crack growth seems an acceptable basis for theoretical predictions of experimental data.

Acknowledgments

The authors appreciate the support of the Nuclear Regulatory Commission and the Oak Ridge National Laboratory, which supplied the material. The authors were supported by the Science and Engineering Research Council (E.R.R.) and the Central Electricity Generating Board (M.W.B.) on research fellowships, and on a visiting fellowship (H.X.P.) from the Chinese Academy of Sciences and The Royal Society of Great Britain.

References

[*1*] Mason, W. and Delaney, W. J., *Reports of the British Association for the Advancement of Science,* 1921, pp. 329–341.
[*2*] Little, R. E., *The Aeronautical Quarterly,* Vol. 20, Part 1, Feb. 1969, pp. 57–60.
[*3*] Kanazawa, K., Miller, K. J., and Brown, M. W., *Transactions,* American Society of Mechanical Engineers, Series H, *Journal of Engineering Materials and Technology,* Vol. 99, July 1977, pp. 222–228.
[*4*] Dietmann, H. and Issler, L. in *Proceedings,* Fifth Conference on Dimensioning and Strength Calculations, and the Fifth Congress of Material Testing, Budapest, Oct. 28–Nov. 1, 1974, Vol. 2, Akademiai Kiado, Budapest, 1974, pp. 388–399.
[*5*] Zamrik, S. Y. and Frishmuth, R. E., *Experimental Mechanics,* Vol. 13, No. 5, 1973, pp. 204–208.
[*6*] Jordan, E. H. in *A Decade of Progress,* American Society of Mechanical Engineers, New York, Section 7.3, 1982.
[*7*] Taira, S., Inoue, T., and Yoshide, T. in *Proceedings,* 11th Japan Congress on Materials Research, 1968, pp. 60–65.
[*8*] Garud, Y. S., *Transactions,* American Society of Mechanical Engineers, Series H, *Journal of Engineering Materials and Technology,* Vol. 103, 1981, pp. 118–125.
[*9*] Brown, M. W. and Miller, K. J. in *Proceedings,* Institution of Mechanical Engineers, Vol. 187, 1973, pp. 745–755.
[*10*] Brown, M. W. and Miller, K. J. in *Low-Cycle Fatigue and Life Prediction, ASTM STP 770,* American Society for Testing and Materials, Philadelphia, 1982, pp. 482–499.
[*11*] de los Rios, E. R., Mohamed, H. J., and Miller, K. J., *Fatigue of Engineering Materials and Structures,* Vol. 8, 1985, pp. 49–63.
[*12*] Forsyth, P. J. E. in *Proceedings,* Crack Propagation Symposium, Cranfield, U.K., 1961, pp. 76–94.
[*13*] Jordan, E. H., Brown, M. W., and Miller, K. J. in *Multiaxial Fatigue, ASTM STP 853,* American Society for Testing and Materials, Philadelphia, 1985, pp. 569–585.
[*14*] Brown, M. W. and Miller, K. J., *Multiaxial Fatigue, ASTM STP 853,* American Society for Testing and Materials, Philadelphia, 1985, pp. 135–152.
[*15*] Brown, M. W. in *Measurement of High Temperature Mechanical Properties of Materials,* M. S. Loveday, M. F. Day, and B. F. Dyson, Eds., Her Majesty's Stationery Office, London 1982, pp. 185–203.
[*16*] Miller, K. J. and Kfouri, A. P., *International Journal of Fracture,* Vol. 10, 1974, pp. 393–404.
[*17*] Meakin, J. D. and Petch, N. J., *Philosophical Magazine,* Vol. 29, 1974, pp. 1149–1156.
[*18*] Al-Haidery, J. T., Petch, N. J., and de los Rios, E. R., *Philosophical Magazine,* Vol. 47, 1983, pp. 891–902.
[*19*] Al-Haidary, J. T., Petch, N. J., and de los Rios, E. R., *Philosophical Magazine,* Vol. 47, 1983, pp. 869–890.
[*20*] Laird, C. and Smith, G. C., *Philosophical Magazine,* Vol. 7, 1962, pp. 847–857.
[*21*] Hertzberg, R. W. in *Fatigue Crack Propagation, ASTM STP 415,* American Society for Testing and Materials, Philadelphia, 1967, pp. 205–225.

[22] McMillan, J. C. and Pelloux, R. M. N. in *Fatigue Crack Propagation, ASTM STP 415,* American Society for Testing and Materials, Philadelphia, 1967, pp. 505–535.
[23] Tomkins, B. and Biggs, W. D., *Journal of Materials Science,* Vol. 4, 1969, pp. 532–553.
[24] Tomkins, B., *Philosophical Magazine,* Vol. 17, 1968, pp. 1041–1068.
[25] Spitzig, W. H., Talola, P. M., and Wei, R. P., *Engineering Fracture Mechanics,* Vol. 1, 1968, pp. 155–166.
[26] Pelloux, R. M. N., *Quarterly Transactions of the American Society for Metals,* Vol. 62, 1969, pp. 281–285.
[27] Bowles, C. Q. and Broek, D., *International Journal of Fracture Mechanics,* Vol. 8, 1972, pp. 75–85.
[28] Knott, J. F., *Fundamentals of Fracture Mechanics,* Butterworths, London, 1973.
[29] Rice, J. R., *Journal of the Mechanics and Physics of Solids,* Vol. 22, 1974, pp. 17–26.
[30] Hutchinson, J. W., *Journal of the Mechanics and Physics of Solids,* Vol. 16, 1968, pp. 13–31.
[31] Rice, J. R. and Rosengren, G. F. *Journal of the Mechanics and Physics of Solids,* Vol. 16, 1968, pp. 1–12.
[32] Hu, W. and Liu, H. W. in *Proceedings,* Second International Conference on Mechanical Behavior of Materials, Boston, MA, 1976, pp. 1058–1062.
[33] Brown, M. W., de los Rios, E. R., and Miller, K. J. in *Basic Questions in Fatigue: Volume I, ASTM STP 924,* J. T. Fong and R. J. Fields, Eds., American Society for Testing and Materials, Philadelphia, 1988, pp. 233–259.

E. K. Tschegg[1] *and S. E. Stanzl*[2]

The Significance of Sliding Mode Crack Closure on Mode III Fatigue Crack Growth

REFERENCE: Tschegg, E. K. and Stanzl, S. E., **"The Significance of Sliding Mode Crack Closure on Mode III Fatigue Crack Growth,"** *Basic Questions in Fatigue: Volume I, ASTM STP 924,* J. T. Fong and R. J. Fields, Eds., American Society for Testing and Materials, Philadelphia, 1988, pp. 214–232.

ABSTRACT: During Mode III loading, the surfaces of a crack move parallel to each other; friction, abrasion, and mutual support of the contacting parts of the microscopically rough crack faces result. These energy-dissipative processes reduce the effective load at the crack tip, a phenomenon known as "sliding mode crack closure" (SMCC). This effect has been studied by performing crack growth experiments on circumferentially notched specimens of high-strength (AISI 4340) and mild steel (AISI C1018) under pure cyclic torsion and under combined loading (cyclic Mode III + static Mode I). Crack growth rates without the influence of sliding mode crack closure ("true" crack growth rates) can be determined using an extrapolation procedure. These crack growth rates are independent of crack depth, specimen diameter, and loading level and are therefore a material characteristic for Mode III fatigue crack growth. With the aid of the "true" crack growth curve the extent of sliding mode crack closure can be quantitatively determined and the changes in fracture mode explained. By superimposing a static tensile load the sliding mode crack closure is reduced whereas the "true" crack growth rates are only slightly influenced.

KEY WORDS: fatigue crack growth, Mode III, complex loading, mixed-mode loading, torsional loading, crack closure, fractography

Nomenclature

a	Uncracked ligament radius of circumferentially notched specimen
b	External radius of circumferentially notched specimen
c	Crack length (depth) emanating from the notch root
Δc	Change in crack length (depth)
$(\Delta c/\Delta N)_i$	Fatigue crack growth rate in Mode i = I, III
$\mathrm{CTD_I}$	Crack-tip displacement in Mode I
$\Delta\mathrm{CTD_{III}}$	Cyclic crack-tip displacement in Mode III
G	Shear modulus
k	Shear yield strength
K_i	Stress-intensity factor for Mode i; i = I, III
K_{iC}	Fracture toughness for Mode i; i = I, III
ΔK_i	Alternating cyclic stress-intensity factor in Mode i; i = I, III
$\Delta K_{\mathrm{IIIeff}}$	Effective alternating stress intensity in Mode III acting at crack tip
$\Delta K_{\mathrm{IIInom}}$	Nominal alternating stress intensity in Mode III

[1] Associate professor, Technical University, Institute of Applied and Technical Physics, Vienna, Austria.
[2] Associate professor, University of Vienna, Institute for Solid State Physics, Vienna, Austria.

R Ratio of minimum to maximum torque or load
r Radial coordinate from specimen axis to plastic zone
r_p Radius from the specimen axis to boundary of plastic zone
R_y Plastic zone size
Δ_γ Plastic shear strain
$\Delta\Gamma_{III}$ Cyclic plastic strain intensity in Mode III
$\Delta\Gamma_{IIIeff}$ Effective cyclic plastic strain intensity in Mode III
$\Delta\Gamma_{IIInom}$ Nominal cyclic plastic strain intensity in Mode III
$\Delta\Gamma_{IIIR}$ Dissipated plastic strain intensity carried by broken ligament
σ Normal stress

During cyclic loading of a crack in the antiplane shear mode (Mode III) or in the shear mode (Mode II) the crack theoretically remains closed in contrast to loading in the crack opening mode (Mode I). The macroscopically flat but microscopically rough fracture surfaces of polycrystalline material slide against each other, thus giving rise to mutual support, friction, abrasion, and similar dissipative processes; therefore, the nominal crack-tip stress values calculated from the externally applied loads are reduced to a lower effective value. This phenomenon is somewhat analogous to fatigue crack closure in Mode I [*1,2*] and is accordingly called "sliding mode crack closure" (SMCC) [*3*].

The fundamental question is how and to what extent this phenomenon influences the crack growth behavior of a material, loaded in the sliding modes. This influence is discussed for the case of Mode III fatigue crack propagation in a high-strength steel and in mild steel. Particular attention is paid to critical experiments, which should help clarify quantitatively the influence of the sliding mode crack closure.

Literature Review

Early analytical work [*4–7*] dealt with longitudinal cracks in torsionally loaded specimens under elastic and fully plastic conditions. Mathematically convenient descriptions of the stress-strain fields in Mode III have also been developed [*8–10*]. The rubbing fracture surfaces were mentioned in this previous work; however, their influence on crack propagation has not been studied, nor has their effect been treated in any way theoretically or experimentally.

More recently, the question of Mode III fatigue crack growth has been reexamined in the context of fracture mechanics. The first crack growth curves of Mode III loading were published on variously tempered EN 16 steel [*11*] and on AISI 4340 steel [*12*]. In both cases the crack growth rates found for Mode III were smaller by a factor of 10 to 50 than for Mode I crack growth at the same nominal stress-intensity range. These measurements were performed on circumferentially notched cylindrical specimens with crack depths of 2 to 4 mm. Mode III threshold values of Ref *11* are larger by a factor of three than Mode I values. The results of Pook [*13–15*] are in contrast to this; he suggested much lower threshold values based on theoretical considerations and experimentally found them to be so.

A possible influence of rubbing fracture surfaces has been mentioned in these recent investigations; however, no further qualitative or quantitative studies have been performed.

Tschegg [*16*] and Tschegg et al. [*17*] were the first to perform qualitative and quantitative investigations in order to determine the influence of the SMCC. In these studies the great influence of the SMCC on Mode III fatigue crack propagation was clearly shown. In general, the SMCC influence increases with increasing crack length, decreases with increasing external loading level, and is strongly influenced by the fracture surface morphology. Based on the results of Refs *16* and *17* the influence of the SMCC was also investigated in AISI 4140 steel [*18*]. In that work, the SMCC was diminished to an unknown extent by the application of a static Mode I load during Mode III fatigue crack propagation.

Obviously the effective loading value at the crack tip is responsible for crack propagation and not the nominal external value which is partly dissipated. In Ref *19* the first attempt was made to estimate the Mode III crack growth rates of AISI 4340 steel based on the effective loading levels using a special measuring procedure. This procedure has also been applied to AISI 4340 steel [*20*] and to AISI C1018 steel [*3,21*]. A "true" crack growth curve without the SMCC influence is obtained for Mode III fatigue crack growth in Refs *3*, *19*, and *20*, showing higher crack growth rates in the case of AISI 4340 steel at high cyclic Mode III loading levels than for equivalent Mode I loading.

To the authors' knowledge, only the micromechanical model of McClintock and Ritchie [*22*] does exist for Mode III fatigue crack propagation. This model is based on the premise that crack advance results from coalescence of Mode II microcracks initiated at inclusions ahead of the main crack front. The success of this model in predicting rates of Mode III crack propagation is discussed in detail for constant-amplitude loading in Ref *23* and for simple variable loading spectra in Ref *24*. Several features of Mode III crack growth behavior after a single overload are described in Ref *25*. The SMCC is not considered in these studies, as it is assumed that a small amount of Mode I crack opening during cyclic Mode III loading reduces the energy-dissipative processes to a negligible level.

In contrast to earlier works [*14,15*], the SMCC is included in a recent investigation by Pook [*26*] in calculations of the stress intensity by using a multiplication factor. The success of this procedure is demonstrated by comparison of these results with data from Ref *19*, among others.

Experimental Procedure and Results

From mechanical tests and the resulting crack growth data [*3,20,21*] the influence of the SMCC effect on crack propagation during pure Mode III loading and combined loading (cyclic Mode III + superimposed static Mode I) will be demonstrated in the following.

Material, Specimens, and Experimental Details

High-strength low-alloy AISI 4340 steel was tested as well as commercial mild steel (AISI C1018) which is softer by comparison. The chemical compositions and heat treatments of both materials are summarized in Table 1. In order to perform crack growth in Mode III, cyclindrical shafts with a circumferential starter notch have been used. Two specimen sizes, namely 25.4 mm diameter (with a 1.06-mm-deep notch and a notch radius of 0.02 mm) and 12.7 mm (with a 0.631-mm-deep notch and a notch radius of 0.015 mm), were machined. The tests were performed on a servohydraulic closed-loop tension-torsion testing machine and in laboratory air at 20°C. The testing frequency was 1 Hz, loading was sinusoidal, and the mean torque was zero ($R = -1$). The rotary and linear motions of the test machine were under torque and load control, respectively. Radial crack extension was continuously monitored using the d-c electrical potential-drop technique [*27*]. According to Ref *27* a small static axial load of 500 N maximum was superimposed during the test in order to prevent any pressure forces (these small loads do not give rise to a mixed mode condition [*17*]). The specimens were precracked with a low torque level. Crack propagation measurements were begun at a crack depth of 0.2 mm and were usually finished at a crack depth of 1 mm. After testing, the specimens were removed from the machine, cooled in liquid nitrogen, and broken

TABLE 1—*Chemical composition, heat treatment, and strength levels of AISI 4340 and C1018 steels.*

	Element	AISI 4340	AISI C1018
Chemical composition (amount in weight%)	C	0.4	0.18
	Si	0.26	0.1
	Mn	0.78	0.6
	P	0.07	0.06
	S	0.013	0.03
	Cu	0.14	0.04
	Cr	0.81	0.04
	Sn	...	0.04
	Ni	1.77	0.025
	Mo	0.25	0.01
	Fe	balance	balance
Heat treatment		austenitized at 870°C, oil-quenched, and tempered at 650°C	1 h at 870°C and furnace-cooled
Strength levels			
Monotonic tensile yield strength, Y		956 MPa	260 MPa
Cyclic tensile yield strength, Y_c		589 MPa	...
Ultimate tensile strength (UTS)		1076 MPa	412 MPa
Monotonic shear yield strength, $k = Y/\sqrt{3}$		552 MPa	150 MPa

by impact bending. The size of the final fracture was used to confirm that the fatigue crack had grown uniformly and that the calibration for measuring the crack depth was valid [*17,19*]. Further experimental details are to be found elsewhere [*3,16,17,19–21*].

Fracture Mechanical Characterization of Mode III Fatigue Crack Growth

Under small-scale yielding conditions the Mode III crack propagation rates may be represented as a function of the cyclic stress intensity (ΔK_{III}) [*11,12,17–20*] or of cyclic crack-tip displacement (ΔCTD_{III}) [*12,17–20*]. This has been shown to be an appropriate technique for characterizing the crack growth behavior of AISI 4340 steel, a high-strength steel showing little plastic deformation before failure. It has been assumed that the nominal small-scale yielding condition prevails up to $\Delta K_{III} = 60$ MPa$\sqrt{m}$. Calculations of the characteristic parameters for circumferentially notched specimens have been summarized in the literature [*18,20,23,24,35*]. The characteristic specimen dimensions are shown in Fig. 1.

If an attempt is made to measure a broad range of crack velocities in soft material such as AISI C1018 steel, the mechanical state varies from small-scale yielding to full-scale yielding. This invalidates the stress-intensity concept and it becomes necessary to use the "plastic strain intensity" concept ($\Delta\Gamma_{III}$) [*22*]. The "plastic strain intensity" is defined in terms of the plastic strain γ in front of the crack, the radius of the plastic zone R_y, the geometrical parameters a, r, r_p as shown in Fig. 1 as

$$\Delta\Gamma_{III} = \lim_{a-r\to 0} \{\Delta\gamma \cdot (a - r)\} = \frac{2 \cdot k}{G} \cdot \Delta \left\{ \left(\frac{a}{r_p}\right)^2 \cdot R_y \right\}$$

where k is the shear flow strength, and G the shear modulus. Calculation of the plastic zone size R_y for cylindrical specimens with a circumferential crack is described in Ref *28* for

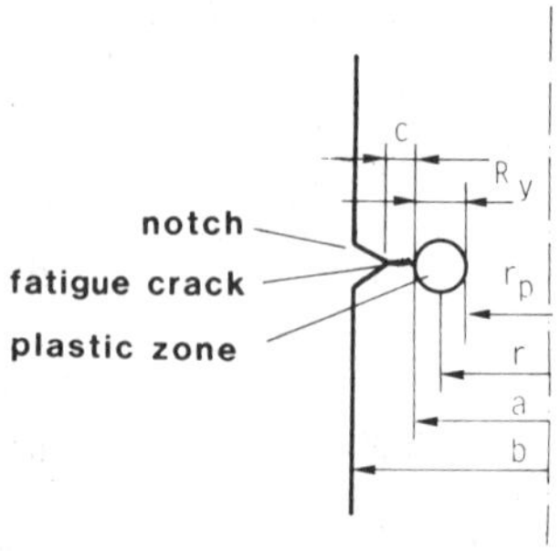

FIG. 1—*Dimensions for the plastic zone and respective radii in circumferentially notched specimen.*

conditions ranging from small-scale yielding to full-scale yielding. For small-scale yielding, the ratio $a/r_p \rightarrow 1$ and the relationship of the plastic strain intensity parameter to the crack-tip displacement becomes

$$\Delta \mathrm{CTD}_{\mathrm{III}} = 2 \cdot \Delta\Gamma_{\mathrm{III}}$$

For large-scale yielding this relationship is only approximately valid. The plastic strain intensity thus seems to be an appropriate parameter for the characterization of the Mode III crack growth results over a wide range of applied torques [*3,18,21,24,25,35*]. Experimental results on AISI C1018 steel will be discussed in terms of this parameter in the following.

Measuring Procedure

The Mode III fatigue crack growth results to date have been obtained using several different measurement procedures. In some studies the torque range ΔT was held constant throughout the test [*11,17*]; in others it was changed periodically [*1,12,16,18,23,24*], in a manner similar to the experimental procedure for Mode I crack growth measurement. In Ref *18* the ratio of torque range to net section limit torque was held constant during measurement. A constant range of crack mouth rotation monitored using a twistmeter has been applied in [*18,24*]. These procedures do not quantitatively account for the SMCC effect and give results which depend on the specimen geometry and thus are not material characteristics. The advantages and disadvantages of all these procedures are discussed elsewhere [*17,20,22*]. The procedures using "stress intensity control" [*17,19,20*] or "plastic strain intensity control" [*3,21*] do not have these drawbacks. In these tests the nominal stress intensity $\Delta K_{\mathrm{IIInom}}$ (calculated from the alternating torque) or the plastic strain intensity $\Delta\Gamma_{\mathrm{IIInom}}$ at the circumferential crack is kept constant. This requires that the applied cyclic torque be readjusted after each crack advance of about 0.05 mm. If the SMCC effect did not exist, the nominal values would be identical with the effective values $\Delta K_{\mathrm{IIIeff}}$ and $\Delta\Gamma_{\mathrm{IIIeff}}$ at the crack tip. If the crack growth rates were plotted versus the crack length (depth) c, a horizontal line would result as shown in Fig. 2 schematically (full line). If, however, SMCC is present, $\Delta K_{\mathrm{IIIeff}} < \Delta K_{\mathrm{IIInom}}$ and $\Delta\Gamma_{\mathrm{IIIeff}} < \Delta\Gamma_{\mathrm{IIInom}}$. These differences increase with increasing crack length. The effective values control the crack growth rates and therefore the curve in Fig. 2 must drop as indicated schematically by a dashed line. The vertical distance between the two crack growth curves for tests with and, hypothetically, without the SMCC indicates quantitatively the influence of the SMCC on the crack growth behavior (Fig. 2).

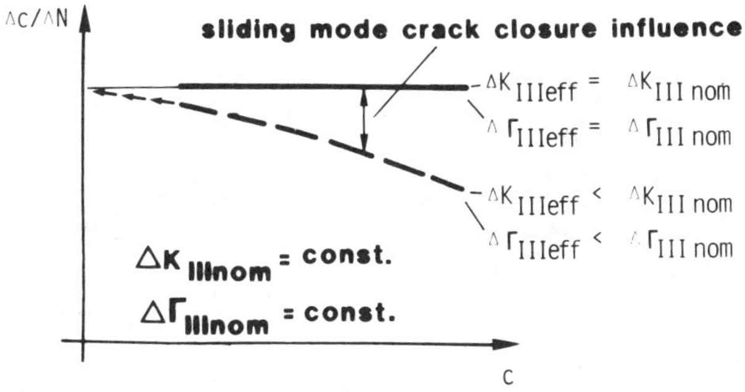

FIG. 2—*Schematic representation of the SMCC influence on the crack growth rates depending on crack depth* c *measured under stress intensity control or plastic strain intensity control.*

A measurement procedure keeping $\Delta c/\Delta N$ constant gives useful results analogous to the above-mentioned procedure [*16*]. However, this procedure has not been applied by the authors as the experimental difficulties to be contended with are substantial.

If the monotonically decreasing curve in Fig. 2 is extrapolated to a crack length $c \rightarrow 0$ the crack growth rate in the absence of sliding mode crack closure may be estimated according to Ref *19*. It is discussed in Ref *19* to what extent extrapolation to zero crack length is justified. The extrapolated crack growth values for crack lengths between c_{minimum} and $c =$ 0 cannot be applied to the short crack problem [*19*]. In order to understand the influence of a superimposed static axial load on Mode III fatigue crack growth and thus, also, the influence of the SMCC, the following procedure was used in Ref *21*. During Mode III loading in "plastic strain intensity" control, an axial load was applied in such a way that a constant Mode I stress intensity value K_I was effective at the crack tip during the whole measurement procedure. In order to study the influence of a negative axial load on the SMCC, a constant pressure on the overall specimen cross section (instead of the constant K_I) was applied.

Mode III Fatigue Crack Growth Results Obtained from "Stress Intensity Control" and "Plastic Strain Intensity Control" Measurements

Preliminary tests on shafts of AISI 4340 steel with a diameter of 25.4 mm, tempered at 650°C, showed that two essentially different fracture modes may occur during pure cyclic torsional loading [*16*]. Faces of these characteristic fracture surfaces are shown in Fig. 3. In Fig. 3*a* the "macroscopically flat" Mode III fracture surface of a shaft, which had been loaded with a high torsional load amplitude, is shown and in Fig. 3*b* a "factory roof" fracture of a shaft, which was loaded with a low torsional amplitude (similar fracture modes were found for mild steel [*35*]). Fracture (*a*) is a Mode III fracture, whereas Fracture (*b*) consists of planes which are oriented at 45 deg to the specimen axis; crack advance occurred in Mode I. The "factory roof" type fracture surfaces will strongly interact with each other during loading, causing interlocking and friction, thus reducing the crack growth rates drastically compared with the case of macroscopically flat fracture surfaces. Macroscopically flat fracture occurs at ΔK_{III} values higher than 20 MPa$\sqrt{m}$ for AISI 4340 steel and $\Delta\Gamma_{III}$ higher than 8×10^{-4} mm for AISI C1018. Below these values a "factory roof" type is observed.

Experimental Mode III fatigue crack growth rates depend on the crack depth c as shown in Fig. 4 for AISI 4340 steel and in Fig. 5 for mild steel AISI C1018. The dashed line in

FIG. 3—*Fatigue fracture surface of AISI 4340 steel produced in* (a) *constant* ΔK_{IIInom} = *45 MPa$\sqrt{m}$, macroscopically flat Mode III fracture type; and* (b) *constant* ΔK_{IIInom} = *12 MPa$\sqrt{m}$, "factory roof" type.*

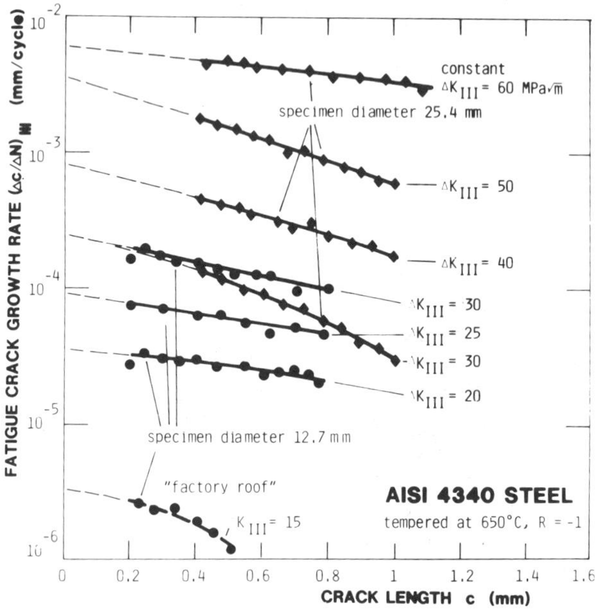

FIG. 4—*Comparison of fatigue crack growth rates in Mode III as a function of crack length* c *(emanating from the notch root) for 12.7-mm- and 25.4-mm-diameter specimens stressed at constant* ΔK_{IIInom} *values.*

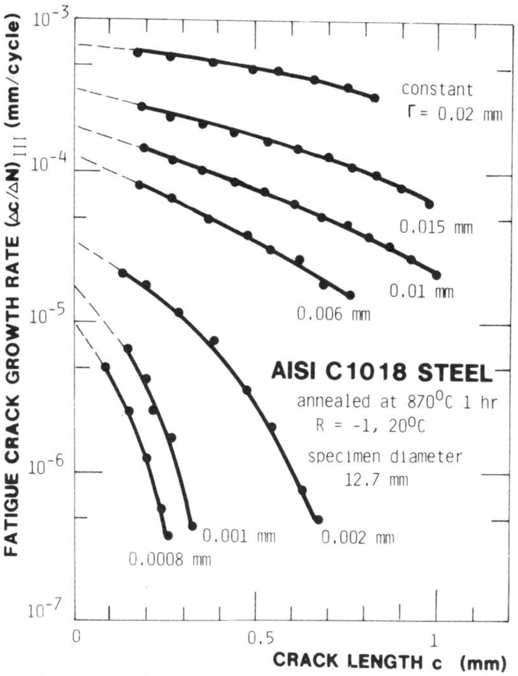

FIG. 5—*Mode III fatigue crack growth rates* $(\Delta c/\Delta N)_{III}$ *in AISI C1018 steel versus crack length* (*depth*) c (*emanating from the notch root*) *loaded under different constant nominal plastic strain intensity* $\Delta\Gamma_{IIInom}$.

Fig. 4 ($\Delta K_{\text{IIInom}} = 15$ MPa$\sqrt{\text{m}}$) points to the fact that crack propagation did not occur in Mode III.

Furthermore, Fig. 4 and especially Fig. 5 show that the slope of the curves increases with decreasing cyclic amplitudes and deviates to the maximum extent from a horizontal line for the smallest amplitudes, indicating qualitatively that the SMCC is more effective at low loading amplitudes than at high ones. Moreover, the slopes of the curves vary systematically with specimen diameter for a given cyclic Mode III stress intensity value (see Fig. 4: $\Delta K_{\text{IIInom}} = 30$ MPa$\sqrt{\text{m}}$). Therefore it may be concluded qualitatively that the SMCC effect decreases in magnitude with decreasing shaft diameter.

In contrast to this, the crack growth rates resulting from extrapolation to zero crack length are identical for the two shaft diameters at $\Delta K_{\text{IIInom}} = 30$ MPa$\sqrt{\text{m}}$. Obviously the "true" crack growth rates are independent of the specimen geometry, at least in the indicated range, and are a material specific property.

"True" Crack Growth Curve

Crack growth rates not influenced by SMCC were obtained by the above-mentioned extrapolation procedure [*19*]. In Figs. 4 and 5 the extrapolations are indicated by dashed lines. If the resulting estimated crack growth rates are plotted versus the cyclic stress intensity or plastic strain intensity, the so-called "true" crack growth curve is obtained which is based on the values of ΔK_{IIIeff} and $\Delta\Gamma_{\text{IIIeff}}$. For AISI 4340 steel this is shown in Fig. 6 and for AISI C1018 steel in Fig. 7. The "true" crack growth values which were obtained for AISI 4340

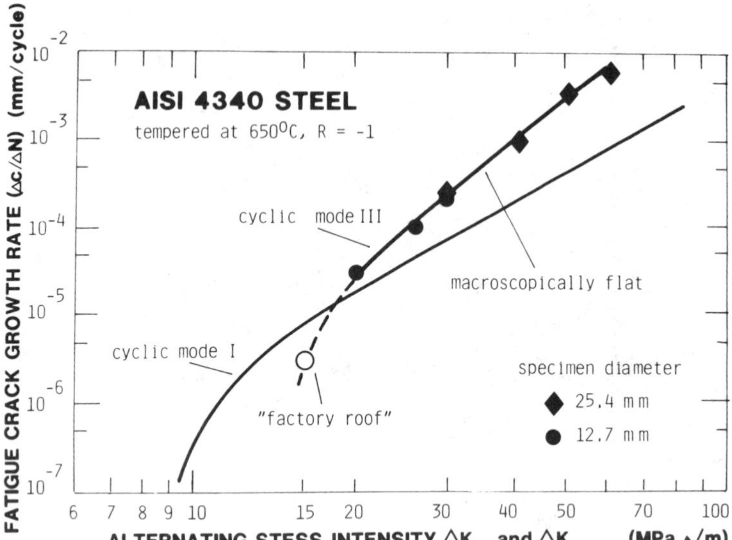

FIG. 6—*Comparison of fatigue crack growth rates in Modes I and III in AISI 4340 steel as a function of* K_I *and* ΔK_{IIIeff}. *Mode III values obtained through an extrapolation procedure from Fig. 4.*

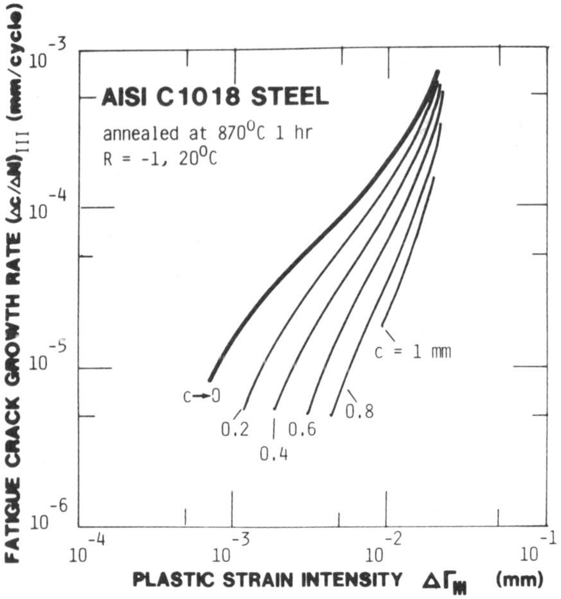

FIG. 7—*Mode III fatigue crack growth rates* $(\Delta c/\Delta N)_{III}$ *versus nominal plastic strain intensity* $\Delta\Gamma_{IIInom}$ *for different crack lengths (depths)* c *in AISI C1018 steel. The curve for* $c \rightarrow 0$ *comes from the extrapolated value of Fig. 5. In absence of dissipation processes, all curves would coincide.*

steel from specimens with different diameter lie on a straight line as shown in Fig. 6. These Mode III crack growth rates may be described by a power law related to the stress intensity, as shown in Refs *12* and *20* for the same material. The following equation is obtained

$$\left(\frac{\Delta c}{\Delta N}\right)_{III} = 8.7 \cdot 10^{-12}(\Delta K_{IIIeff})^5$$

where $\Delta c/\Delta N$ is in mm per cycle and ΔK_{IIIeff} in MPa$\sqrt{m}$.

In an analogous manner the "true" crack growth rates may be correlated with ΔCTD_{IIIeff} for AISI C1018 steel [*19,20*].

In order to compare the "true" fatigue crack growth behavior in Mode III with that in Mode I, the Mode I fatigue crack growth behavior [*12*] is also shown in Fig. 6. The "true" Mode III curve lies above the Mode I curve and has a larger slope. This result is in contrast with the results which have been published to date, for example, Refs *11* and *12*, and which show much slower fatigue crack growth rates in Mode III compared to Mode I for the same material. These different results can now be explained in light of SMCC as this has not been considered in Refs *11* and *12*; there the crack growth rates were related to the nominal loads. The higher crack growth rates for Mode III than for Mode I loading and the steeper $(\Delta c/\Delta N)_{III}$ versus ΔK_{IIIeff} curve may be explained according to Ref *20* by the fact that at high load levels crack-tip blunting is substantial in Mode I loading. This is not the case for Mode III loading as the crack theoretically remains closed during loading.

Another possibility for presenting Mode III crack growth results is shown in Fig. 7 for mild steel. Here the crack growth curves are plotted with crack depth c as parameter. The "true" crack growth rates ($c \rightarrow 0$) are highest; with increasing crack length the curves shift to lower crack growth rates and the slope of the curves increases. With the aid of this set of plots the reason for the discrepancy between the "true" crack growth is found, and the results of Refs *11* and *12* become quite obvious, as these measurements were performed at crack depths of 2 to 4 mm. Again the great importance of considering the SMCC is demonstrated.

Analogous to the equation for the "true" crack growth behavior of AISI 4340 steel a power-law fit may be given for the central region of the "true" crack growth curve for AISI C1018 steel in Fig. 7

$$(\Delta c/\Delta N)_{III} = 0.035 \cdot (\Delta\Gamma_{IIIeff})^{1.13}$$

where $(\Delta c/\Delta N)_{III}$ is in mm/cyclic and $\Delta\Gamma_{IIIeff}$ in mm.

With the aid of the "true" crack growth curves it is possible to characterize the influence of the SMCC quantitatively; this will be shown in the next chapter.

Quantitative Characterization of the Sliding Mode Crack Closure

One possibility for characterizing SMCC is to plot the effective cyclic stress or plastic strain intensity values versus crack length. The difference between effective and nominal values gives a measure of SMCC. Such a plot is used in the following to show how the SMCC effect is influenced by specimen diameter. From the measured crack growth rates for AISI 4340 steel specimens with diameters of 12.7 and 25.4 mm, fatigued at ΔK_{IIInom} = 30 MPa$\sqrt{m}$ (see Fig. 4), and from the "true" crack growth curve (Fig. 6), the effective ΔK values can be determined, as crack propagation is promoted by the effective stress intensity alone. In Fig. 8 the result of this procedure demonstrates that the effective stress intensity

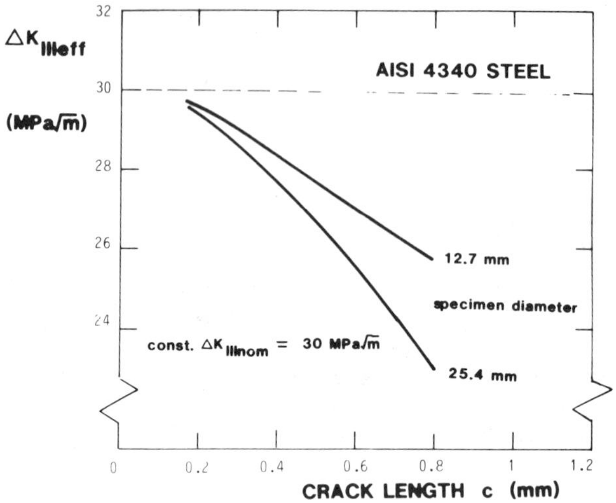

FIG. 8—*Effect of specimen diameter on the SMCC influence. In absence of the SMCC, both curves would coincide with the dashed line.*

values decrease with increasing crack length. This difference to the nominal stress intensity values (dashed line) is more pronounced for specimens with the greater diameter than for those with the smaller diameter. The reduction of nominal values to smaller effective values is a quantitative measure of the SMCC. This interesting result has been explained by a model assuming a set of inclined planes for representing the microscopically rough fracture surfaces [*20*]. During loading these planes rub against each other and give about 40% higher fracture surface pressure for specimens with a large diameter than with a smaller diameter. A qualitative experimental verification of this model is the observation that much more ferric oxide (Fe_2O_3) powder is squeezed out from the crack in thick specimens during fatigue than in thin ones. Production of Fe_2O_3 powder is a typical result of metallic abrasion, wear, and friction [*30*].

Another possibility for describing SMCC quantitatively is to plot the difference between the nominal and effective cyclic plastic strain (or stress) intensities versus the nominal values with crack depth as parameter. This is shown in Fig. 9 for mild steel. The difference between $\Delta\Gamma_{IIInom}$ and $\Delta\Gamma_{IIIeff}$ (effective at the crack tip) is called $\Delta\Gamma_{IIIR}$

$$\Delta\Gamma_{IIIR} = \Delta\Gamma_{IIInom} - \Delta\Gamma_{IIIeff}$$

The numerical values for $\Delta\Gamma_{IIIR}$ can be determined from Fig. 7. A given plastic strain intensity value and a given crack length correspond to a certain fatigue crack growth rate as seen in Fig. 7. The horizontal distance between the associated point in Fig. 7 and the "true" crack growth curve ($c \to 0$) gives $\Delta\Gamma_{IIIR}$. These distances are plotted in Fig. 9 for crack lengths between 0.2 and 1 mm versus $\Delta\Gamma_{IIInom}$. $\Delta\Gamma_{IIIR}$ increases for all crack depths with increasing nominal plastic strain intensity until a maximum is reached at a $\Delta\Gamma_{IIInom}$ of approximately 0.015 mm. This increase shows that the energy-dissipative processes (friction, abrasion, and mutual support) increase in extent with increased loads and ΔCTD_{III} values. The existence of a maximum and the following drop of $\Delta\Gamma_{IIIR}$ is explained in Ref *3* as the result of Mode I crack opening occurring at higher Mode III cyclic loads which, reduces SMCC according

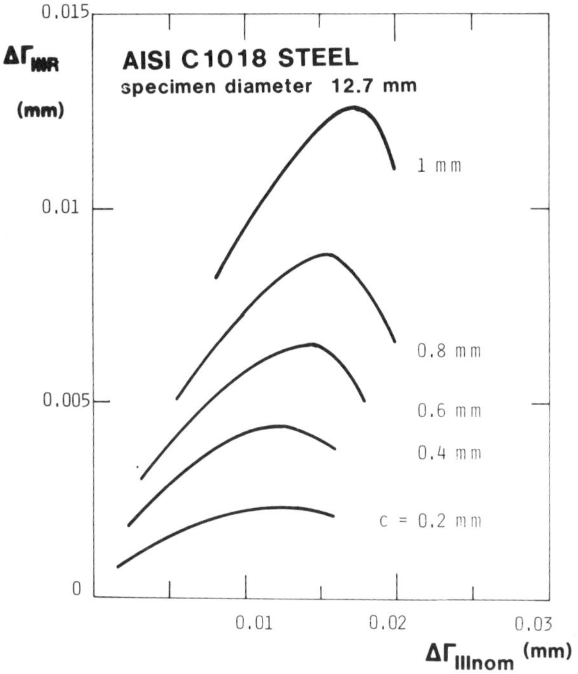

FIG. 9—*Dissipated strain intensity* $\Delta\Gamma_{IIIR}$ *versus nominal plastic strain intensity* $\Delta\Gamma_{IIInom}$ *for different crack lengths (depths) in AISI C1018 steel specimens with a diameter of 12.7 mm.*

to McClintock.[3] The Mode I crack opening is brought about by large Mode III plastic zones. The work of Rie and Stüwe [*31*] support this assumption. They found that length changes of up to 20% occur during rotational fatigue of metal shafts with torsional amplitudes chosen to give substantial plasticity. The vertical shift of the curves in Fig. 9 for different crack lengths (depths) c is approximately linearly related to crack length, which indicates that $\Delta\Gamma_{\mathrm{IIIR}}$ per unit crack length is approximately the same for each $\Delta\Gamma_{\mathrm{IIInom}}$ value. The values in Fig. 9 are valid only for mild steel rods with a diameter of 12.7 mm. However, it is the authors' belief that the general features of the SMCC discussed here are of general validity.

Results of Cyclic Mode III + Static Mode I Fatigue Crack Growth Measurements

A static axial load superimposed on cyclic Mode III loading causes Mode I crack opening, which reduces the sliding mode crack closure. Such superposition of a static axial load must therefore be studied in order to understand the nature of sliding mode crack closure. Using the measurement procedure described earlier, crack growth rates have been measured at a low cyclic stress ($\Delta\Gamma_{\mathrm{IIInom}} = 0.002$ mm) and substantially higher value ($\Delta\Gamma_{\mathrm{IIInom}} = 0.015$ mm) with a superimposed stress intensity K_{I} taking on several values between $K_{\mathrm{I}} = 0$ and $K_{\mathrm{I}} = 10.5\ \mathrm{MPa}\sqrt{\mathrm{m}}$ [*21*]. These results are shown in Fig. 10 in relation to the crack length (depth) c. Because of sliding mode crack closure, the crack growth rates decrease in all cases with increasing crack length. By means of the superimposed static load the sliding mode crack

[3] McClintock, F. A., personal communication, May 1981.

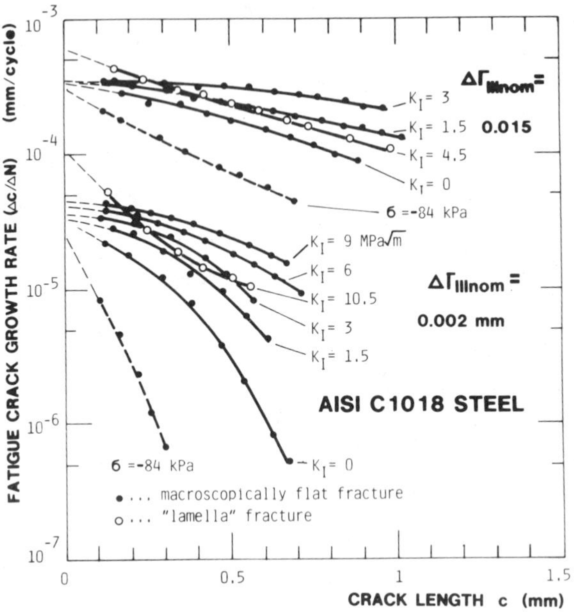

FIG. 10—*Mode III fatigue crack growth rates* $(\Delta c/\Delta N)_{III}$ *in AISI C1018 steel versus crack length (depth)* c *(emanating from the notch root) fatigued in plastic strain intensity control* $(\Delta\Gamma_{IIInom}$ *= constant) with different superimposed static axial loads* $(K_I$ *= constant).*

closure influence is reduced. This reduction is smaller at high cyclic Mode III loads than at low ones. The complete elimination of the influence of SMCC using static axial loads does not seem possible in the cyclic Mode III load range studied, as the curves in Fig. 10 still show a negative slope for $K_I = 3$ MPa$\sqrt{m}$ at $\Delta\Gamma_{IIInom} = 0.02$ mm (which is equivalent to $CTD_I = 3.2$ μm) and for $K_I = 9$ MPa$\sqrt{m}$ at $\Delta\Gamma_{IIInom} = 0.002$ mm ($CTD_I = 30$ μm). This result indicates that the fracture surface roughnesses of the macroscopically flat fractures are greater than 3.2 and 30 μm, respectively; microscopic measurements of metallographic sections confirm this result. The fracture surface roughness depends not only on the material but also on the torque level [*3*].

Further increases of K_I to 4.5 MPa$\sqrt{m}$ and 10.5 MPa$\sqrt{m}$, respectively, cause a change of the fracture mode and change the crack growth behavior completely, as shown in Fig. 10. The fracture changes from a macroscopically flat mode to a so-called "lamellar" fracture [*21*]. This fracture consists of neighboring lamellae which are separated by Mode III + II cracks lying in planes parallel to the specimen axis. A typical micrograph is shown in Fig. 11.

Another interesting result is obtained by extrapolating the curves in Fig. 10 to a crack depth $c \rightarrow 0$ as in Figs. 4 and 5. Identical "true" crack growth rates result for the curves with $K_I = 0$, 1.5, and 3 MPa$\sqrt{m}$ and $\sigma = -84$ kPa (pressure) at high and low Mode III load levels. Further increase of K_I to 6 to 9 MPa$\sqrt{m}$ results in a minor increase of the "true" crack growth rates at the low Mode III load level.

In summary, the following results have been obtained: The fatigue crack growth rates increase if an axial load is applied by reducing the SMCC. The "true" crack growth rates are, however, almost independent of the static axial load. The SMCC effect cannot be completely removed by superimposing a static axial load. A quantitative characterization

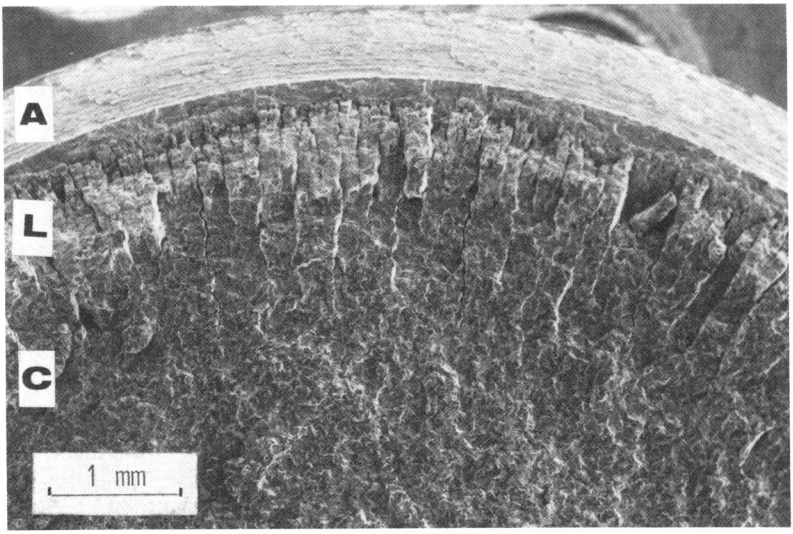

FIG. 11—*Fracture surface after fatigue with* $\Delta\Gamma_{IIInom}$ = *0.015 mm and a superimposed static axial load* K_I = *4.5* $MPa\sqrt{m}$ *showing a typical "lamella" fracture* (A = *notch,* L = *lamella fracture,* C = *impact fracture at* LN_2).

of the sliding mode crack closure influence during combined loading is possible and is given in Ref *36*.

Scientific Relevance of the Results

1. Crack growth models are usually based on material properties alone and do not account for SMCC under Mode III or Mode II loading (for example, Ref *22*). In view of this, the "true" crack growth curves described above, which do not depend on crack depth and specimen diameter, seem to serve as an important basis for examining the validity of models for Mode III, Mode II, and combined-mode fatigue crack propagation. The reduction of SMCC at high $\Delta\Gamma_{IIInom}$ values (Fig. 9) has been explained earlier as resulting from Mode I opening accompanying large Mode III plastic zones. The fundamental question arises as to which forces lead to the Mode I crack opening displacement during Mode III loading.

Witzel [*32*] has studied longitudinal elongations at forces accompanying unidirectional torsional loading of polycrystalline cylinders and tubes. He showed that the specimens become longer or pressure forces arise within the specimens if the applied torque is increased. After a maximum is reached the elongations or pressure forces decay and negative elongations or tensile forces result, if the applied torque is further increased. These measured tensile and pressure forces are small compared with the applied forces. If these results on unidirectional [*32*] and cyclic [*31,33*] torsional loading of cylindrical and tubular specimens are qualitatively applied to the Mode III plastic zone of polycrystalline material, it may be surmised that tensile forces arise which may tear apart the material at the crack tip (high-load situation). Farther away from the crack tip the plastic zone expands and pressure forces arise (low-load situation) as shown schematically in Fig. 12. The local forces in the plastic zone will have significant fluctuating components depending on the applied torsional loads. By means of this second-order effect the development of small Mode I crack opening

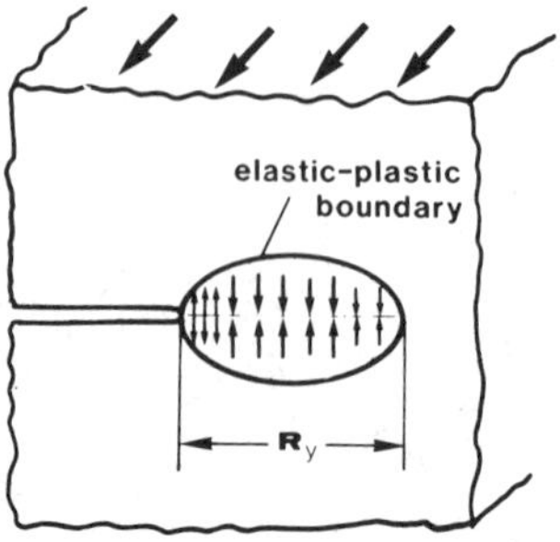

FIG. 12—*Schematic illustration of proposed model for Mode I crack opening produced by pure cyclic Mode III loading.*

displacements could perhaps be explained as a result of the above-mentioned force distribution within the plastic zone. These forces mitigate against complete rewelding of the newly created fracture surface, thus encouraging further crack propagation. If the plastic zone size is a large fraction of the specimen diameter the pressure forces within the plastic zone may be large, thus wedging open the crack tip and reducing sliding mode crack closure dramatically.

2. Fractographic investigation has given the following important results:

(i) A fracture-mode transition from a macroscopically flat Mode III fracture at high cyclic torque levels to a "factory roof" type fracture at low cyclic torque levels is observed (Fig. 3). Several models for this change are discussed in Ref *20* and compared with results from the literature. In general, a change in the fracture mode occurs near the point of intersection of Mode I and Mode III fatigue crack growth curves (Fig. 6). The change of the fracture mode is probably initiated by branch cracks which form at inclusions within the plastic zone during the process of plastic deformation. At low load levels the Mode I branch cracks grow faster than the Mode III crack and lead to a "factory roof" fracture. For high load levels the Mode III + II branch cracks dominate crack growth in the longitudinal or transverse planes of the specimen and a macroscopically flat Mode III fracture occurs.

(ii) A second kind of fracture-mode change is the change from a macroscopically flat Mode III fracture to a "lamellar" fracture, which occurs during combined cyclic Mode III + static Mode I loading (see Fig. 11). The mechanism for this fracture mode change is not yet fully understood. Several simple models have been discussed in Ref *21*. These models are based on the assumption that shear cracks are formed predominantly in the axial direction, if the axial static load which is superimposed on the cyclic torsional load is large enough.

(iii) In several scanning electron microscopic (SEM) studies of macroscopically flat Mode III fracture surfaces, fine striations parallel to the crack front have been found in the close vicinity of the crack tip, where they have not been entirely destroyed by rubbing of the fracture surfaces. They are called "Mode III striations" according to Ref *3* and are characteristic of Mode III fractures. The Mode III striation distances are bigger for high cyclic loading levels than for low ones. However, the striation spacings may be identified with the crack growth increments only qualitatively, in contrast to the Mode I striations which coincide quantitatively with the crack growth distances in a defined stress intensity range. Another difference between Mode I and Mode III striations is that Mode III striations are limited in their length by "balconies," which is not the case for Mode I striations. A typical scanning electron micrograph of a Mode

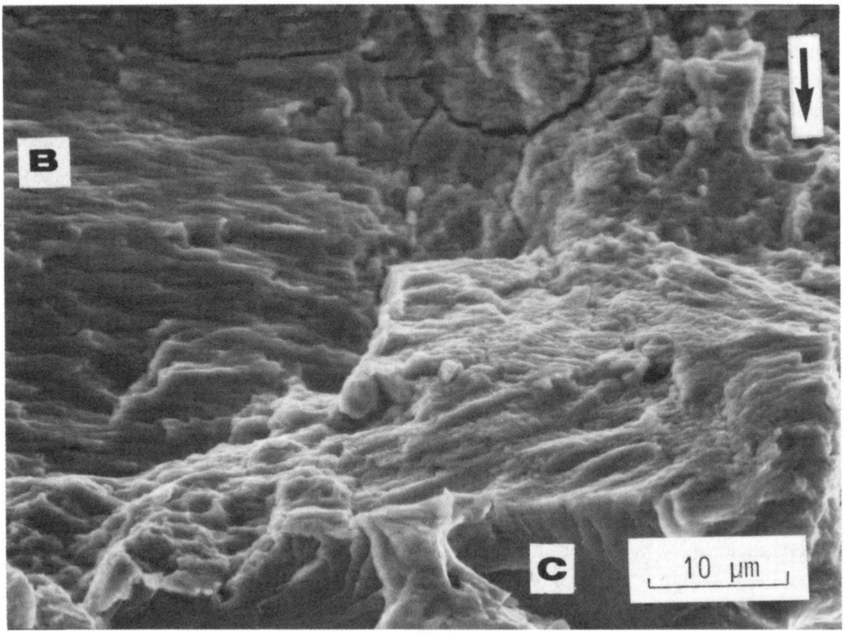

FIG. 13—*Scanning electron micrograph of a Mode III fracture surface of AISI C1018 steel fatigued at* $\Delta\Gamma_{IIInom}$ = *0.002 mm,* K_I = *6* $MPa\sqrt{m}$ *(refers to* $\Delta\Gamma_{IIIeff} = 8 \times 10^{-4}$ *mm). Arrow shows crack growth direction* (B = *typical Mode III striations,* C = *final fracture at liquid nitrogen temperature*).

III fracture surface of AISI C1018 steel is shown in Fig. 13. Clear Mode III striations may be seen in (B). How these striations are formed is not yet known.

3. Friction and wear of rubbing fracture surfaces occur during crack propagation under Mode II loading similar to that seen under Mode III loading. Therefore it would be of interest to know whether Mode II crack growth rates are also influenced by sliding mode crack closure. The experimental and analytical techniques reported in studies for Mode III may be of relevance in answering this question. Parallels between Mode III and Mode II loading have been discussed previously [*19,29,34*].

Practical Relevance of the Results

Cyclic torsional loading frequently occurs in parts of rotating power trains and force transmission systems of machines and other mechanical systems (such as motor vehicles, power plants, elevators, and trains). Knowledge of crack propagation behavior in structural materials can be used to help prevent unexpected failures.

1. Sliding mode crack closure has to be considered in lifetime predictions and failure risk estimations of torsionally loaded parts, as the extent of sliding mode crack closure will influence the crack growth behavior and lifetime of the part drastically. If lifetime calculations are based on the "true" crack growth curves (where the SMCC is zero) conservative results will be obtained. If, however, SMCC is taken into consideration, several important points must be kept in mind. Two such points of practical importance are:
   (i) A superimposed static Mode I load reduces the SMCC during cyclic Mode III loading (as mentioned earlier). Thus the effective stress intensity at the crack tip is increased in spite of constant nominal loading and more rapid crack propagation results.

(ii) SMCC is also influenced by the applied cyclic Mode III level, (as mentioned earlier). At high torsional load levels it is assumed that Mode I crack opening takes place, reducing the SMCC influence. This is especially important for multistage loading conditions. For example, SMCC is reduced after an overload. This is likely due to the overload-induced Mode I crack opening displacement which increases the effective stress or plastic strain intensity at the crack tip. The crack growth rates are therefore higher after an overload sequence, although the nominal load is being held constant. This behavior is in contrast to Mode I crack growth behavior where crack growth retardation occurs after an overload.

2. The various fracture modes may allow a failure analyst to readily determine whether a part failed under high cyclic torque levels (macroscopically flat fracture) or under low cyclic torque levels ("factory roof" fracture). If "lamellar" fracture is found, this indicates that combined loading conditions (cyclic Mode III + static Mode I) prevailed.

3. The fracture mode change from a macroscopically flat to a "factory roof" fracture restricts "true" Mode III fatigue crack propagation to higher load levels. The minimum load level consistent with a macroscopically flat fracture surface may be viewed as a "Mode III threshold value." This is not a threshold value in the classic sense, in that crack propagation occurs below this value, though with a different macroscopic fracture mode.

4. If a crack under Mode III or Mode II loading is monotonically stressed to final fracture, microscopically rough fracture surfaces slide over each other. The authors do not know any study of the influence of SMCC on the results of K_{IIIC} or K_{IIC} tests. This would be of some significance in reference to the ASTM draft standard E 24.01.05 (1981) on the measurement of K_{IIIC}.

5. To date, results on the influence of the sliding mode crack closure exist only for steels. The authors believe, however, that the sliding mode crack closure occurs in other materials such as cements, concrete, plastics, minerals, and composite materials and is of great practical importance in these, too.

Summary

The significance of sliding mode crack closure during pure cyclic Mode III and during combined loading (cyclic Mode III + static Mode I) has been demonstrated for a high-strength steel (AISI 4340) and a mild steel (AISI C1018). The results show that the extent of SMCC depends primarily on crack depth, material properties, specimen diameter, and loading level. Crack growth rates without the SMCC influence (called "true" crack growth rates) can be determined by an extrapolation procedure which uses results obtained from crack growth measurements under "stress intensity" or "plastic strain intensity" control. These crack growth rates give the "true" crack growth curve which is obviously independent of crack length (depth) and specimen diameter and is therefore a material characteristic for Mode III fatigue crack growth. With the aid of the "true" crack growth curve the extent of SMCC can be quantitatively determined and the change in fracture mode can be explained.

During combined loading, SMCC is drastically reduced by the applied Mode I opening; the "true" crack growth rates are, however, only slightly influenced.

Finally, the technical and scientific importance of the SMCC is demonstrated for Mode III (and Mode II) loading.

Acknowledgment

Some results reported here were obtained during a sabbatical stay at Massachusetts Institute of Technology. We want to thank Professor A. S. Argon, who initiated this work.

Thanks are due as well to Professors F. A. McClintock and R. O. Ritchie for making these experiments possible. Special thanks are due to Mr. W. Moffatt for numerous helpful discussions.

References

[1] Elber, W. in *Damage Tolerance in Aircraft Structures, ASTM STP 486,* American Society for Testing and Materials, Philadelphia, 1971, pp. 230–242.
[2] Ritchie, R. O. and Suresh, S., *Metallurgical Transactions,* Vol. 13A, 1982, p. 937.
[3] Tschegg, E. K., *Acta Metallurgica,* Vol. 31, 1983, pp. 1321–1330.
[4] McClintock, F. A. in *Proceedings,* International Conference on Fatigue of Metals, Institution of Mechanical Engineers, London, 1956, pp. 538–542.
[5] Hult, J. A. H., *Transactions,* Royal Institute of Technology, Stockholm, No. 119, 1958, pp. 1–47.
[6] Walsh, J. B. and Mackenzie, A. C., *Journal of the Mechanics and Physics of Solids,* Vol. 7, 1959, pp. 247–257.
[7] Kayan, I., *Bulletin of the Technical University of Istanbul,* Vol. 14, 1969, pp. 45–49.
[8] Hult, J. A. H. and McClintock, F. A. in *Proceedings,* International Congress on Applied Mechanics, University of Brussels, Vol. 8, 1957, pp. 51–58.
[9] Rice, J. R. in *Fracture,* Volume 2, H. Liebowitz, Ed., Academic Press, New York, 1968, pp. 191–311.
[10] Rice, J. R. in *Fatigue Crack Propagation, ASTM STP 415,* American Society for Testing and Materials, Philadelphia, 1967, pp. 247–311.
[11] Hurd, N. J. and Irving, P. E. in *Design of Fatigue and Fracture Resistant Structures, ASTM STP 761,* American Society for Testing and Materials, Philadelphia, 1967, pp. 212–233.
[12] Ritchie, R. O., McClintock, F. A., Nayeb-Hashemi, H., and Ritter, M. A., *Metallurgical Transactions,* Vol. 13A, 1982, pp. 101–110.
[13] Pook, L. P. and Sharples, J. K., *International Journal of Fracture,* Vol. 15, 1979, pp. R223–R225.
[14] Pook, L. P., NEL-Report No. 667, National Engineering Laboratory, East Kilbride, Glasgow, U.K., 1980. pp. 1–11.
[15] Pook, L. P. in *Fatigue Thresholds, Fundamentals and Engineering Applications,* J. Bäcklund, A. Blom, and J. C. Beevers, Eds., Engineering Materials Advisory Services Ltd., Warley, West Midland, U.K., Vol. 2, 1982, pp. 1007–1032.
[16] Tschegg, E. K., "Results of Torsional Tests on AISI 4340 Steel at Various Heat Treatments," Research Note, Massachusetts Institute of Technology, Cambridge, MA, Feb. 1981.
[17] Tschegg, E. K., Ritchie, R. O., and McClintock, F. A., *International Journal of Fatigue,* Jan. 1983, pp. 29–35.
[18] Nayeb-Hashemi, H., McClintock, F. A., and Ritchie, R. O., *Metallurgical Transactions,* Vol. 13A, 1982, pp. 2197–2204.
[19] Tschegg, E. K., *Materials Science and Engineering,* Vol. 54, 1982, pp. 127–136.
[20] Tschegg, E. K., *Journal of Materials Science,* Vol. 18, 1983, pp. 1604–1614.
[21] Tschegg, E. K., *Materials Science and Engineering,* Vol. 59, 1983, pp. 127–137.
[22] McClintock, F. A. and Ritchie, R. O. in *Mechanics of Fatigue,* Mura T., Ed., American Society of Mechanical Engineers, AMD, Vol. 47, 1982, pp. 1–11.
[23] Nayeb-Hashemi, H., McClintock, F. A., and Ritchie, R. O., *International Journal of Fracture,* Vol. 23, 1983, pp. 163–185.
[24] Nayeb-Hashemi, H., McClintock, F. A., and Ritchie, R. O., *Engineering Fracture Mechanics,* Vol. 18, 1983, pp. 763–783.
[25] Nayeb-Hashemi, H., Suresh, S., and Ritchie, R. O., *Materials Science and Engineering,* Vol. 59, 1983, pp. L1–L5.
[26] Pook, L. P. in *International Symposium on Biaxial/Multiaxial Fatigue, ASTM STP 853,* American Society for Testing and Materials, Philadelphia, 1985, pp. 249–263.
[27] Ritter, M. A. and Ritchie, R. O., *Fatigue of Engineering Materials and Structures,* Vol. 5, 1982, pp. 91–104.
[28] McClintock, F. A. and Tschegg, E. K., *International Journal of Fracture,* Vol. 26, 1984, pp. R11–R14.
[29] *Metals Handbook,* American Society for Metals, Metals Park, OH, Vols. 9 and 10, 1974.

[30] Waterhouse, R. B. in *Corrosion Fatigue,* A. J. McEvily and R. W. Staehle, Eds., *Proceedings,* National Association of Corrosion Engineers Conference, University of Connecticut, Storrs, CT, June 14–18, 1971, pp. 608–616.
[31] Rie, T. K. and Stüwe, H. P., *Zeitschrift für Metallkunde,* Vol. 64/1, 1973, pp. 37–40.
[32] Witzel, W. in *Verformung und Bruch,* H. P. Stüwe, Ed., Verlag der österreichischen Akademie der Wissenschaft, Vienna, 1979, pp. 103–115.
[33] Rose, W. and Stüwe, H. P., *Zeitschrift für Metallkunde,* Vol. 59, 1968, p. 369.
[34] Otsuka, A., Mori, K., Oshima, T., and Tsumaya, S. in *Advances in Fracture Research, Proceedings,* Fifth International Conference on Fatigue, D. Francois, Ed., Cannes, France, 1981, pp. 1851–1858.
[35] Tschegg, E. K. and Stanzl, S. E., DGM-Hauptversammlung (ER02), Stuttgart, May 1985.
[36] Tschegg, E. K., *Theoretical and Applied Fracture Mechanics,* Vol. 3, 1985, pp. 157–178.

Michael W. Brown,[1] *Eduardo R. de los Rios,*[1] *and Keith J. Miller*[1]

A Critical Comparison of Proposed Parameters for High-Strain Fatigue Crack Growth

REFERENCE: Brown, M. W., de los Rios, E. R., and Miller, K. J., **"A Critical Comparison of Proposed Parameters for High-Strain Fatigue Crack Growth,"** *Basic Questions in Fatigue: Volume I, ASTM STP 924,* J. T. Fong and R. J. Fields, Eds., American Society for Testing and Materials, Philadelphia, 1988, pp. 233–259.

ABSTRACT: A number of elastic-plastic fracture mechanics parameters have been suggested for the description of fatigue crack growth, of which three are considered in this paper: the J-integral, crack-tip opening displacement, and plastic zone size. A critical comparison of the effectiveness of each parameter may be devised from a study of Mode I fatigue crack growth rates under various states of biaxial stress. Data for one heat of AISI 316 austenitic stainless steel at 20°C and 550°C showed that, if crack closure is taken into account, the cyclic plastic zone size is the most suitable correlating parameter for stress levels up to the yield value. A second heat of 316 stainless steel showed a preference for crack opening displacement control of crack growth. The analysis is based on the Dugdale model for crack-tip plasticity.

KEY WORDS: biaxial stresses, crack propagation, fatigue (materials), plastic zone size, crack opening displacement, J-integral, stress intensity, stainless steels, crack closure, stress ratio

Nomenclature

A	Constant defined by Eq 15
a	Crack length
b	Crack length plus cyclic plastic zone
C	Constant
c	Crack length plus monotonic plastic zone
CFD	Displacement between crack faces at $x = 0$
E	Young's modulus
E'	$E/(1 - \nu^2)$
F	Stress intensity geometry factor
f	Angular function for stress intensity factor
h	Distance of point load from crack centerline
J	J-integral
K	Stress intensity factor
k	a/c
M	Constant
N	Number of cycles
P	Applied load
R	Ratio of minimum and maximum loads
r	Distance from crack tip
r_p	Plastic zone size

[1] Lecturer, senior research fellow, and professor, University of Sheffield, Mechanical Engineering Department, Mappin Street, Sheffield, United Kingdom.

T T-stress
W Half specimen width
W^* Modified specimen width
x,y Cartesian coordinates
Y Flow stress
Z Function of complex variable
z $x + iy$
α Weighting factor
Δ Range
δ Crack-tip opening displacement
ϵ Strain
θ Polar coordinate
Λ σ_x/σ_y, biaxiality ratio
ν Poisson's ratio
σ Normal stress
τ Shear stress
ϕ Function defined by Eq 32
ψ Crack face displacement

Subscripts

c Compressive or cyclic
d Crack-tip closure
e Effective closure stress
f Crack face closure
o Tensile cohesive stress
p Plastic
th Threshold
x,y Cartesian coordinate
ϵ Strain intensity factor
I Mode I opening
1,2,3 Principal values
∞ Infinite plate value

For a number of years, linear elastic fracture mechanics (LEFM) has been used to provide a succinct description of fatigue crack growth, where the speed of a crack is related to the stress intensity factor at the crack tip. Whether in terms of the cyclic range or a combination of range and peak values, this relationship between stress intensity factor and the crack growth rate, da/dN, provides not only a useful tool for the designer but also a basis for understanding the crack extension process for the scientist. However, its applicability is limited by the necessary condition for valid LEFM analyses that the stress far from the crack should be well below yield. A corrollary to this condition is that cracks must be "long" when LEFM is applicable, that is, the crack length

$$a > (\Delta K_{th}/F\Delta\sigma)^2/\pi \quad (1)$$

where

$$\Delta\sigma < Y/3 \quad (2)$$

if the crack is to grow. For many steels, ΔK_{th}, the threshold stress intensity factor range, is typically 5 MPa $\sqrt{m}$, the geometry factor F is of order 1.0, and for LEFM the applied stress range $\Delta\sigma$ should be less than about 80 MPa, giving $a > 1$ mm.

But frequently cracks are "short," either physically short because a is less than 1 mm and becomes comparable to typical microstructural dimensions, or short in comparison with the plastic zone at the crack tip because the stress range, $\Delta\sigma$, exceeds $Y/3$ where Y is the yield stress. This paper is concerned with "short" cracks of the second type, since stresses are of the order of the proof stress with crack lengths in the range of 2 to 30 mm. It considers the development of possible laws to describe their growth.

A number of experimental studies have been performed in recent years [*1–4*] where extensive local plasticity at the crack tip has implied that elastic-plastic fracture mechanics (EPFM) was required to analyze the results. A corresponding variety of EPFM parameters has been proposed, each parameter intended to provide a unique description of fatigue crack growth in the same way as ΔK governs crack propagation through the Paris law. Therefore the fundamental question arises: Which is the best characterizing parameter to quantify high-strain fatigue crack growth? The answer is important for two reasons: (*a*) its practical implications in designing against fatigue, and (*b*) the fundamental guidance that may be derived concerning the mechanics of crack extension.

Previously the answer has been sought through standard fracture mechanics tests, using for example the compact tension specimen (CTS) or center-cracked plate (CCP). Since only one independent variable was available (applied load), differentiation between parameters was not easy, and frequently two theories may give equally good representations of the data [*1*]. With biaxial stress fatigue tests, however, a second independent variable is introduced since the stresses on two axes may be controlled. This paper explores the use of biaxial stress tests for the verification of EPFM crack growth laws, where the stresses normal to and parallel to a Mode I crack are cycled in phase with each other.

EPFM Parameters in Fatigue

A number of EPFM parameters have been proposed for high-strain fatigue crack growth, some common examples being the strain intensity factor [*2,3*], the cyclic J-integral [*1,4*], crack-tip opening displacement (CTOD) [*5,6*], and cyclic plastic zone size [*7,8*]. An extensive review has been written recently by Skelton [*9*] on the propagation of short fatigue cracks, describing the use of such parameters when specifically applied to fatigue. Although there is some doubt whether a single parameter can adequately describe crack tip plasticity under large-scale yielding [*10*], empirical high-strain fatigue data have been always presented in terms of such parameters, and therefore it is worthwhile to investigate the efficacy of one-parameter correlations.

Strain-based approaches pose a problem in biaxial stress investigations in that, as a rule, they have been developed for uniaxial loading alone, and have not been defined for more general multiaxial situations. If the strain range used is defined as the maximum principal strain range as measured in a uniaxial test, that is, for isotropic elastic conditions

$$\Delta\epsilon_1 = (\Delta\sigma_1 - \nu\Delta\sigma_2 - \nu\Delta\sigma_3)/E \tag{3}$$

where E is Young's modulus and ν is Poisson's ratio, then crack propagation predictions from the strain intensity factor within the LEFM regime will be at variance with the Paris law for Mode I crack propagation. This is because the Paris law implies that crack propagation is predicted from maximum principal stress through the stress intensity factor

$$\Delta K = F \cdot \Delta\sigma_1 \sqrt{\pi a} \tag{4}$$

and Eq 3 shows that stress and strain intensity factors in terms of the principal values $\Delta\sigma_1$ and $\Delta\epsilon_1$, respectively, cannot be equivalent for general multiaxial loading. Within the restrictions of LEFM, however, biaxial stress tests on steels show that the Paris law is indeed obeyed [*11–13*], which implies that the only valid definition for strain intensity factor is

$$\begin{aligned} \Delta K_\epsilon &= F(\Delta\sigma_1/E)\sqrt{\pi a} \\ &= F[\Delta\epsilon_1 + \nu(\Delta\epsilon_1 + \Delta\epsilon_2 + \Delta\epsilon_3)/(1 - 2\nu)]\sqrt{\pi a}/(1 + \nu) \end{aligned} \tag{5}$$

and therefore, for steels

$$da/dN = C(\Delta K_\epsilon)^m \tag{6}$$

The definition of ΔK_ϵ in Eq 5 may be also used beyond the realm of LEFM if the elasto-plastic value for ν is employed.

The J-integral [*14*] has been used to predict the initiation of subcritical crack growth with a large degree of success, because it characterizes the strength of the crack-tip singularity in EPFM problems in the same manner as *K* for LEFM. The extension of its use to cyclic loading [*1,4*] is therefore to be expected, and here also it has proved a useful parameter even though there is no physical basis for this approach where unloading or load reversal occurs. This is because *J* has been defined for a nonlinear elastic material, whereas in high-strain fatigue, metals exhibit nonlinear elastic-plastic behavior. Indeed the width of the hysteresis loop is seen as a fundamental control variable in many philosophies of fatigue, whereas nonlinear elastic materials do not admit such behavior.

The CTOD has been always considered as an obvious choice for a crack growth parameter, because the new surface created on opening a crack is available for extending the crack during the compressive half-cycle [*6*]. Hence crack growth rate is equal to half the CTOD. Formulas for CTOD usually are derived for local plasticity around the crack tip, as is the case in the present paper, but Tomkins [*5*] has extended the approach to include the contribution of plastic strain range in high-strain fatigue situations.

Plastic zone size [*7,8*] has been less widely used than CTOD or *J*, but the concept of plastic zone size is applicable as a parameter characterizing the local stress strain field at the crack tip. Note this is the same argument used for *J* previously, but plastic zone size has two additional advantages: (*a*) being related directly to the Tomkins crack growth equation, which should facilitate correlation of fast growing cracks; and (*b*) having an obvious physical interpretation under cyclic conditions.

The plastic zone size, J-integral, and CTOD may be determined by finite-element methods for general loading in the elastic-plastic range. However, such numerical methods are not usually extended to cyclic loading. For the analysis of extensive data generated in fatigue experiments, a closed-form solution is desirable, and therefore the Dugdale-Bilby-Cottrell-Swinden (DBCS) model is an attractive if simplified approach to deriving EPFM parameters [*15,16*]. Based on the strip yield model of crack-tip plasticity, the length of plastic zone, r_p, at the tip of a Mode I crack of length $2a$ in an infinite plate under an equibiaxial tensile stress σ is given by

$$r_p = a[\sec(\pi\sigma/2\sigma_o) - 1] \tag{7}$$

where σ_o is the cohesive stress in the plastic zone, usually taken to be the yield stress, Y. The analysis was extended by Burdekin and Stone [17] to derive the CTOD, δ, with

$$\delta = (8\sigma_o\, a/\pi E') \cdot \ln[\sec(\pi\sigma/2\sigma_o)] \tag{8}$$

where $E' = E$ for plane stress and $E' = E/(1 - \nu^2)$ for plane strain. The J-integral may be derived by taking an integration path around the edge of the plastic zone, where the stress is equal to the cohesive stress, to obtain [18]

$$J = \sigma_o\, \delta \tag{9}$$

For small-scale yielding, Eqs 7, 8, and 9 can be approximated by the first term of their respective series expansions, since $\sigma << \sigma_o$, to give (for $K = \sigma\sqrt{\pi a}$)

$$r_p = (\pi/8)(K/\sigma_o)^2$$

$$\delta = (K^2/\sigma_o\, E')$$

$$J = (K^2/E')$$

These expressions show that each parameter depends on K, so that each one may be used with equal validity compared to the Paris law within the LEFM regime. Even beyond the bounds of LEFM given by Eqs 1 and 2, J and δ are equivalent, and not markedly different from r_p, so long as σ_o is constant.

However, for biaxially stressed center-cracked panels, the term σ_o can be altered by the application of a stress parallel to the crack [11]. This stress is directly related to the T-stress, which is the nonsingular component of stress at the crack tip, parallel to the crack plane. Thus at the crack tip [19,20]

$$\begin{bmatrix} \sigma_{xx} & \sigma_{xy} \\ \sigma_{yx} & \sigma_{yy} \end{bmatrix} = \frac{K_I}{\sqrt{2\pi r}} \begin{bmatrix} f_{xx}(\theta) & f_{xy}(\theta) \\ f_{yx}(\theta) & f_{yy}(\theta) \end{bmatrix} + \begin{bmatrix} T & 0 \\ 0 & 0 \end{bmatrix} \tag{10}$$

and it has been shown that for plane stress the cohesive stress and T-stress are linked through the yield criterion [11]

$$\sigma_o = \tfrac{1}{2}(T + \sqrt{4Y^2 - 3T^2}) \tag{11}$$

Biaxial Stress Fatigue Data

A limited number of fatigue crack propagation studies conducted on steels, aluminum alloys, and polymers have been reviewed thoroughly elsewhere [21–23]. This paper concentrates on data generated for Type 316 stainless steel, using biaxially stressed cruciform specimens [11]. Two heats of this material were examined, at 20 and 550°C in a solution-annealed condition. The composition and mechanical properties are given in Tables 1 and 2. Details concerning the test procedure and the specimen may be found in Ref 11.

Figure 1 shows some typical data, plotted as crack growth rate against the function ΔK, defined by Eq 4 where

$$F = \sqrt{[\sec(\pi a/2W^*)]} \tag{12}$$

TABLE 1—*Composition of Type 316 stainless steel,[a] weight %.*

Material	C	Mn	P	S	Si	Cr	Ni	Mo	Grain Diameter, μm
A	0.06	1.88	0.023	0.020	0.62	17.30	13.40	2.34	52
B	0.049	1.36	0.023	0.018	0.54	17.26	11.20	2.15	40

[a] Solution treated at 1070°C and water quenched.

for $2W^* = 101.9$ mm, being the specimen width, but modified to allow for the stiffening effect of the edges on the K calibration [*11*]. Clearly there is no correlation of crack growth rate with respect to stress state in terms of ΔK, which is not entirely surprising in view of the high stress, well outside the LEFM condition imposed by Eq 2.

In all cases tested, the shear loading gave the highest crack growth rate, and the equibiaxial loading gave the lowest. The stress state, defined by Λ, is the ratio of applied loads parallel to and normal to the crack (Fig. 1), giving $\Lambda = -1$ for shear, $\Lambda = +1$ for equibiaxial loads. At room temperature, three stress levels were studied, as shown in Table 3, with R ratios of 0 and -1 where R is the ratio of minimum to maximum stress normal to the crack during a cycle. During cycles at $R = -1$, crack closure occurred, but the stress range $\Delta\sigma$ quoted is the full stress range, irrespective of closure effects.

The Dugdale Model Under Biaxial Cyclic Stress (DBCS)

Although the strip yield model is applicable to ideal elastic-plastic metals with no strain hardening, the availability of closed-form solutions makes the DBCS approach useful even under plane-strain conditions. The results obtained from the strip yield model have been compared with finite-element solutions for monotonic loading [*24*], and show reasonable agreement for a strain hardening material with large-scale yielding. The influence of biaxial stress is also well represented by the DBCS model when compared with numerical solutions for small-scale yielding [*24*].

Clearly the strip yield zone is physically unrealistic for strain hardening materials, in that all other solutions show two inclined shear ears for the crack-tip plastic zone at an angle between 45 and 90 deg to the crack plane. However, the expressions derived in this section agree well with those due to Rice [*19*] for the effect of T-stress on inclined plastic zones, apart from the constant of proportionality. In this paper, only Mode I fatigue crack growth will be considered, and the conclusions drawn should not be extended arbitrarily to Mode II or Mode III cracks, where a significant shear component dominates fatigue processes at the crack tip.

For monotonic loading, the EPFM analysis of crack-tip plasticity can be reduced to an elastic problem with a notional crack length $2c$ replacing the physical crack of length $2a$, following Dugdale's analysis [*15*]. This has been extended to biaxial loading, as shown in

TABLE 2—*Mechanical properties.*

Material	Temperature, °C	0.2% Proof Stress, MPa	Tensile Strength, MPa	Reduction in Area, %	E, GPa	ν
A	20	395	611	71	198	0.29
	550	268	489	54	155.1	0.332
B	20	243	597	71	198	0.29
	550	133	474	55	155.1	0.332

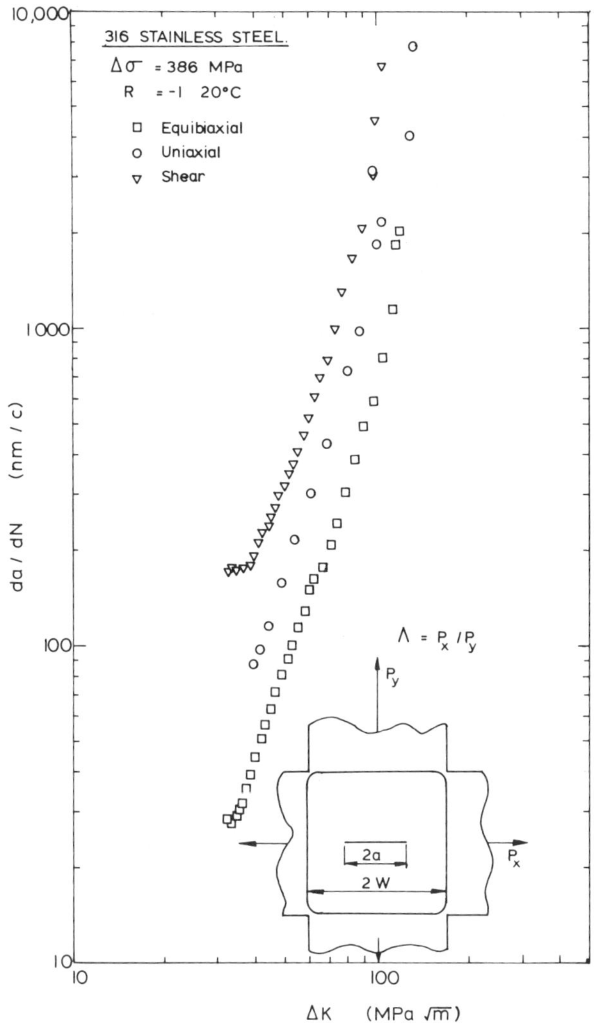

FIG. 1—*Typical biaxial fatigue crack propagation data.*

Fig. 2*a*, by using the principle of superposition for the elastic problem [*11*]. Here the Mode I crack under biaxial stress has been replaced by three separate load systems:

1. an elastic crack under equibiaxial tension, a problem already solved by Westergaard [*25*];
2. a crack loaded by the *T*-stress, where $T = (\Lambda - 1)\sigma$ is not influenced by the crack-tip singularity; and
3. a crack loaded by a cohesive stress, σ_o, within the plastic zone, $r_p = c - a$.

Equations 7, 8, and 9 give the solution to this problem where the value of the cohesive stress should be determined by the yield criterion in the plastic zone. Load System 1 gives zero stresses along the crack face, and therefore contributes nothing to the yield calculation.

TABLE 3—*Closure stress ranges and plastic zone sizes for short cracks* (a << W).

$\Delta\sigma$, MPa	R	Λ	r_p/a, $\times 10^3$	$\Delta\sigma_f$, MPa	[a]$(r_{pc}/a)_f$, $\times 10^3$	$\Delta\sigma_e$, MPa	[a]$(r_{pc}/a)_e$, $\times 10^3$	σ_o, MPa	σ_c, MPa
				MATERIAL A, 20°C					
62.7	0.0	1.00	10.4	62.7	2.6	62.7	2.6	686	1371
66.6	0.0	−0.01	13.3	66.6	3.1	66.6	3.1	646	1332
63.1	0.0	−0.99	13.3	63.1	2.9	63.1	2.9	610	1296
70.8	0.0	1.00	13.3	70.8	3.3	70.8	3.3	686	1372
191.3	−1.0	1.00	24.5	96.3	6.1	115.9	8.9	686	1372
193.0	−0.99	−0.03	30.5	97.9	6.9	119.1	10.5	626	1299
195.2	−1.01	−1.00	37.7	98.3	7.7	121.0	12.2	566	1224
387.9	−1.0	1.00	107.4	198.9	26.5	237.8	38.2	686	1372
383.0	−0.96	−0.01	170.1	203.9	34.1	247.9	53.4	563	1218
385.7	−1.0	−1.00	310.4	207.7	45.8	257.3	80.8	431	1039
				MATERIAL A, 550°C					
199.0	−1.0	1.00	41.1	100.5	10.2	120.8	14.8	555	1109
199.4	−1.0	−0.03	53.3	101.0	11.7	123.0	17.9	490	1030
198.8	−0.99	−1.01	73.0	101.8	13.6	126.0	22.4	423	945
				MATERIAL B, 550°C					
204.0	−1.0	0.99	46.3	103.2	11.5	124.0	16.7	537	1074
204.3	−1.0	−0.02	60.6	103.6	13.2	126.2	20.2	472	994
203.9	−1.0	−0.51	70.9	103.9	14.2	127.7	22.6	438	952
204.7	−1.0	−1.00	85.5	104.8	15.6	129.8	25.9	403	907

[a] $(r_{pc}/a)_f$ is derived from Eq 16 for $\Delta\sigma = \Delta\sigma_f$, and $(r_{pc}/a)_e$ is for $\Delta\sigma = \Delta\sigma_e$.

Load System 2 gives a stress of T parallel to the crack, and Load System 3 gives σ_o normal to the crack, in the plastic zone. Thus for a von Mises yield criterion and plane-stress loading

$$\sigma_o^2 + T^2 + (\sigma_o - T)^2 = 2Y^2$$

whence

$$\sigma_o = \tfrac{1}{2}(T \pm \sqrt{4Y^2 - 3T^2}) \tag{13}$$

Clearly for tensile loading with $T = 0$, $\sigma_o = Y$, and so one takes the positive sign, but for compressive loading one takes the negative sign. So for proportional loading, as used in the experiments reported here, Eq 13 may be written

$$\sigma_o = Y[(T/2Y) + \sqrt{1 - 3(T/2Y)^2}] \tag{14}$$

where $T = (\Lambda - 1)\sigma$. The corresponding equation for plane strain is derived in Appendix I, giving

$$\sigma_o = (Y/\sqrt{1 - \nu + \nu^2})\,[A\alpha T/Y + \sqrt{1 - (\alpha T/Y)^2(1 - A^2)}] \tag{15}$$

where

$$A = [(1 + \nu)/2]/\sqrt{1 - \nu + \nu^2}$$

and

$$\alpha = 1/\sqrt{2}$$

For cyclic loading, the cyclic plastic zone size may be derived by taking the stresses at maximum load as an initial or prestress condition, and superposing a compressive unloading stress of $\Delta\sigma$, as shown in Fig. 2*b*. The solution to this problem follows directly by analogy with the monotonic loading case, whence

$$r_{pc} = b - a$$

$$= a[\sec(\pi\Delta\sigma/2\sigma_c) - 1] \quad (16)$$

$$\delta_c = (8\sigma_c\, a/\pi E')\, \ln[\sec(\pi\Delta\sigma/2\sigma_c)] \quad (17)$$

$$J_c = \sigma_c\, \delta_c \quad (18)$$

where for plane stress, from Eq 13

$$\sigma_o - \sigma_c = Y[(T_c/2Y) - \sqrt{1 - 3(T_c/2Y)^2}] \quad (19)$$

with σ_o given by Eq 14. Similarly, for plane strain

$$\sigma_o - \sigma_c = (Y/\sqrt{1 - \nu + \nu^2})[A\alpha T_c/Y - \sqrt{1 - (\alpha T_c/Y)^2(1 - A^2)}] \quad (20)$$

In each case, T_c is the value of T-stress after unloading, that is, at minimum load or at the closure load if crack closure occurs, giving

$$T_c = (\Lambda - 1)(\sigma - \Delta\sigma) \quad (21)$$

Note that for equibiaxial loading, $\sigma_o = Y$ and $\sigma_c = 2Y$, giving the usual formula for cyclic plastic zone size from Eq 16 for this mode of stressing only [*18*]. For $\Lambda = 0$ and -1, the plastic zone size and CTOD will be increased above the corresponding equibiaxial values, but J is almost insensitive to the T-stress [*19*]. Figure 2*c* shows a schematic of the stresses ahead of the crack tip corresponding to maximum and minimum load in the cycle.

Note that no account has been taken for the finite width of the specimen in Eqs 16–21. The Dugdale model may be extended to CCP specimens using the result of Bilby et al. [*26*], who considered a periodic array of cracks in an infinite plate. They found that

$$r_p = \frac{2W}{\pi} \text{ arc sin} \left[\sin \frac{\pi a}{2W} \sec \frac{\pi\sigma}{2\sigma_o}\right] - a \quad (22)$$

The CCP problem has also been analyzed by Chell [*27*], who integrated an approximate Green's function solution for point loading of a crack to obtain

$$r_p = a \left\{\left[\left(1 - \frac{a}{w}\right) \cos \left(\frac{\pi\sigma}{2\sigma_o\,(1 - a/w)}\right) + \frac{a}{w}\right]^{-1} - 1\right\} \quad (23)$$

Figure 3 shows the plastic zone size, normalized to the value for the infinite plate, according to these formulas.

Another finite-width correction may be derived by noting that for small-scale yielding, Eq 8 and 9 give $J_\infty = K_\infty^2/E'$ for the infinite plate, since $K_\infty = \sigma\sqrt{\pi a}$. But for any other

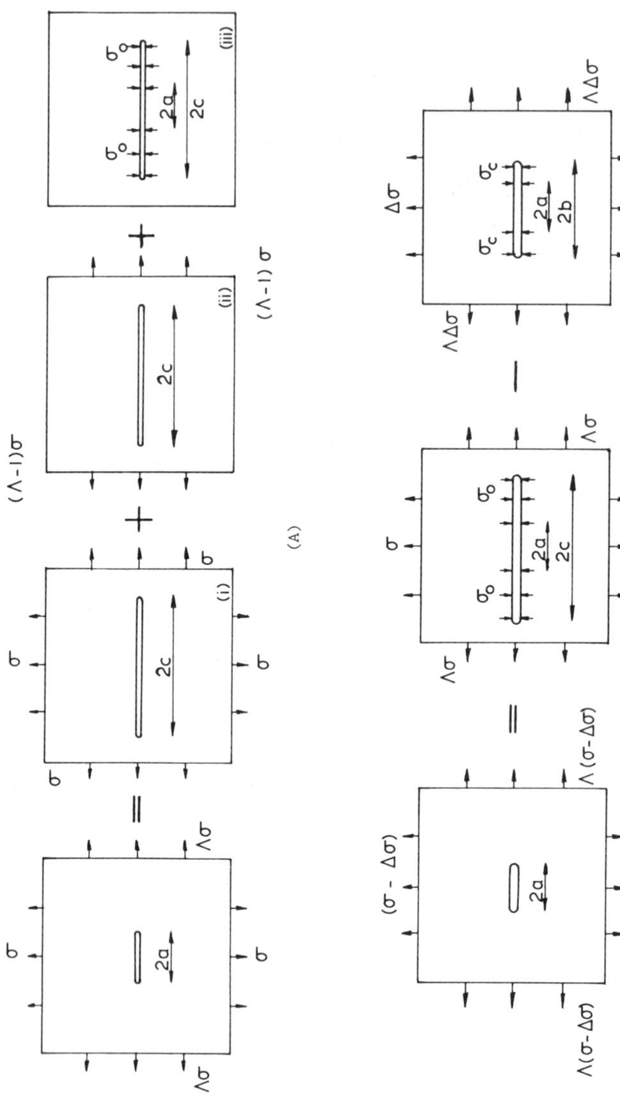

FIG. 2—Superposition of loads for the Dugdale model: (a) monotonic load, (b) cyclic load, (c) cyclic stress distribution ahead of crack tip.

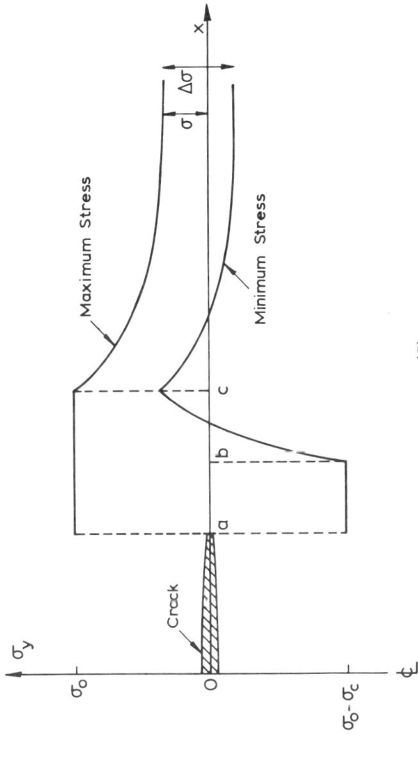

(C)

FIG. 2—*Continued.*

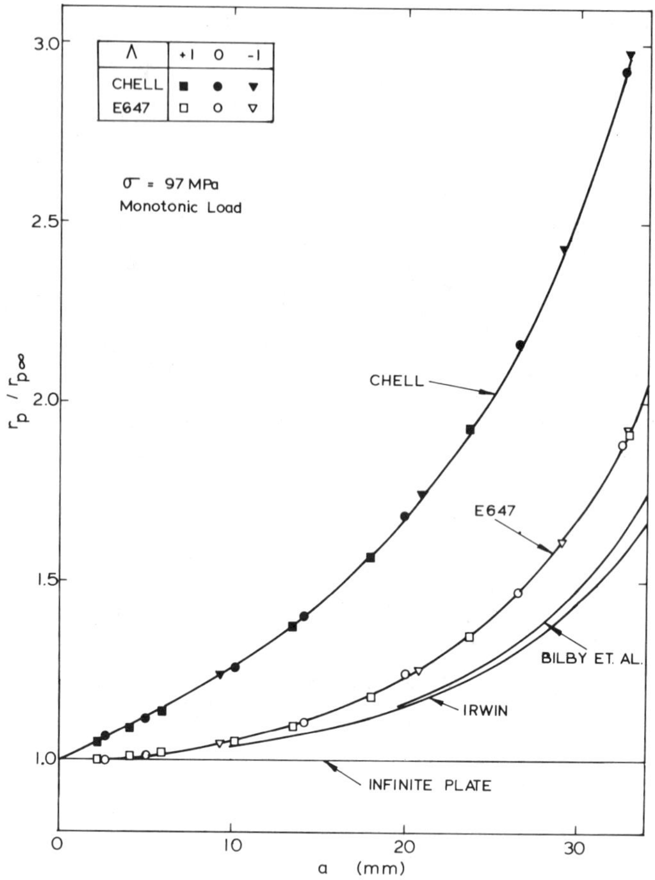

FIG. 3—*Finite-width corrections for plastic zone size* (2W = *101.9 mm*).

geometry, $K = F\sigma\sqrt{\pi a}$, so one should find that

$$J = F^2 \sigma^2 \pi a / E'$$

or

$$J = F^2 J_\infty \tag{24}$$

From Eqs 7, 8, and 9

$$J = (8\sigma_o^2 a / \pi E') \ln[1 + r_p / a]$$

which, on substitution in Eq 24, yields

$$\ln(1 + r_p/a) = F^2 \ln(1 + r_{p\infty}/a)$$

or

$$r_p = a[(1 + r_{p\infty}/a)^{F^2} - 1] \tag{25}$$

Equation 25 is also plotted in Fig. 3 for two values of F, firstly Eq 12, which is based on the ASTM Test Method for Constant-Load-Amplitude Fatigue Crack Growth Rates Above 10^{-8} m/Cycle (E 647-83), and secondly the expression

$$F = \sqrt{\left[\frac{2W}{\pi a}\tan(\pi a/2W)\right]} \quad (26)$$

due to Irwin [28]. The latter is in close agreement with the Bilby formula as both apply to a periodic array of cracks. Since it is generally believed that the CCP solution falls between the Chell and Bilby formulas [29], the E 647 result is employed in this paper. To avoid the possibility of errors due to this assumption, crack lengths were generally kept below 20 mm.

An equation for crack face displacement is derived in Appendix II. This enables the crack profile to be examined during a load cycle, as shown in Fig. 4. For this example, the maximum applied stress of 193 MPa generated a CTOD δ = 6.86 μm and a crack face displacement (CFD) = 19.5 μm, in the center of the crack (a = 5 mm). On unloading, crack closure was first observed in the center of the crack, at a compressive stress of 15 MPa for a plane-strain calculation. If a correct evaluation of cyclic plasticity is required, closure should be

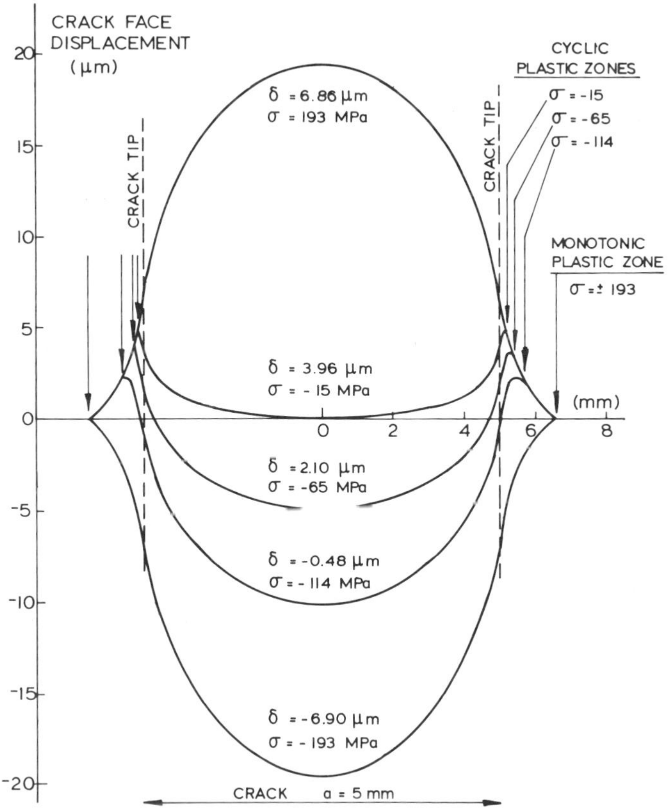

FIG. 4—*Calculated displacements of the crack face during a load cycle* (Λ = -1, R = -1, $\Delta\sigma$ = *386 MPa*).

taken into account, since large "negative" CFD values arise below 15 MPa, as shown in Fig. 4 from calculations with Eq 35 in Appendix II.

Crack Propagation Results

Figure 5 shows the collected data for three stress levels and three biaxial stress states, plotting crack growth rate versus cyclic plastic zone size derived from Eq 16, including the width correction (Eq 25), in plane strain (Eq 20). In this figure five points only are plotted for each test conducted, to avoid undue emphasis being placed on any one experiment. A plane-strain calculation was used since cyclic plastic zone sizes were generally less than the specimen thickness of 4 mm. The flow stress, Y, was taken as the tensile strength, to incorporate the cyclic behavior of the material [*5,11*]. The use of yield or proof stress for the flow stress is not admissible since the applied loads can and do exceed the yield condition without causing instantaneous failure, in strain hardening materials.

The shaded points in Fig. 5 correspond to $R = -1$, and clearly suffer from a closure effect because of the lack of correlation to $R = 0$ data. Taking a traditional approach to closure problems [*30*], the open points were plotted by assuming a reduced stress range such that the effective minimum stress corresponded to zero CFD, that is, the first closure in the center of the crack. It is apparent that an overcorrection is made by this procedure, illustrated by the open points in Fig. 5.

An examination of Fig. 4 shows that although the crack initially closes at $\sigma = -15$ MPa,

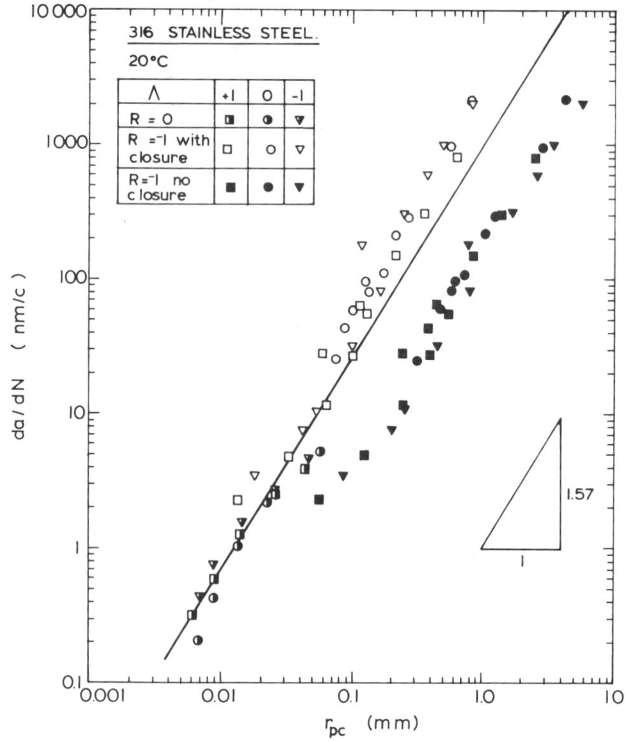

FIG. 5—*Dependence of crack growth rate on cyclic plastic zone size for Elber's approach to closure, and for ignoring closure effects.*

the residual CTOD (δ = 3.96 μm, that is, 58% of the monotonic CTOD) has not been fully reversed. Appendix II shows that for this example a stress of approximately −114 MPa is required for crack-tip closure, leaving very considerable scope for further crack-tip plasticity after initial closure has occurred. In the absence of better analyses, it was assumed that effective crack closure occurred halfway between the stress levels for zero CFD and zero CTOD as determined in Appendix II. Figure 6 shows da/dN versus r_{pc} plotted on this basis, with a good agreement between data from each stress level and R-ratio.

In this analysis of closure, no account is taken of the possible effect of oxide wedging cracks open. Indeed for stainless steel this appears to be an unlikely mechanism at reasonable crack growth rates. Secondly, being plane strain, there is no possibility of a plastic wake producing early closure due to the constant-volume condition for plastic flow. A plastic wake could be produced only if deep striations were formed, the striation depth being of the order of 4 μm (that is, the CTOD). However, striations are observed to be shallow, with a depth of not more than one-quarter the crack growth increment per cycle, and very much smaller than the CTOD. Thus closure in plane strain appears to occur below zero load, and this point of view is substantiated by the data correlation for R = 0 and R = −1 tests.

Also included in Fig. 6 are data for δ_c, the cyclic CTOD, and J_c (Eqs 17, 18), allowing for closure in each case. There is little difference in the degree of correlation of data by each parameter when da/dN is less than 10 nm/cycle, although r_{pc} is slightly better. Since this corresponds to the region of low stress tests, one would expect only a moderate de-

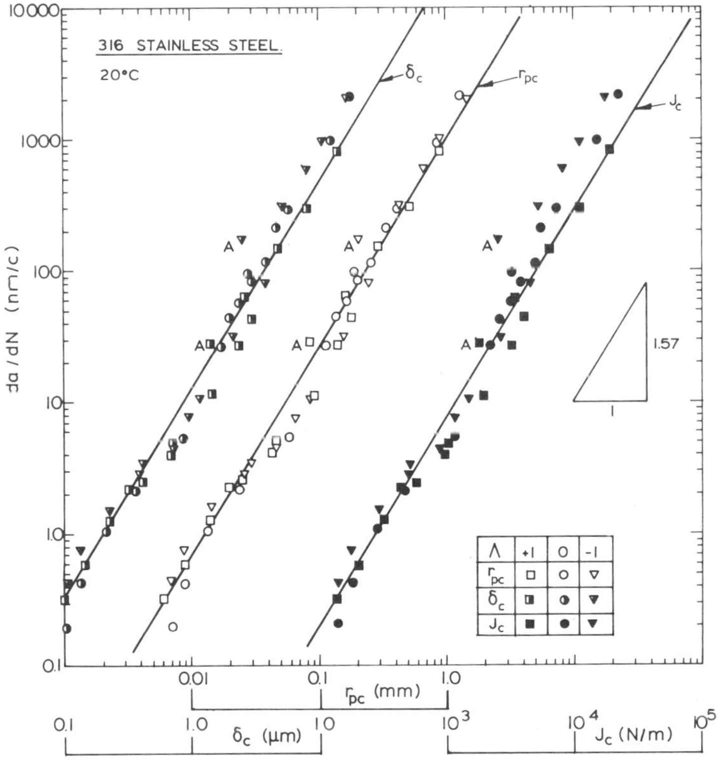

FIG. 6—*Crack growth rate dependence on cyclic plastic zone size, CTOD, and J-integral.*

pendence on biaxiality as discussed above for LEFM behavior. However, if *da*/*dN* exceeds 100 nm/cycle, a much greater dependence on stress state can be seen for the J_c data. The data for shear loading exceeds the equibiaxial line by a factor of 3 in propagation rate, which corresponds closely to the divergence in Fig. 1. This agrees with the observation of Rice [*19*] that *J* (like *K*) is insensitive to the *T*-stress. (Note that each line is drawn through the equibiaxial data in the figure, and in no sense is the line a "best fit.") However, the shear data for δ_c is rather closer to the equibiaxial line compared with that for J_c, but the closest agreement of uniaxial, equibiaxial, and shear results is obtained with the r_{pc} parameter.

The upper portion of Fig. 6 is reproduced on an expanded scale in Fig. 7, showing the results of tests at high stress alone. The correlation with r_{pc} is good except at the start of the crack growth tests (Region A). The abnormal high growth rates are probably due to the soft form of the annealed metal before achieving a cyclically stable state, in view of the high degree of cyclic hardening found in austenitic steels. Therefore Region A should be ignored in Figs. 6 and 7 when assessing biaxial loading effects.

Figures 8 and 9 present data at a single stress range for two heats of 316 stainless steel at 550°C. In both cases, the early parts of the tests reveal an uncharacteristically low slope, giving a Paris law exponent of around 1. Previously such low slopes have been attributed to aging effects [*11,31,32*], indicative of some complex interactive mechanisms in the crack tip region. This appears to be reflected in the marked lack of correlation with any of the parameters plotted. But once the material becomes stable, the slope changes abruptly, giving a rapidly accelerating crack on a slant plane, including a Mode III component. Here Material A is best correlated by r_{pc} once again, although agreement is not ideal. But Material B, which had a very much lower yield stress due to more effective annealing (Table 2), conforms to a CTOD criterion more closely (Fig. 9).

Discussion

Although the general trend of results presented in this paper points to r_{pc}, the cyclic plastic zone size, as the ideal crack growth parameter, this has been demonstrated for only one material. Further work is necessary to show which classes of material follow this propagation criterion. Even for Material B in Fig. 9, r_{pc} is an acceptable parameter in that the scatter of data is less than that observed in Fig. 6. In Fig. 8 the equibiaxial test results have been extended to a crack length of 25.5 mm to enable comparable crack growth rates to be obtained. As noted above, this implies a considerable contribution from the plate width correction factor, and therefore the data from Fig. 8 are replotted in Fig. 10 but assuming that specimen width was infinite (that is, $F = 1$). The improvement in correlation with respect to r_{pc} implies that even at 550°C the plastic zone size provides a good description of fatigue crack growth, and also that the effect of the side arms on the *K* calibration of the cruciform specimen is of some significance. The finite-width correction factors plotted in Fig. 3 assume that the side arms may be ignored on the basis that the vertical stress is undisturbed by the application of a horizontal load. But the constraint obtained in the horizontal direction from the side arms may shift the curve for cruciform specimens on Fig. 3 toward, and probably beyond, the results for a periodic array of cracks, that is, the Bilby and Irwin curves.

The use of Eq 9 to derive the J-integral for plane-strain situations has been justified by a wide range of EPFM studies, although the equation is usually written as

$$J = MY\delta$$

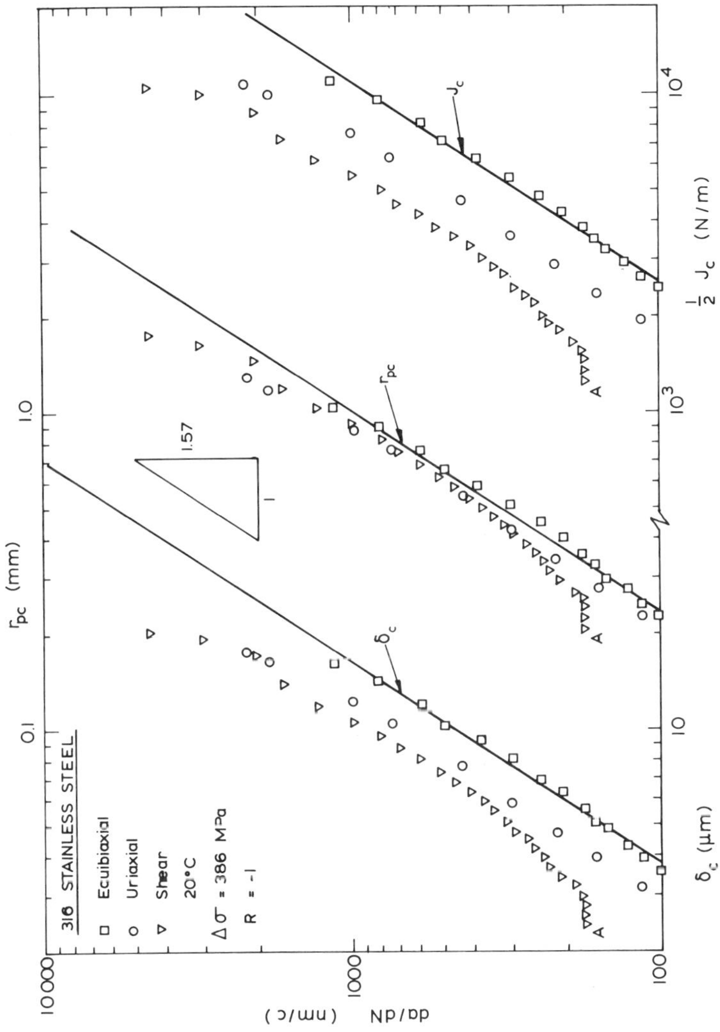

FIG. 7—*Crack growth rate correlations for Material A, 20°C.*

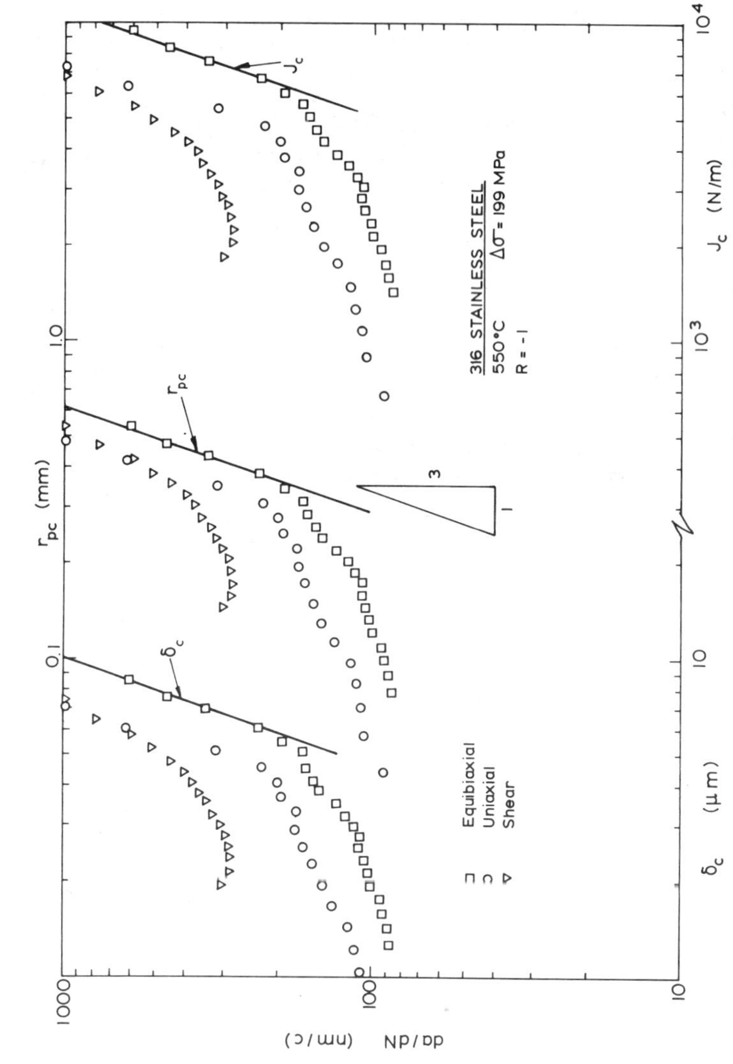

FIG. 8—*Crack growth rate correlations for Material A, 550°C.*

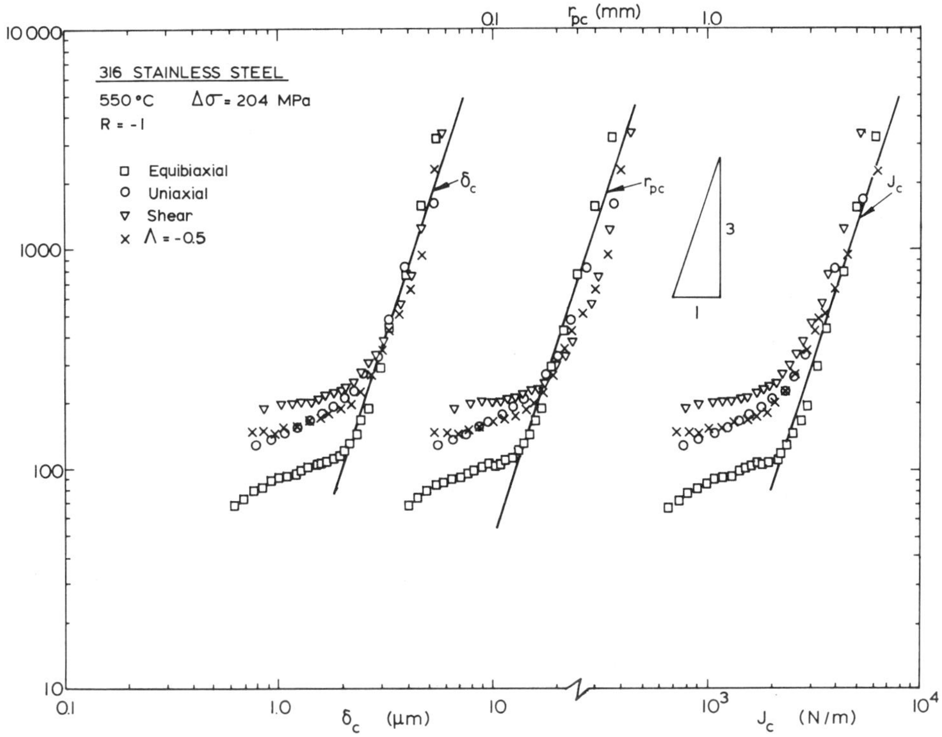

FIG. 9—*Crack growth rate correlations for Material B, 550°C.*

where M is a constant. Although the DBCS model gives M equal to unity for plane stress, in plane-strain values between 1 and 3 are generally found; for example, 2.1 was derived by Rice [*19*]. However, the actual value has been shown to depend on the ratio (Y/E) and the strain hardening exponent, so that for any given material M may be treated as a constant [*33*]. When T is taken as zero in Eq 15, one obtains for the strip yield case

$$\sigma_o = Y/\sqrt{1 - \nu + \nu^2}$$

which implies that M is 1.125 for $\nu = 0.3$, or 1.155 for $\nu = 0.5$. These values compare favorably with 1.33 for an ideal plastic material, derived by Marston [*33*] for the Hutchinson-Rice-Rosengren singularity [*34*], even though this relates to a very different shape of plastic zone. Whatever constant value M takes, the degree of correlation shown in Figs. 6–10 will be unaffected.

The results presented here show that biaxial stress fatigue crack growth experiments are able to provide an answer to the fundamental question posed in the introduction. A more convincing proof could be provided by using a wider range of T-stresses; indeed the Λ-values used here were chosen arbitrarily on the basis that previous studies used similar values. Higher compressive stresses may be applied subject to the limitation provided by buckling of the working section. However, increasing the T-stress above zero will lead to crack bifurcation [*35*], the crack path changing through 90 deg to seek a route with lower T-stress and the maximum stress range opening the crack. A statistical analysis of Figs. 6–10 is given in Appendix III.

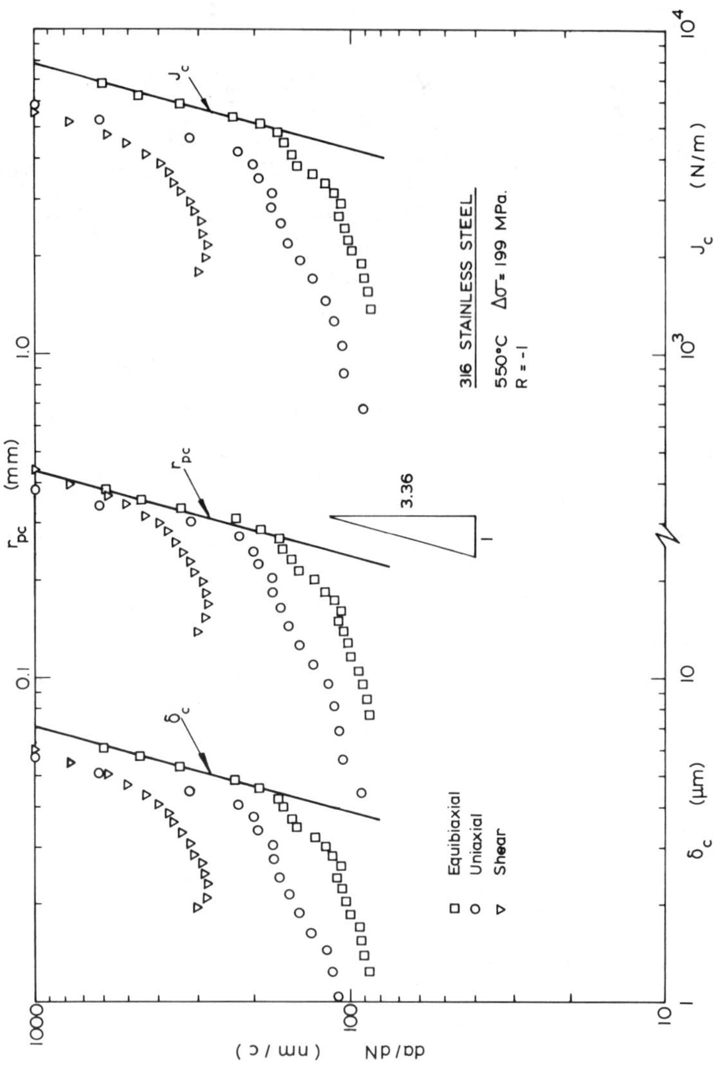

FIG. 10—*Crack growth rate correlations for Material A at 550°C using the infinite plate formulas.*

Simpler tests to answer the critical question can be conducted by comparing the results of tests on compact tension (CT) and center-cracked (CCP) specimens, without resorting to the cruciform shape [*36*], since the CT specimen exhibits a positive T-stress, and the CCP shows a negative value [*20*]. However, a full calibration for T as a function of (a/w) is required [*37*].

The results presented here do not extend very far into the region of general yield. On the basis of Tomkins' theory, one would expect a contribution to crack propagation from the applied plastic strain range in high-strain fatigue [*5*], which should be added to the crack extension increment due to cyclic plastic zone size. Similarly, there will be plastic strain contributions for inclusion in both J-integral and CTOD calculations.

Conclusions

1. Biaxial stress fatigue crack propagation tests may be used to give a critical comparison between crack growth rate predictions for various fatigue criteria.
2. Tests on Type 316 stainless steel show that cyclic plastic zone size is a better correlating parameter for Mode I fatigue crack growth rate than crack-tip opening displacement or the cyclic range of the J-integral. In some instances crack opening displacement is applicable.

Acknowledgments

The authors would like to thank the Central Electricity Generating Board and the Science Research Council for financial support, the Electric Power Research Institute and Oak Ridge National Laboratory for providing the materials, and the United States Nuclear Regulatory Commission for support of the project.

APPENDIX I

Dugdale Model Under Plane Strain

Under plane-strain conditions, the plastic zone size and CTOD are given by Eqs 7 and 8, with $E' = E/(1 - \nu^2)$, the derivation being given by Burdekin and Stone [*17*]. After dividing the loads into three components, as illustrated in Fig. 2*a*, the principal stresses in the cohesive stress region may be determined as follows.

1. For the elastic crack under equibiaxial stress, the stresses on the crack face must be zero in all directions, since for a function of the complex variable, Z, the stress field around a crack lying in the plane $y = 0$ is given by [*25*]

$$\sigma_x = \mathrm{Re}(Z) - y\,\mathrm{Im}(Z') \tag{27}$$

$$\sigma_y = \mathrm{Re}(Z) + y\,\mathrm{Im}(Z') \tag{28}$$

$$\tau_{xy} = -y\,\mathrm{Re}(Z') \tag{29}$$

$$Z' = dZ/dz \text{ where } z = x + iy$$

Clearly $\sigma_y = 0$ on the crack face, being the stress normal to the crack, and therefore $\sigma_x = 0$ from Eqs 27 and 28, with $y = 0$. From symmetry, $\tau_{xy} = 0$ when $y = 0$, as given by Eq 29.

2. For the applied stress parallel to the crack, no singularity is formed, and a uniform stress field results, with $\sigma_x = (\Lambda - 1)\sigma = T$. Although this is a plane-strain calculation, the plane-strain condition arises only from the constraint around the crack-tip stress field, generating stresses σ_z. Since the T-stress produces a uniform stress field, no such constraint can arise, and therefore we take $\sigma_z = 0$ for a plate of uniform thickness.

3. For the cohesive stresses, the stress function is [*17*]

$$Z = -\int_a^c \frac{2\sigma_o z\sqrt{c^2 - h^2}}{\pi\sqrt{z^2 - c^2}\,(z^2 - h^2)}\,dh$$

which, for z real and $a \leqslant z < c$, is an imaginary number. Thus Eq 27 can give only $\sigma_x = 0$ in the cohesive zone, since $y = 0$. For plane strain

$$\sigma_z = \nu(\sigma_x + \sigma_y)$$

and for $c \geqslant x \geqslant a$, σ_y must be σ_o, the cohesive stress, giving

$$\sigma_z = \nu\sigma_o$$

Applying the von Mises yield criterion for these stresses, 1 to 3

$$(\sigma_o - T)^2 + (T - \nu\sigma_o)^2 + (\nu\sigma_o - \sigma_o)^2 = 2Y^2$$

which, being a quadratic equation in σ_o, gives after some algebra

$$\sigma_o = (Y/\sqrt{1 - \nu + \nu^2})[AT/Y \pm \sqrt{1 - (T/Y)^2(1 - A^2)}] \qquad (30)$$

$$A = [(1 + \nu)/2]/\sqrt{1 - \nu + \nu^2}$$

As a concession to strain hardening, a weighting factor α has been added to the T-stress in Eq 15, since the presence of a nonzero strain hardening exponent will make the strip yield zone more diffuse, and therefore of finite width. This implies that the T-stress will not in fact be uniform at the crack face, but reduced somewhat by the proximity of the plastic zone. Thus $0 < \alpha \leqslant 1$, where $\alpha = 1$ for an ideal plastic material.

Comparison of plastic zone size determined from the full LEFM stress field with small-scale yielding [*38*] shows that for a Mode I crack, to second-order accuracy

$$\alpha = [2(1/8 + \nu - \nu^2)/(1 + \nu)]\sqrt{\{(1 - \nu + \nu^2)/(5/8 - \nu + \nu^2)\}} \qquad (31)$$

Therefore

$$\alpha = 1/\sqrt{2} \qquad \text{for } \nu = 0.5$$

$$= 0.733 \qquad \text{for } \nu = 0.4$$

$$= 0.711 \qquad \text{for } \nu = 0.3$$

$$= 0.681 \qquad \text{for } \nu = 0.25$$

Taking $\alpha = 1/\sqrt{2}$ gives close agreement with the small-scale yield solution for Mode I cracks over the useful range of values for Poisson's ratio in isotropic metals [*24*]. (Note for a Mode

II crack, one obtains $\alpha = \sqrt{3}/2$ for small-scale yielding, giving the divergence of curves noted in Fig. 4 of Ref *24*.)

APPENDIX II

Displacement of the Crack Face

Burdekin and Stone [*17*] have derived expressions for the displacements associated with the DBCS model. The relative displacement of the two crack faces at a distance x from the centerline was given, for $k = a/c$ and equibiaxial loading, as

$$\psi = \frac{8Y}{\pi E'}\left\{a \coth^{-1}\left(\frac{1}{c}\sqrt{\frac{c^2 - x^2}{1 - k^2}}\right) - x \coth^{-1}\left(\frac{k}{x}\sqrt{\frac{c^2 - x^2}{1 - k^2}}\right)\right\}$$

for $x \leqslant a$. Since

$$2 \coth^{-1}(x) = \ln[(x + 1)/(x - 1)]$$

if one defines ϕ as

$$\phi = \sqrt{[(c^2 - x^2)/(c^2 - a^2)]} \tag{32}$$

then $\phi \geqslant 1$ and

$$\psi = \frac{4Y}{\pi E'} \ln\left[\left(\frac{\phi + 1}{\phi - 1}\right)^a \cdot \left(\frac{a\phi - x}{a\phi + x}\right)^x\right] \tag{33}$$

But for $c \geqslant x > a$, $0 \leqslant \phi \leqslant 1$, and it may be shown that

$$\psi = \frac{8Y}{\pi E'}\{a \operatorname{th}^{-1}(\phi) - x \operatorname{th}^{-1}(a\phi/x)\}$$

$$= \frac{4Y}{\pi E'} \ln\left[\left(\frac{1 + \phi}{1 - \phi}\right)^a \cdot \left(\frac{x - a\phi}{x + a\phi}\right)^x\right] \tag{34}$$

Therefore for $c \geqslant x \geqslant 0$, combining Eqs 33 and 34 for ψ, one obtains

$$\psi = \frac{4Ya}{\pi E'} \ln\left[\left|\frac{\phi + 1}{\phi - 1}\right| \cdot \left|\frac{a\phi - x}{a\phi + x}\right|^{x/a}\right] \tag{35}$$

For $x = 0$, the crack face displacement may be derived

$$\text{CFD} = \frac{4Ya}{\pi E'} \ln[(\phi + 1)/(\phi - 1)] \tag{36}$$

For $x \rightarrow a$, the CTOD is given by the limit value of ψ

$$\delta = \frac{8Ya}{\pi E'} \ln(1/k) \tag{37}$$

These formulas have been used to plot the crack and plastic zone profiles in Fig. 4.

Under cyclic loading, the condition for initial crack closure can be obtained from Eq 36. At the maximum load in a cycle, let ϕ take the value ϕ_o derived from Eqs 7 and 15. For closure in the center of the crack, the stress range applied must produce an equal but opposite CFD. Thus for $\phi = \phi_f$ for the crack face closure stress, Eq 36 gives

$$\frac{4\sigma_o a}{\pi E'} \ln[(\phi_o + 1)/(\phi_o - 1)] = \frac{4\sigma_c a}{\pi E'} \ln[(\phi_f + 1)/(\phi_f - 1)]$$

where σ_c is derived from Eqs 15 and 20. This condition reduces to

$$\Delta\sigma_f = \frac{2\sigma_c}{\pi} \sin^{-1} \left\{ \frac{X - 1}{X + 1} \right\} \tag{38}$$

where

$$X = \{[1 + \sin(\pi\sigma/2\sigma_o)]/[1 - \sin(\pi\sigma/2\sigma_o)]\}^{\sigma_o/\sigma_c}$$

and $\Delta\sigma_f$ is the unloading required to produce closure. Since σ_c is a function of $\Delta\sigma_f$ through Eq 21, a numerical solution is required for Eq 38. However, very rapid convergence is found when using a repeated substitution procedure.

For equibiaxial loading, $\Lambda = +1$ and $\sigma_c = 2\sigma_o = 2Y$, enabling Eq 38 to be simplified, giving

$$\Delta\sigma_f = \frac{4\sigma_o}{\pi} \sin^{-1} \left[\tan \left(\frac{\pi\sigma}{4\sigma_o} \right) \right] \tag{39}$$

which also provides a reasonable initial value for repeated substitution in Eq 38. Note that for small-scale yielding, Eq 39 gives $\Delta\sigma_f = \sigma$.

To determine a crack-tip closure condition, one may follow the same procedure as above, but based on Eq 37 for CTOD, to obtain

$$\Delta\sigma_d = \frac{2\sigma_c}{\pi} \cos^{-1}\{[\cos(\pi\sigma/2\sigma_o)]^{\sigma_o/\sigma_c}\} \tag{40}$$

and for equibiaxial loading, this may be simplified as above to get

$$\Delta\sigma_d = \frac{2\sigma_o}{\pi} \cos^{-1}[2 \cos(\pi\sigma/2\sigma_o) - 1] \tag{41}$$

Note that for small-scale yielding, $\Delta\sigma_d = \sqrt{2}\sigma$. Clearly a zero value for CTOD cannot be achieved in practice because the calculated crack face displacements, ψ, will all be negative, and ψ cannot be less than zero for a real crack. Therefore an effective closure stress was assumed to be given by

$$\Delta\sigma_e = (\Delta\sigma_f + \Delta\sigma_d)/2 \tag{42}$$

since the effective crack closure should be bounded by Eqs 38 and 40. Results are given in Table 3.

APPENDIX III

Statistical Analysis

The results have been presented graphically in Figs. 6–10, with a visual appreciation of the accuracy of fit for each parameter. A statistical analysis of the full data set is given in Tables 4 and 5.

The crack growth rates were determined in accordance with ASTM E 647-83, using a five-point least-squares fit to a parabola. The data for each test may be fitted to the Paris law

$$da/dN = C(\Delta K)^m$$

by using a linear fit on a logarithmic graph, using the procedure recommended in the ASTM Practice for Statistical Analysis of Linear or Linearized Stress-Life (S-N) and Strain-Life (ϵ-N) Fatigue Data (E 739-80).

The results for each test are given in Table 4, for crack growth rates in the range 1 to 3000 nm/cycle. For the elevated-temperature tests, the initial data have been excluded where strain aging effects dominate the crack growth process. The variance gives a figure for accuracy of prediction of $\log_{10}(da/dN)$ from the Paris law, and is therefore a measure of the degree of scatter. Exclusion of data below 1 nm/cycle avoids the region of curvature close to threshold, where the Paris law is not acceptable.

The average variance in Table 4 is 0.049, and the maximum value is 0.091. When combining the data for all tests by using an EPFM parameter, and excellent fit would be indicated by obtaining a variance of similar magnitude.

In Table 5, the results are given for collected data of all tests, in terms of the three EPFM

TABLE 4—*Paris law equations.*

$\Delta\sigma$, MPa	Λ	Log C, nm/cycle	m	Variance	No. of Data Points
		MATERIAL A, −20°C			
62.7	1.00	−1.79	2.08	0.023	10
66.6	−0.01	−1.99	2.29	0.025	9
63.1	−0.99	−1.42	1.80	0.022	9
70.8	1.00	−2.11	2.37	0.023	8
191.3	1.00	−2.73	2.50	0.059	27
193.0	0.03	−4.60	3.91	0.058	10
195.2	−1.00	−3.13	2.91	0.066	23
387.9	1.00	−3.00	2.90	0.026	27
383.0	−0.01	−3.40	3.31	0.038	10
385.7	−1.00	−1.97	2.66	0.055	24
		MATERIAL A, −550°C			
199.0	1.00	−5.01	4.24	0.047	6
199.4	−0.03	−6.34	5.25	0.080	6
198.8	−1.01	−5.66	5.04	0.067	7
		MATERIAL B, −550°C			
204.0	0.99	−7.09	6.05	0.062	10
204.3	−0.02	−4.88	4.78	0.038	6
203.9	−0.51	−5.51	5.19	0.091	8
204.7	−1.00	−3.87	4.22	0.048	9

TABLE 5—*Correlations for EPFM parameters.*

Material	Temperature, °C	Parameter	Log *C*, nm/cycle	m	Variance	No. of Data Points
A	20	r_{pc}	3.04	1.68	0.133	157
(*a* < 20 mm)		δ_c	1.06	1.76	0.169	157
		J_c	−4.56	1.81	0.218	157
A	550	r_{pc}	3.78	2.82	0.140	19
		δ_c	0.71	2.52	0.215	19
		J_c	−4.18	1.81	0.278	19
B	550	r_{pc}	4.19	2.41	0.088	33
		δ_c	1.30	2.64	0.078	33
		J_c	−6.46	2.60	0.127	33
A	20	r_{pc}	3.12	1.72	0.121	172
infinite		δ_c	1.09	1.80	0.150	172
plate		J_c	−4.64	1.84	0.197	172
A	550	r_{pc}	4.44	3.73	0.092	19
infinite		δ_c	0.03	3.79	0.176	19
plate		J_c	−7.25	2.68	0.269	19

parameters. The fitted lines correspond to Figs. 6, 8, 9, and 10, respectively. In each case, the variance values indicate that the plastic zone size correlation is the best one for Material A, whereas for Material B the cyclic CTOD is marginally better. But even for plastic zone size with Material B, the variance of 0.088 is acceptable, and comparable to the values in Table 4 which indicate the general scatter of data.

The use of the infinite plate formula for the geometry factor F clearly improves both sets of Material A data, where crack lengths up to 35 mm have been included. The values of variance for plastic zone size again approach the reference values in Table 4, which indicate a reasonable correlation from this EPFM parameter.

References

[*1*] Starkey, M. S. and Skelton, R. P., *Fatigue of Engineering Materials and Structures,* Vol. 5, 1982, pp. 329–341.
[*2*] Haigh, J. R. and Skelton, R. P., *Materials Science and Engineering,* Vol. 36, 1978, pp. 133–137.
[*3*] Boettner, R. C., Laird, C., and McEvily, A. J., *Transactions,* The Metallurgical Society of the American Institute of Mining, Metallurgical and Petroleum Engineers, Vol. 233, 1965, pp. 379–387.
[*4*] Dowling, N. E. in *Cyclic Stress-Strain and Plastic Deformation Aspects of Fatigue Crack Growth, ASTM STP 637,* American Society for Testing and Materials, Philadelphia, 1977, pp. 97–121.
[*5*] Tomkins, B., *Philosophical Magazine,* Vol. 18, 1968, pp. 1041–1066.
[*6*] Pook, L. P. and Frost, N. E., *International Journal of Fracture,* Vol. 9, 1973, pp. 53–61.
[*7*] Hu, W. and Liu, H. W. in *Proceedings,* Second International Conference on Mechanical Behavior of Materials, Boston, 1976, pp. 1058–1062.
[*8*] Brown, M. W., Liu, H. W., Kfouri, A. P., and Miller, K. J. in *Advances in Fracture Research,* D. Francois, Ed., Pergamon, Oxford, U.K., Vol. 2, 1980, pp. 891–898.
[*9*] Skelton, R. P. in *Low-Cycle Fatigue and Life Prediction, ASTM STP 770,* American Society for Testing and Materials, Philadelphia, 1982, pp. 337–381.
[*10*] Wang, T. C. and Miller, K. J., *Fatigue of Engineering Materials and Structures,* Vol. 5, 1982, pp. 249–263.
[*11*] Brown, M. W. and Miller, K. J. in *Multiaxial Fatigue, ASTM STP 853,* American Society for Testing and Materials, Philadelphia, 1985, pp. 135–152.
[*12*] Kitagawa, H., Yuuki, R., and Tohgo, K., *Fatigue of Engineering Materials and Structures,* Vol. 2, 1979, pp. 195–206.

[13] Kitagawa, H., Yuuki, R., Tohgo, K., and Tanabe, M. in *Multiaxial Fatigue, ASTM STP 853,* American Society for Testing and Materials, Philadelphia, 1985, pp. 164–183.
[14] Rice, J. R., *Transactions,* American Society of Mechanical Engineers, Series E, *Journal of Applied Mechanics,* Vol. 35, 1968, pp. 379–386.
[15] Dugdale, D. S., *Journal of the Mechanics and Physics of Solids,* Vol. 8, 1960, pp. 100–104.
[16] Bilby, B. A., Cottrell, A. H., and Swinden, K. H. in *Proceedings of the Royal Society,* London, Series A, Vol. 272, 1963, pp. 304–314.
[17] Burdekin, F. M. and Stone, D. E. W., *Journal of Strain Analysis,* Vol. 1, 1966, pp. 145–153.
[18] Rice, J. R. in *Fracture, An Advanced Treatise,* H. Liebowitz, Ed., Academic Press, New York, Vol. 2, 1968, pp. 191–311.
[19] Rice, J. R., *Journal of the Mechanics and Physics of Solids,* Vol. 22, 1974, pp. 17–26.
[20] Larsson, S. G. and Carlsson, A. J., *Journal of the Mechanics and Physics of Solids,* Vol. 21, 1973, pp. 263–277.
[21] Smith, E. W. and Pascoe, K. J., *Fatigue of Engineering Materials and Structures,* Vol. 6, 1983, pp. 201–224.
[22] Charvatt, I. M. I. and Garrett, G. G., *Journal of Testing and Evaluation,* Vol. 8, 1980, pp. 9–17.
[23] Brown, M. W. and Miller, K. J. in *Defects and Fracture,* G. C. Sih and H. Zorski, Eds., Martinus Nijhoff, The Hague, 1982, pp. 29–38.
[24] Brown, M. W., Miller, K. J., and Walker, T. J. in *Transactions,* Seventh International Conference on Structural Mechanics in Reactor Technology, North Holland, Amsterdam, 1983, Vol. L, pp. 143–149.
[25] Westergaard, H. M., *Journal of Applied Mechanics,* Vol. 6, 1939, pp. 49–53.
[26] Bilby, B. A., Cottrell, A. H., Smith, E., and Swinden, K. H. in *Proceedings of the Royal Society,* London, Series A, Vol. 279, 1964, pp. 1–9.
[27] Chell, G. G., *International Journal of Fracture,* Vol. 12, 1976, pp. 135–147.
[28] Irwin, G. R., *Transactions,* American Society of Mechanical Engineers, Series E, *Journal of Applied Mechanics,* Vol. 24, 1957, pp. 361–364.
[29] Larsson, L. H. in *Advances in Elasto-Plastic Fracture Mechanics,* L. H. Larsson, Ed., Applied Science, London, 1980, pp. 261–278.
[30] Elber, W. in *Damage Tolerance in Aircraft Structures, ASTM STP 486,* American Society for Testing and Materials, Philadelphia, 1971, pp. 230–242.
[31] Shahinian, P., Smith, H. H., and Watson, H. E., *Transactions,* American Society of Mechanical Engineers, Series B, *Journal of Engineering for Industry,* Vol. 93, 1971, pp. 976–980.
[32] James, L. A., *Metallurgical Transactions,* Vol. 5, 1974, pp. 831–838.
[33] Marston, T. U. in *Advances in Elasto-Plastic Fracture Mechanics,* L. H. Larsson, Ed., Applied Science, London, 1980, pp. 191–235.
[34] Hutchinson, J. W., *Journal of the Mechanics and Physics of Solids,* Vol. 16, 1968, pp. 13–31.
[35] Pook, L. P. and Holmes, R. in *Proceedings,* Society of Environmental Engineers International Conference on Fatigue Testing and Design, London, 1976, Vol. 2, Paper 36.
[36] Rhodes, D. and Radon, J. C. in *Multiaxial Fatigue, ASTM STP 853,* American Society for Testing and Materials, Philadelphia, 1985, pp. 153–163.
[37] Cardew, G. E., Goldthorpe, M. R., Howard, I. C., and Kfouri, A. P. in *Fundamentals of Deformation and Fracture,* International Union of Theoretical and Applied Mechanics Symposium, Sheffield, U.K., B. A. Bilby, K. J. Miller, and J. R. Willis, Eds., Cambridge University Press, U.K., 1984, pp. 465–476.
[38] Gao, H., Alagok, N., Brown, M. W., and Miller, K. J. in *Multiaxial Fatigue, ASTM STP 853,* American Society for Testing and Materials, Philadelphia, 1985, pp. 184–202.

Michael C. Smith[1] *and Roderick A. Smith*[2]

Toward an Understanding of Mode II Fatigue Crack Growth

REFERENCE: Smith, M. C. and Smith, R. A., **"Toward an Understanding of Mode II Fatigue Crack Growth,"** *Basic Questions in Fatigue: Volume I, ASTM STP 924,* J. T. Fong and R. J. Fields, Eds., American Society for Testing and Materials, Philadelphia, 1988, pp. 260–280.

ABSTRACT: The Mode II crack flank displacement and crack growth responses of three precracked specimens made from structural steel were measured, using plastic replicas and a crack-tip compliance gage. Crack surface interaction was found to dominate behavior: at low stress intensity range (ΔK_{II}^{nom} = 9 MPa $\sqrt{m}$) the precracks did not suffer reversed slip to their tips and no crack growth occurred, while at high stress intensity range (ΔK_{II}^{nom} = 19 MPa $\sqrt{m}$) the effective stress intensity range was less than half that nominally applied. Three sources of crack flank frictional attenuation were identified: compressive residual stresses due to precracking, Mode I wedging over asperities, and gross plastic deformation of interlocking asperities. The measured unlocking response was modeled successfully by assuming that crack flank frictional stresses obeyed a constant interfacial shear stress friction law.

KEY WORDS: fatigue crack growth, Mode II, shear mode, steels, crack flank locking, crack flank slip, friction

Nomenclature

a_e	Instantaneous effective crack length (locked/unlocked boundary)
a_e^{max}	Maximum extent of reversed slip during a load cycle
a_m	Total crack length (precrack + notch)
a_n	Notch length
CTG	crack-tip clip gage
K_{II}^{nom}	Nominal stress intensity (no friction, unlocking to a_m)
K_{II}^{eff}	Effective stress intensity developed at the precrack tip
K_{II}^{a}	Applied stress intensity (no friction, unlocking to a_e)
K_{II}^{f}	Attenuation in applied stress intensity due to crack flank frictional shear stresses
K_{II}^{a-f}	Stress intensity developed at a_e
K_{op}	Mode I crack opening stress intensity
P	Applied load
r_f	Critical distance ahead of a_e over which $\tau(x')$ must be exceeded to cause further unlocking
R	Load ratio
SEM	scanning electron microscope
Δu	Relative Mode II displacement of crack flanks
Δv	Relative Mode I displacement of crack flanks
x'	Distance along precrack

[1] Lead stress engineer, Mechanical Design and Development Department, British Aerospace PLC, Dynamics Group, Space and Communications Division, Argyle Way, Stevenage SG1 2AS, United Kingdom.

[2] Lecturer, Cambridge University Engineering Department, Trumpington St., Cambridge CB2 1PZ, United Kingdom.

x'_s Distance along precrack to instantaneous effective crack tip
δ CTG anvil displacement
μ Friction coefficient
$\sigma(x')$ Compressive normal stress distribution acting across the precrack flanks
$\tau(x')$ Limiting frictional shear stress distribution
τ_ℓ Linear component of $\tau(x')$
τ_s Constant component of $\tau(x')$
τ_{xy} Shear stress ahead of crack tip

The overwhelming majority of fatigue failures are due to the growth of cracks in the tensile opening mode, Mode I. However, there are a number of areas where crack growth in Mode II or Mode III may be important; for example:

1. Contact fatigue failure of rails [*1,2*] or bearings [*3*] in which cracks grow in a region of compressive Mode I and fully reversed Mode II loading.
2. Fatigue failure of shafts loaded in torsion [*4*].
3. Wear theories based on "delamination" [*5*].

It is well known that premature contact between the flanks of cracks subjected to Mode I loading (crack closure) can have important effects on crack growth. If a crack is subjected to pure Mode II loading its flanks are always in contact; we might expect that such contact has effects that are at least as large as those of closure in Mode I. These surface interactions have never been studied explicitly, although published data suggest that they are indeed important, for the following reasons:

1. Coplanar Mode II growth nearly always ends in crack arrest, unless crack flanks are separated by a Mode I load [*6,7*].
2. Specimens containing slits, with no crack flank contact, always exhibit lower Mode II thresholds for branch crack growth than do precracked specimens, which suffer crack flank contact [*7,8*].
3. The crack surfaces of precracked specimens tested in Mode II show evidence of surface rubbing and wear [*9,10*].

Studies of Mode III fatigue crack growth have revealed similar evidence of the importance of crack surface interaction [*11,12*].

Service fatigue problems in which Mode II loading is important often also involve compressive normal stresses acting across crack flanks (for example, contact fatigue cracks in rails or bearings). The effects of this compressive load are usually modeled by assuming Coulomb frictional behavior of the crack surfaces [*13*]. No experimental data exist to support this assumption. Thus we must answer the question:

—What form is taken by the interaction between the flanks of a crack under cyclic Mode II loading, and what effect does this interaction have on crack growth?

Experimental

The Mode II crack flank displacement and crack growth responses were measured for three specimens manufactured from 3.2-mm-thick hot-rolled BS 4360 50B structural steel; material composition and mechanical properties are given in Tables 1 and 2. The specimen

TABLE 1—*Chemical composition of 3.2-mm BS 4360 50B sheet steel.*

C	Si	Mn	P	S	Fe
0.130	0.008	1.200	0.014	0.014	balance

geometry of Richard [*14*] was used (see Fig. 1); each specimen contained a 50-mm-deep machined notch and a 5-mm Mode I precrack which was inserted at constant stress intensity range (ΔK_I = 15 MPa $\sqrt{m}$ and R = 0.05).[3] The following Mode II test conditions were used: one specimen was tested at a nominal stress intensity range, ΔK_{II}^{nom}, of 19 MPa $\sqrt{m}$, and R = 0.05, while the second and third were both tested at ΔK_{II}^{nom} = 9 MPa $\sqrt{m}$, one at R = 0.05 and the other at R = 0.5. Test details are given in Table 3.

Mode II testing was carried out under load control, each test being performed at a constant Mode II load range. The crack flank displacement response of each specimen was monitored at intervals during cyclic loading, using both plastic replicas and a crack-tip clip gage (CTG); details of these techniques are given below. Crack growth was monitored using a traveling microscope. After testing, the crack surfaces were separated and examined in the SEM.

Plastic Replication Technique

Before testing, a set of fine parallel lines spaced at 200 μm was scribed perpendicular to the precrack on one face of each specimen. Plastic replicas of the precrack and lines were then taken at intervals during Mode II testing. Positive replicas made from these negatives were examined in the SEM, and crack flank displacements measured directly; a resolution of ±0.25 μm was attainable and Mode I, Mode II and Mode III components of displacement could be separated. The process is shown diagrammatically in Fig. 2.

Crack-Tip Clip Gage

A twin-cantilever clip gage [*15*] was mounted at the crack tip to record the near-tip load-displacement response, using mounting anvils that responded to Mode II displacements

TABLE 2—*Mechanical properties of 3.2-mm BS 4360 50B sheet steel.*

	Yield Strength, σ_y, MPa	Ultimate Tensile Strength, σ_u, MPa	Elongation, %
Longitudinal	398	481	27
Transverse	385	468	35

[3] The conditions chosen for precracking will influence subsequent Mode II behavior in two major ways:

1. By changing the magnitude and distribution of the residual compressive stresses left along the crack flanks (in effect changing the closure response of the Mode I crack).
2. By affecting the surface roughness of the precrack. The ideal precrack is smooth, frictionless, and free of residual compressive stress, a situation which is unattainable in practice. In BS 4360 50B both the magnitude of compressive residual stress (due to plasticity-induced closure) and the CLA roughness decrease with the precracking stress intensity range ΔK_I, while the distribution of stress depends on the stress intensity gradient $(1/K)(dK/da)$ (see Refs *17* and *18*). The conditions chosen gave the minimum practicable center-line average (CLA) roughness and residual stress, and allowed precracks to be grown rapidly and simply. Precracking philosophy is discussed in detail in Ref *17*.

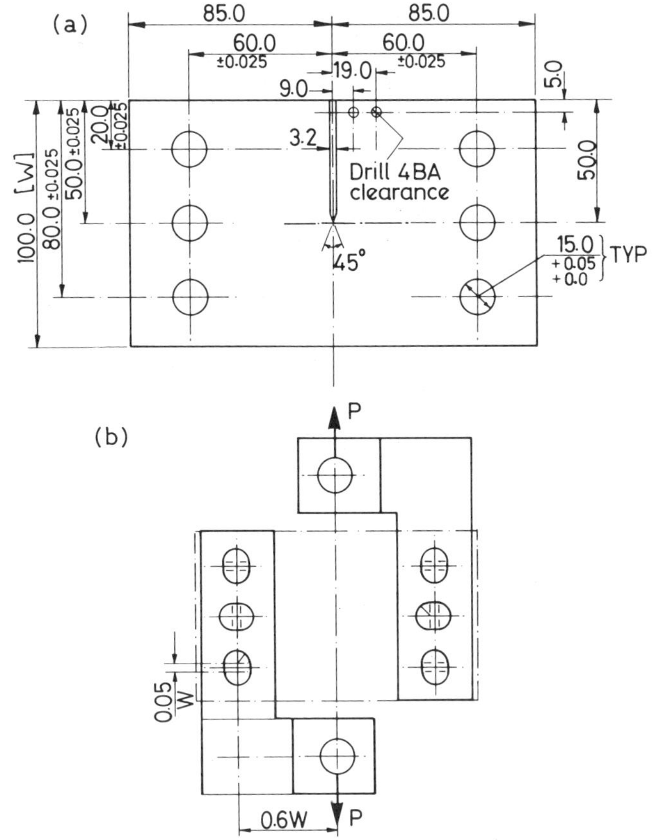

FIG. 1—*New compact shear (NCS) specimen geometry and load introduction; dimensions in mm.*

(Fig. 3). During testing, the CTG response was recorded at a test frequency of 0.05 Hz, using the instrumentation of Fig. 4. Both load and displacement signals were filtered before storage to remove high-frequency electrical noise.

Results

Definitions

It is important to have a precise terminology to describe the state of the flanks of a crack under Mode II loading. Tschegg [*16*] uses the phrase "sliding mode crack closure" to describe

TABLE 3—*Test details.*

Specimen	Applied Loads, kN	Load Ratio	ΔK_{II}^{nom}, MPa $\sqrt{m}$	Duration, cycles	Outcome
1	9.72 to 0.48	0.05	19	1×10^5	Mode I branch crack growth
2	4.84 to 0.24	0.05	9	2×10^6	no growth
3	9.22 to 4.61	0.5	9	1×10^6	no growth

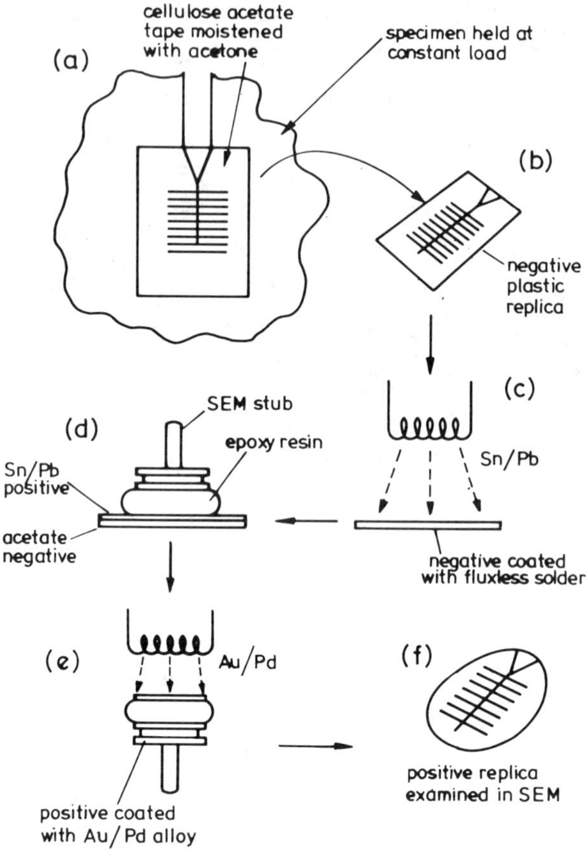

FIG. 2—*Surface replication technique.*

the interaction between crack surfaces undergoing sliding contact. This convention is not used in the current study, because it causes confusion with Mode I crack closure. Instead, the following terminology is adopted, using the crack-tip coordinate system in Fig. 5*a*.

1. A crack is "open" if Mode I displacements are present along its flanks ($\Delta v > 0$; see Fig. 5*b*). It is "fully open" if Mode I displacements reach the crack tip, in which case $K_I \geqq K_{op}$.
2. A crack is "closed" at any point if no Mode I displacements are present ($\Delta v = 0$; see Fig. 5*b*).
3. A crack is "slipped" at any point if Mode II displacements are present ($\Delta u \neq 0$; see Fig. 5*c*). If slip does not reach the crack tip, then $\Delta u = 0$ at the crack tip and the crack is "partially slipped." If $\Delta u \neq 0$ at the crack tip, then the crack is "fully slipped."
4. A crack is "locked" at any point if Δu does not change as the applied load changes $[\partial(\Delta u)/\partial P = 0]$.
5. A crack is "unlocked" at any point if its flanks are *free* to slide over one another as the load is changed $[\partial(\Delta u)/\partial P \neq 0]$. Unlocking may be full or partial in the same manner as slip.

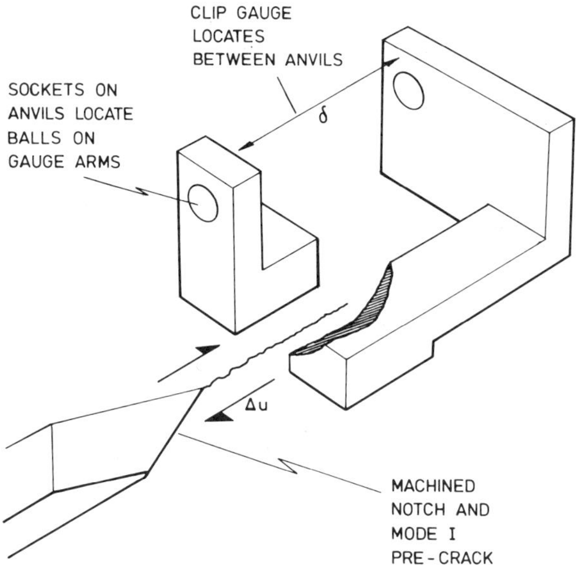

FIG. 3—*CTG anvil geometry.*

6. Slip is said to be "static" if it is unidirectional, and only takes place on first loading, or during the loading portion of a load cycle (see Fig. 5*d*).

7. Slip is said to be "reversed" if it takes place during both loading and unloading portions of a load cycle, and $\partial(\Delta u)/\partial P$ changes sign between loading and unloading (see Fig. 5*d*)

Measurements from Replicas

Unlocking Behavior on First Loading—The unlocking response on first loading was examined by taking a series of replicas at intervals during the first load application. The crack flank displacement profile for Specimen 3 is shown in Fig. 6. Slip spread progressively from the notch toward the precrack tip as the load was increased. The magnitude of slip was very much less than would have been the case if no frictional attenuation were present, and in

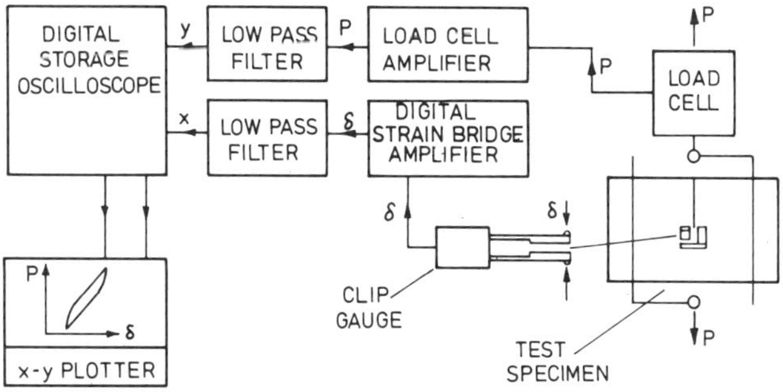

FIG. 4—*Recording instrumentation for crack-tip clip gage.*

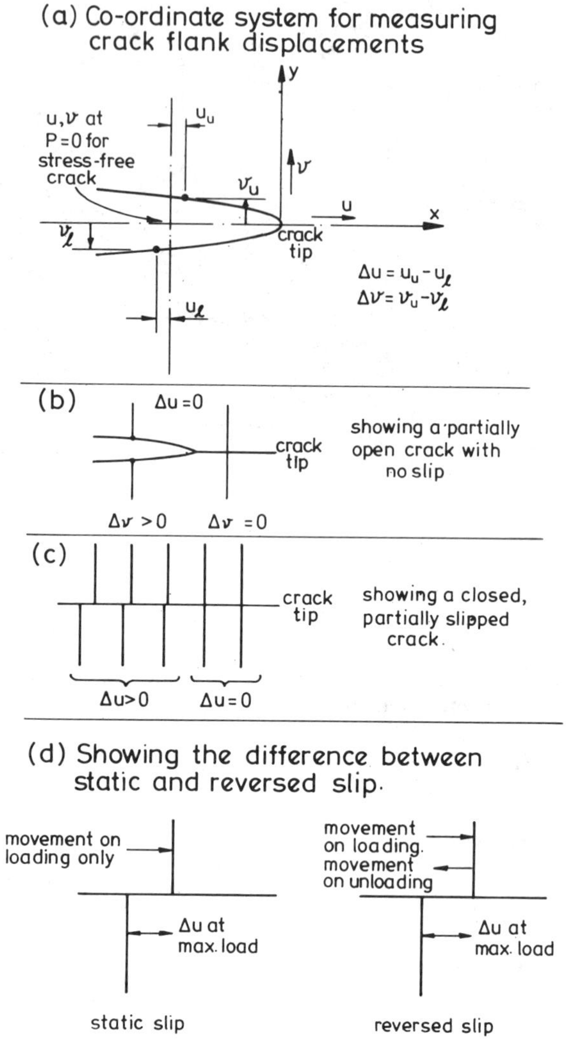

FIG. 5—*Coordinate system and notation for describing the crack flank displacement response under Mode II loading.*

this example no slip at all was observed until K_{II}^{nom} exceeded 4.5 MPa $\sqrt{m}$. During loading, the crack flanks were separated into two regions:

1. Behind the instantaneous effective crack tip, a_e, the crack flanks were unlocked and measurable slip was present.
2. Ahead of a_e, the crack flanks were locked, with zero slip.

If loading stopped before unlocking reached the crack tip, then no stress intensity was developed there ($K_{II}^{eff} = 0$). This was the case in Specimen 2 ($\Delta K_{II}^{nom} = 9$ MPa $\sqrt{m}$, $R = 0.05$).

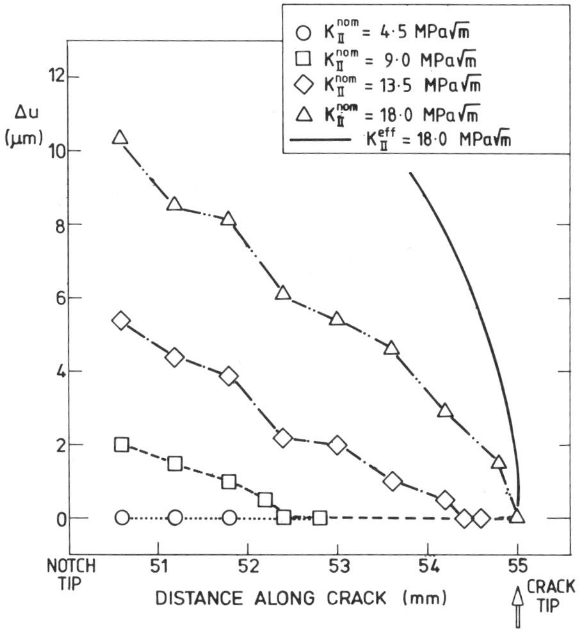

FIG. 6—*Crack flank sliding response on first loading for Specimen 3, compared with predicted response at* K_{II}^{eff} = *18 MPa*$\sqrt{m}$.

Cyclic Unlocking and Slip Behavior, Specimen 1—The cyclic crack flank displacement profile of Specimen 1, tested at ΔK_{II}^{nom} = 19 MPa $\sqrt{m}$ and R = 0.05, is shown in Fig. 7. In this specimen static slip spread to the precrack tip on first loading. On unloading, reversed slip also spread from the notch to the precrack tip. Hence a cyclic Mode II stress intensity was developed there ($\Delta K_{II}^{eff} > 0$) and crack growth subsequently took place. From Fig. 7 it can be seen that crack flank displacements at maximum load were less, and at minimum load were more, than predicted by assuming that the full nominal stress intensity range was developed at the precrack tip (ΔK_{II}^{eff} – ΔK_{II}^{nom}). The *effective* load ratio at the precrack tip was thus higher than the applied load ratio of 0.05, and the *effective* stress intensity was less than that applied.

Cyclic Unlocking and Slip Behavior, Specimen 2—The cyclic crack flank displacement profile of Specimen 2, tested at ΔK_{II}^{nom} = 9 Mpa $\sqrt{m}$ and R = 0.05, is shown in Fig. 8. In this specimen, static slip extended only for about 2 mm from the notch on first loading, the remainder of the precrack remaining locked with zero slip. Unloading resulted in reversed slip over only 1 mm of the precrack. The remainder of the crack remained locked, either with no slip, or with some positive amount, and a small residual stress intensity was left at the boundary between slipped and unslipped regions of the crack. As the test progressed, both static and reversed slip spread toward the precrack tip. It can be seen from Fig. 8 that the magnitudes of static and reversed slip were low. Because slip did not reach the precrack tip, no crack growth took place.

Cyclic Unlocking and Slip Behavior, Specimen 3—The cyclic crack flank displacement profile of Specimen 3, tested at ΔK_{II}^{nom} = 9 MPa $\sqrt{m}$ and R = 0.5, is shown in Fig. 9. In

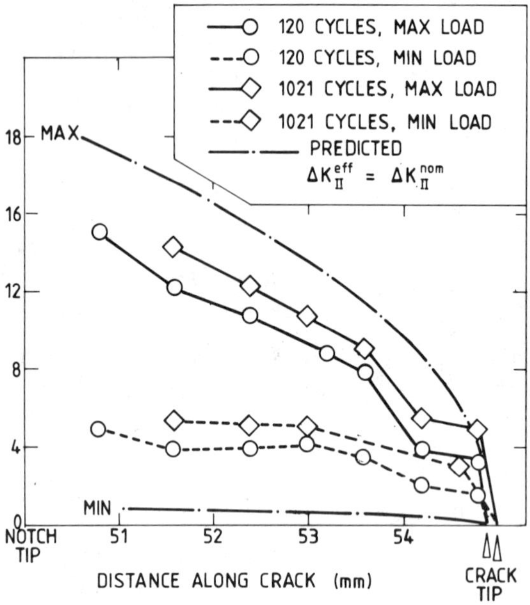

FIG. 7—*Crack flank displacements measured from replicas for Specimen 1, tested at* ΔK_{II}^{nom} = *19 MPa* $\sqrt{m}$ *and* R = *0.05, compared with predicted displacements for* ΔK_{II}^{eff} = ΔK_{II}^{nom} *(no crack flank frictional attenuation).*

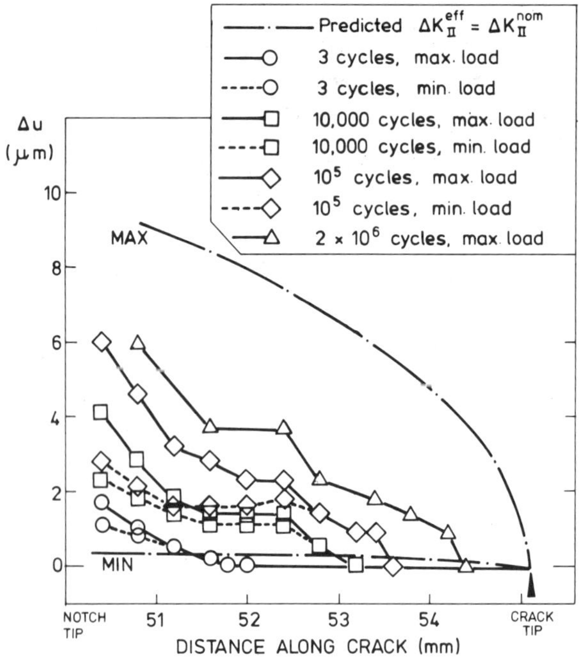

FIG. 8—*Crack flank displacements measured from replicas for Specimen 2, tested at* ΔK_{II}^{nom} = *9 MPa* $\sqrt{m}$ *and* R = *0.05, compared with predicted displacements for* ΔK_{II}^{eff} = ΔK_{II}^{nom}.

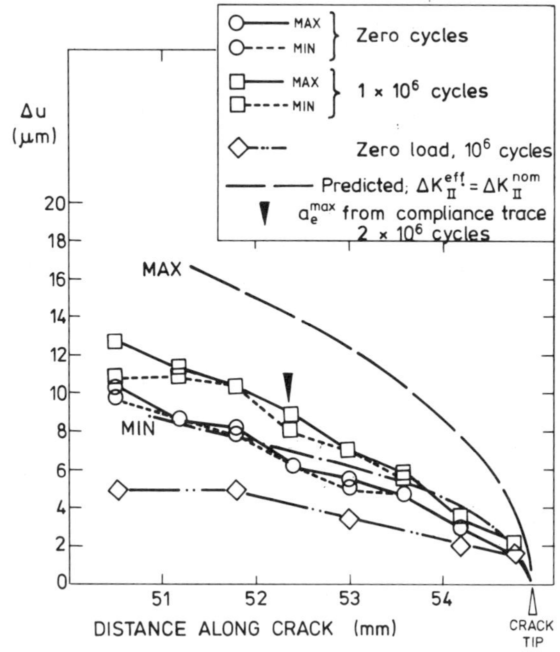

FIG. 9—*Crack flank displacements measured from replicas for Specimen 3, tested at* ΔK_{II}^{nom} = *9 MPa* $\sqrt{m}$ *and* R = *0.5, compared with predicted displacements for* $\Delta K_{II}^{eff} = \Delta K_{II}^{nom}$.

this specimen static slip spread to the precrack tip on first loading. Its magnitude at maximum load was similar to that of Specimen 1. On unloading to minimum load (K_{II}^{nom} = 9 MPa $\sqrt{m}$) reversed slip spread only 1 mm from the notch, and its magnitude was low. Large static residual Mode II displacements remained locked into the precrack beyond a_e^{max}, the maximum extent of reversed slip, and a residual Mode II stress intensity of −9 MPa $\sqrt{m}$ was locked into the precrack tip. Cyclic loading resulted in a slow spread of reversed slip toward the precrack tip, although reversed slip never reached the precrack tip and no crack growth took place. Large residual displacements remained along the entire precrack when unloaded to zero load at the end of the test (Fig. 9). The behavior of this specimen indicates that crack growth cannot occur at the precrack tip unless a *cyclic* Mode II stress intensity is developed there.

Crack-Tip Compliance Measurements

It is clear from the replica measurements that considerable frictional shear stresses act across the crack flanks to resist slip in a precrack loaded in Mode II. If these shear stresses obey a constant interfacial shear stress (CSS) friction law, then the load-displacement ($P - \delta$) response of a CTG takes the form shown in Fig. 10. The trace form is quite different from a Mode I closure trace, a consequence of the differences between unlocking behavior in Mode II and crack closure in Mode I [*17*]. However, the principles of interpretation remain the same: when the trace is linear, the effective crack length, a_e, is constant; when the trace is curved, then a_e is changing. The frictional attenuation in applied stress intensity, K_{II}^f, is also constant when a_e is constant, and only varies when a_e varies. It can be seen from

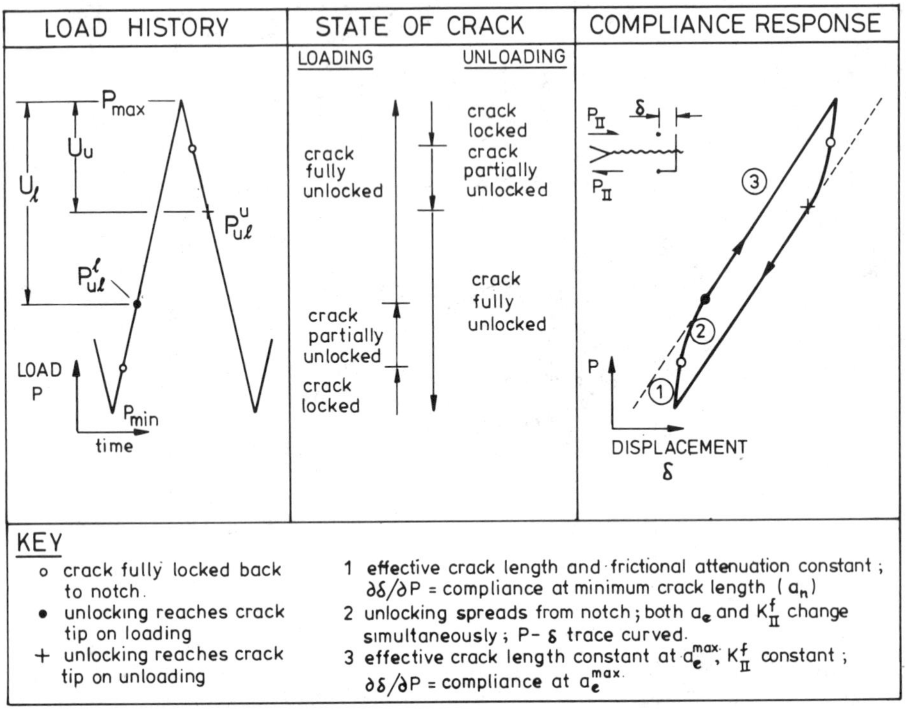

FIG. 10—*Idealized near-tip compliance response of a crack under Mode II loading.*

Fig. 10 that the maximum extent of reversed slip, a_e^{max}, may be deduced from a CTG trace if a CSS friction law is obeyed. Deviations from the idealized unlocking behavior will cause deviations in the trace shape.

Figure 11 illustrates the general form of CTG traces obtained in the current study, and compares predictions of a_e^{max} made from compliance traces with both replica measurements and visual observation. The following results can be seen:

1. The trace shapes obtained conform closely to the idealized response of Fig. 10.
2. Excellent agreement is obtained between replica measurements of a_e^{max} and CTG measurements.
3. The extent of reversed slip in Specimens 2 and 3 is virtually identical, although the extent of static slip is very different. It appears that reversed slip is controlled by the applied stress intensity range, while static slip is determined by the maximum applied stress intensity (these specimens were tested at the same ΔK_{II}^{nom}, but different R).
4. The increased extent of both static and reversed slip in Specimen 2 during testing can be clearly seen. A similar increase in the extent of reversed slip took place in Specimen 3.

These measurements indicate that the frictional shear stresses resisting slip in a precrack do indeed obey a CSS friction law, to a good approximation. They also indicate that the frictional resistance to slip decreases as a test proceeds and the crack surfaces suffer wear.

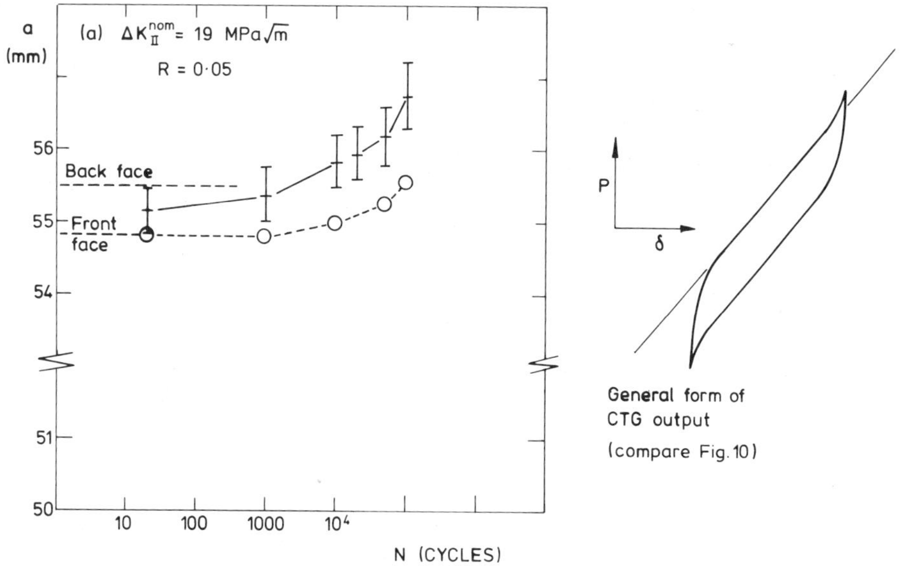

FIG. 11*a*—*The extent of static and reversed slip measured from replicas, compared with the effective crack length* a_e^{max} *deduced from the slope of a crack tip gage load-displacement trace (region 3 in Fig. 10), for Specimen 1. The general form of the crack tip gage load-displacement trace is also shown. For key see Fig. 11*b.

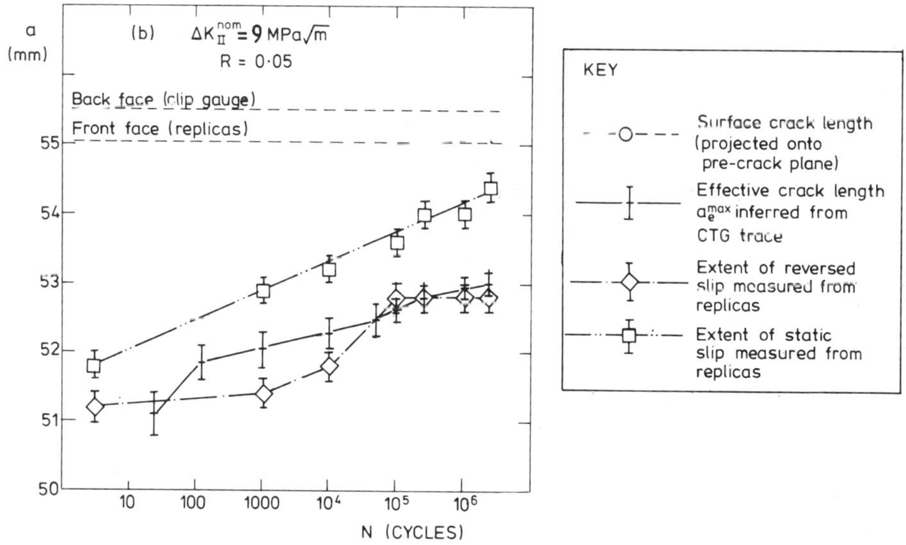

FIG. 11*b*—*The extent of static and reversed slip measured from replicas, compared with the effective crack length* a_e^{max} *deduced from the slope of a crack tip gage load-displacement trace, for Specimen 2.*

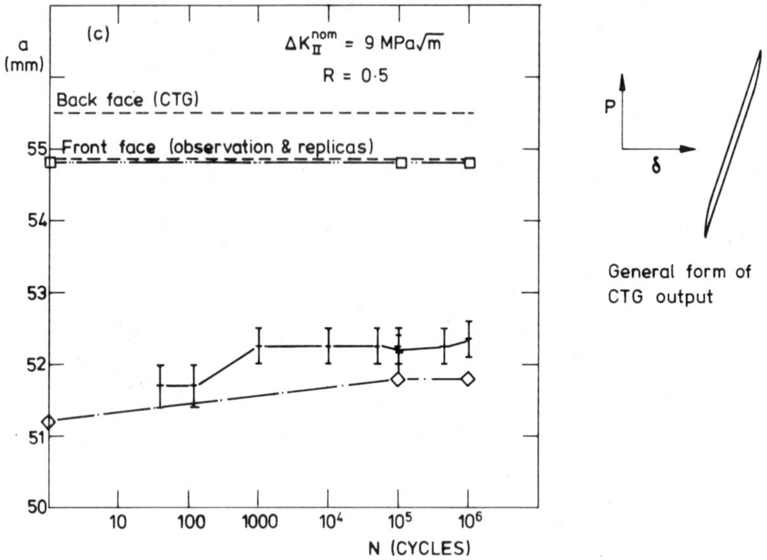

FIG. 11c—*The extent of static and reversed slip measured from replicas, compared with the effective crack length* a_e^{max} *deduced from the slope of a crack tip gage load-displacement trace, for Specimen 3. The general form of the crack tip gage load-displacement trace is also shown. For key see Fig. 11*b.

Summary

The unlocking response of precracks tested in Mode II shows the following features:

1. Unlocking is progressive from notch tip to precrack tip on loading.
2. Statically slipped regions of the precrack lock on unloading, retaining residual static stress intensities at the precrack tip.
3. The extent and magnitude of reversed slip are smaller than those of static slip, even at load ratios approaching zero.
4. Reversed slip appears to be controlled by the applied stress intensity range, while static slip is determined by the maximum applied stress intensity.
5. The applied load ratio has an exaggerated effect on the difference between static and reversed slip.
6. Large residual displacements remain on unloading to zero load.
7. Crack growth from the precrack tip can occur only if reversed slip reaches the precrack tip.

Mechanisms of Crack Surface Interaction

The specimens of the current study suffered pure Mode II loading. At first sight the lack of an externally applied compressive normal load across the crack flanks suggests that no frictional resistance to slip should exist. However, there are several possible sources for this resistance:

1. Compressive residual stresses exist behind the precrack tip due to plasticity-induced crack closure experienced during precracking [*18*]. These cause frictional shear stresses opposing slip along the crack flanks. Stress-relieving is often used to reduce these stresses. Unfortunately, such heat treatment also causes some rewelding of the crack surfaces, thereby

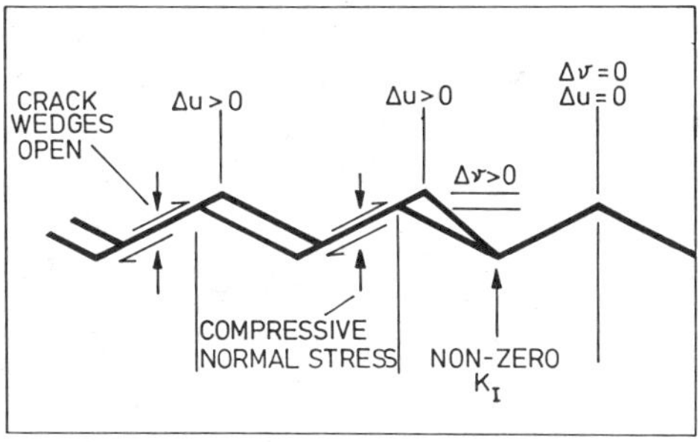

FIG. 12—*Mode I wedging behavior of crack flanks of a precracked specimen tested in Mode II.*

increasing the severity of crack surface interaction [*17*]. Thus heat treatment was not performed in the present study.

2. Examination of replicas showed that, under Mode II loading, the crack surfaces wedge open over surface irregularities (Fig. 12), thus generating a positive Mode I stress intensity at the effective crack tip, a_e^{max}, and compressive normal stresses along the crack flanks. These in turn cause frictional shear stresses opposing slip.

3. Fractographic examination by SEM after testing revealed extensive plastic deformation of interlocking surface asperities even in areas of solely static slip (see Fig. 13). Such deformation also results in resistance to slip.

FIG. 13*a*—*Worn precrack surface on Specimen 2 after testing, in a region which suffered reversed slip throughout testing. Note smeared worn surface and extensive fretting debris.*

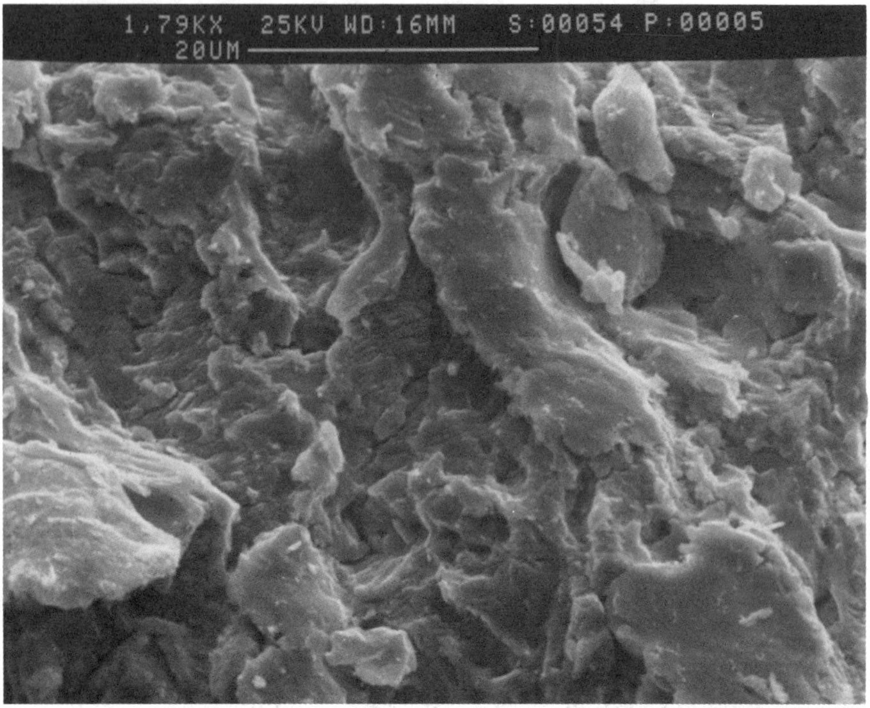

FIG. 13*b*—*Precrack surface on Specimen 3 in a region which suffered slip on first loading and post-test unloading only. Note extensive smearing of high spots.*

The compressive stresses due to precracking and wedging will increase as the precrack tip is approached, and hence the frictional shear stresses will follow the same pattern. The shear stresses due to asperity deformation will probably be reasonably constant along the crack flanks, as the surface roughness is constant in the tests conducted.

These frictional shear stresses appear to control the behavior of cracks under Mode II loading, so it is important to be able to model their effects on slip between crack flanks.

Modeling the Unlocking Behavior of Cracks Under Mode II Loading

Description of the Model

The notch and precrack of the test specimens are idealized as a semi-infinite crack in an infinite plate; the notch is considered to be frictionless while the precrack is treated as an elastic line crack with a limiting frictional shear stress distribution, $\tau(x')$, acting across unlocked or previously unlocked portions of its flanks to resist slip. The frictional shear stress at any point, x', is independent of the magnitude of slip, Δu, once the crack flanks are unlocked at that point. Thus the frictional stresses obey a CSS friction law. This assumption leads to the same unlocking behavior as would an assumption of Coulomb frictional behavior. It merely replaces two variables, the friction coefficient μ and compressive normal crack flank stress distribution $\sigma(x')$, with one, $\tau(x')$. Such simplification is necessary because

μ for a Mode I precrack is unknown. A simple linear distribution is used for $\tau(x')$, such that

$$\tau(x') = \tau_t x' + \tau_s \quad (1)$$

where x' is the distance along the precrack from its origin at the notch tip. A linear distribution is chosen because the crack flank compressive stresses (and hence frictional shear stresses) due to precracking and Mode I wedging both rise as the precrack tip is approached. A linear distribution for $\tau(x')$ is the simplest that will model these phenomena.

At a given *monotonic* load, the precrack is assumed to unlock over a distance x'_s such that (Fig. 14)

$$K_{II}^a = K_{II}^f + K_{II}^{a-f} \quad (2)$$

where K_{II}^a is the applied stress intensity at the instantaneous effective crack tip, a_e ($a_e = a_n + x'_s$). It is calculated by assuming zero friction along the unlocked portion of the precrack, and by assuming that the locked portion does not exist. K_{II}^f is the attenuation in applied stress intensity caused by the frictional shear stresses, $\tau(x')$, acting across the unlocked

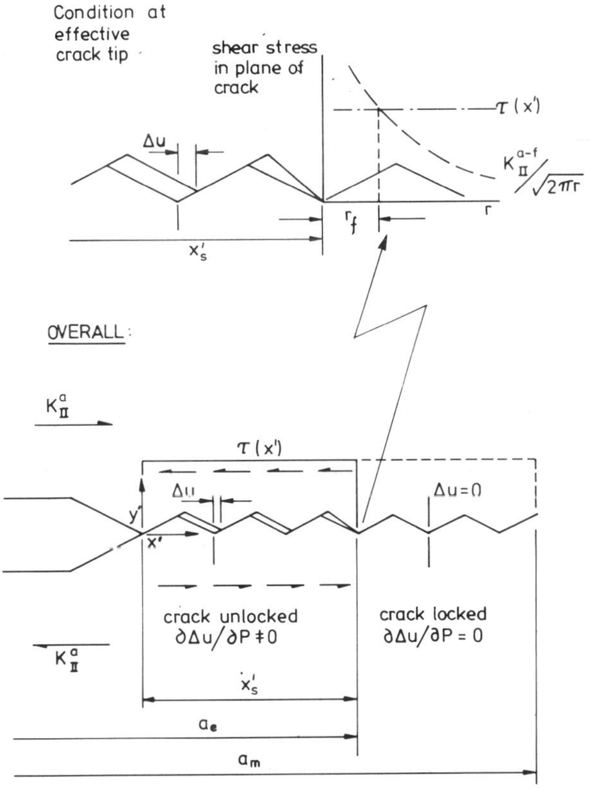

FIG. 14—*Unlocking behavior of a precrack on first loading, showing notation and conditions at the instantaneous effective crack tip.*

portion of the precrack. It is calculated using a weight function for a point loaded semi-infinite crack in an infinite plate [*19*]

$$K_{II}{}^{f} = \int_0^{x'_s} \frac{2\tau(x')dx'}{\sqrt{(2\pi x')}} \tag{3}$$

$K_{II}{}^{a-f}$ is the effective stress intensity developed at the instantaneous effective crack tip, a_e. For a crack unlocked to its tip ($a_e = a_m$), $K_{II}{}^{a-f}$ is the effective stress intensity $K_{II}{}^{eff}$. For a partially unlocked crack ($a_e < a_m$), $K_{II}{}^{a-f}$ is the stress intensity required to cause further unlocking beyond a_e. It is assumed that, for unlocking to spread further, the applied shear stress, τ_{xy}, ahead of the instantaneous effective crack tip in the precrack plane must exceed $\tau(x')$ over a critical distance r_f (Fig. 14). Thus

$$\tau_{xy} = \frac{K_{II}{}^{a-f}}{\sqrt{(2\pi r)}} = \tau(x'_s + r_f) = \tau_l(x'_s + r_f) + \tau_s \tag{4}$$

r_f may be rationalized as the average asperity size on the precrack surface (typically 10 to 30 μm).[4]

Model Behavior

The response of the model to constant and variable-amplitude loading is detailed elsewhere [*17,20*]. Briefly, the model behaves as follows:

During monotonic loading (Fig. 14) the precrack unlocks progressively from the notch (a_n). Ahead of the instantaneous effective crack tip, a_e, the crack flanks are locked with zero slip, and the material behaves as if no crack is present. Behind a_e, the crack is unlocked with the limiting frictional shear stresses, $\tau(x')$, acting across its flanks.

Load reversal at maximum load causes the *entire* precrack to lock, since reversed slip requires unlocking in the opposite sense to first loading. Thus reduction in the applied load causes unlocking to again spread progressively from the notch toward the precrack tip (Fig. 15). However, because the crack flanks in the statically slipped region of the precrack are already locked with the limiting frictional shear stresses acting to resist slip in the loading direction, the frictional resistance to slip in the unloading direction is doubled. Similar behavior occurs on reloading after load reversal at minimum load. This feature of unlocking behavior is the key to understanding many of the characteristic features of the crack flank displacement response of precracks. It accounts for the decreased magnitude and extent of reversed slip compared with static slip, for the exaggerated effect of applied load ratio on the differences between static and reversed slip, and for the large residual displacements left at zero load. Because the precrack locks fully at each load reversal, CTG compliance traces adopt the characteristic form of Fig. 10.

Comparison with Experiment

The CSS unlocking model was used to model the crack flank displacement responses of Specimens 1 to 3 (Fig. 16). It can be seen that excellent qualitative agreement between

[4] The precise value chosen for $K_{II}{}^{a-f}$ has little effect on the unlocking behavior predicted by the model. A finite value was chosen partly to aid visualization of the behavior of crack-tip clip gages and unlocking crack flanks, and partly because experimental evidence [*17*] supported the concept. No attempt was made to vary asperity sizes on real precracks and thereby validate the assumption made in the model, since experimental scatter would swamp such second-order effects.

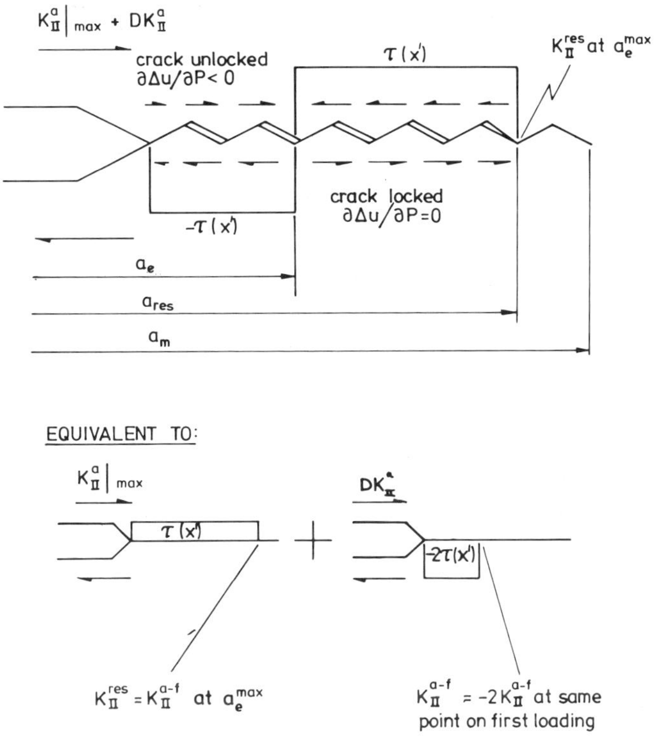

FIG. 15—*Unlocking behavior of a precrack on unloading from maximum load, showing doubled frictional resistance to slip.*

model and experiment was obtained, if reasonable values for the frictional shear stresses are assumed.

Implications

The results of this study show that crack surface interaction has a dominating influence on crack growth under Mode II loading. Crack growth occurred in only one of the three tests, and in that test the CSS model predictions (Fig. 16) show that ΔK_{II}^{eff} was less than 50% of ΔK_{II}^{nom}. This has the following implications for Mode II studies:

1. Common mixed-mode threshold criteria compare Mode II behavior with Mode I thresholds [*7,8,21*]. Such comparisons will be of little value unless they compare effective rather than nominal Mode I and Mode II thresholds.
2. In previous work on Mode II crack growth a variety of preparation routes has been used to generate sharp cracks for Mode II testing (these routes are reviewed by Smith [*17*]). The mechanisms of surface interaction identified in this study are all sensitive to preparation route. Thus considerable experimental scatter is likely to result from variations in specimen preparation.
3. Attempts to model the behavior of Mode II cracks in a compressive applied stress field have always assumed Coulomb frictional behavior [*13,22,23*]. The results of this study indicate that such an approach is valid. However, it is likely to underestimate the severity

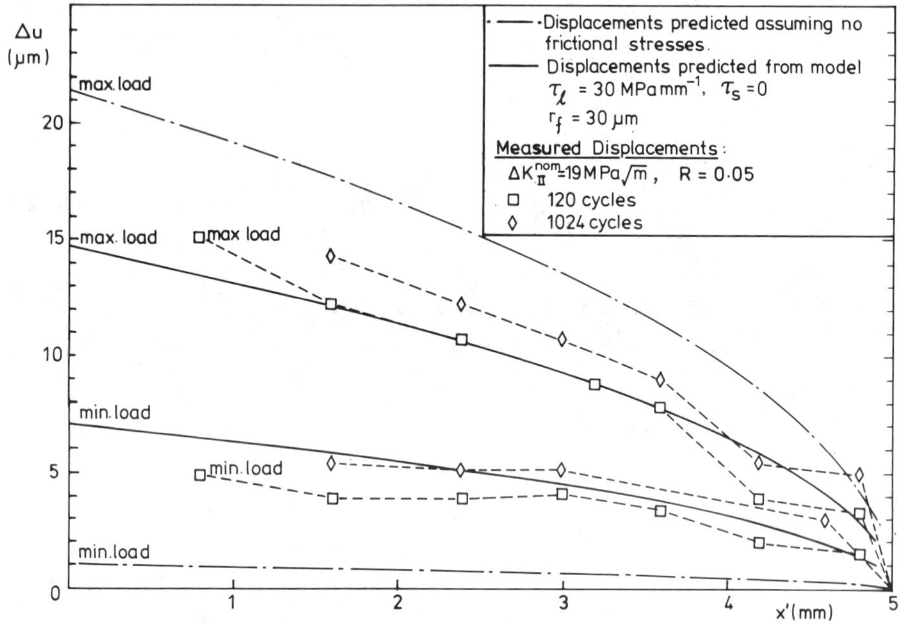

FIG. 16*a*—*Comparison of experimental crack flank displacement profiles with the predictions of the CSS unlocking model: Specimen 1.*

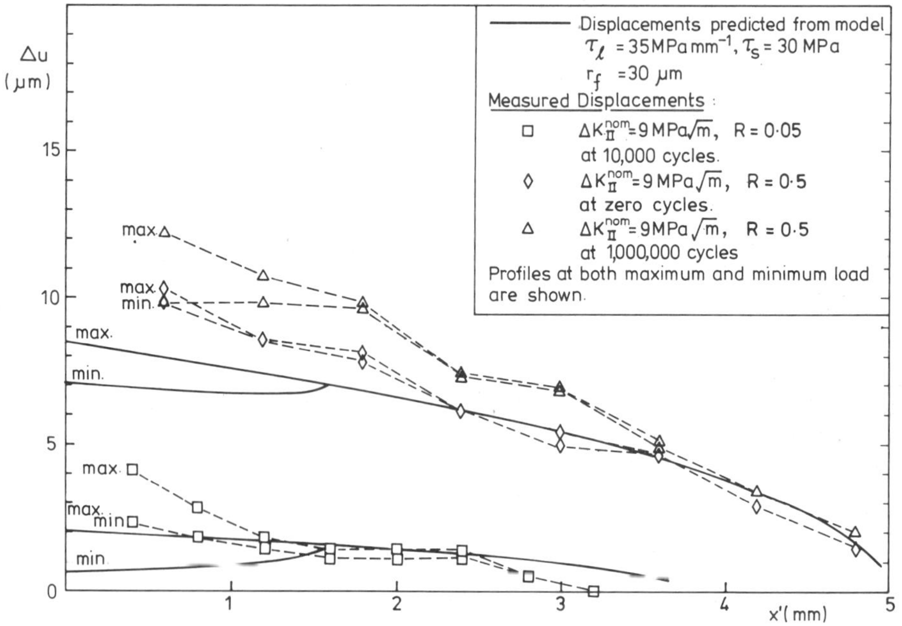

FIG. 16*b*—*Comparison of experimental crack flank displacement profiles with the predictions of the CSS unlocking model: Specimens 2 and 3.*

of crack flank frictional attenuation unless due account is taken of the possible contributions of Mode I wedging and plastic deformation of the mating surfaces

Both the compliance measurement technique and the modeling approach used in this study are likely to be of use in Mode III studies, since the mechanical environment of sliding crack surfaces is similar.

Conclusions

In Mode I precracks under Mode II loading, crack surface interaction takes the form of shear stresses acting across the crack flanks to resist slip. These shear stresses obey a constant interfacial shear stress friction law. Crack growth from the precrack tip can take place only if sufficient cyclic stress intensity is developed there; and crack flank shear stresses exert a major influence on the magnitude, and indeed the presence, of such cyclic stress intensities.

Acknowledgment

The first author (M.C.S.) gratefully acknowledges the support of a Science and Engineering Research Council studentship for the period during which this work was performed.

References

[*1*] Clayton, P. and Allery, M. B. P., "Metallurgical Aspects of Surface Damage Problems in Rails," MR MF1/264-384-1, British Rail Research, Derby, U.K., 1981.

[*2*] Skinner, D. H. and Judd, P. A., "A Metallographic Study of Fatigue in Rails" in *Proceedings,* AIM 34th Annual Conference, Queensland, Australia, 1981, pp. 68–72.

[*3*] Scott, D., "Rolling Contact Fatigue," *Treatise on Materials Science and Technology,* Vol. 13, 1979, pp. 321–361.

[*4*] Hurd, N. J. and Irving, P. E., "Factors Influencing Propagation of Mode III Fatigue Cracks Under Torsional Loading," *Design of Fatigue and Fracture Resistant Structures, ASTM STP 761,* American Society for Testing and Materials, Philadelphia, 1982, pp. 212–233.

[*5*] Fleming, J. R. and Suh, N. P., "Mechanics of Crack Propagation in Delamination Wear," *Wear,* Vol. 44, 1977, pp. 39–56.

[*6*] Otsuka, A., Mori, K., Ohshima, T., and Tsuyama, S., "Mode II Fatigue Crack Propagation in Aluminium Alloys and Mild Steel," *Proceedings,* 5th International Conference on Fracture, Cannes, France, Vol. 4, 1980, pp. 1851–1858.

[*7*] Hua, G., Brown, M. W., and Miller, K. J., "Mixed-Mode Fatigue Thresholds," *Fatigue of Engineering Materials and Structures,* Vol. 5, 1982, pp. 1–17.

[*8*] Pook, L. P. and Greenan, A. F., "Mode II Fatigue Crack Growth Threshold in Mild Steel," NEL Report No. 654, National Engineering Laboratory, East Kilbride, Glasgow, U.K., 1978.

[*9*] Robin, C., Jodin, P. and Pluvinage, G., "Essai de Fatigue en Cisaillement sur une Eprouvette a Double Entaille (Shear Fatigue Test in a Double Notched Specimen)," *Comptes Rendus de l'Academie des Sciences,* Series B, Vol. 286, Paris, 1978, pp. 77–80.

[*10*] Smith, M. C. and Smith, R. A., "The Formation of Spherical Wear Debris in Mode II Fatigue Cracks," *Wear,* Vol. 76, 1982, pp. 105–128.

[*11*] Nayeb-Hashemi, H., McClintock, F. A., and Ritchie, R. O., "Effects of Friction and High Torque on Fatigue Crack Propagation in Mode III," *Metallurgical Transactions,* Vol. 13A, 1982, pp. 2197–2204.

[*12*] Tschegg, E. K., "A Contribution to Mode III Fatigue Crack Propagation," *Materials Science and Engineering,* Vol. 54, 1982, pp. 127–136.

[*13*] Hearle, A. D. and Johnson, K. L., to be published in *Journal of the Mechanics and Physics of Solids.*

[*14*] Richard, H. A., "A New Compact Shear Specimen," *International Journal of Fracture,* Vol. 17, 1981, pp. R105–R107.

[*15*] Fleck, N. A., "Some Aspects of Clip Gauge Design," *Strain,* Feb. 1983, pp. 17–22.

[*16*] Tschegg, E. K., "The Influence of the Static Mode I Load and R Ratio on Mode III Fatigue Crack Growth Behaviour in Mild Steel," *Materials Science and Engineering,* Vol. 59, 1983, pp. 127–137.
[*17*] Smith, M. C., "Some Aspects of Mode II Fatigue Crack Growth," Ph.D. thesis, University of Cambridge, Cambridge, U.K., 1984.
[*18*] Fleck, N. A., "An Investigation of Fatigue Crack Closure," Ph.D. Thesis, University of Cambridge, Cambridge, U.K., 1983.
[*19*] Tada, H., Paris, P. C., and Irwin, G.R., *The Stress Analysis of Cracks Handbook,* DEL Research Corp., Hellertown, PA, 1973.
[*20*] Smith, M. C. and Smith, R. A. "Modelling the Unlocking and Slip of Crack Surfaces Under Mode II Loading," *Proceedings,* 6th International Conference on Fracture, Delhi, Dec. 1984.
[*21*] Otsuka, A., Mori, K., and Miyata, T., "The Condition of Fatigue Crack Growth in Mixed Mode Condition," *Engineering Fracture Mechanics,* Vol. 7, 1975, pp. 429–439.
[*22*] Chipperfield, C. G., "The Interaction of Inclusions and Service Stresses in the Initiation of Fatigue Damage in Rails," BHP Melbourne Research Laboratory Report No. MRL/PS13/80/006, Broken Hill Proprietary Co., Melbourne, Victoria, Australia, 1980.
[*23*] Hills, D. A. and Ashelby, D. W., "On the Application of Fracture Mechanics to Wear," *Wear,* Vol. 54, 1979, pp. 321–330.

Keisuke Tanaka[1] *and Yoshiaki Akiniwa*[1]

The Propagation of Short Fatigue Cracks at Notches

REFERENCE: Tanaka, K. and Akiniwa, Y., **"The Propagation of Short Fatigue Cracks at Notches,"** *Basic Questions in Fatigue: Volume I, ASTM STP 924,* J. T. Fong and R. J. Fields, Eds., American Society for Testing and Materials, Philadelphia, 1988, pp. 281–298.

ABSTRACT: Center-notched plate specimens and single-edge-notched specimens of a low-carbon steel were fatigued under axial and bending loads. The growth behavior of a short fatigue crack near the notch tip was analyzed based on the measurement of crack closure. The crack closure develops as a crack grows; the effective fraction of the applied stress decreases. This accounts for the decrease of the crack growth rate with increasing crack length. The relation between growth rate and the effective stress intensity range is unique, and nearly identical to that for a long crack, meaning that the effective stress intensity range is a good measure of the crack driving force. The results on the threshold of crack growth are compared with the model proposed by Tanaka and Nakai. The variations of the threshold values of the stress intensity range and the effective stress intensity range, and the crack closure at the threshold with the length of nonpropagating cracks, agree well with the model prediction.

KEY WORDS: fatigue (materials), notch, short crack, crack closure, fracture mechanics, low-carbon steel, grain size, threshold condition, fatigue limit

Nomenclature

ρ	Notch tip radius
t	Notch length for single-edge-notched (SEN) specimen and half notch length for center-notched plate (CNP) specimen
W	Specimen width for SEN specimen and half specimen width for CNP specimen
c	Crack length from notch tip
c_{np}	Nonpropagating crack length from notch tip
c'_0, c_0	Intrinsic crack length
R	Ratio of minimum load to maximum load
σ_a	Nominal (net-) stress amplitude
σ_{w0}, $\Delta\sigma_{w0}$	Stress amplitude and range[2] at fatigue limit of smooth specimen
σ_{w1}, $\Delta\sigma_{w1}$	Nominal stress amplitude and range[2] at threshold for crack initiation in notched specimen
σ_{w2}, $\Delta\sigma_{w2}$	Nominal stress amplitude and range[2] at threshold for fracture in notched specimen
$\Delta\sigma_{th}$, $\Delta\sigma_{effth}$	Threshold stress range[2] and effective threshold stress range for crack growth
K_t	Elastic concentration factor
K	Stress intensity factor (SIF)
F	Correction fractor for SIF
K_{max}, K_{min}	Maximum and minimum SIF values

[1] Associate professor and graduate student, respectively, Department of Engineering Science, Kyoto University, Kyoto 606, Japan.

[2] The range of each quantity is taken for the nominal tensile portion only; the range equals its maximum value for the case of negative R.

K_{op}	SIF at crack opening
ΔK, ΔK_{eff}	SIF range² and effective SIF range
ΔK_{th}, ΔK_{effth}	SIF range² and effective SIF range at threshold for crack growth (suffix ∞ indicates the value for a long crack)
U_{th}, $U_{th\infty}$	Effective fractions for a short crack and for a long crack

The reduction of fatigue strength due to notches is a classical problem for fatigue engineers. One of the fundamental questions of this problem is "Why does a fatigue crack decelerate or stop at the root of a sharp notch?" Since the stress intensity factor (SIF) for a crack emanating from a notch tip is a monotonic increasing function of the crack length under constant-amplitude loading, the conventional technique of fracture mechanics (dealing with long cracks) fails to predict the growth behavior of short cracks at sharp notches.

Taira et al. [*1*] observed that the size of the slipband zone ahead of the crack tip diminished with increasing crack length at a sharp notch. They found that the diminishing size of the slipband zone corresponded to the decreasing rate of a short fatigue crack. Smith et al. [*2,3*] and Hammounda et al. [*4*] ascribed the deceleration of a short crack to the falling notch plasticity effect. El Haddad et al. [*5,6*] explained the growth dip by combining the anomalous crack growth behavior due to the smallness of a crack with the notch plasticity effect. A recent study of Tanaka and Nakai [*7*] shows that the development of crack closure is primarily responsible for the decreasing rate of a short crack in notch fatigue. Similar results have been reported by Ogura et al. [*8*] and Schijve [*9*].

Questions to be answered in relation to the crack closure of a short crack are the following two: (1) "Is the closure of a short crack a function of crack length and stress level?" (2) "Is the effective stress intensity range a good measure of the driving force for short crack growth?" Stress ratio, notch geometry, loading mode, and material grain size are those variables that influence the closure and propagation of a short crack. The results of previous exploratory experiments conducted under a limited experimental condition support an affirmative answer to the foregoing two questions [*7–9*].

In the present study, to investigate the generality of the answer and to quantify the crack closure development, the method for measuring the crack closure of a short crack is improved, and the measurement is conducted under a wide range of experimental variables of stress level, material grain size, notch geometry, and loading mode. Quantitative discussion is centered on the threshold of short crack growth. The model proposed by Tanaka and Nakai [*10*] is evaluated based on the experiments.

Experimental Procedure

Material and Test Specimen

The experimental material used is a structural low-carbon steel (JIS SM41B) with the following chemical composition in weight percent: 0.17C-0.19Si-0.79Mn-0.016P-0.020S. The specimens were annealed under two different conditions to have different (ferrite) grain sizes. The material with a grain size of 64 μm is denoted Material L, and that with a grain size of 14 μm Material S. The heat-treatment condition and the mechanical properties derived from tension tests are summarized in Table 1.

Figure 1 shows two types of notched specimens. Center-notched plate (CNP) specimens are subjected to axial tension compression; single-edge-notched (SEN) specimens to in-plane bending. Most of the specimens were made of Material L. Only CNP specimens having notch-tip radius $\rho = 0.16$ mm and notch depth $t = 3$ mm were made of Material S as well as Material L to study the grain size effect. The influence of notch-tip radius is studied using

TABLE 1—*Heat-treatment and mechanical properties of test materials.*

Material	Heat Treatment	Grain Size, d, μm	Yield Strength σ_Y, MPa	Tensile Strength σ_B, MPa
L	1200°C 5 h anneal	64	194	423
S	900°C 30 min anneal	14	280	445

CNP specimens with ρ = 0.16, 0.39, and 0.85 mm. SEN specimens were used to study the effects of loading mode and notch depth. To study the notch-size effect, SEN specimens are preferable because of easy machining of a small notch and of easy measurement of crack closure as described later. The elastic stress concentration factors K_t are also indicated in the figure. Ishida's solution [*11*] is used to calculate K_t for CNP specimen and Neuber's solution [*12*] for SEN specimen.

Fatigue Testing and Stress Intensity Factor

The fatigue tests under axial loading were conducted in a closed-loop servo hydraulic testing machine. The specimen was gripped by a ram with wood's metal chuck to obtain good alignment. The ratio of minimum to maximum load R was -1. The frequency of stress cycling is 30 Hz. The crack length was measured by traveling microscope. In-plane bending

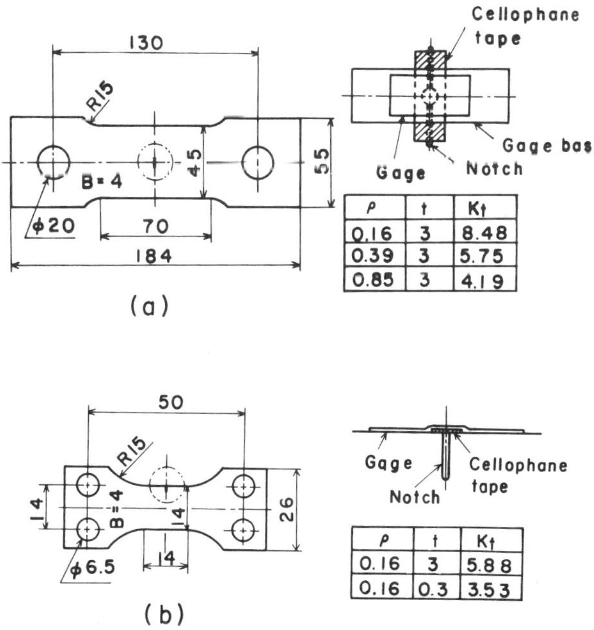

FIG. 1—*Test specimens (dimensions are in mm).* (a) *Center-notched plate (CNP) specimen.* (b) *Single-edge-notched (SEN) specimen.*

fatigue tests were performed in a resonance-type bending testing machine operated at a test speed of 33.3 Hz. The specimen was subjected to a fully reversed bending moment, that is, $R = -1$. With SEN specimens under bending, only the threshold condition of crack growth was studied. All the tests were conducted under constant amplitude of applied load or moment.

The range of SIF for a crack emanating from a notch tip is generally expressed by

$$\Delta K = \Delta\sigma(\pi c)^{1/2} \cdot F(c) \tag{1}$$

where

$\Delta\sigma$ = nominal (net-) stress range,
c = crack length, and
F = correction factor.

The F term is a function of notch geometry and loading mode. When c is small compared with ρ, F is given by

$$F_A = 1.122K_t/(1 + 4.5c/\rho)^{1/2} \tag{2}$$

for both tension and bending [*13*]. When c becomes large, the effect of a notch can be neglected to compute SIF. In this case, the F term for CNP specimen under tension is given by [*14*]

$$F_B = (1 - t/W)(a/c)^{1/2} \cdot [1 - 0.25(a/W)^2 + 0.06(a/W)^4] \cdot [\sec(\pi a/2W)]^{1/2} \tag{3}$$

and that for SEN specimen under bending is [*15*]

$$F_C = (1 - t/W)^2(a/c)^{1/2} \cdot [1.122 - 1.40(a/W) + 7.33(a/W)^2 - 13.08(a/W)^3 + 14.0(a/W)^4] \tag{4}$$

For the CNP specimen, SIF is calculated using F_A when $c \leq c^*$ (c^* is the crack length at $F_A = F_B$), and by F_B at $c > c^*$ [*7*]. For the SEN specimen under bending, F_A is used when $c \leq \rho$, and F_C is used when $c > \rho$.

Crack Closure Measurement

The crack closure of a short crack at the notch was measured by an improved compliance method. For CNP specimens, the cellophane tape of about 1.5 mm width was first put over the center notch, and then a foil strain gage of length 5 mm and width 1.9 mm (KYOWA KFC-5-C1-11) was glued on across the tape along the centerline of the specimen (see Fig. 1*a*). The output of the strain gage was used as the displacement signal. The hysteresis loop of load versus displacement was recorded intermittently during the test by manual loading of one cycle, and the point of the crack opening was determined as the inflection point of compliance magnified by a circuit subtracting the elastic compliance from the loading curve [*7*]. An example of the compliance curves obtained is presented in Figs. 2*a* and 2*b*, where 2*a* and 2*b* are before and after subtraction. In Fig. 2*b*, the *Y*-axis is magnified by about nine times from that in Fig. 2*a*. The arrow in the figure indicates the crack opening point.

For SEN specimens, a similar technique was used. A strain gage is glued across a taped notch on the upper face of the specimen (see Fig. 1*b*). The applied moment was monitored by the output of the strain gage glued on the transmission torque bar. The output of both

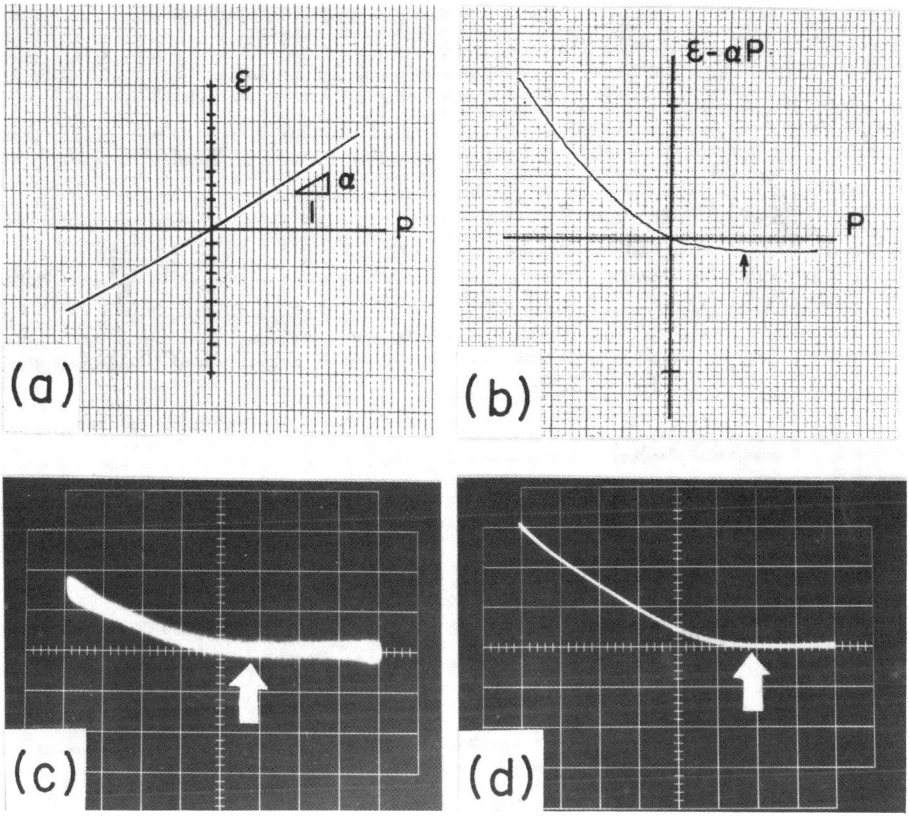

FIG. 2—*Compliance curves.* (a) *Material L, CNP* (ρ = *0.16 mm,* t = *3 mm*), σ_a = *55 MPa,* c_{np} = *0.584 mm.* (b) *After subtraction of the elastic compliance from* (a). (c) *Material L, SEN* (ρ = *0.16 mm,* t = *3 mm*), σ_a = *60 MPa,* c_{np} = *0.181 mm.* (d) *Material L, SEN* (ρ = *0.16 mm,* t = *3 mm*), σ_a = *80 MPa,* c_{np} = *1.29 mm.*

strain gages was amplified by the same type of amplifiers, and the opening point was determined from the compliance curve (where the elastic compliance is subtracted) on a synchroscope during the test. Figures 2*c* and 2*d* show two examples of the curve. When the crack length is short (Fig. 2*c*), the magnification of the *Y*-axis is large.

The above compliance method was capable of measuring the opening point of crack length down to about 30 μm. For a crack longer than about 2 mm, alternative methods were used to measure the displacement because of the break of strain gages. A differential transformer with needle-pointed pickup [*16*] was used to measure the crack opening displacement of the CNP specimen. The strain gage was glued on the back face (the edge face without notch) for SEN specimens. Though the sensitivity was reduced in those two methods, they are enough for longer cracks.

Relation Between Crack Propagation Rate and Stress Intensity Range

The rate of a short crack nucleated at the tip of a sharp notch first decreases as the crack propagates. The crack then becomes nonpropagating under low stresses, while it begins to accelerate after taking a minimum growth rate under high stresses. Since the SIF range is a monotonic increasing function of crack length under constant-amplitude loading, a dip of growth rate is also seen when the rate is correlated with the SIF range.

Figure 3 shows an example of the relation between crack growth rage dc/dN and SIF range ΔK, where ΔK equals K_{max} in the present test with $R = -1$. The specimen is CNP with $\rho = 0.16$ mm and $t = 3$ mm, and is made of Material L. The minimum crack length measured was about 50 μm. The solid line in the figure indicates the relation obtained for long cracks [*17*]. At nominal stress amplitudes, σ_a, between 40 and 60 MPa, the nucleated crack becomes nonpropagating. (The data at $\sigma_a = 60$ MPa are not shown in the figure.) The threshold value of the SIF range, ΔK_{th}, increases with increasing stress amplitude. Most of the growth data lie above the relation of long cracks; ΔK_{th} is lower than the threshold SIF range for long cracks. At a stress amplitude of 65 MPa, a crack propagates to cause the final fracture of the specimen. At $\sigma_a = 35$ MPa, neither crack nor slip was formed at the notch. The fatigue limit for crack initiation, σ_{w1}, is between 35 and 40 MPa; that for fracture, σ_{w2}, is between 60 and 65 MPa.

The opening point of a short crack was measured by the compliance method and the effective stress intensity range ΔK_{eff} was calculated by

$$\Delta K_{eff} = K_{max} - K_{op} \quad (5)$$

where K_{max} is the maximum stress intensity and K_{op} the SIF value at the crack opening.

In Fig. 4, the rate is correlated to ΔK_{eff}. A unique relation can be seen between dc/dN and ΔK_{eff}, and it agrees fairly well with the rate versus ΔK_{eff} relation (dashed line) for long cracks [*17*].

Crack Closure Development

In Fig. 5*a*, ΔK_{eff} is plotted against crack length for the case of a CNP specimen ($\rho = 0.16$ mm, $t = 3$ mm) of Material L. The ΔK_{eff}-value decreases with increasing crack length, like

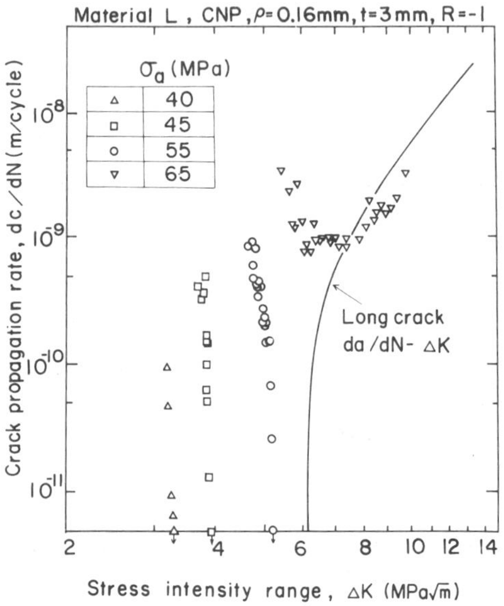

FIG. 3—*Relation between crack propagation rate and stress intensity range. Material L, CNP specimen (ρ = 0.16 mm, t = 3 mm).*

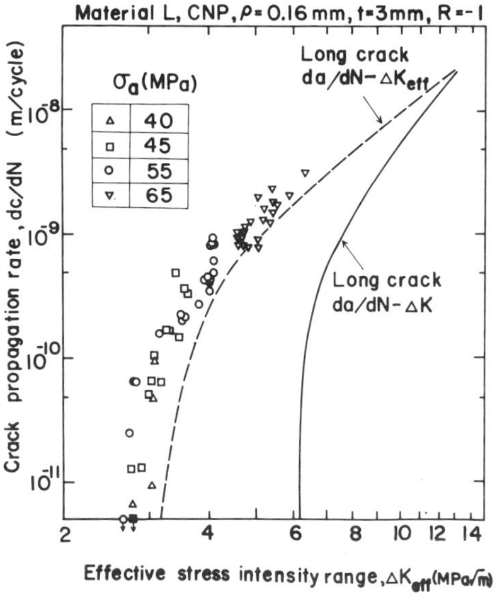

FIG. 4—*Relation between crack propagation rate and effective stress intensity range. Material L, CNP specimen* (ρ = *0.16 mm,* t = *3 mm).*

the growth rate. At the threshold (the last data points for σ_a = 40, 45, and 55 MPa), ΔK_{eff} drops down to the level of the effective threshold stress intensity for a long crack, $\Delta K_{\text{effth}\infty}$ (=3.06 MPa $m^{1/2}$). The effective fraction U of the applied stress is calculated by

$$U = (K_{\max} - K_{\text{op}})/K_{\max} = \Delta K_{\text{eff}}/\Delta K \tag{6}$$

Its change with crack length is shown in Fig. 5*b*. U decreases with crack length until the nonpropagation for σ_a = 40 to 55 MPa. At σ_a = 65 MPa, U gradually decreases and then becomes nearly constant. The U-value is not only a function of c, but also of σ_a. The dashed line in the figure is the semi-empirical formula for U at the threshold, which will be explained later.

Model for Notch Fatigue Threshold

Tanaka-Nakai Model

Tanaka and Nakai [*10*] have proposed a predictive model for the threshold of the initiation and propagation of a short crack in notch fatigue by combining the crack closure with their blocked slipband model. A simplified version of their model is briefly described below. Figures 6*a* and 6*b* illustrate the threshold model for crack initiation and propagation. For the initiation of a fatigue crack at the notch tip, the slip deformation spreading over one or two grains is responsible. In Fig. 6*a*, the threshold condition for crack initiation is given by the condition of the propagation of an intrinsic microcrack. The critical value of the SIF range for the propagation of an intrinsic crack is $\Delta K_{\text{effth}\infty}$. The length of an intrinsic crack c'_0 is determined from the condition that the fatigue limit of smooth specimens $\Delta\sigma_{w0}$ is to be derived from the nonpropagation criterion for an edge crack of length c'_0 in a semi-infinite plate:

$$c'_0 = (\Delta K_{\text{effth}}/1.122\Delta\sigma_{w0})^2/\pi \tag{7}$$

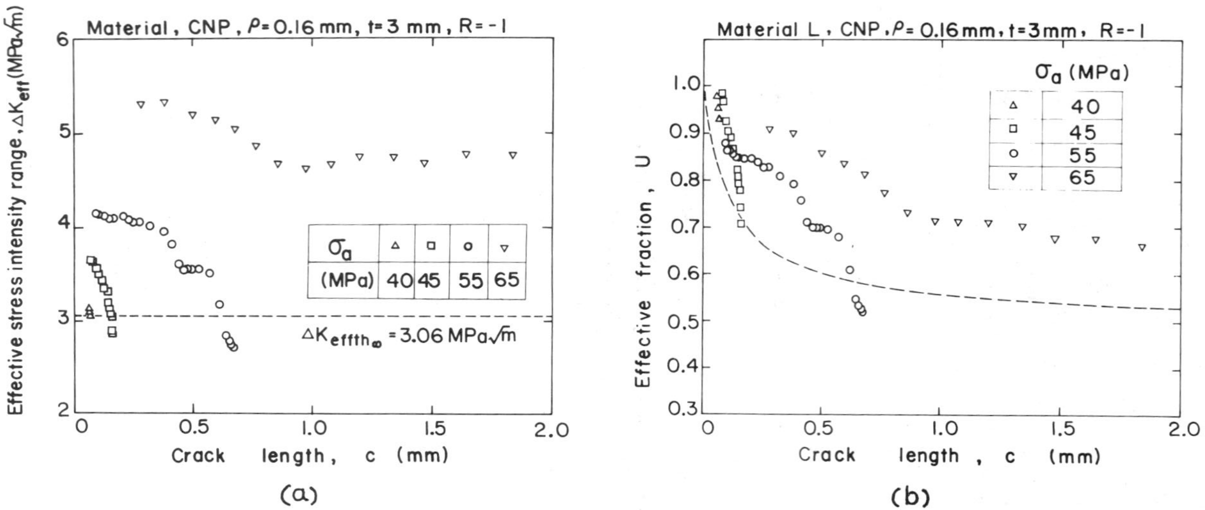

FIG. 5—*Development of crack closure. Material L, CNP specimen* (ρ = *0.16 mm*, t = *3 mm*). (a) *Change of effective stress intensity range with crack length.* (b) *Change of effective fraction with crack length.*

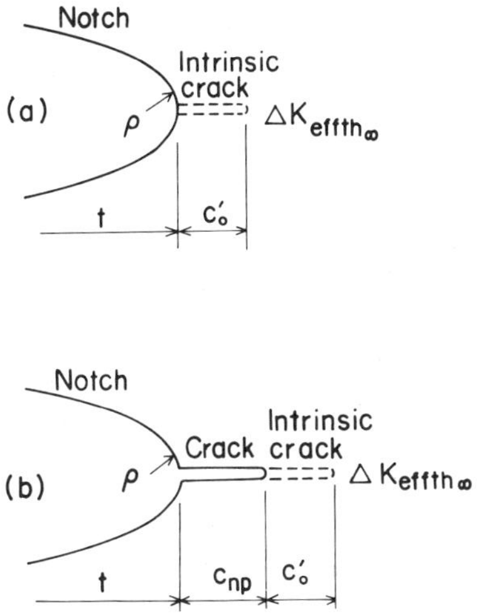

FIG. 6—*Models of thresholds of notch fatigue.* (a) *Crack initiation threshold.* (b) *Crack propagation threshold.*

For the propagation of a short crack of length c_{np} as shown in Fig. 6*b*, the SIF range for a fictitious crack of length c_{np} plus c'_0 must exceed $\Delta K_{effth\infty}$. It should be noted that the model of crack growth at $c_{np} = 0$ is identical to the model of crack initiation. The two models are consistent.

Because crack closure is not yet included, the condition that the SIF range at the tip of a fictitious crack of length c_{np} plus c'_0 equals $\Delta K_{effth\infty}$ gives the effective threshold stress range $\Delta\sigma_{effth}$:

$$\Delta K_{effth\infty} = \Delta\sigma_{effth}[\pi(c_{np} + c'_0)]^{1/2} \cdot F(c_{np} + c'_0) \tag{8}$$

The value of ΔK_{effth} for a small crack at the notch tip is calculated from $\Delta\sigma_{effth}$ and c_{np} as

$$\Delta K_{effth} = \Delta\sigma_{effth}(\pi c_{np})^{1/2} \cdot F(c_{np})$$

Then we have

$$\frac{\Delta K_{effth}}{\Delta K_{effth\infty}} = \left[\frac{c_{np}}{c_{np} + c'_0}\right]^{1/2} \cdot \frac{F(c_{np})}{F(c_{np} + c'_0)} \tag{9}$$

Since the crack closure is caused by the touching of crack faces, it may depend on the distance of the crack face made by fatigue, or on the crack length. Based on the previous results of small fatigue cracks without the influence of a notch [*16*], Tanaka and Nakai assumed the following equation for the effective fraction of a short crack at the notch tip at the threshold:

$$U_{th}/U_{th\infty} = [(c_{np} + c_0)/(c_{np} + c'_0)]^{1/2} \tag{10}$$

where c_0 is determined by

$$c_0 = (\Delta K_{th\infty}/1.122\Delta\sigma_{w0})^2/\pi \tag{11}$$

and $\Delta K_{th\infty}$ is the threshold SIF range for a long crack. (In the original equation proposed by Tanaka and Nakai, c'_0 and c_0 are defined by dropping a factor of 1.122 in Eqs 7 and 11.) U_{th} is unity at $c_{np} = 0$ because

$$U_{th\infty} = \Delta K_{effth\infty}/\Delta K_{th\infty} = (c'_0/c_0)^{1/2}$$

With increasing crack length, U_{th} gradually decreases down to $U_{th\infty}$.

The threshold stress intensity range ΔK_{th} is given from Eqs 9 and 10 as

$$\frac{\Delta K_{th}}{\Delta K_{th\infty}} = \left(\frac{\Delta K_{effth}}{\Delta K_{effth\infty}}\right) \Big/ \left(\frac{U_{th}}{U_{th\infty}}\right) = \left[\frac{c_{np}}{c_{np} + c_0}\right]^{1/2} \cdot \frac{F(c_{np})}{F(c_{np} + c'_0)} \tag{12}$$

The corresponding stress range $\Delta\sigma_{th}$ is calculated from ΔK_{th} through

$$\begin{aligned} \Delta\sigma_{th} &= \Delta K_{th}/[(\pi c_{np})^{1/2} \cdot F(c_{np})] \\ &= \Delta K_{th\infty}/[\sqrt{\pi}(c_{np} + c_0)^{1/2} \cdot F(c_{np} + c'_0)] \end{aligned} \tag{13}$$

The fatigue limit for crack initiation, $\Delta\sigma_{w1}$, is determined by substituting $c_{np} = 0$ into Eq 8 or 13 as

$$\Delta\sigma_{w1} = \Delta K_{effth\infty}/[(\pi c'_0)^{1/2} \cdot F(c'_0)] \tag{14}$$

The material parameters required for the above calculation are three: $\Delta K_{th\infty}$, $\Delta K_{effth\infty}$, and $\Delta\sigma_{w0}$. The F-term in those equations accounts for the influences of notch geometry and loading mode on the fatigue threshold. For the case of negative R as in the present experiment, $\Delta K_{th\infty}$ and $\Delta\sigma_{w0}$ are taken to be equal to their maximum value.

Evaluation of Model

The fatigue data of Materials L and S required for prediction have been reported elsewhere [*17*] and are summerized in Table 2. The value of σ_{w0} of Material L is smaller than that of Material S, while the grain size effect is opposite with $\Delta K_{th\infty}$. The value of $\Delta K_{effth\infty}$ of Material L is slightly larger than that of Material S. The values of c'_0 and c_0 calculated by Eqs 7 and 11 are also given in the table. They are larger for Material L than for Material S.

When $\Delta K_{th\infty}$, $\Delta K_{effth\infty}$, and σ_{w0} are given for a particular material, the variations of ΔK_{th} and ΔK_{effth} with crack length can be calculated for any case of specimen geometry and loading mode by Eqs 9, 10, and 12, provided that the SIF equation is known. Figure 7 shows the

TABLE 2—*Fatigue threshold and intrinsic crack length* (R = −*1*).

Material	Smooth Specimen Fatigue Limit, σ_{w0}, MPa	Long Crack Threshold SIF Range, $\Delta K_{th\infty}$, MPa$\sqrt{m}$	Long Crack Effective Threshold SIF Range, $\Delta K_{effth\infty}$, MPa$\sqrt{m}$	Intrinsic Crack Length, c'_0, μm	Intrinsic Crack Length, c_0, μm
L	163	6.11	3.06	89	355
S	198	5.09	2.77	50	167

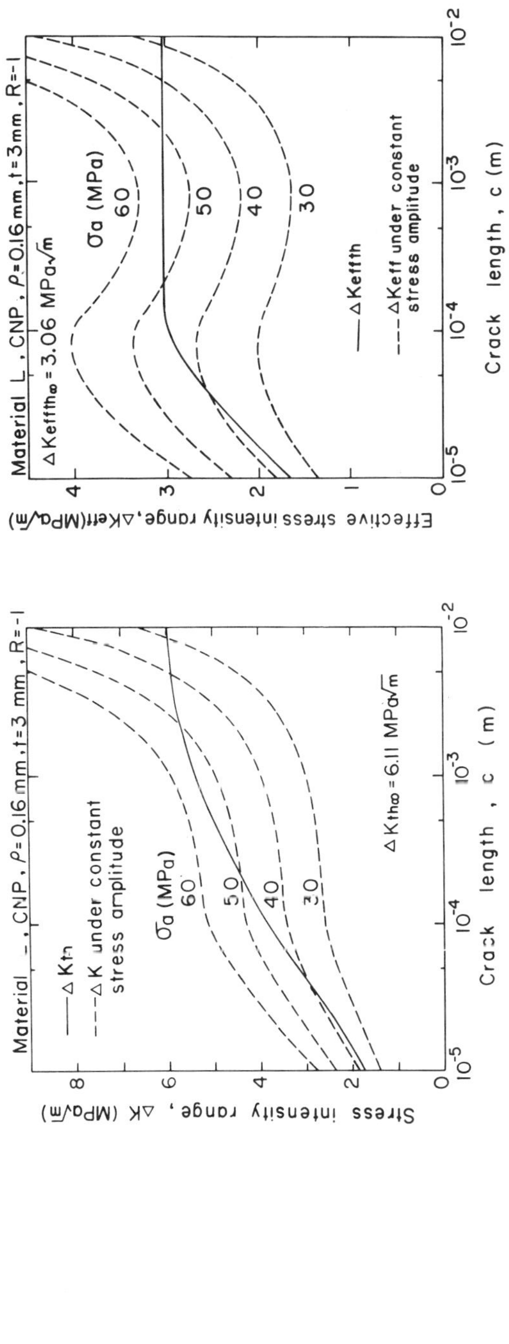

FIG. 7—*Driving force versus resistance. Material L, CNP specimen* (ρ = *0.16 mm,* t = *3 mm*). (a) *Change of stress intensity range with crack length.* (b) *Change of effective stress intensity range with crack length.*

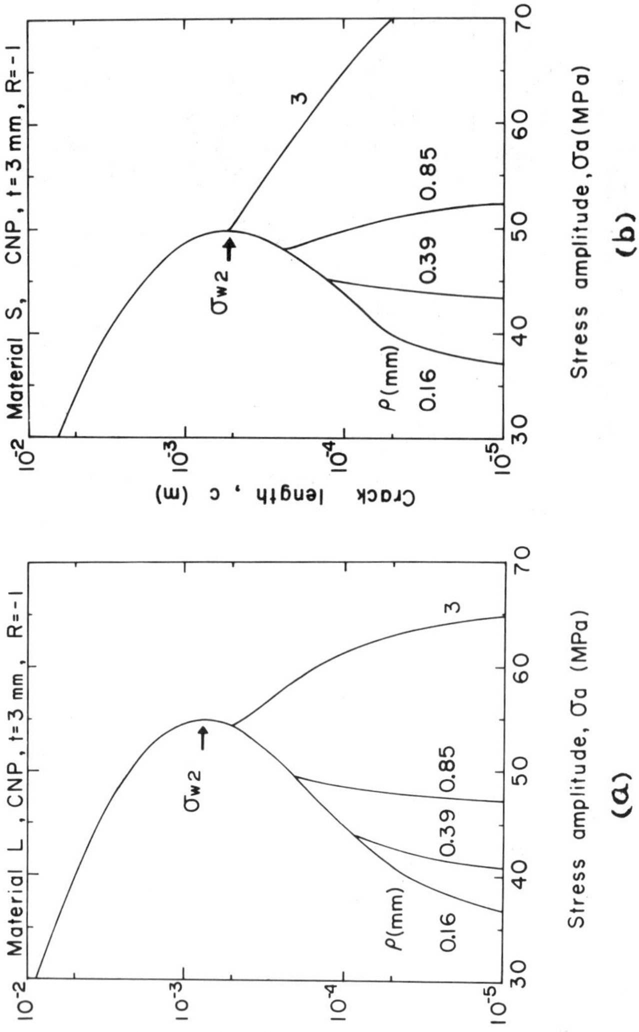

FIG. 8—*Limiting curve for nonpropagation of fatigue crack in CNP specimen with notch depth* t = 3 *mm.* (a) *Material L.* (b) *Material S.*

prediction for a CNP specimen (ρ = 0.16 mm, t = 3 mm) of Material L. In Fig. 7*a*, the dashed lines indicate the change of SIF calculated by Eq 1 for several values of the applied stress amplitude. The value of ΔK_{th} increases with crack length, approaching $\Delta K_{th\infty}$ at large crack length. A fatigue crack can propagate as far as the applied SIF value is above the threshold. Under a given amplitude, the crack becomes nonpropagating when it grows up to the length at the intersection of ΔK-curve and ΔK_{th}-curve. In Fig. 7*b*, the effective SIF range is taken as the ordinate. The effective SIF range ΔK_{eff} is calculated by

$$\Delta K_{eff} = U\Delta K = U_{th\infty}[(c + c_0)/(c + c'_0)]^{1/2} \cdot \Delta K \tag{15}$$

where Eq 10 is assumed applicable for a propagating crack. At crack lengths longer than about 100 μm, the intrinsic resistance for crack growth, ΔK_{effth}, is nearly constant, equal to $\Delta K_{effth\infty}$. The applied value of ΔK_{eff} drops with crack growth, and the crack stops at the length where ΔK_{eff} reaches ΔK_{effth}.

By using Eq 13, the threshold value of the stress amplitude (= range for $R = -1$) for Materials L and S is calculated as a function of crack length for the cases of several notch-tip radii of the CNP specimen (t = 3 mm). Figures 8*a* and 8*b* show the threshold relation between crack length and stress amplitude for Matcrials L and S, respectively. In the figures, when a certain combination (c, σ_a) lies to the right of the threshold line for each ρ-value, the crack can propagate. The stress at the diminishing crack length corresponds to the fatigue limit of crack initiation σ_{w1}. The stress at the nose indicated with the arrow means the fatigue limit for fracture σ_{w2}. The value of σ_{w2} is independent of notch-tip radius. It is larger with Material L than with Material S. The value of σ_{w1} is very much dependent on the ρ-value and material. While σ_{w1} of Material L is nearly equal to that of Material S at ρ = 0.16 mm, it is lower at larger ρ. The notch-tip radius at the branch point, below which a nonpropagating crack can be formed, is 1.57 mm for Material L and 0.67 for Material S.

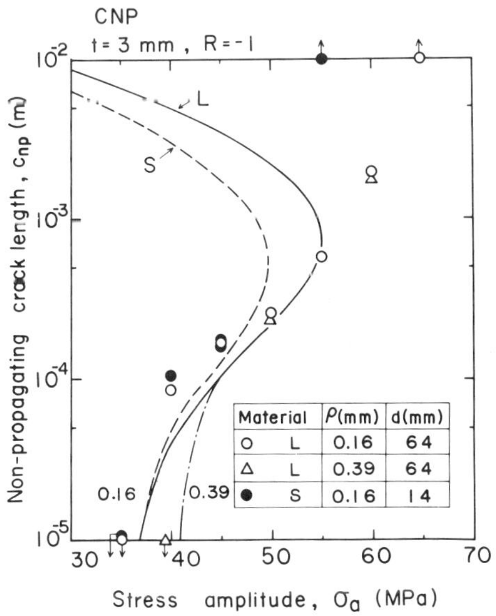

FIG. 9—*Variation of nonpropagating crack length with stress amplitude. CNP specimen* (ρ = *0.16 mm*, t = *3 mm*).

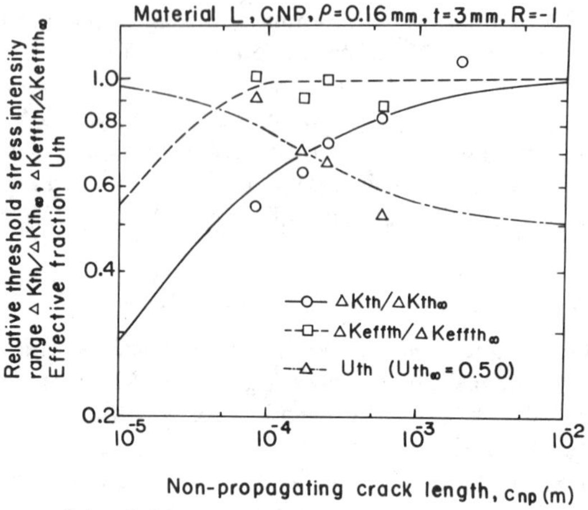

FIG. 10—*Changes of threshold stress intensity ranges and crack closure with crack length at the threshold. Material L, CNP specimen* (ρ = *0.16 mm,* t = *3 mm).*

A general feature of the mechanism of the nonpropagation of a notch-tip short crack observed is well described by the model. Quantitative comparison will be made next.

One significant assumption made in the model is that the change of crack closure with crack length is given by Eq 10. This equation is drawn with the dashed line in Fig. 5*b* for the comparison with experiments. Near the threshold, the data are close to the dashed line. On the other hand, the U-value of a propagating crack is larger than that predicted by Eq 10. In Fig. 7*b*, indicating clearly the reduction of ΔK_{eff} due to crack closure development to be responsible for the nonpropagation of a crack, the real value of ΔK_{eff} will be larger than the predicted line when it is above ΔK_{effth}.

In Fig. 9, the nonpropagating crack length, c_{np}, measured with CNP specimens of Materials L and S is compared with the prediction. For the case of CNP specimens with $\rho = 0.16$ mm, the fatigue limits, σ_{w1} and σ_{w2}, predicted are 36.0 and 55.0 MPa for Material L, and 36.1 and 49.7 MPa for Material S. The experimental value of σ_{w1} is between 35 and 40 MPa for both materials, which agrees with the prediction. At $\sigma_a = 40$ and 45 MPa, the c_{np}-values of both materials are about the same. At $\sigma_a = 50$ MPa, a crack propagates to fracture in Material S, while it becomes nonpropagating in Material L. At $\sigma_a = 60$ MPa in Material L, the nonpropagating crack is formed, contrary to the predicted continuous growth. Similar conclusions can be drawn for the case of CNP specimens with $\rho = 0.39$ mm. The value of σ_{w1} is predicted very well, while the prediction of σ_{w2} is slightly conservative.

Equations 9, 10, and 12 are directly compared with experimental data in Fig. 10. The ΔK_{th}-value decreases as a crack becomes shorter, following the theoretically predicted line. The crack closure was measured in four cases. The value of U_{th} decreases with crack length, while ΔK_{effth} is nearly constant and equal to $\Delta K_{effth\infty}$.

Effect of Notch Geometry and Loading Mode on Fatigue Thresholds

In the Tanaka-Nakai model, the effects of notch geometry and loading mode on the fatigue threshold are included as the F-term of the SIF expression. The relation between effective fraction and crack length is independent of notch geometry and loading mode because Eq 10 does not include the F-term. On the other hand, the changes of ΔK_{th} and ΔK_{effth} with crack length depend on notch geometry and loading mode (see Eqs 9 and 12).

The variable which predominantly controls those changes is the notch-tip radius. For the case of $\rho = 0.16$ mm, for example, the difference in ΔK_{th} or ΔK_{effth} calculated for CNP and SEN specimens having $t = 0.3$ and 3 mm is less than 10%. For further evaluation of the model, the prediction is compared with the results obtained with differently sized specimens of Material L under different loading mode.

Two types of SEN specimens ($\rho = 0.16$ mm; $t = 0.3$ and 3 mm) were fatigued under several stress amplitudes, and nonpropagating cracks with various crack lengths were obtained. Figure 11 shows the data of ΔK_{th} and ΔK_{effth} measured with those cracks, including the data of CNP specimens with $\rho = 0.16$ mm. The lines drawn in the figure are the prediction for the case of CNP specimen with $\rho = 0.16$ mm and $t = 3$ mm. As seen in the figure, the effects of loading mode and notch length on ΔK_{th} and ΔK_{effth} are minimal as predicted. The value of ΔK_{th} decreases as the crack length becomes smaller, nearly following the predicted curve. At crack lengths longer than about 1 mm, ΔK_{th} is slightly larger than $\Delta K_{th\infty}$. This may result from the macroscopic plastic deformation made near the notch which is predominant at high stresses (equivalent to at longer nonpropagating cracks). The ΔK_{eff}-value is nearly constant at the crack length longer than 0.1 mm, which is in agreement with the prediction. When the crack length is very small, ΔK_{effth} tends to decrease. More experimental data of cracks shorter than 0.1 mm are necessary for confirmation.

In Fig. 12, all the measured values of U_{th} for a nonpropagating crack in Material L are plotted as a function of crack length. The solid line in the figure corresponds to Eq 10. All the data roughly follow the solid line. Strictly speaking, the data at large crack lengths are below the line, and those at small crack lengths are above the line. Because the number of the data with notch-tip radius different from 0.16 mm is only two, we need more data for other notch-tip radii to prove the independence of the effective fraction against notch geometry change.

The fatigue limits, σ_{w1} and σ_{w2}, predicted from the model are compared with the experimental data in Table 3. The experimental data are shown by two values. The smaller value means the stress where no crack was observed for the case of σ_{w1}, and the stress where the specimen was not fractured for the case of σ_{w2}. The larger value means the stress where the specimen was cracked or fractured for σ_{w1} or σ_{w2}. For all the cases examined, the prediction

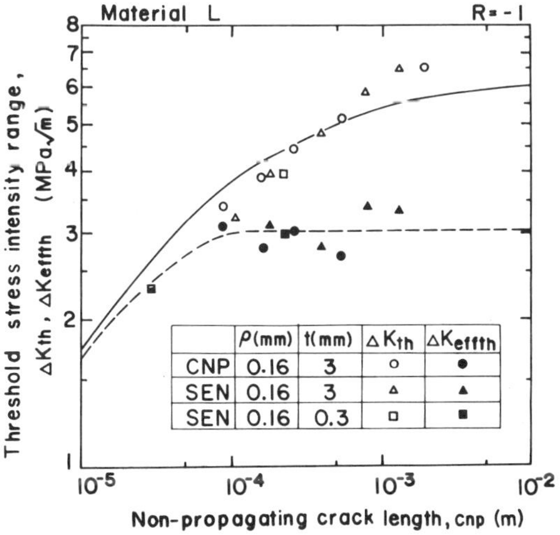

FIG. 11—*Changes of threshold stress intensity ranges with crack length of Material L.*

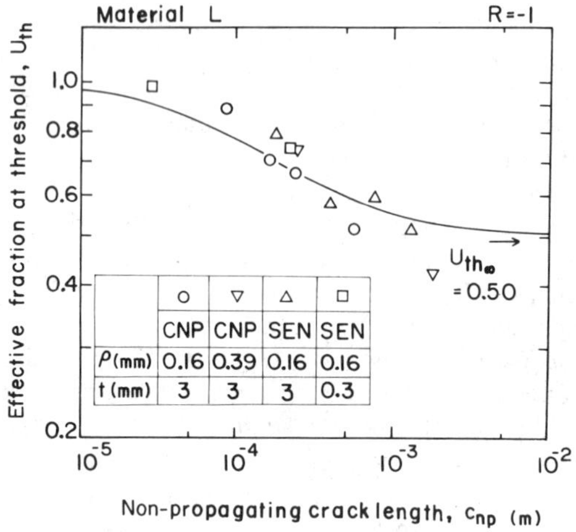

FIG. 12—*Change of effective fraction with crack length at the threshold of Material L.*

of σ_{w1} agrees very well with the experimental result. With respect to σ_{w2}, the predicted value is smaller than the experiment in the three cases of Material L (CNP: ρ = 0.16 mm, t = 3 mm; CNP: ρ = 0.39 mm, t = 3 mm; and SEN: ρ = 0.16 mm, t = 3 mm). This discrepancy may result from the macroscopic plastic deformation at high stresses which is not considered in the model. The macroscopic plastic deformation at the notch root may increase the crack closure of the nonpropagating crack (see large U-value for long nonpropagating cracks). For other cases, good agreement is obtained.

Conclusions

Based on the experiments conducted in the present study, the answers to two questions posed in Introduction, about short crack growth at a notch, are given as follows: To the first question, our answer is "Yes." The crack closure develops as a crack grows; the effective fraction of the applied stress decreases. The effective fraction tends to be smaller at lower stress level when compared at the same crack length. Especially near the threshold, the effect of stress level is large. To the second question about the effective SIF range as a measure of the crack driving force, our answer is "Yes, but it has some limitations." When the crack length is shorter than about 100 μm, the effective stress intensity range as a

TABLE 3–*Effect of loading mode and notch size on fatigue thresholds of notched specimens.*

Specimen			Fatigue Limit for Crack Initiation, σ_{w1}, MPa		Fatigue Limit for Fracture, σ_{w2}, MPa	
Type	Dimension, mm	Material	Predicted	Experimental	Predicted	Experimental
CNP	ρ = 0.16, t = 3	L	36.0	35 to 40	55.0	60 to 65
	ρ = 0.16, t = 3	S	36.1	35 to 40	49.7	45 to 55
	ρ = 0.39, t = 3	L	39.6	39 to 50	55.0	60 to 65
	ρ = 0.85, t = 3	L	47.2	50 to 51.5	55.0	51.5 to 53
SEN	ρ = 0.16, t = 3	L	51.9	50 to 55	71.6	80 to 90
	ρ = 0.16, t = 0.3	L	86.4	80 to 85	99.5	95 to 105

measure of the crack driving force is questioned. A crack of microstructural size may grow under mixed-mode loading, so the characterization by Mode I loading only will not be enough. Also, the reversed plastic zone may become large compared with crack length; the size is no longer controlled by ΔK_{eff}. Another limitation of ΔK_{eff} as the crack driving force is due to the macroplasticity effect which will be predominant under a high stress level. Although the experimental data obtained in the present test do not show it clearly, it is expected that a short crack grows faster in the macroplasticity field when the rate is correlated to ΔK_{eff} [*19*]. To clarify those limitations of the ΔK_{eff} approach, further experiments are necessary. A new mechanical parameter is required to express the driving force for the growth of very short cracks under plasticity condition.

With respect to the threshold condition of crack growth, the Tanaka-Nakai model is evaluated based on the present experiment. The change of the threshold SIF range and the effective threshold SIF range with crack length predicted agree fairly well with the experimental results obtained with specimens having several kinds of notches under axial and bending loads. The relation between the effective fraction and crack length assumed in the model is well supported by the data. The prediction of the fatigue limit for crack initiation agrees very well with the experiment. The predicted value of the fatigue limit for fracture is smaller than the experiment only for the specimen of large-grain-sized material with a long notch. To prove further the usefulness and the limitation of the model, experiments using specimens with different notch-tip radii under different stress ratio are necessary. For predicting the finite life of notched components, the modeling of crack closure development for a *growing* crack is essential.

References

[*1*] Taira, S., Tanaka, K., and Ryu, J. G. in *Proceedings,* 16th Japan Congress of Materials Research, 1973, pp. 36–44.
[*2*] Smith, R. A. and Miller, K. J., *International Journal of Mechanical Science,* Vol. 20, 1978, pp. 201–206.
[*3*] Smith, R. A. in *Fatigue Mechanisms: Advances in Quantitative Measurement of Physical Damage, ASTM STP 811,* American Society for Testing and Materials, Philadelphia, 1983, pp. 264–279.
[*4*] Hammouda, M. M., Smith, R. A., and Miller, K. J., *Fatigue of Engineering Materials and Structures,* Vol. 2, 1979, pp. 139–154.
[*5*] El Haddad, M. H., Smith, K. N., and Topper, T. H. in *Fracture Mechanics, ASTM STP 677,* American Society for Testing and Materials, Philadelphia, 1979, pp. 274–289.
[*6*] El Haddad, M. H., Topper, T. H., and Smith, K. N., *Engineering Fracture Mechanics,* Vol. 11, 1979, pp. 573–584.
[*7*] Tanaka, K. and Nakai, Y., *Fatigue of Engineering Materials and Structures,* Vol. 6, 1983, pp. 315–327.
[*8*] Ogura, K., Miyoshi, Y., and Nishikawa, I. in *Proceedings,* 26th Japan Congress on Materials Research, 1983, pp. 91–96.
[*9*] Schijve, J. in *Fatigue Thresholds,* Vol. 2, Engineering Materials Advisory Service Ltd., Warley, U.K., 1982, pp. 881–908.
[*10*] Tanaka, K. and Nakai, Y., *Journal of Engineering Materials and Technology, Transactions,* American Society of Mechanical Engineers, Vol. 106, 1984, pp. 192–199.
[*11*] Ishida, M., *Transactions,* Japan Society of Mechanical Engineers, Vol. 21, 1955, pp. 507–513.
[*12*] Neuber, H., *Kerbspannungslehre,* 2nd ed., Springer-Verlag, Berlin, 1958, p. 51.
[*13*] Lukas, P. and Klesnil, M., *Materials Science and Engineering,* Vol. 34, 1978, pp. 61–66.
[*14*] Tada, H., Paris, P. C., and Irwin, G. R., *The Stress Analysis of Cracks Handbook,* Del Research Corp., Hellertown, PA, 1973, p. 22.
[*15*] Brown, W. E., Jr., and Srawley, J. in *Plane Strain Crack Toughness Testing of High Strength Metallic Materials, ASTM STP 410,* American Society for Testing and Materials, Philadelphia, 1966, p. 13.
[*16*] Nakai, Y., Tanaka, K., and Nakanishi, T., *Engineering Fracture Mechanics,* Vol. 15, 1981, pp. 291–302.

[17] Akiniwa, Y. and Tanaka, K., *Transactions*, Japan Society of Mechanical Engineers, Vol. 53, 1987, pp. 393–400.
[18] Tanaka, K., Nakai, Y., and Yamashita, M., *International Journal of Fracture,* Vol. 17, 1981, pp. 519–533.
[19] Hoshide, T., Tanaka, K., and Nakata, M., *Journal of the Society of Materials Science* (Japan), Vol. 31, 1982, pp. 566–572.

Fatigue of Aluminum and Other Alloys—Mechanical Loading

Ouyang Jie[1] *and Yan Minggao*[1]

A Study of the Mechanism of Striation Formation and Fatigue Crack Growth in Engineering Alloys

REFERENCE: Ouyang, J. and Yan, M., **"A Study of the Mechanism of Striation Formation and Fatigue Crack Growth in Engineering Alloys,"** *Basic Questions in Fatigue: Volume 1, ASTM STP 924,* J. T. Fong and R. J. Fields, Eds., American Society for Testing and Materials, Philadelphia, 1988, pp. 301–311.

ABSTRACT: A large number of fatigue cracks and fractures of test specimens and components of various engineering alloys were investigated by means of the fractographic technique. In particular, critical experiments were performed on the observation of natural profiles of fracture plateaus and mating fracture surfaces using the scanning electron microscope. It was found that the profile of a fatigue crack appears to be a tunnel with numerous steps. In consequence, the striations on the matching surfaces are basically in register other than any kinds of symmetrical configurations such as peaks to peaks and valleys to valleys. Furthermore, the appearance of them on one surface is usually more regular and distinct than that on the matching surface due to their greater slip length and, sometimes, slipband notches or cracks. The process of fatigue crack growth is just like alternatively walking up or down the steps of a path in different areas or grains.

A more general and practical model schematically illustrating the mechanism of fatigue crack growth and striation formation, which is one of the most fundamental questions on fatigue, is proposed. According to this model, the process of fatigue crack growth under a loading cycle includes four steps: (i) crack-tip opening; (ii) shear crack growth in a very short distance by single system slip decohesion; (iii) normal crack growth by alternative conjugate slip decohesion; and (iv) crack-tip closing, resharpening and, sometimes, inducing one or two slipband notches or cracks at the very tip. It differs markedly from the previous models which, in general, describe the crack advancing under conditions of symmetric deformation at the crack tip and growing straight along the crack plane in a loading cycle. Based on the present model, the markings on the fracture surfaces and phenomena associated with the process of crack growth, such as the secondary shear cracks accompanying by the striations, the slipbands between striations, and the simultaneous growth of branching cracks, can properly be explained.

KEY WORDS: fatigue (materials), fractography, fatigue crack growth, fatigue striations, slipbands, slip decohesion, secondary crack, engineering alloys

It is well known that fatigue striation is the most significant and dominant microscopic feature of fatigue fracture surfaces in ductile metals and alloys. It runs parallel to the crack front and perpendicular to the direction of crack growth, hence indicating the successive position of the advancing crack front [*1*]. Some researchers [*2–6*] have also shown that striations can both identify fatigue fracture as such and offer significant quantitative information concerning fatigue crack growth rates. Furthermore, the striations can be used to seek out the original site and propagation direction of the crack, and the relations between growth rate and microstructures, load history, environment, etc. Thus, they are quite beneficial to service failure analysis. Despite all this, the fundamental question, that is, the

[1] Senior research engineer and professor/technical director, Institute of Aeronautical Materials, Beijing, China.

process and mechanism by which striations are formed, is not very clear up to the present. A study of this question makes it possible to clarify the mechanism of fatigue crack growth and to render information for increasing fatigue resistance of metallic materials and structures. Many investigators, therefore, have paid a great deal of attention to it, and several different models of striation formation and fatigue crack growth have been proposed during the past 25 years [*7–21*]. Although these models are based on different materials, loading conditions, and interpretations, it is generally presumed that the mode, process, and area of slip deformation at the crack tip are symmetrical about the crack plane and that the crack should bisect the angle between two operable slip systems so that it extends straight forward and there is a commonly characteristic morphology of the profile of fatigue crack with striations; that is, the striations on both sides of the crack are in antiregister, but symmetrical in the sorts of peaks to peaks and valleys to valleys. As a result, the striations on the matching fracture surfaces are considered to have basically the same appearance. A few of the models, to a certain extent, may depict the processes of fatigue crack propagation in some single crystals or pure metals under rather high strain cycles. However, quite a few of the markings concerned on the fracture surfaces and phenomena associated with the process of crack growth, such as the secondary shear cracks accompanying the striations and the simultaneous growth of branching cracks, cannot be rationally interpreted.

Since the 1960's, a large number of fatigue cracks and fracture surfaces of test specimens and structural components of various alloys such as aluminum, copper, magnesium, nickel, iron-nickel and titanium base alloys, as well as structural steels of different strength levels under different loading modes, have been observed and analyzed by means of the fractographic technique. The critical and unique experiments conducted are: (1) stereoscopic observation of the mating features on the matching fracture surfaces by using the scanning electron microscope (SEM) through turning and tilting the stage; and (2) stereoscopic observation of the natural profiles of the mating fracture plateaus by the same means. The chemical compositions of several alloys reported in this paper are listed in Table 1. The configurations of the fatigue crack profile and the mating fracture surfaces observed have not been reported by the earlier authors. Based upon fractographic observation, it is intended in this paper to find a more general and practical model to describe the mechanism of striation formation and fatigue crack propagation in engineering alloys by which almost all of the configurations on the fracture surface and the phenomena associated with the crack propagation can reasonably be explained.

Results and Discussion

Figure 1 provides three pairs of matching surfaces of fatigue fractures in aluminum, iron-nickel and titanium alloys. It is shown that the striations on both sides are not of the same appearance, that is, they do not match symmetrically in any patterns of peaks to peaks and valleys to valleys, but are basically in register stepwise with slopes to slopes and planes to planes. It is also seen that the striations on each side of the matching surfaces are not equally distinct; they are usually distinguished on one side but not on the other. Furthermore, they are not always distinct on the same side but alternatively change from one side to the other in different areas or grains. So we can always observe some distinct striations on one area but indistinct ones in the neighborhood. In the meantime it can be seen that there are a lot of secondary cracks associated with the striations. It is interesting to note that the secondary cracks are usually created only on one of the matching surfaces and not always on the same side, either. It is obvious that this kind of secondary crack is produced between slipbands which generally make an angle of about 45 deg to the main crack plane (see Fig. 1*b*) instead of being perpendicular to it [*22*]. It is remarkable that these secondary cracks are different

TABLE 1—*Chemical composition of the alloys investigated (weight %).*

Material	C	Al	Cu	Mg	Mn	Si	Zn	Fe	Ni	Cr	Co	Ti	Mo	Nb	V	W
LY12	...	bal	4.4	1.5	0.6	...	0.1	...	...	...	...	...	...	...	...	...
LC4	...	bal	1.7	2.3	0.4	0.3	6.0	0.2	...	0.18	...	...	...	...	...	...
GH36	0.37	...	...	...	8.5	0.6	...	bal	8.0	12.5	...	0.10	1.2	0.35	1.35	...
GH49	0.05	4.1	...	...	...	...	...	1.3	bal	10.3	15	1.6	5.0	...	0.35	5.5
K5	0.14	5.5	...	...	...	...	...	...	bal	10.5	10	2.5	3.8	...	...	4.9
TC4	...	6.0	...	...	...	...	...	...	...	...	...	bal	...	...	3.9	...

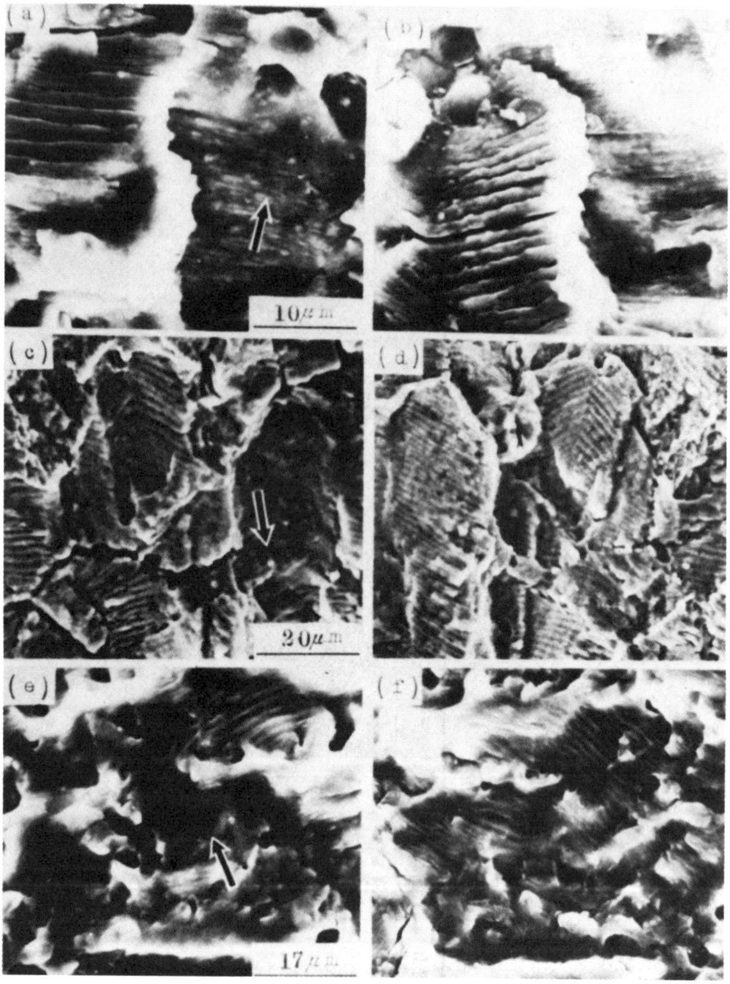

FIG. 1—*Matching fracture surfaces of fatigue specimens under tension-tension cycles:* (a) *and* (b) *LY12 aluminum alloy;* (c) *and* (d) *GH36 iron-nickel alloy;* (e) *and* (f) *TC4 titanium alloy.*

from another kind of secondary crack that usually accompanies boundaries of subgrains, grains, twins, large ridges, and secondary-phase particles or inclusions.

The slipbands shown on the natural profile of the fracture plateaus in Fig. 1*b* are more typical and distinguished. It is exhibited that they are connected with those slipbands appearing on the surfaces of the neighboring plateaus. Some very distinct slipbands associated with striations are shown in Fig. 2. It is evident that the slipbands are parallel (Fig. 2*a*) or make an angle of about 45 deg (Fig. 2*b*) with the striations and should be formed during cyclic loading, by which the striations are induced. These further manifest that the formation of striations and secondary cracks during crack growth is in connection with the process of shear decohesion. It is noticeable that the slipbands are more distinct near the place where the crack front rested due to unloading or compressive loading when the crack tip suffered, in turn, the maximum tensile and compressive stresses.

Figure 3 further provides various striations from very narrow to pretty wide ones on some

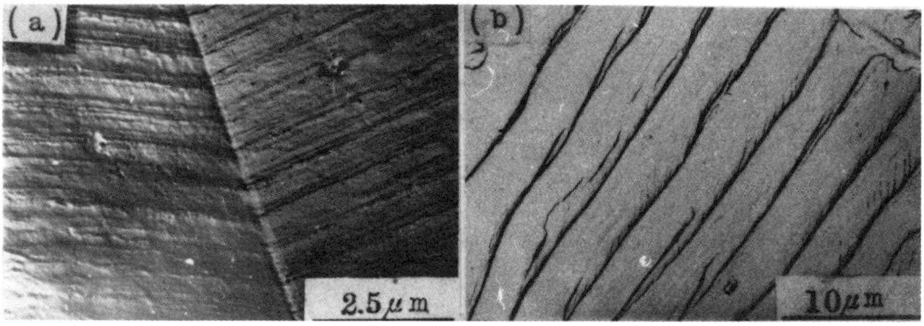

FIG. 2—*Slipbands associated with striations under rotating bending cycles:* (a) *GH49 nickel alloy;* (b) *LC4 aluminum alloy.*

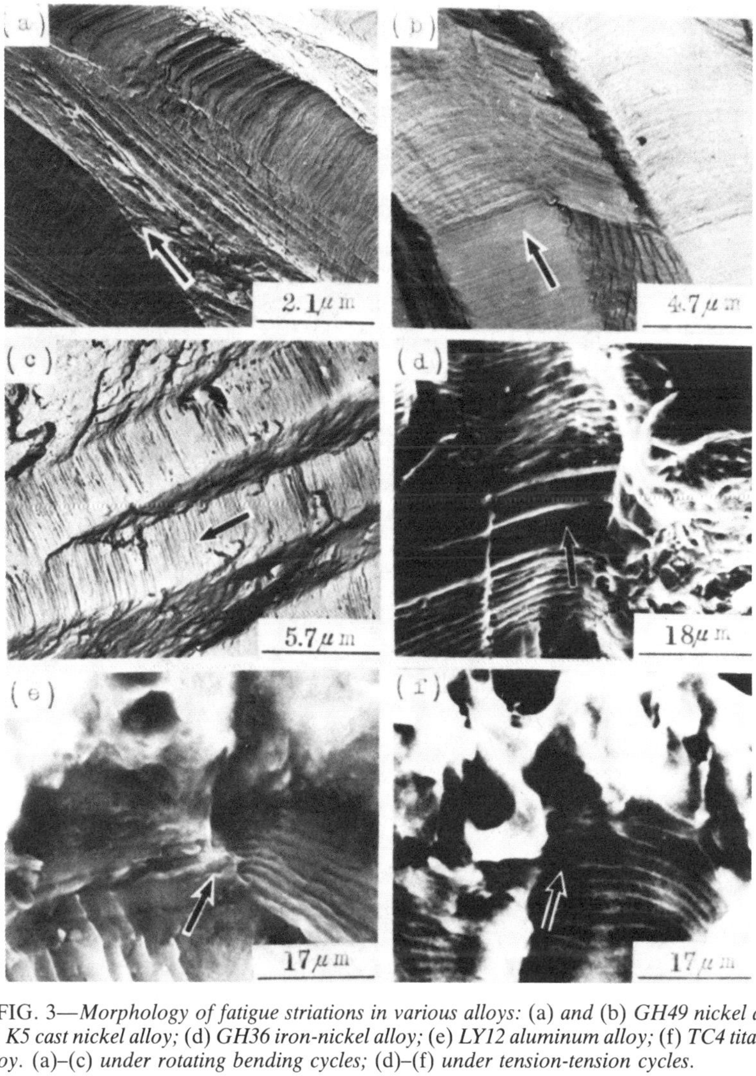

FIG. 3—*Morphology of fatigue striations in various alloys:* (a) *and* (b) *GH49 nickel alloy;* (c) *K5 cast nickel alloy;* (d) *GH36 iron-nickel alloy;* (e) *LY12 aluminum alloy;* (f) *TC4 titanium alloy.* (a)–(c) *under rotating bending cycles;* (d)–(f) *under tension-tension cycles.*

natural profiles of the fracture plateaus in different alloys. It is obvious that the appearances of striations in these alloys are step-like. It implies that the crack propagation is just like walking up and down steps at different gradients and spaces instead of straight going. Also, these configurations are observed most clearly in some combined fatigue specimens of tension, bending, and torsion.

From those exhibited above it is not difficult to think how the fatigue crack grows practically. Figure 4 illustrates schematically the process of fatigue crack growth in detail during a loading cycle in face-centered-cubic (fcc) metals as an example. The following scenario is proposed:

1. As the tensile load increases, the closure stress will be progressively overcome and the crack tip will be fully open correspondingly (Fig. 4*b*).
2. As the load further increases to a certain level, because of relatively large shear stress near the surface, the slip of crystal planes of the more favorable slip system will preferentially occur along one side at about 35 to 55 deg to the main crack plane just ahead of the crack tip under maximum shear stress; then the slipband crack will grow from a slipband intrusion, notch, or crack which occurred during unloading or com-

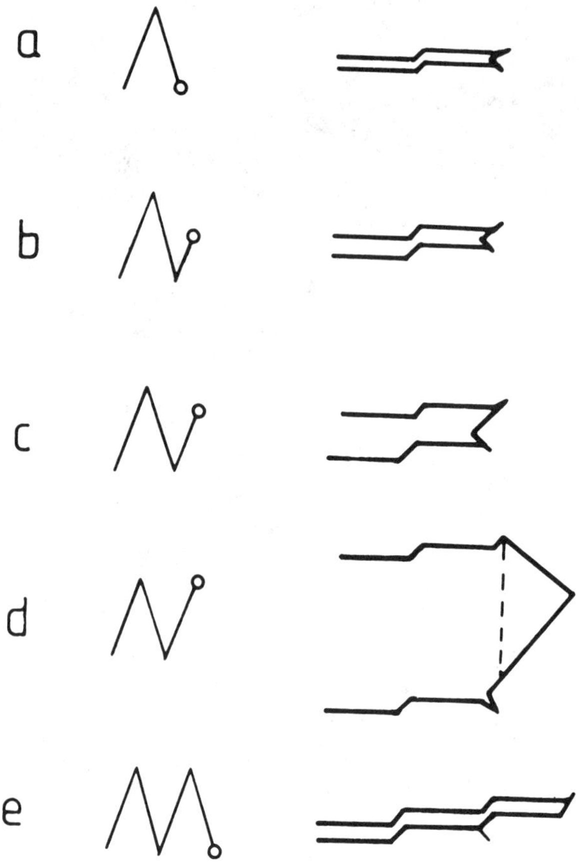

FIG. 4—*A scheme of the process and mechanism of fatigue crack growth and striation formation.*

pressive loading of the last cycle by reverse shear decohesion, that is, somewhat similar to the crack growth of Stage I. The process may also react successively on the conjugate slip system (Fig. 4*c*).

3. When the load increases to a higher level, both systems of intersecting conjugate slip would alternatively be activated at the main shear crack tip and the crack will begin to grow by alternative shear decohesion in the direction perpendicular to the nominal stress axis under normal stress until the load increases to a maximum at which the crack tip is blunted and a tensile plastic zone will be created around it, which is unfavorable for further extension during the next cycle. In the meanwhile, a slipband crack may also grow a very short distance on the other side (Fig. 4*d*).
4. After the foregoing the load begins to decrease, some asymmetrical and inhomogeneous reverse slip occurs, the crack begins to close at the very tip, and the slipbands, intrusions, and extrusions yield simultaneously. Finally, a slipband notch or crack may initiate at one or both sides of the crack tip after unloading or compressive loading; the wake of the crack tip will be fully closed by residual tensile plastic deformation and compressive stresses; and a reverse plastic zone will yield around it, which is favorable for further extension during the next cycle (Figs. 4*a* and *e*). It can be seen that if the slipband notches or cracks initiate at both sides of the crack tip, there will be a secondary crack associated with a striation left on one of the mating fracture surfaces.

Figure 5 shows schematically the profile of the step-like fatigue crack (*a*) and the different appearances of striations with secondary cracks and slipbands on the matching fracture surfaces (*b*). It can be seen that the crack surfaces are matched basically in register and that the striations associated with longer shear decohesion slopes and slipband notches or cracks are more regular and distinct. Moreover, it is shown in Fig. 5*a* that the crack does not always

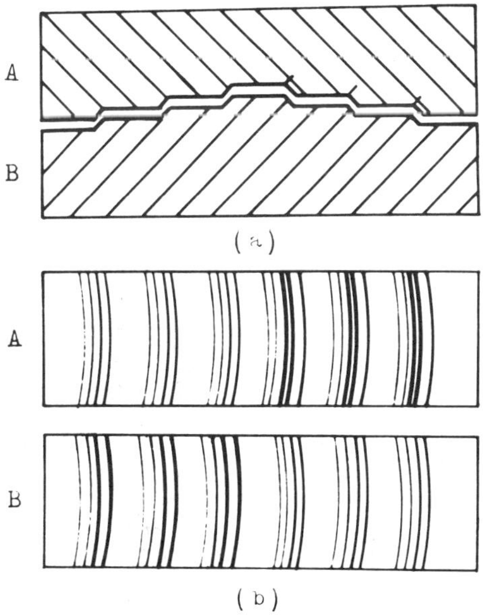

FIG. 5—*A representation of the profile of step-like fatigue crack* (a) *and the different appearances on the matching surfaces* (b).

go up or down steps but changes alternatively and frequently in different areas or grains. That is why the macrocrack and fracture surfaces are always nearly perpendicular to the nominal stress axis.

However, lots of very fine striations formed by the identical mechanism under very low value of stress-intensity range, Δk, in the early growth period of Stage II are usually invisible. This is because the resolving power of either the SEM or the transmission electron microscope (TEM) is not enough to separate them. Therefore, more often, we can only observe some flat narrow plateaus divided by ridges that are similar to river patterns.

As we know, it is rather difficult to slip in steels and alloys of high strength. There may be very small shear decohesion in those areas with unfavorable crystal orientations. So it is hard to form any regular and distinct striations. In some other favorable areas the striations may be formed, but they are too shallow and blurred to be resolved. Besides, some very fine and indistinct striations may be fretted out during a long period of cyclic loading. Only on some of the most favorable patches can a few regular and distinguished striations be exhibited.

Some researchers reported that the striations are either nonexistent or rare on fracture surfaces fatigued in vacuum conditions [*23–26*]. According to the present model, the single system slip and alternative slip deformation may be reversible in vacuum; therefore, it is more difficult to form slipband notches or cracks at the crack tip during unloading or compressive loading. Hence, it is hard to cause striations in vacuum as regular and distinct as in air. Certain evidence was reported by Wanhill [*23*] that there are a number of striations on fracture surfaces both fatigued in air and in vacuum. The only difference is that in the latter condition the striations are much less regular and distinct, and the spaces between them are smaller than those in air. As a result, the crack growth rate must be less in vacuum than in air, too.

On the other hand, the morphology of matching surfaces may occasionally be interrupted by the inhomogeneity of microstructures along the crack path as shown in Fig. 6, and, sometimes, the matching features of striations may be broken by branching cracks and their simultaneous growth. It is shown on the lower part of the matching surfaces that the striations are quite different from each other (Figs. 7*a* and *b*). It shows that a two-or-more-layer crack growth has happened by crack branching and a piece of material between them has dropped

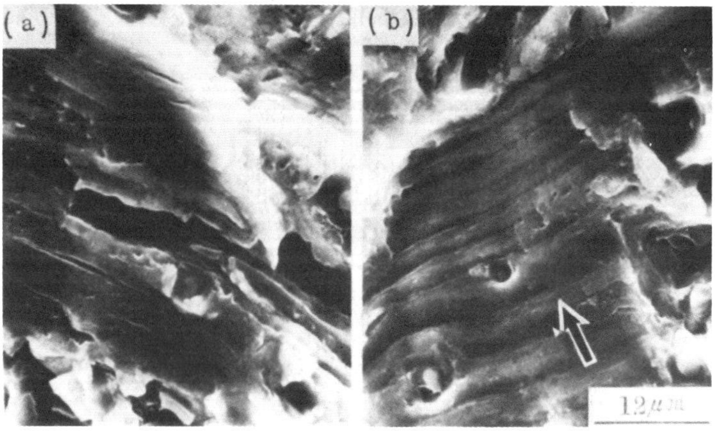

FIG. 6—*Effect of microstructure on the matching fracture surfaces in LY12 aluminum alloy under tension-compression and tension-tension cycles.*

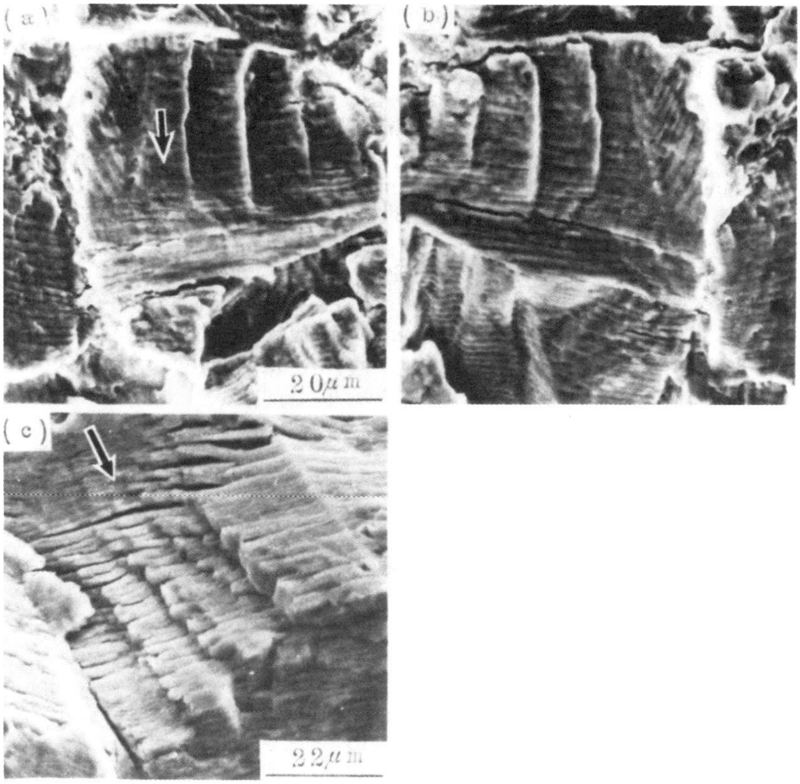

FIG. 7—*Fractographs showing crack branching and simultaneous growth of the branches in GH36 iron-nickel alloy under tension-tension cycles:* (a) *and* (b) *matching surfaces;* (c) *multiple crack branching.*

away. Figure 7*c* provides multilayered fracture surfaces induced by multibranched cracks. It is rather difficult to explain how this occurs by mechanisms proposed previously. However, the present model makes it easier to interpret. Figure 8 shows schematically the process of crack branching and simultaneous growth according to the present model. Doubtless the simultaneous growth of two or multilayered cracks may happen under certain favorable conditions. Of course, the broken matching texture may also result from intersecting of the main crack and the other crack nucleated ahead of it under a few particular conditions.

It often may be seen that some separated particles of secondary phase or inclusions interrupt or bend temporarily the advancing crack fronts, hence the striations. In general, as the particle is much less than the plastic zone size (PZS) in front of the crack tip, there

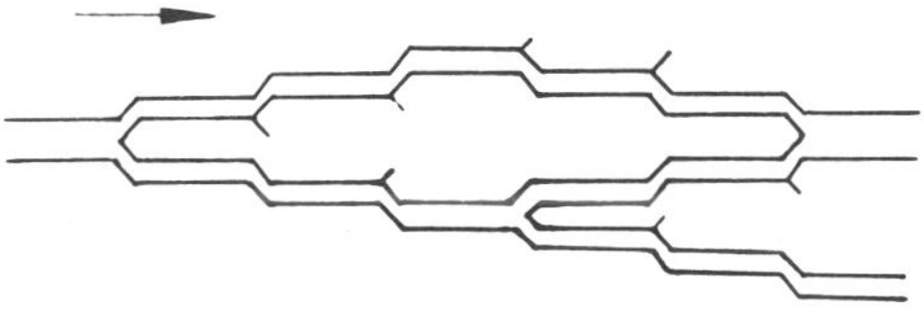

FIG. 8—*A scheme of crack branching and simultaneous growth of the branches.*

is little influence on the crack growth and striation shape. While it is comparable to the PZS, the crack growth by shear decohesion may be prevented or even stopped temporarily for a certain number of cycles. The particle will not separate until the crack grows gradually through both sides of it and the stress intensity grows progressively to a certain level to pull it out or cleave it. Then a elongated shallow dimple and a pair of microridges or a cleavage plane appear. After that the crack will jump forward cycle by cycle in striation mechanism to catch up to the neighboring crack fronts. Some striations may thereby be cut off or curved by it. If the particle is big enough or some particles concentrate somewhere, no striations will form; instead, a large dimple or a group of dimples will appear. As a result, the number of striations is not in accordance with the loading cycles in these areas. As the maximum stress or amplitude of stress intensity increases to a rather high level and the plastic zone size in front of crack tip increases gradually to equal or larger than those of the most numerous particles in the material, the mechanism of striation extension will be progressively replaced by a mechanism of dimple rupture. If the grain boundary is weaker than the inside of the grain, the intergranular fracture will occur instead.

Conclusions

1. The profiles of fatigue cracks of the engineering alloys investigated appear to be tunnels with numerous steps. In consequence, the appearance of striations induced by slip decohesion on the matching surfaces is basically in register; that is, slopes match slopes and planes match planes other than any symmetrical patterns of peaks to peaks and valleys to valleys. The configurations of the striations on one surface accompanied by higher steps, slip notches, or secondary cracks are usually more regular and distinct than those on the matching surface.

2. The process of fatigue crack growth is just like walking up or down the steps of a path in different areas or grains. Therefore, the macrocracks and fracture surfaces are always nearly perpendicular to the nominal stress axis. Likewise, the level direction of crack growth in the individual area or grain may be different from the general growth direction depending mainly on its crystal orientation, microstructure, stress condition, etc.

3. A model illustrating the mechanism of fatigue crack growth and striation formation is proposed. According to the present model, the process of fatigue crack growth under a loading cycle may be divided into four steps: (i) crack-tip opening after the closure stress is overcome as the tensile load rises; (ii) shear crack growth in a very short distance along a slipband notch or crack at the very tip by single system slip decohesion as the tensile load continues to increase; (iii) normal crack growth by alternative conjugate slip decohesion under higher tensile stresses up to a maximum; and (iv) crack-tip closing, resharpening, and slipband notch or crack may be induced due to compressive shear deformation at one or both sides of the very tip during unloading or compressive loading.

4. Based on the present mechanism, other markings on the fracture surface and some phenomena during crack growth, such as secondary shear cracks associated with striations, crack branching, and simultaneous growth of multilayered cracks, could reasonably be interpreted.

References

[1] Forsyth, P. J. E. and Ryder, D. A., "Fatigue Fracture," *Aircraft Engineering,* Vol. 32, No. 374, 1960, pp. 96–99.

[2] Hertzberg, R. W., "Application of Electron Fractography and Fracture Mechanics to Fatigue Crack Propagation in High Strength Aluminum Alloys," Ph.D. Dissertation, Lehigh University, Bethlehem, PA, June 1965.

[3] Piper, D. E., Quist, W. E., and Anderson, W. E., "The Effect of Composition on the Fracture Properties of 7178-T6 Aluminum Alloy Sheet," presented at the Fall Meeting of the American Institute of Mining, Metallurgical and Petroleum Engineers, Philadelphia, PA, Oct. 1964.
[4] Carmen, C. and Shuler, M., "Low Cycle Fatigue Properties of 18 Ni-Co-Mo 250 Maraging Steel," presented at ASTM Subcommittee on Electron Fractography, Schenectady, NY, American Society for Testing and Materials, Philadelphia, Sept. 1964.
[5] Laird, C. and Smith, G. C., "Crack Propagation in High Stress Fatigue," *Philosophical Magazine,* Vol. 7, No. 77, 1962, pp. 847–857.
[6] Pelloux, R. M. H., "Fractographic Analysis of the Influence of Constituent Particles on Fatigue Crack Propagation in Aluminum Alloys," *Transactions,* American Society for Metals, Vol. 57, No. 2, 1964, p. 511.
[7] Forsyth, P. J. E. and Ryder, D. A., "Some Results of the Examination of Aluminum Alloy Specimens Fracture Surfaces," *Metallurgia,* Vol. 63, 1961, pp. 117–127.
[8] Schijve, J., "Analysis of the Fatigue Phenomenon in Aluminum Alloys," NLR-TR, M2212, National Aerospace Laboratory, Amsterdam, April 1964.
[9] McMillan, J. C. and Pelloux, R. M. N., "Fatigue Crack Propagation under Program and Random Loads," *Fatigue Crack Propagation, ASTM STP 415,* Philadelphia, 1967, pp. 505–535.
[10] Laird, C., "The Influence of Metallurgical Structure on the Mechanisms of Fatigue Crack Propagation," *Fatigue Crack Propagation, ASTM STP 415,* Philadelphia, 1967, pp. 131–180.
[11] Tomkins, B., "Fatigue Crack Propagation—An Analysis," *Philosophical Magazine,* Vol. 18, No. 155, 1968, pp. 1041–1066.
[12] Lindley, T. C., Richards, C. E., and Ritchie, R. O., "The Mechanics and Mechanisms of Fatigue Crack Growth in Metals," Paper 24, Conference on the Mechanics and Physics of Fracture, Cambridge, U.K., Jan. 1975.
[13] Pelloux, R. M. N., "Mechanisms of Formation of Ductile Fatigue Striations," *Transactions,* American Society for Metals, Vol. 62, 1969, pp. 281–284.
[14] Neumann, P., "Buldung und Ausbreitung von Rissen bei der Wechselverformung," *Zeitschrift für Metallkunde,* Vol. 11, 1967, pp. 780–789.
[15] Liu, H. W., "Analysis of Fatigue Crack Propagation," National Aeronautics and Space Administration Contract Report, NASA CR-2032, Washington, DC, May 1972.
[16] Hahn, G. T., Hoagland, R. G., and Rosenfield, A. R., "Local Yielding Attending Fatigue Crack Growth," *Metallurgical Transactions,* Vol. 3, May 1972, pp. 1189–1202.
[17] Neumann, P., "A Rupture Criterion for Ductile Fatigue Crack Advance," *Proceedings,* 3rd International Conference on Fracture, Munich, Germany, Vol. 2, 1974, p. 1232.
[18] Neumann, P., "New Experiments Concerning the Slip Process at Propagation Fatigue Cracks—I, II," *Acta Metallurgica,* Vol. 22, 1974, p. 1155.
[19] Tomkins, B., "Fatigue Failure," Paper 23, Conference on Mechanics and Physics of Fracture, Cambridge, U.K., Jan. 1975.
[20] Wanhill, R. J. H., "Fatigue Striation Mechanisms in Metals" in *Proceedings,* International Conference on the Mechanical Behavior of Materials, ICM-II, Boston, 1976, p. 558.
[21] Kuo, A. S. and Liu, H. W., "An Analysis of Unzipping Model for Fatigue Crack Growth," *Scripta Metallurgica,* Vol. 10, 1976, p. 723.
[22] Gauthier, P., De Rabaudy, H., and Auvinet, J., "Secondary Cracking Process During Fatigue Crack Propagation," *Engineering Fracture Mechanics,* Vol. 5, No. 4, 1973, pp. 977–984.
[23] Mein, D. A., "Comparison Between Air and Vacuum Fatigue Fracture in 7A1-2Nb-1Ta Titanium Alloy," U.S. Naval Research Laboratory Reports, May 1966, pp. 20–21.
[24] Wanhill, R. J. H., "Fractography and Fatigue Crack Propagation in 2024-T3 and 7075-T6 Aluminum Alloys in Air and Vacuum," *Metallurgical Transactions,* Vol. 6A, 1975, p. 1587.
[25] Ritter, D. L. and Wei, R. P., "Fractographic Observations of Ti-6A1-4V Alloy Fatigue in Vacuum," *Metallurgical Transactions,* Vol. 2, 1971, pp. 3229–3230.
[26] Grinberg, N. M., "The Effect of Vacuum on Fatigue Crack Growth," *International Journal of Fatigue,* April 1982, p. 83.

Toshihiko Hoshide,[1] *Mitsuo Miyahara,*[1] *and Tatsuo Inoue*[1]

Elastic-Plastic Behavior of Short Fatigue Crack in Smooth Specimen

REFERENCE: Hoshide, T., Miyahara, M., and Inoue, T., **"Elastic-Plastic Behavior of Short Fatigue Crack in Smooth Specimen"** *Basic Questions in Fatigue: Volume I, ASTM STP 924,* J. T. Fong and R. J. Fields, Eds., American Society for Testing and Materials, Philadelphia, 1988, pp. 312–322.

ABSTRACT: The fundamental problem posed in this study is that associated with the behavior of short surface crack in low-cycle fatigue, which plays a significant role in predicting fatigue life. In the investigation, low-cycle fatigue tests of smooth specimens were conducted under push-pull loading for several materials, and behaviors of cracks on the specimen surface were observed by means of a plastic replication technique. When the growth rate of surface cracks was correlated to J-integral range, the data for cracks longer than about three times grain-size without coalescence growth were found to coincide with the relation obtained for large through-thickness cracks. Other data, even when excluding the results with coalescence, shifted toward the higher-rate region than the large-crack relation. Such an acceleration was suggested to be due to the difference in growth mechanism.

The prediction of fatigue life was investigated on the basis of the J-integral approach, and it was found that some restrictions remained in the applicability of the crack growth law of ordinary fracture mechanics type to the life prediction for smooth specimens under low-cycle fatigue.

KEY WORDS: fatigue, short crack, crack growth, elastic-plastic fracture mechanics, J-integral, life prediction

Quantitative evaluation of short crack growth behavior is of importance for the prediction of fatigue life, since the large part of the fatigue process is covered with the growth process of the short crack. Several investigators [*1–5*] have treated the subject, especially the derivation of Coffin-Manson's law representing an empirical relation between plastic strain and fatigue life. In the treatments, a fundamental question associated with the behavior of short fatigue crack is posed from an engineering viewpoint: "Is the crack growth law of ordinary fracture mechanics type directly applicable to the life prediction for smooth specimens under low-cycle fatigue?" The growth behavior of short cracks in the low-cycle-fatigue regime should be analyzed through the elastic-plastic fracture mechanics (EPFM) because of gross plasticity. Although a candidate parameter in the EPFM approach is suggested to be the range of J-integral (ΔJ) [*6–10*], the applicability of J-integral range to short crack growth, which is directly related to the life prediction, has not been sufficiently clarified [*6,11*].

In the present investigation, the growth behaviors of short cracks in smooth specimen of some metals were observed in the low-cycle fatigue process, and the applicability of ΔJ to short crack growth was examined experimentally. In order to answer the question posed in this study, the prediction of fatigue life was investigated based on the ΔJ analysis.

[1] Research associate, graduate student, and professor, respectively, Department of Mechanical Engineering, Kyoto University, Kyoto, Japan.

Experimental Procedure

The data of materials employed in the experiments are listed in Table 1. The chemical composition (in weight percent) of the medium-carbon steel (JIS S35C steel) is 0.37C, 0.24Si, 0.77Mn, 0.019P, 0.023S, 0.01Cu, 0.02Ni, and 0.04Cr. The steel with the mean grain-size $d = 9.7$ μm is denoted as S35C(S), and the larger grain-sized one ($d = 31$ μm) as S35C(L). The aluminum alloy contains 2.60Mg, 0.12Si, 0.28Fe, 0.04Cu, 0.06Mn, 0.25Cr, 0.02Zn, and 0.01Ti in weight percent. The copper (oxygen-free high conductivity [OFHC] copper) has 99.99% purity and contains oxygen of 3 ppm.

Round-bar-type specimens (6 mm in diameter and 10 mm in gage length) were employed throughout the whole fatigue tests, and the fully reversed push-pull loading was imposed under displacement-controlled conditions by a closed-loop servo-hydraulic testing machine. In this case, the relation between the displacement and the strain measured with the foil strain gage attached on the surface of gage section was first determined for the calibration. The growth behavior of fatigue crack on specimen surface was observed by a plastic replication technique in a series of fatigue tests. The crack length $2a$ on the surface was measured as the component of the actual length to the direction perpendicular to the loading axis. The crack shape was also observed by means of electropolishing and replication in another series of fatigue tests.

Fatigue Strength Properties and Crack Growth Behaviors

Figure 1 represents the empirical relation between the total range of applied strain $\Delta\epsilon$ given by the controlled displacement through the calibrated relation and the failure life N_f for four kinds of material.

The relation between the strain range $\Delta\epsilon$ and the stress range $\Delta\sigma$ is determined from the hysteresis loop at half of the fatigue life in each fatigue test and is formulated by

$$\Delta\epsilon = \Delta\epsilon_e + \Delta\epsilon_p = \Delta\sigma/E + (\Delta\sigma/k')^{1/n'} \tag{1}$$

where

$\Delta\epsilon_e, \Delta\epsilon_p$ = elastic and plastic components of $\Delta\epsilon$, respectively,
E = Young's modulus,
k' = cyclic strength coefficient, and
n' = cyclic strain hardening exponent.

TABLE 1—*Heat-treatment conditions, grain size, and mechanical properties of materials tested.*

Material	Heat Treatment	Grain Size, d, μm	Yield Strength, σ_Y, MPa	Tensile Strength, σ_B, MPa	Elongation, %
S35C(S)	annealed at 865°C for 0.5 h in vacuum and cooled in air	9.7	382	668	37
S35C(L)	annealed at 940°C for 4 h in vacuum and cooled in furnace	31	267	607	37
Al alloy	annealed at 400°C for 1 h in vacuum and cooled in furnace	47	134	239	31
Copper	annealed at 600°C for 1 h in vacuum and cooled in furnace	97	40.6	216	80

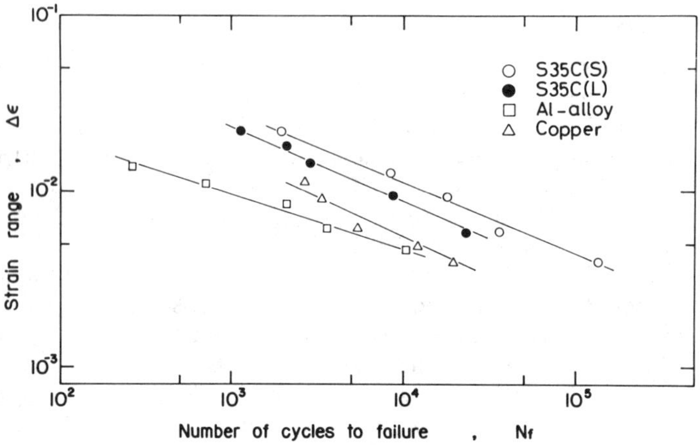

FIG. 1—*Fatigue life curves of four types of material.*

Values of the parameters are as follows: E = 206 GPa, k' = 2.49 GPa, and n' = 0.23 for S35C(S) and S35C(L); E = 76 GPa, k' = 580 MPa, and n' = 0.036 for aluminum alloy; E = 122 GPa, k' = 1.28 GPa, and n' = 0.27 for copper.

Dominant fatigue cracks in each material were found to initiate along or near the boundary of grains, the size of which was observed to be larger than the mean grain size of the material.

The interaction between crack growth and metallurgical microstructure such as pearlite phase, grain boundary, and so on, and the coalescence of cracks for the carbon steel S35C(S) with small grains were detected especially in the early growth stage of cracks of length less than 50 μm, whereas cracks in the subsequent growth process, except for the final fracture stage, were accompanied by little coalescence. Crack coalescence of large-grained steel S35C(L) occurred more frequently than that in S35C(S), and cracks whose growth was affected by the microstructure similarly to S35C(S) were also observed in the early stage of growth, where such an interaction was less than in S35C(S). Cracks in copper and aluminum-alloy joined each other more frequently than in both types of carbon steel over a wide range of crack sizes.

Assuming the shape of surface crack to be semi-elliptic, based on the observation, the aspect ratio λ of the crack is defined as the ratio of the depth c to half-length a on the surface. The ratio λ for a single crack of the steels decreased from 1.0 to 0.7 as the crack grew from the grain-size order to a few millimetres, which means the change of crack shape from semi-circular to shallower semi-ellipse, and the value of λ in the case of joining was apparently smaller than that expected from the single-crack relation and nearly equal to 0.5. The aspect ratio in copper was somewhat larger than that in the steels for the same crack length. The relation for aluminum alloy, however, shifted toward the lower side than the cases of other materials; λ was 0.7 ~ 0.5 for single cracks and 0.4 ~ 0.3 for joint cracks. The empirical relation of λ against crack length will be used in the following analysis.

Mechanics and Mechanism of Short Crack Growth

The crack growth rate da/dN on the specimen surface is discussed in this section with respect to the range of J-integral (ΔJ). A procedure of estimating ΔJ for surface cracks proposed by Dowling [*6*] was modified by Hoshide and Tanaka [*8*] and Tanaka et al. [*9*] to take into account the effects of crack shape, loading mode, crack closure, and so on, and the following equation based on the formula indicated in Refs *8* and *9* was suggested to be available for such a crack on a round-bar-type specimen in Ref *10*

$$\Delta J = 2\pi M_J \Delta \tilde{W} a \tag{2}$$

where

$$M_J = (M_K/\Phi)^2 \lambda \tag{3}$$

$$\Delta \tilde{W} = \Delta\sigma^2/2E + f(n')\Delta\sigma\Delta\epsilon_p/(1 + n') \tag{4}$$

$$f(n') = (1 + n')[3.85(1 - n')/\sqrt{n'} + \pi n']/2\pi \tag{5}$$

In Eq 3, M_K is the correction factor for the stress-intensity factor K of the surface crack; $M_K = K/(\sigma\sqrt{\pi\lambda a/\Phi})$ [*12*]; and Φ is the complete elliptic integral of the second kind with the modulus $k = \sqrt{1-\lambda^2}$. Therefore, M_J is reduced to the function of crack shape. When evaluating ΔJ, use is made of the value of λ obtained experimentally and, as seen in Eq 4, the full ranges of stress and plastic strain are supposed to be effective for the crack growth due to the direct observation of crack opening and closing behavior [*10*].

The growth rate da/dN for each material is plotted against ΔJ in Fig. 2, where the strain ranges adopted are also indicated. The central bold lines with formulas of growth rate in the figures indicate the ΔJ representation of growth law for large through-thickness cracks, which was obtained in the range $10^{-9} \sim 10^{-5}$ m/cycle for S35C steel, $5 \times 10^{-8} \sim 10^{-5}$ m/cycle for aluminum alloy, and $5 \times 10^{-9} \sim 5 \times 10^{-5}$ m/cycle for copper. Two fine lines are the boundaries of scatter for the case of long cracks with a factor of four in growth rate. It is noted that, as for the medium-carbon steel, the growth equation for long cracks was not affected by the difference of grain size, that is a result similar to that reported by a few investigators [*13,14*] (noting that ΔJ is identical to $(\Delta K_{\text{eff}})^2/E$, where ΔK_{eff} is the effective range of stress-intensity factor). In the figures, the slashed marks are data corresponding to coalescence. Although such a coalescence was hardly observed in S35C(S), the growth by coalescence occurred more frequently in the order of S35C(L), aluminum alloy, and copper, which corresponds to the order of grain size.

As seen in Fig. 2, some data for surface-cracks are not predictable from the relation for through-thickness cracks; cracks with short length particularly, as well as those with coalescence, show the higher rate.

For brevity and clarity, the data for the highest and the lowest strain ranges tested in each material are summarized in Fig. 3. The abscissa in the figure represents the relative crack length normalized by the mean grain size, and the ordinate is the ratio of the measured crack growth rate $(da/dN)_{\text{exp}}$ to the rate $(da/dN)_{\Delta J}$ predicted from the growth law for large through-thickness cracks in the form

$$da/dN = C_J(\Delta J)^{m_J} \tag{6}$$

It follows that, if the ratio $(da/dN)_{\text{exp}}/(da/dN)_{\Delta J}$ takes unity, the growth behavior of a short crack can be described by the same relation as that of a large crack. As seen from the figure, $(da/dN)_{\text{exp}}/(da/dN)_{\Delta J}$ for each material is larger than one for smaller values of $2a/d$ and approaches one after the transition region of $2a/d$ ranging from 3 to 5. The order of such a deviation from unity in the region of small values of $2a/d$ is S35C steel, aluminum alloy, and copper. The solid marks indicating coalescence growth lie in the region of higher growth rate, as easily anticipated.

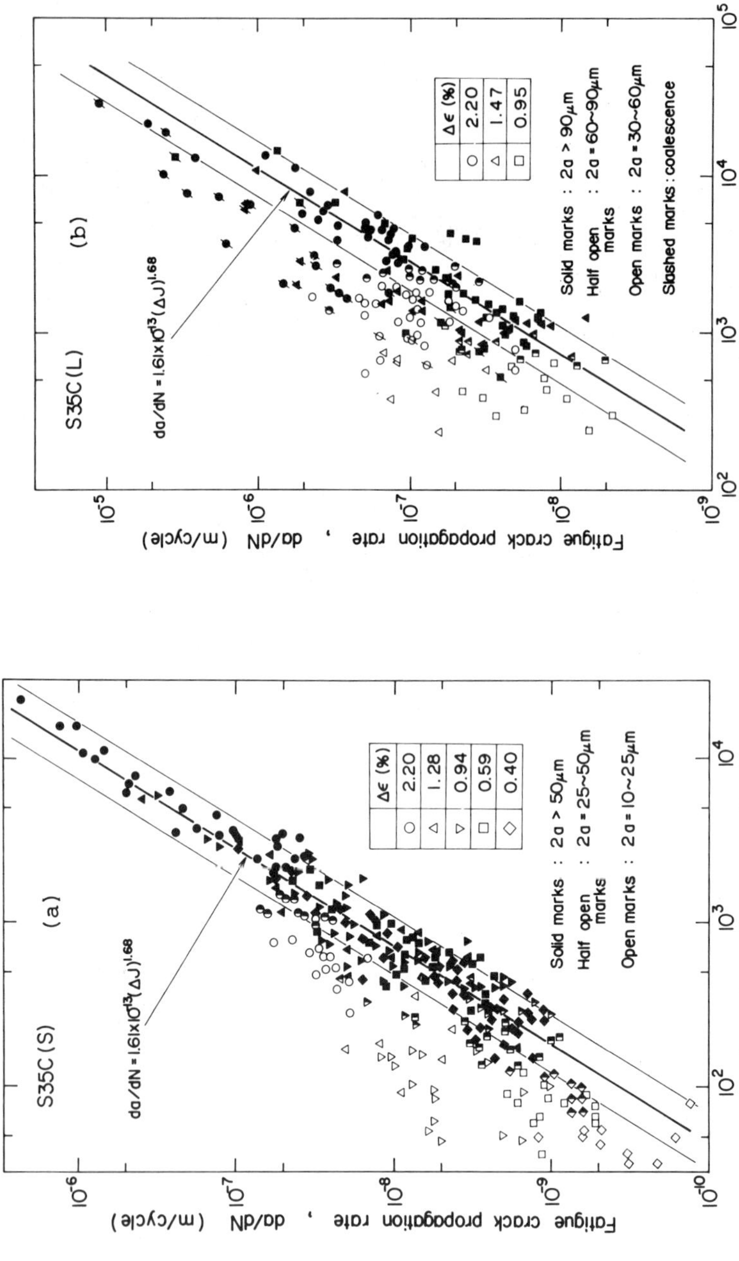
S35C (L)
(b)
da/dN = 1.6(x10^{13})(ΔJ)$^{1.68}$
Δε (%)
2.20
1.47
0.95
Solid marks : 2a > 90μm
Half open marks : 2a = 60~90μm
Open marks : 2a = 30~60μm
Slashed marks : coalescence
Fatigue crack propagation rate , da/dN (m/cycle)
J integral range , ΔJ (N/m)
S35C (S)
(a)
da/dN = 1.6(x10^{13})(ΔJ)$^{1.68}$
Δε (%)
2.20
1.28
0.94
0.59
0.40
Solid marks : 2a > 50μm
Half open marks : 2a = 25~50μm
Open marks : 2a = 10~25μm
Fatigue crack propagation rate , da/dN (m/cycle)
J integral range , ΔJ (N/m)

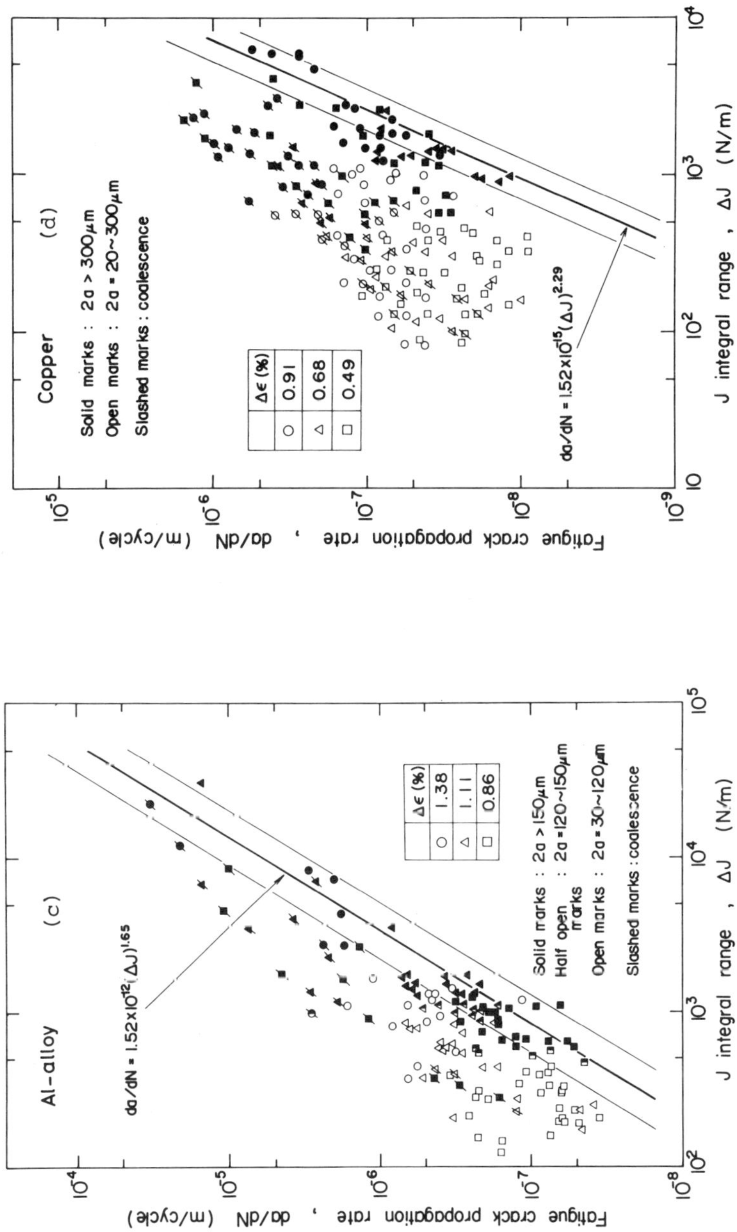

FIG. 2—*Relation between crack propagation rate and J-integral range:* (a) *S35C(S),* (b) *S35C(L),* (c) *aluminum alloy,* (d) *copper.*

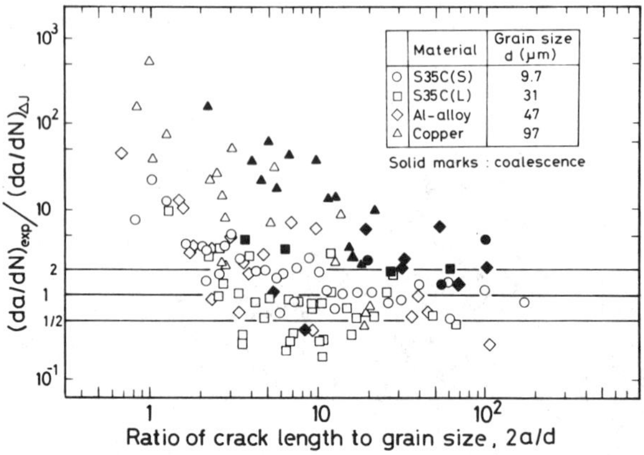

FIG. 3—*Limitations in* ΔJ *representation of small-crack growth behaviors.*

It has been suggested in some works [*9,10*] that, by analogy to J_{Ic} testing [*15*], the applicability of the J-integral range to short fatigue cracks might be restricted by the minimum size requirement for valid ΔJ testing

$$a \geq \alpha(\Delta J/2\sigma_Y) \tag{7}$$

where $\alpha = 50$ for a low-carbon steel [*9*] and $\alpha = 10$ for a medium-carbon steel [*10*]. As seen in Fig. 2, however, the shift toward the higher rate region for each material occurs at much the same crack length in spite of a wide variation in ΔJ-value. This fact as well as the result in Fig. 3 implies that the acceleration might be explained by the size of crack relative to the metallurgical microstructure rather than the mechanical size requirement.

The observations of crack growth behaviors related to metallurgical microstructure will be stated in more detail. An example of S35C(S) is shown in Fig. 4, which indicates the growth behavior of a main crack under the applied strain range of $\Delta\epsilon = 0.4\%$. In the figure, an illustration of the crack with the surrounding structure in the early growth stage is also presented, where the slashed parts correspond to the pearlite phase and the other to the ferrite. Point A with arrows means a situation where the crack was disturbed to grow by the pearlite grains as indicated by State A in the illustration. That leads to a decrease in growth rate, followed by the recovery of the rate at Point B, where the crack breaks through the blocking as also indicated by State B in the illustration. In the region of $2a \leq 50$ μm, excluding the data during interruption of growth, the growth rate is seen to be higher than that expected from the growth law of large cracks, as observed in other materials. It is worth noting that, in all materials tested, the growth of a crack of length less than a few times grain-size is affected strongly by the microstructure, and the growth rate is higher than that of longer cracks.

From fractographic observation of copper indicating the most significant acceleration in growth rate, the crack nucleation part was found to show a morphology of grain boundary cracking of the facet type corresponding to the size of an individual grain followed by striation or a striation-like pattern. This suggests that a cracking pattern of Mode I combined with Mode II or Mode III might change into the dominant Mode I state as the crack grows, and that the acceleration behavior of shorter cracks and its dependence on grain size might be attributed mainly to such a difference in growth mechanism.

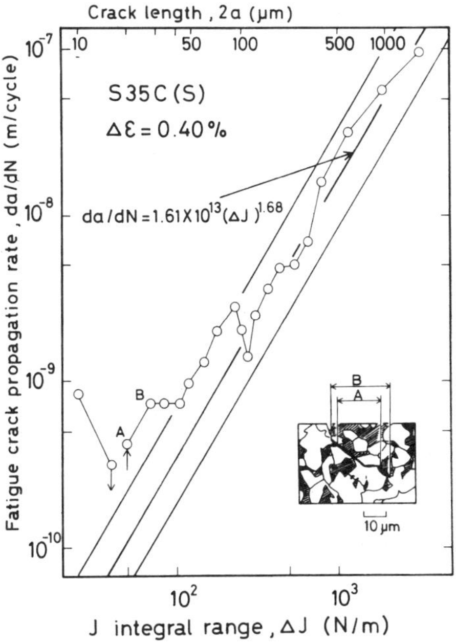

FIG. 4—da/dN *versus* ΔJ *relation of individual crack in S35C(S).*

J-Integral Approach to Life Prediction

From the previous discussion it is found that the growth of a crack of length less than about three times grain size cannot be represented directly by the growth law of through-thickness cracks due to a mixed-mode growth mechanism. This implies the first limitation in the applicability of the crack growth law of ordinary fracture mechanics type to life prediction. For the more appropriate prediction of failure life, the growth process of the smaller cracks should be treated as a crack initiation stage through other procedures, since the fraction of the number of strain cycles, when the total length of the crack becomes about three times that of grain size, against the failure life is large enough to be 0.3 ~ 0.4 for S35C(S), 0.1 ~ 0.2 for S35C(L), 0.2 ~ 0.4 for aluminum alloy, and 0.35 ~ 0.45 for copper (the fraction showed a tendency to increase for a smaller applied strain).

In the following, the evaluation of propagation life of cracks larger than such size is examined in terms of a new parameter derived from ΔJ. The number of cycles necessary for crack growth from initial value a_i to final value a_f is obtained by substituting Eq 2 into Eq 6 and executing the integration, giving the expression for crack propagation life N_p as

$$(\Delta \tilde{W})^{m_J} N_p = C_W \tag{8}$$

where

$$C_W = [(a_i)^{1-m_J} - (a_f)^{1-m_J}]/[C_J(m_J - 1)(2\pi M_J)^{m_J}] \tag{9}$$

The value C_W can be regarded as a constant because m_J and C_J are determined as material constants by growth tests of through-thickness type cracks.

Figures 5*a* and 5*b* illustrate examples of the relation between the energy density parameter

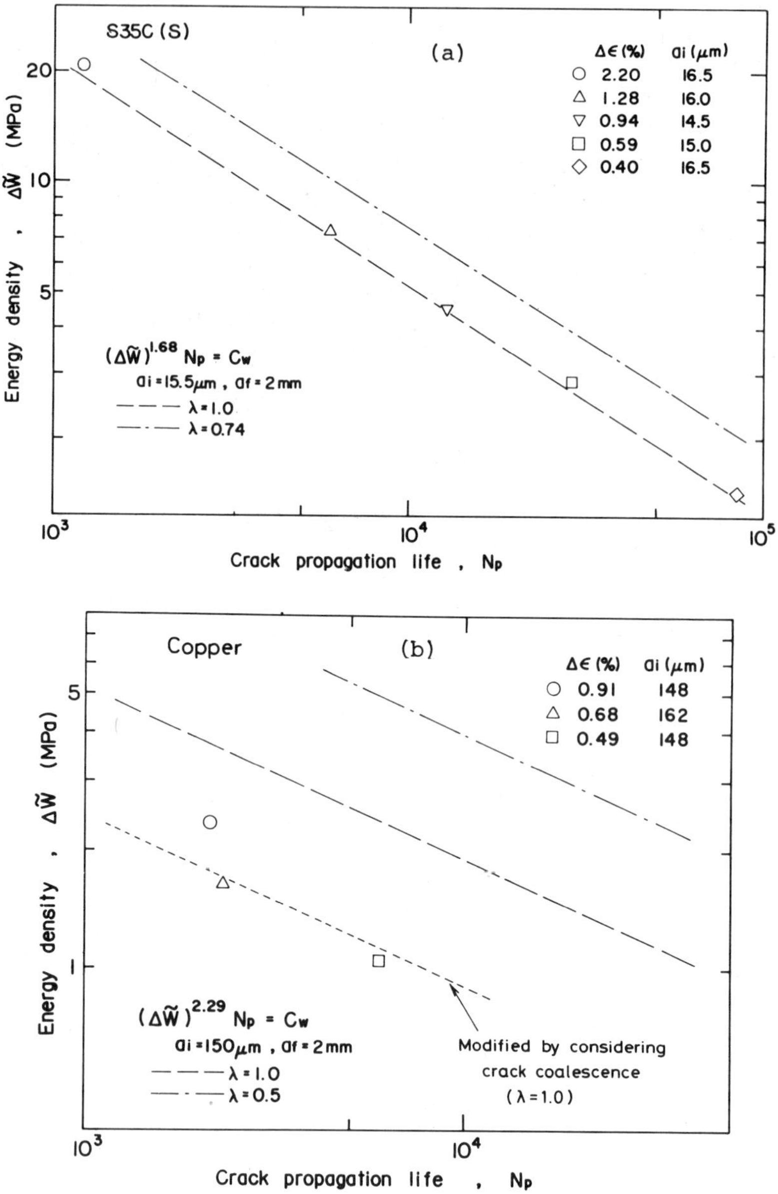

FIG. 5—*Relation between crack propagation life and strain energy parameter:* (a) *S35C(S)*, (b) *copper.*

$\Delta \tilde{W}$ shown in Eq 4 and the crack propagation life N_p after a crack grows to length longer than three times grain size, especially for S35C(S) and copper, which indicate two extreme cases of crack growth behavior. In the calculation, $a_f = 2$ mm was adopted, because failures in most cases occurred immediately after a main crack grew to that length. The two extreme values of M_f in Eq 9 are calculated from Eq 3, corresponding to the maximum and minimum values of aspect ratio λ which lead to the shortest and the longest lives in the prediction, respectively, and the broken and dot-dash lines in Fig. 5, which indicate the predicted crack propagation life by Eq 8 for these values of λ. The experimental data of S35C(S), in which coalescence of cracks scarcely occurred, agree with the predicted relation for $\lambda = 1$ as shown in Fig. 5*a*. This is attributed to the fact that the growth period of a very short crack with $\lambda \fallingdotseq 1$ is dominant in fatigue life for the material. In the case of copper, in which crack joinings were often observed, the predicted life tends to shift toward the longer-life region than the experimental results (see Fig. 5*b*). So, the problem to be settled is how the fatigue life for the case of coalescence can be estimated by the crack growth equation of long cracks. Such an evaluation is attempted in the following. The dotted line in Fig. 5*b* presents the relation modified by considering crack coalescence, which shows good agreement with experimental data. In the modification, the reduction in crack propagation life due to coalescence is presumed to be equivalent to the number of cycles during which a single crack propagates by increments of its length corresponding to jumps in crack length due to coalescences.

The crack growth law of the fracture mechanics type is concluded to be directly applicable to the prediction of the life of a smooth specimen with cracks longer than about three times grain size, although some modification considering crack coalescence needs a further refinement for special material.

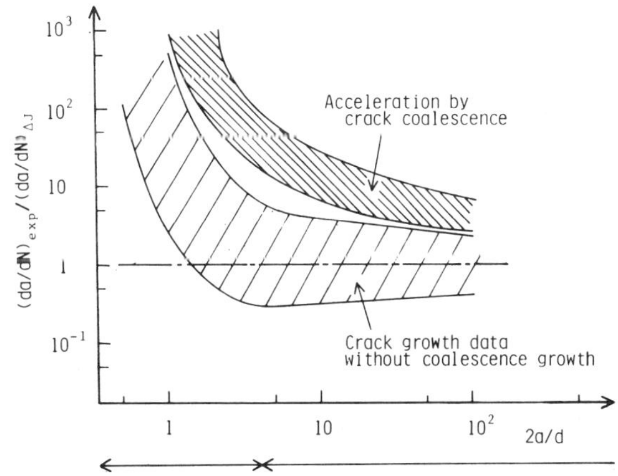

Microstructure effect	large	small	little
Morphology of fracture surface	grain-boundary cracking (facet-type)	striation-like or striation pattern	
Cracking mode	Mode I + Mode II or Mode III	Mode I	

FIG. 6—*Growth rate and mechanism of short crack.*

Conclusions

Low-cycle fatigue tests of a smooth specimen were conducted under push-pull loading of displacement-controlled conditions for four kinds of materials: two medium-carbon steels with distinct grain-sizes, copper, and aluminum alloy. The crack growth behavior was observed by a plastic replication technique and analyzed in the framework of elastic-plastic fracture mechanics.

The growth rate of cracks was correlated to the J-integral range ΔJ, and a summary of the results obtained is illustrated in Fig. 6. For cracks larger than a few times mean grain-size, except for the case of coalescence growth, the rate of propagation could be estimated by the growth law for through-thickness cracks, whereas the rate of smaller cracks became higher. Such results were attributed to the differences in growth mechanism. This suggests that there are some restrictions in applicability of the crack growth law of the fracture mechanics type to the life prediction of smooth specimens in low-cycle fatigue.

In spite of the fact that the fracture mechanics type of crack growth law gave a good estimation of the propagation life of cracks larger than the size of a few grains, some modifications were, in fact, necessary for considering the coalescence growth of cracks in the cases of copper and aluminum alloy. For the shorter cracks, the ordinary fracture mechanics type of analysis was not directly applicable to the prediction of their propagation lives.

References

[*1*] Boettner, R. C., Laird, C., and McEvily, A. J., *Transactions,* The Metallurgical Society of the American Institute of Mining, Metallurgical and Petroleum Engineers, Vol. 233, 1965, pp. 379–387.

[*2*] Tomkins, B., *Philosophical Magazine,* Vol. 18, 1968, pp. 1041–1066.

[*3*] Mowbray, D. F. in *Cracks and Fracture, ASTM STP 601,* American Society for Testing and Materials, Philadelphia, 1976, pp. 33–46.

[*4*] Radhakrishnan, V. H., *Fatigue of Engineering Materials and Structures,* Vol. 3, 1980, pp. 75–84.

[*5*] Murakami, Y., Harada, S., Endo, T., Tani-Ishi, H., and Fukushima, Y., *Engineering Fracture Mechanics,* Vol. 18, 1983, pp. 909–924.

[*6*] Dowling, N. E. in *Cyclic Stress-Strain and Plastic Deformation Aspects of Fatigue Crack Growth, ASTM STP 637,* American Society for Testing and Materials, Philadelphia, 1977, pp. 97–121.

[*7*] Douglas, M. J. and Plumtree, A. in *Fracture Mechanics, ASTM STP 677,* American Society for Testing and Materials, Philadelphia, 1979, pp. 68–84.

[*8*] Hoshide, T. and Tanaka, K., *Transactions,* Japan Society of Mechanical Engineers, Vol. 48A, 1982, pp. 1102–1110.

[*9*] Tanaka, K., Hoshide, T., and Maekawa, O., *Engineering Fracture Mechanics,* Vol. 16, 1982, pp. 207–220.

[*10*] Yamada, T., Hoshide, T., Fujimura, S., and Manabe, M., *Transactions,* Japan Society of Mechanical Engineers, Vol. 49A, 1983, pp. 441–451.

[*11*] Hoshide, T., Fujimura, S., and Yamada, T., *Transactions,* Japan Society of Mechanical Engineers, Vol. 50A, 1984, pp. 320–329.

[*12*] Raju, I. S. and Newman, J. C., Jr., *Engineering Fracture Mechanics,* Vol. 11, 1979, pp. 817–829.

[*13*] Taira, S., Tanaka, K., and Hoshina, M. in *Fatigue Mechanisms, ASTM STP 675,* American Society for Testing and Materials, Philadelphia, 1979, pp. 135–173.

[*14*] Nakai, Y., Tanaka, K., and Yamashita, M., *Journal of the Society of Materials Science (Japan),* Vol. 31, 1982, pp. 1121–1127.

[*15*] Landes, J. D. and Begley, J. A. in *Developments in Fracture Mechanics Test Methods Standardization, ASTM STP 632,* American Society for Testing and Materials, Philadelphia, 1977, pp. 57–81.

Michael T. Resch,[1] D. V. Nelson,[1] H. H. Yuce,[1] and B. D. London[1]

Use of Nondestructive Evaluation Techniques in Studies of Small Fatigue Cracks

REFERENCE: Resch, M. T., Nelson, D. V., Yuce, H. H., and London, B. D., **"Use of Nondestructive Evaluation Techniques in Studies of Small Fatigue Cracks,"** *Basic Questions in Fatigue: Volume I, ASTM STP 924,* J. T. Fong and R. J. Fields, Eds., American Society for Testing and Materials, Philadelphia, 1988, pp. 323–336.

ABSTRACT: This paper reports progress to date in the development of an ultrasonic surface acoustic wave (SAW) technique to monitor *in situ* the depth of small cracks as they grow, as well as the stresses required to cause the cracks to begin to open and to fully open.

Two SAW transducers (transmitter and receiver) are used to monitor crack reflection signals. The surface length is measured by conventional methods. Based on new acoustic theory, the reflection signal and length data are used to infer crack depth below the surface. Acoustically determined depth is compared with that obtained by destructive examination of specimens, and agreement is good.

The crack opening stress determined acoustically is compared with opening stress determined by scanning electron microscope (SEM) measurements of crack mouth opening displacement versus applied stress. This comparison requires specimens small enough to fit inside an SEM chamber. Preliminary results demonstrate the ability of the acoustic measurements of crack opening to detect behavior below the surface not detected by SEM measurements. Acoustically determined changes in crack opening behavior as a function of constant-amplitude and variable-amplitude stress histories are also presented.

KEY WORDS: nondestructive testing, small crack growth, crack opening stress, closure, load sequence effects, fatigue, acoustic technique

Various nondestructive evaluation (NDE) techniques have been used over the past two decades in the study of fatigue of metals. For example, bulk wave ultrasonic reflection techniques have been used [*1*] to detect the formation of fatigue cracks in thin center-notched specimens of aluminum alloys and steel. Multiple surface acoustic wave (SAW) transducers have detected microcracking while scanning the surface of round bar fatigue specimens [*2*]. SAW transducers have been utilized to observe crack opening behavior and to monitor the accumulation of "fatigue damage" [*3*]. Direct observations of crack mouth opening displacement (CMOD) and crack tip opening displacement (CTOD) as a function of applied stress have been made [*4*] on small surface cracks in specimens mounted within a scanning electron microscope (SEM) in an effort to determine opening stress.

The use of acoustic NDE techniques for studying the opening behavior of "large" through-thickness fatigue cracks has also received considerable attention. Experimental studies [*5*] have been made of the transmission and reflection characteristics of bulk ultrasonic waves at an interface which simulates the acoustic response of a true fatigue crack. Detailed theoretical calculations have been performed [*6*] which demonstrate the nonlinear depend-

[1] Research affiliate/Materials Science and Engineering Department, associate professor/Mechanical Engineering Department, research assistant/Mechanical Engineering Department, and research assistant/Materials Science and Engineering Department, Stanford University, Stanford, CA 94305.

ence on contact stress of transmission and reflection of bulk waves at an irregular interface. Transmission and reflection as a function of the geometry of the contacting surface area at an interface have also been analyzed [*7*]. In addition, a method has been developed [*8*] to calculate the scattering of surface acoustic, or Rayleigh, waves from partly closed surface breaking cracks. In principle, this method can be utilized to model cracks which are not simply connected (for example, which have islands of closure). These studies provide a theoretical basis for the nonlinear behavior of crack acoustic signals versus applied stress observed experimentally by a number of investigators [*2,3,9*].

Heretofore, such acoustic techniques have been limited to empirical correlations between crack size and acoustic signals. Optical observations of crack opening behavior are incapable of determining directly changes in opening behavior below the surface of small surface cracks. Consequently, there is considerable room for improvement in the utilization of NDE techniques to quantitatively characterize the growth behavior of small surface cracks.

The important questions to be addressed by these techniques include: What is the relation between surface crack length and depth during growth? When does a crack begin to open? When is it completely open? How does stress history influence the opening of small cracks? How is the opening of a small crack affected by residual stresses? Is there an effective stress range for small crack growth based on the range of stress above the opening level?

This paper describes the use of SAW and SEM techniques to provide quantitative information about the growth behavior of small surface fatigue cracks. In particular, the applicability of the techniques for monitoring crack depth and the variation of opening stress below the surface with growth are discussed. Also, the potential use of a SAW NDE technique for determining residual stresses at and just below the surface of specimens is summarized.

Acoustic Measurement of Crack Size

Based on advances in SAW transducers and acoustic scattering theory [*10–12*], it has recently been demonstrated [*9*] that it is possible to predict the depth of a small surface fatigue crack by making an acoustic measurement of the reflection coefficient of Rayleigh waves from the crack and optically measuring the length of the crack on the surface. A brief summary of the acoustic theory, including the definition of the reflection coefficient, is included in the Appendix. In addition, if the crack is known to be growing at an equilibrium value of crack depth-to-length ratio, or aspect ratio, a/c, then measurement of the reflection coefficient alone allows the depth to be predicted. Alternatively, if the aspect ratio is not at equilibrium, and the surface length is not known, the effective radius of a half-penny-shaped crack with the same reflection coefficient can still be calculated, yielding a conservative estimate of the crack depth. This variant of the technique is most useful for remote monitoring of crack growth when surface length monitoring is impractical or undesirable.

The primary limitation of the technique is that the frequency of the Rayleigh waves must be adjusted during fatigue crack growth so that the maximum crack depth normalized to the wavelength, κa (where $\kappa = 2\pi/\lambda$) is less than unity. In the experiments to be described shortly, a frequency of 3 MHz was used, resulting in a wavelength of approximately 1 mm in each material tested. Also, a tensile stress must be applied to a specimen during an acoustic measurement which is large enough to completely separate the crack faces in order to ensure that the true reflection coefficient of the crack is measured. This effect of crack closure below the surface on the reflection coefficient lends itself to characterization of crack opening behavior.

In order to investigate the validity of the preceding method, experiments have been performed [*9,13–16*] on specimens of quenched and tempered 4340 steel, 7075-T651 aluminum, and Pyrex glass. In these experiments predictions of crack depth obtained from

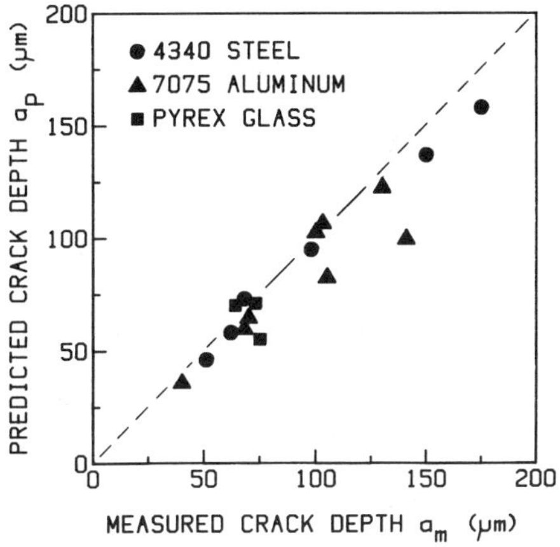

FIG. 1—*Acoustic predictions of crack depth versus post-fracture measurements of crack depth.*

SAW scattering measurements from small surface cracks were compared with post-fracture measurements of crack depth. In Fig. 1 the acoustic prediction of crack depth, a_p, is plotted versus the post-fracture measurement of crack depth, a_m. Good agreement is observed between the acoustic predictions and post-fracture measurements with the predicted length equal to or slightly less than the measured depth. In these tests, aspect ratios varied between 0.4 and 1.0. Knowledge of growth rate in depth is important and the SAW technique allows that to be obtained during the course of a fatigue test.

Acoustic Measurement of Crack Opening

For the case of SAW scattering under the conditions that the normalized crack depth, κa, is less than 1, it has been observed experimentally that measurement of the reflection coefficient, S_{21}, versus applied stress, σ, facilitates the detection of two important physical events during crack opening. The first (see Fig. 2) is the approximate point at which the

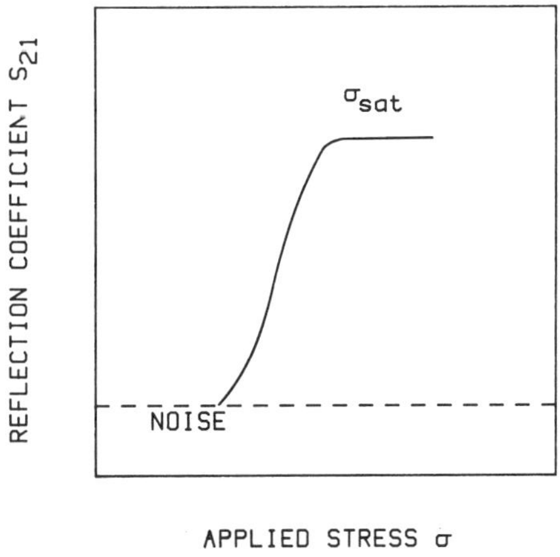

FIG. 2—*Typical response of the reflection coefficient versus applied tensile stress.*

crack faces just begin to separate. Unfortunately, the presence of backscattered signals (noise) from microstructural features such as grain boundaries and inclusions partially obscures the acoustic reflection from the crack, denoting the beginning of the opening of adjacent crack faces. The stress at which the reflection coefficient emerges from the noise actually denotes separation of a small portion of the previously closed adjacent crack faces. The second is the point at which the crack faces completely separate, so that increasing the applied stress produces no subsequent increase in the reflection coefficient. The stress at which the reflection coefficient "saturates" is denoted σ_{sat}. It has been observed experimentally that σ_{sat} can change significantly during fatigue cycling as a consequence of changes in the applied stress range, $\Delta\sigma$, and stress ratio, $R = \sigma_{min}/\sigma_{max}$, during fatigue tests. This type of measurement technique permits experimental investigations of the relation between crack growth rate and effective stress range for growth (difference between maximum applied stress and the stress required to fully open a crack). It is currently of interest to understand the significance of these acoustic measurements with respect to previous methods for measuring the opening behavior of the crack at the surface.

Microscopic Measurement of Crack Opening

A conventional method to determine the opening stress of a small surface crack measures the crack opening behavior under load. This may be accomplished through measurement of either crack mouth opening displacement (CMOD) or crack tip opening displacement (CTOD) (at either crack tip) in a SEM. The results of this type of measurement are depicted schematically in Fig. 3. There is often a small amount of residual opening at zero stress, which remains approximately linear with applied stress up to the stress at which the crack just begins to open at the particular point of inspection along the crack mouth. Above this stress, the opening is linear with increasing applied stress, but with a different slope of CMOD or CTOD versus stress. The marked change in slope is interpreted as defining the opening stress, σ_{op}. It has been observed experimentally [*17*] that the effective stress range deduced from measurements of the crack opening stress at individual crack tips correlates fairly well with observed crack growth rates on the surface. It is not clear, however, whether or not these measurements correlate with crack growth below the surface.

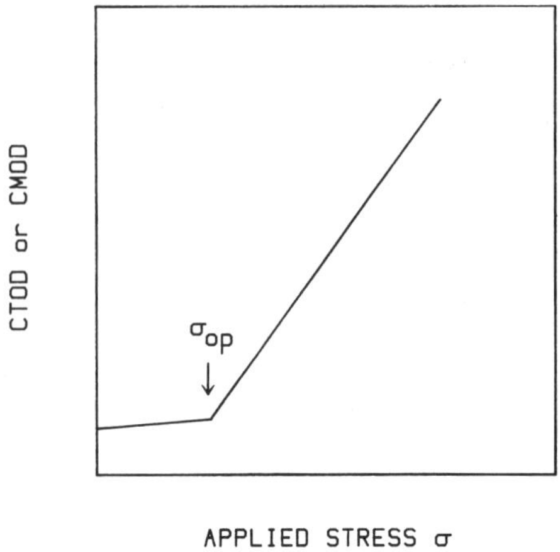

FIG. 3—*Typical response of surface crack opening displacement versus applied tensile stress.*

An analytical relationship [*18*] has been obtained between CMOD, the crack aspect ratio, a/c, and applied stress, σ, for semi-elliptical surface cracks in a finite plate in tension or pure bending. This result could allow SEM measurements of CMOD versus σ, combined with a measurement of crack length at the surface, to be used to estimate corresponding crack depth. However, it is unclear how to apply this result in the event that σ_{op} at either of the crack tips is greater than at the crack mouth, which can occur for small cracks. The combined use of acoustic (SAW) and optical (SEM) measurements of crack opening behavior provides a more complete picture of the physical processes which affect the crack growth behavior than either method alone.

In the next section, some preliminary experimental observations of small crack growth behavior under constant- and variable-amplitude loading are presented. These observations utilize several variants of the SAW-SEM techniques discussed previously, and provide background for a more general discussion concerning current important questions in small crack growth research.

Experimental Work

Preliminary experiments have been conducted in order to demonstrate the utility of the acoustic NDE techniques to characterize the growth behavior of small surface fatigue cracks [*13,14*]. In these experiments, a single fatigue crack is initiated from a defect caused by a pinpoint laser burn or a spark discharge, grown away from the damaged region, and then the damaged surface layer is carefully removed until the acoustically perceived crack depth is at the minimum detectable size. Fatigue testing can then be conducted, and subsequent growth monitored using the SAW-SEM technique with periodic adjustment of the frequency to ensure that the normalized crack depth, κa, remains less than unity.

Constant-Amplitude Stress History

Results of crack growth tests for a small surface crack in 7075-T651 aluminum are shown in Figs. 4*a*–4*c*. The applied stress history for this crack consisted of fully reversed cycling, with a stress amplitude = ±75% of the cyclic yield stress. In this experiment the crack length at the surface was measured optically at the same intervals at which the SAW-SEM opening behavior of S_{21} versus σ and CMOD versus σ were measured. The subsequent history of optical measurements of half the surface crack length, c, and the acoustic measurements of crack depth, a, are plotted in Fig. 4*a*. Knowledge of variation in crack aspect ratio, a/c, with growth is useful for interpretation of growth behavior and could not be obtained from either the SEM or SAW technique by itself. In Fig. 4*b*, the stress intensity factor range [*19*] evaluated at the surface and at the maximum crack depth are plotted versus the number of cycles. Finally, in Fig. 4*c* the opening history for this surface crack is presented combining the acoustic measurements of complete opening below the surface, σ_{sat}, with SEM measurements of opening stress at the surface, σ_{op}. This opening behavior will be discussed shortly.

Variable-Amplitude Stress History

The crack growth history for a small surface crack in quenched and tempered 4340 steel is shown in Figs. 5*a*–5*d*. The variable-amplitude stress history for this crack depicted in Fig. 5*a* consisted of increments of cycling at two values of stress range and stress ratio, $R = 0.8$; then $R = 0.05$. The simplifying assumption was made in this test that the crack would be growing at an equilibrium value of $a/c = 1$, so that remote monitoring of the fatigue test

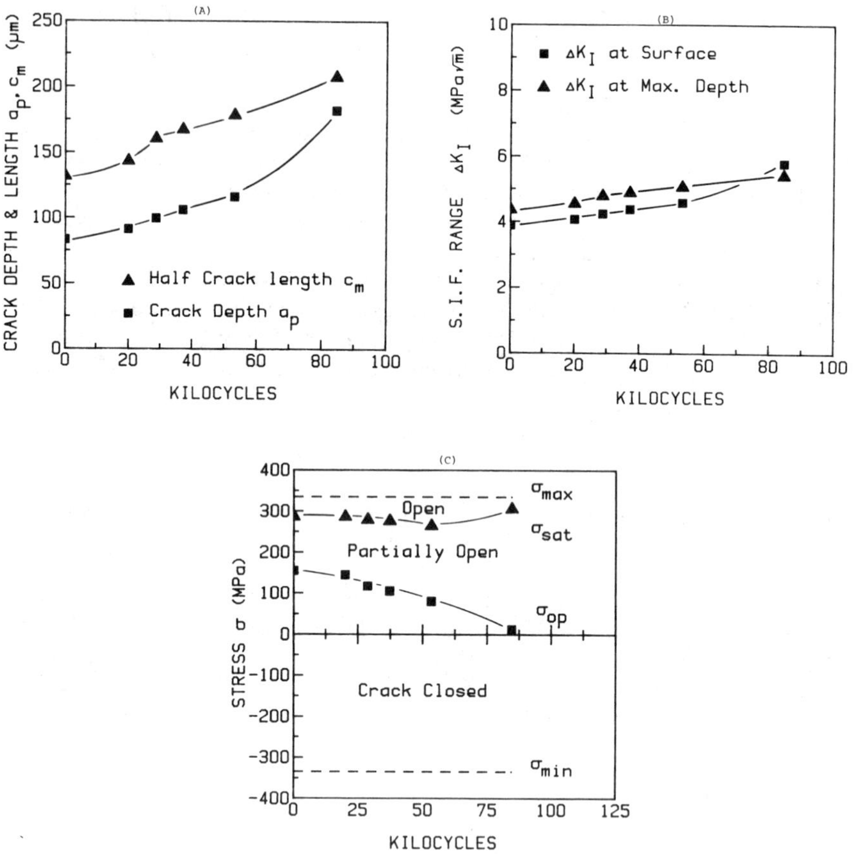

FIG. 4—(a) *Crack half length and depth history,* (b) *stress intensity factor range history, and* (c) *opening history for a small surface crack in 7075-T651 aluminum specimen.*

could be performed without visual observation of the crack. The subsequent crack growth history for this assumed half-penny-shaped crack is shown in Fig. 5*b*. Dramatic changes in the crack growth rate are evident immediately after each change in the stress range $\Delta\sigma$. In Fig. 5*c*, the maximum stress intensity factor range evaluated at the maximum crack depth [*19*] is plotted versus number of cycles. At the point in this fatigue test where the stress ratio was changed, the reflection coefficient was measured versus applied stress in order to determine the saturation stress, σ_{sat}. These measurements of σ_{sat} versus number of cycles are shown in Fig. 5*d*. Here the stress needed to fully open the crack is observed to be highly dependent on stress history. SEM measurements of CMOD versus σ were not performed on this specimen.

SAW and SEM Measurements of Crack Opening

Combined SAW and SEM measurements of the changes in crack opening behavior for the previously described constant-amplitude small crack growth experiment in 7075-T651 aluminum are shown in Figs. 6 and 7. In Fig. 6, the measurements of the reflection coefficient versus the applied stress demonstrate that although the stress required to completely open

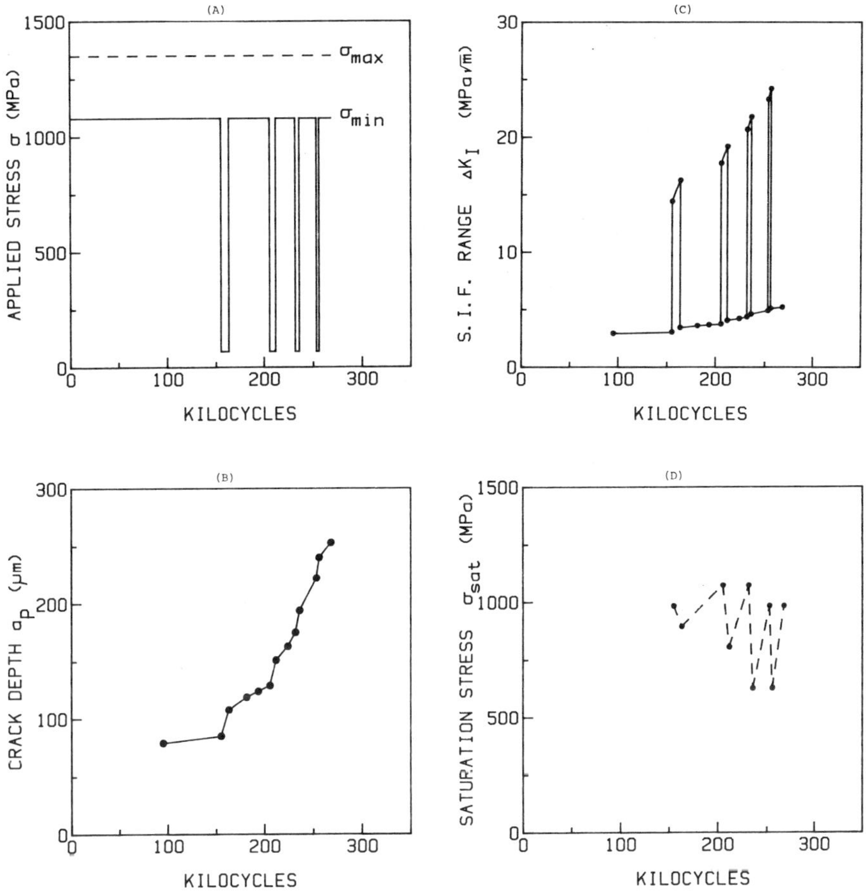

FIG. 5—(a) *Applied stress history,* (b) *crack depth history,* (c) *stress intensity factor range history, and* (d) *opening history for a small surface crack in quenched and tempered 4340 steel specimen.*

the crack does not change appreciably with crack growth, the approximate opening stress at which the acoustic signal emerges from the noise decreases significantly with increasing number of cycles. The corresponding SEM measurements of crack opening in Fig. 7 demonstrate the same trend of decreasing opening stress. Measurements of CMOD versus σ fail to detect the behavior in crack opening below the surface, which the acoustic technique does. These preliminary experiments were designed to investigate the application of various aspects of the SAW-SEM techniques described previously. In the next section questions concerning the relationships between the important physical events which occur during small crack growth experiments and the information provided by the acoustic NDE technique are discussed.

Discussion

The preliminary experiments demonstrate the utility of SAW reflection coefficient measurements performed as a function of applied stress to augment CMOD versus σ measure-

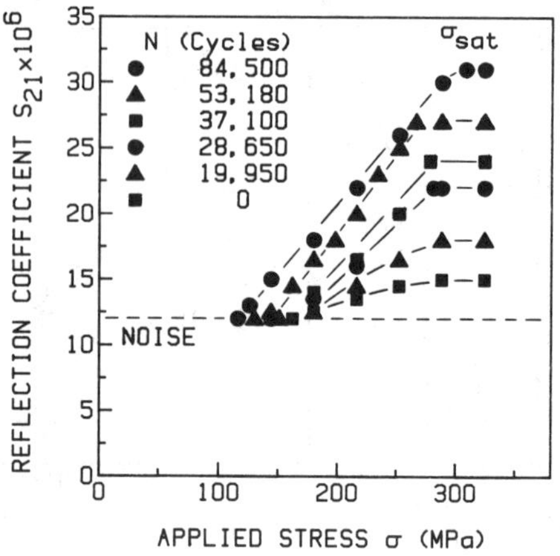

FIG. 6—*Reflection coefficient versus applied stress for a small surface crack in 7075-T651 specimen.*

ments. Comparison of these measurements demonstrates that opening stress determined from the behavior between CMOD and σ does not guarantee that adjacent crack faces of a small surface crack are completely separate, and that a significant increase in stress above the surface opening stress may be necessary to ensure that the crack is completely open. Currently, tests are underway to investigate the relation between acoustically determined opening stress and that determined from CTOD versus σ measurements, to see what, if any, trends exist.

A number of investigators [*20,21*] have documented the efficacy of utilizing the concept of an effective stress range for crack growth to explain the effects of the stress ratio, *R*, and loading sequence on the growth rate of fatigue cracks. However, these studies deal with through-cracked specimens where only the growth rate at the surface is studied. An important question to be asked is, How can the concept of effective stress range be applied to small

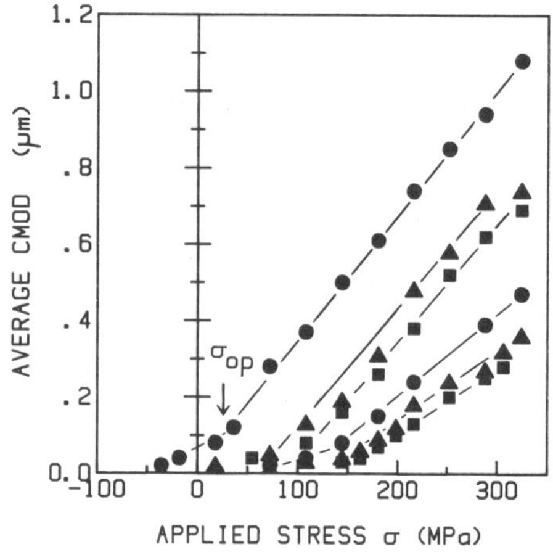

FIG. 7—*Crack mouth opening displacement versus applied stress corresponding to measurements in Fig. 6.*

surface cracks when the opening stress appears to vary along the crack front? These preliminary experiments have demonstrated that the opening stress evaluated at the surface can be considerably less than the stress necessary to completely open the crack at its maximum depth. There is no obvious answer to this question yet. However, if the opening stress determined acoustically proves to be useful in defining an effective stress range for crack growth, the SAW technique would be much faster and more convenient than the SEM techniques.

One of the more important yet least characterized factors involved in a rigorous study of small crack growth is the state of residual stress in the vicinity of growing cracks. The typical experimental approach is to perform stress-relieving heat treatments on specimens combined with careful machining and subsequent metallographic surface treatments (such as polishing) designed to minimize the initial residual stresses. If subsequent cycling is below the elastic limit, the simplifying assumption is made that the specimens begin free of residual stress, and remain free of residual stress during crack growth.

This technique presents two potential difficulties. First, manufactured components will almost always have surface residual stresses, and thus small crack growth data generated with laboratory specimens presumably free of such stresses may not be directly applicable in evaluating small crack behavior in components. Second, even laboratory specimens which start with minimal surface residual stresses may develop shallow residual stresses during the course of fatigue cycling [*22*] which will affect small crack growth behavior. A rigorous study of small crack growth behavior must therefore consider possible changes in the state of residual stress at intervals during fatigue testing.

The most direct technique for quantitative measurements of residual stress involves the use of X-ray diffraction to detect the state of strain in the region within a few micrometres of the surface. In the present work, specimens with surface treatments identical to those used in the fatigue crack growth experiments were examined using X-ray diffraction to determine the initial state of residual stress. It was found that residual stresses were negligible. In future tests the state of residual stress will be monitored periodically during fatigue tests in order to determine if and how the state of residual stress may change due to small-scale yielding.

It is well known that surface treatments such as peening can introduce a compressive state of surface residual stress with a gradient below the surface. High loads applied to notched specimens can produce tensile or compressive residual stresses at the notch, with a gradient below the notch root. Small surface cracks residing in the gradients of these residual stress fields can be expected to exhibit opening behavior which is strongly dependent on depth below the surface. SAW techniques described previously could be used to investigate the influence of residual stresses in order to evaluate the correlations between effective stress range (or stress intensity range) with observed crack growth rates at the surface and in depth.

Recently it has been demonstrated [*23,24*] that time-of-flight measurements of Rayleigh waves may be utilized to infer relative changes in residual stress. This technique utilizes coupling of the SAW phase velocity to the state of stress through the acoustoelastic effect. Although X-ray measurements of the initial state of residual stress are currently necessary at the beginning of a test, relative changes in time of flight of Rayleigh waves could be used to monitor *in situ* possible changes in residual stress during fatigue testing.

Another potential advantage of SAW NDE measurements over X-ray diffraction is that the Rayleigh waves penetrate much more deeply below the surface (that is, a millimetre or so), allowing a characterization of the stresses which interact with the entire crack, instead of just at the surface. In addition, the profile of the residual stress beneath the surface may be inferred by measuring changes in time of flight as a function of the acoustic wavelength, which can be varied. This SAW NDE technique offers potential for monitoring the stress at and beneath the surface during the course of a fatigue test, which is not possible by any other technique.

Conclusions

1. The SAW technique allows the depth of small surface fatigue cracks to be monitored during testing, providing information needed to better understand and characterize growth behavior.
2. A combined SAW-SEM technique gives information about crack opening stress variation during crack growth, which should also prove helpful in evaluating growth behavior.
3. Development of SAW techniques for monitoring surface residual stresses and their variation in depth during testing offers the promise of a useful new experimental tool for fatigue studies.

Acknowledgment

The authors wish to thank the United States Department of Energy, Office of Basic Energy Sciences, Division of Materials Sciences (Dr. S. Wolf) for supporting this research.

APPENDIX

Scattering Theory

A general scattering theory [*10,11*] has been developed which describes the relative signal amplitude of an ultrasonic wave scattered from one transducer to another by a void of arbitrary shape. The geometry considered is shown in Fig. 8. The reflection coefficient, S_{21}, is defined as the amplitude ratio of the maximum reflected signal from the flaw, A_2, received by Transducer 2, to the maximum incident signal, A_2, transmitted by Transducer 1, measured at the terminals of the transducers. The general relationship which defines the reflection coefficient is given by

$$S_{21} = \frac{A_2}{A_1} = \frac{\omega}{4P} \int_{S_f} \sigma_{ij} u_j n_i \, dS \tag{1}$$

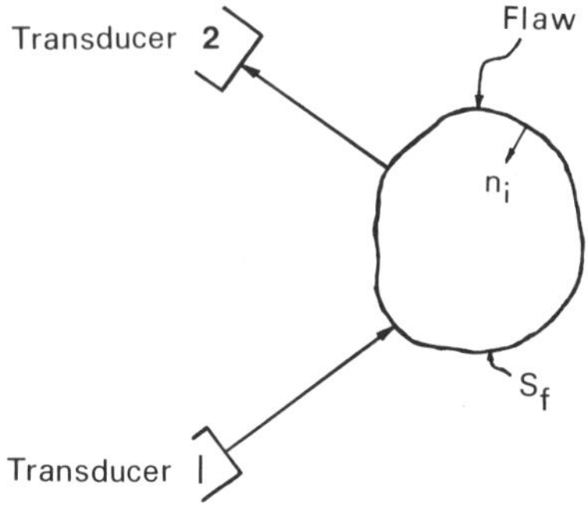

FIG. 8—*Schematic of generalized scattering geometry.*

where

P = power input to the transmitting transducer,
u_j = acoustic displacement field at the flaw surface evaluated from the acoustic stress field which would exist if Transducer 2 were the transmitter,
σ_{ij} = acoustic stress field evaluated in the vicinity of the flaw as if no flaw were present,
n_i = inward directed normal to the flaw surface, S_f, surrounding the scatterer, and
ω = angular frequency of the acoustic wave.

Consider the case of a surface acoustic (Rayleigh) wave propagating in the z-direction normally incident to the crack plane. There are three components of stress associated with the displacement fields of a Rayleigh wave: a longitudinal component, σ_{zz}, parallel to the direction of propagation, a shear component, σ_{xz}, and a normal component, σ_{xx}. The amplitudes of these stresses normalized by the amplitude of the longitudinal component evaluated at the surface, σ_{zz}^0, are plotted as a function of depth below the surface in Fig. 9 in normalized distance units, κx, for $\nu = 1/3$ [25]. Here κ is defined as the wave number $2\pi/\lambda$, where λ is the wavelength. It is apparent from this figure that σ_{xx} and σ_{xz} are zero at the surface, while the longitudinal component of stress σ_{zz} has its maximum value at the surface. Equation 1 can be evaluated for the case of a Rayleigh wave propagating along a surface normal to the crack plane by choosing the frequency of the acoustic wave such that the crack depth, a, will be much smaller than λ. In the long-wavelength limit with $a << \lambda$, the crack resides in a SAW stress field with σ_{zz} approximately uniform over the crack surface, S_C, and with the other components of stress approximately equal to zero. For these conditions, Eq 1 reduces to

$$S_{21} = \frac{\omega}{4P} \int_{S_C} u_z \sigma_{zz} \, dS \qquad (2)$$

The integral of the displacement times stress evaluated over the surface of a crack can be expressed as a line integral of the stress intensity factor squared times a factor ρ evaluated

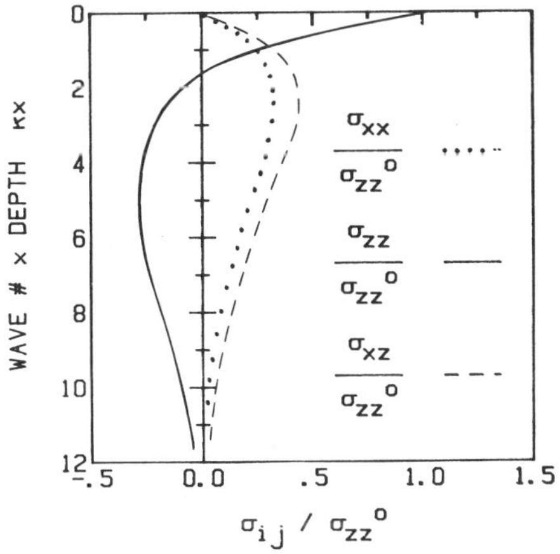

FIG. 9—*Components of stress in a surface acoustic wave normalized to the value of longitudinal stress at the surface.*

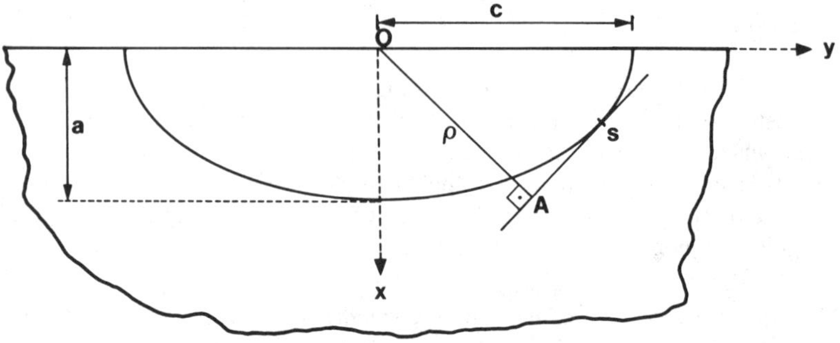

FIG. 10—*Schematic of semi-elliptical surface crack.*

around the crack front, *C* [*26*]. Utilizing this result the reflection coefficient may be expressed as

$$S_{21} = \frac{\omega(1 - \nu^2)}{6PE} \int_C \rho K_I^2 \, ds \tag{3}$$

In Fig. 10 the factor ρ is depicted as the distance between the origin and the tangent line to the crack front at the point of inspection, *s*. K_I is the Mode I stress intensity factor associated with the SAW acoustic stress field σ_{zz}.

In order to evaluate Eq 3, the distribution of K_I around the crack tip due to the SAW stress field must be known. For the case of a surface crack in a finite plate in pure bending for a material with Poisson's ratio, ν, of 0.3 and 1/3, respectively, the distribution in K_I has been evaluated numerically [*19,27*]. An approximation technique has been developed which facilitates the adaptation of these numerical results to the problem of a small surface crack in a SAW stress field, with the limitation that the wavelength normalized crack depth, κ*a*, must be less than unity [*28*]. This technique has recently been used [*14*] to evaluate the reflection coefficient for a material with ν = 0.3 using K_I from Ref *19.* In Fig. 11, the

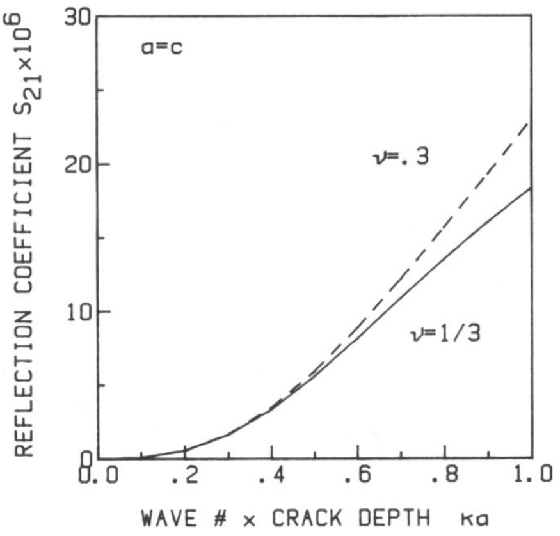

FIG. 11—*Reflection coefficient versus normalized crack depth of half-penny-shaped cracks.*

reflection coefficient, S_{21}, is plotted versus the normalized crack depth, κa, for half-penny-shaped cracks with $a = c$. In addition, the result for a material with $\nu = 1/3$ using K_I from Ref *27* is included for comparison. Similar calculations may be performed to determine the relationship between S_{21} and κa at different constant values of the crack aspect ratio, a/c, or alternatively at constant values of the wavelength normalized half-crack length at the surface, κc. Thus, measurement of reflection coefficient and surface crack length allows depth to be inferred.

References

[*1*] Klima, S. J., Lesco, D. J., and Freche, J. C., *Experimental Mechanics,* Vol. 6, March 1966, pp. 154–161.
[*2*] Teller, C. M., Barton, J. R., Matzkanin, G. A., Kusenberger, F. N., and Beissner, R. E., *Journal of Engineering Materials and Technology,* Vol. 102, Jan. 1980, pp. 50–55.
[*3*] Thompson, R. B., Buck, O., and Thompson, D. O., *Journal of the Acoustical Society of America,* Vol. 59, No. 5, 1976, pp. 1087–1094.
[*4*] Morris, W. L., *Metallurgical Transactions,* Vol. 11A, No. 7, July 1980, pp. 1117–1123.
[*5*] Buck, O., Fiedler, C. J., Reed, L. K., Lakin, K. M., and Thompson, R. B., *Review of Progress in Quantitative Nondestructive Evaluation,* Vol. 3A, D. O. Thompson and D. E. Chimenti, Eds., Plenum Press, New York, 1984, pp. 199–206.
[*6*] Achenbach, J. D. and Norris, A. N., *Review of Progress in Quantitative Nondestructive Evaluation,* Vol. 3A, D. O. Thompson and D. E. Chimenti, Eds., Plenum Press, New York, 1984, pp. 163–174.
[*7*] Thompson, R. B. and Fiedler, C. J., *Review of Progress in Quantitative Nondestructive Evaluation,* Vol. 3A, D. O. Thompson and D. E. Chimenti, Eds., Plenum Press, New York, 1984, pp. 207–216.
[*8*] Visscher, W., *Review of Progress in Quantitative Nondestructive Evaluation,* Vol. 3A, D. O. Thompson and D. E. Chimenti, Eds., Plenum Press, New York, 1984, pp. 187–198.
[*9*] Resch, M. T., "Non-Destructive Evaluation of Small Surface Cracks Using Surface Acoustic Waves," Ph.D. Dissertation, Stanford University, Stanford, CA, 1982.
[*10*] Kino, G. S., *Journal of Applied Physics,* Vol. 49, No. 6, 1978, pp. 3190–3199.
[*11*] Auld, B. A., *Wave Motion,* Vol. 1, No. 1, 1979, pp. 3–10.
[*12*] Fraser, J. D., Khuri-Yakub, B. T., and Kino, G. S., *Applied Physics Letters,* Vol. 32, No. 11, 1978, pp. 698–700.
[*13*] Resch, M. T., London, B. D., Ramusat, G. F., Yuce, H. H., Nelson, D. V., and Shyne, J. C., *Review of Progress in Quantitative Nondestructive Evaluation,* Vol. 3A, D. O. Thompson and D. E. Chimenti, Eds., Plenum Press, New York, 1984, pp. 217–228.
[*14*] Yuce, H. H., Nelson, D. V., and Resch, M. T., *Review of Progress in Quantitative Nondestructive Evaluation,* Vol. 4A, D. O. Thompson and D. E. Chimenti, Eds., Plenum Press, New York, 1985, pp. 103–113.
[*15*] Resch, M. T., Khuri-Yakub, B. T., Kino, G. S., and Shyne, J. C., *Applied Physics Letters,* Vol. 34, No. 3, 1979, pp. 182–184.
[*16*] Resch, M. T., Nelson, D. V., Yuce, H. H., and Ramusat, G. F., *Journal of Nondestructive Evaluation,* Vol. 5, No. 1, 1985, pp. 1–7.
[*17*] Morris, W. L., James, M. R., and Buck, O., *Metallurgical Transactions,* Vol. 12A, No. 1, 1981, pp. 57–63.
[*18*] Kobayashi, A. S. in *Proceedings,* Second International Conference on Mechanical Behavior of Solids, 1976, pp. 1073–1077.
[*19*] Newman, J. C. and Raju, I. S., "Analyses of Surface Cracks in Finite Plates under Tension or Bending Loads," NASA Technical Paper 1578, National Aeronautics and Space Administration, Washington, DC, 1979.
[*20*] Suresh, S., Zamiski, G. F., and Ritchie, R. O., *Metallurgical Transactions,* Vol. 12A, No. 8, 1981, pp. 1435–1443.
[*21*] Newman, J. C. in *Design of Fatigue and Fracture Resistant Structures, ASTM STP 761,* American Society for Testing and Materials, Philadelphia, 1982, pp. 255–277.
[*22*] Field, J. L., Behnaz, F., and Pangborn, R. N. in *Fatigue Mechanisms: Advances in Quantitative Measurement of Physical Damage, ASTM STP 811,* American Society for Testing and Materials, Philadelphia, 1983, pp. 71–94.

[*23*] Husson, D. E., "Measurement of Stress with Acoustic Waves," Ph.D. Dissertation, Stanford University, Stanford, CA, 1983.
[*24*] Liang, K., Bennett, S. D., Khuri-Yakub, B. T., and Kino, G. S., *Applied Physics Letters,* Vol. 41, No. 12, 1982, pp. 1124–1125.
[*25*] Viktorov, I. A., *Rayleigh and Lamb Waves,* Plenum Press, New York, 1967.
[*26*] Knowles, J. K. and Sternberg, E., *Archives for Rational Mechanics and Analysis,* Vol. 44, 1971–1972, pp. 187–211.
[*27*] Smith, F. W., Emery, A. F., and Kobayashi, A. S., *Journal of Applied Mechanics,* Vol. 34, No. 4, 1967, pp. 953–959.
[*28*] Tien, J., Khuri-Yakub, B. T., and Kino, G. S. in *Proceedings,* DARPA/AFML Review of Progress in Quantitative Nondestructive Evaluation, Report No. AFWAL-TR-80-4078, 1980, pp. 671–677.

Robert O. Ritchie,[1] Eliot Zaiken,[1] and Anders F. Blom[2]

Is the Concept of a Fatigue Threshold Meaningful in the Presence of Compression Cycles?

REFERENCE: Ritchie, R. O., Zaiken, E., and Blom, A. F., **"Is the Concept of a Fatigue Threshold Meaningful in the Presence of Compression Cycles?"** *Basic Questions in Fatigue: Volume I, ASTM STP 924,* J. T. Fong and R. J. Fields, Eds., American Society for Testing and Materials, Philadelphia, 1988, pp. 337–356.

ABSTRACT: The fatigue threshold ΔK_{TH}, represents the nominal stress intensity range below which fatigue cracks are presumed not to propagate. Its magnitude, however, has been linked to the degree of crack closure. Since the extent of closure may vary, at fixed nominal ΔK, with such variables as crack length and loading history, the concept of a unique threshold as a material constant for a specific material/microstructure/environment system may become questionable. Of particular importance here for aerospace applications is whether crack growth remains dormant below ΔK_{TH} in the presence of variable-amplitude loading, specifically during the occurrence of single compression cycles. In the present paper, a series of critical experiments on the effect of single compression overloads on the behavior of fatigue cracks arrested at ΔK_{TH} is examined. Based on tests on I/M 7150-T6, 7150-T7, and 7475-T7351 aluminum alloys and 17-4 PH stainless steel, it was found that the occurrence of a single compressive cycle, of magnitude five times the maximum tensile load, caused immediate propagation of previously arrested threshold cracks at ΔK_{TH}. Such observations are interpreted in terms of the degree of compressive residual stress in the cyclic plastic zone ahead of the crack and a measured reduction in crack closure following the compression cycle. The results serve to confirm the dependence of the threshold on crack closure and furthermore to illustrate the danger of utilizing nominal ΔK_{TH} threshold values in damage-tolerant designs to predict fatigue life.

KEY WORDS: fatigue, crack growth, variable-amplitude loading, compression cycles, fatigue threshold, crack closure, microstructure

Over the past 20 years, it has become customary in fracture mechanics methodology to characterize the rate of propagation of fatigue cracks (da/dN) in terms of experimentally determined relationships involving primarily the linear elastic stress intensity range, ΔK [*1*]. Such relationships form the basis of damage-tolerant lifetime predictions through integration over a range of flaw sizes from the assumed largest undetected defect to the critical crack length. In general, fatigue behavior is bounded at very high growth rates (for example, above $\sim 10^{-4}$ m/cycle) by the onset of instability or catastrophic failure, which is characterized by the limit load or fracture toughness. Conversely, at very low growth rates (for example, approaching 10^{-11} m/cycle), it is bounded by the threshold stress intensity range, ΔK_{TH}, representing the nominal value of ΔK below which crack growth is presumed dormant or is experimentally undetectable [2]. Of these limits, the threshold perhaps has attracted the most attention recently as lifetimes clearly will be dominated by behavior at the lower growth

[1] Professor of materials science and research student, respectively, Department of Materials Science and Mineral Engineering, University of California, Berkeley, CA 94720. Coauthor Zaiken is currently with Shiley Inc., Irvine, CA.
[2] Head of Fatigue and Fracture Section, Structures Department, The Aeronautical Research Institute of Sweden, P.O. Box 11021, S-161 11, Bromma, Sweden.

rates and, furthermore, the application of threshold values offers the potential of design for infinite life.

Near-threshold crack growth behavior and the value of the fatigue threshold have been shown to be markedly dependent on microstructural and environmental factors, and to be sensitive to mechanical factors such as the load ratio (that is, the ratio, R, of minimum to maximum load) (for example, Refs *2–6*). As such, it is tempting to consider that the threshold has a unique value, at fixed R, for a given material, microstructure, environmental combination, such that it can be regarded as a "material constant." However, recently it has become clear that the physical rationale behind the existence of experimentally obtained fatigue thresholds, particularly at low load ratios, is intimately associated with the phenomenon of crack closure, arising from interference between mating fracture surfaces principally in the immediate wake of the crack tip [*6–13*]. For example, experiments involving the careful removal of material left in the wake of fatigue cracks arrested at ΔK_{TH} clearly show recommencement of crack growth *at the threshold,* concomitant with a measured reduction in closure [*10–13*]. The closure acts to reduce near-tip driving forces for crack advance from nominal levels, based on global measurements of applied load and crack size, for example, $\Delta K = K_{max} - K_{min}$, to lower effective levels actually experienced at the crack tip, for example, $\Delta K_{eff} = K_{max} - K_{cl}$, where K_{cl} is the stress intensity on first contact of the crack surfaces during unloading [*14*]. Due to its many origins (reviewed in Fig. 1), including cyclic plasticity [*14*], which predominates under plane stress conditions, and crack surface corrosion deposits [*15–17*] and irregular crack path morphologies [*18–20*] (coupled with significant Mode II crack tip displacements), which predominate under plane strain conditions, the magnitude of the closure may depend, *at a fixed nominal* ΔK, on such variables as crack size and loading history [*6*]. Thus, due to this inherent variability in closure, the uniqueness of fatigue thresholds must be brought into question. In fact, it is well known that short cracks, when they are small compared with the scale of microstructure or local plasticity or simply when they are physically small (that is, less than ~0.5 mm), can initiate and grow at nominal ΔK levels *well below* ΔK_{TH}, again consistent with reduced closure in the more limited wake behind the crack tip (for example, Ref *21*).

Of equal concern to questions surrounding the meaning of the fatigue threshold in the presence of small flaws is the significance of the threshold during variable-amplitude loading. This is clearly of great importance in aerospace applications where defect-tolerant fatigue design invariably is used and where loading spectra generally are variable amplitude. Although much information exists on the effect of loading histories involving cycles in tension, such as single or multiple tensile overloads and block loading sequences (reviewed recently in Ref *22*), comparatively few results are available on the effect of compression overloads [*23–31*]. This perhaps is surprising as such excursions can occur frequently under service conditions, such as during air-to-ground cycles during aircraft flight loading histories. However, although the importance of compressive loads in the process of crack initiation is well known, their effect on crack propagation behavior, at least at intermediate to high ΔK levels, has long been considered minimal, except where they follow tensile overloads [*23*]. This has been rationalized in terms of the crack being closed during the compressive portion of the cycle, implying a stress intensity of zero at the tip. Very recent studies involving periodic compression overloads at lower growth rates, conversely, have indicated significantly accelerated growth rates and a reduction in measured thresholds [*27–31*]. Although precise explanations are uncertain, such results are consistent with experimentally monitored changes in crack closure [*31*] and analyses showing a reduction in the residual compressive stresses within the cycle plastic zone at the crack tip [*30*].

In light of the prior studies described above, the objective of the present work is to pose the question, *"Is the concept of a fatigue threshold still meaningful in the presence of loading*

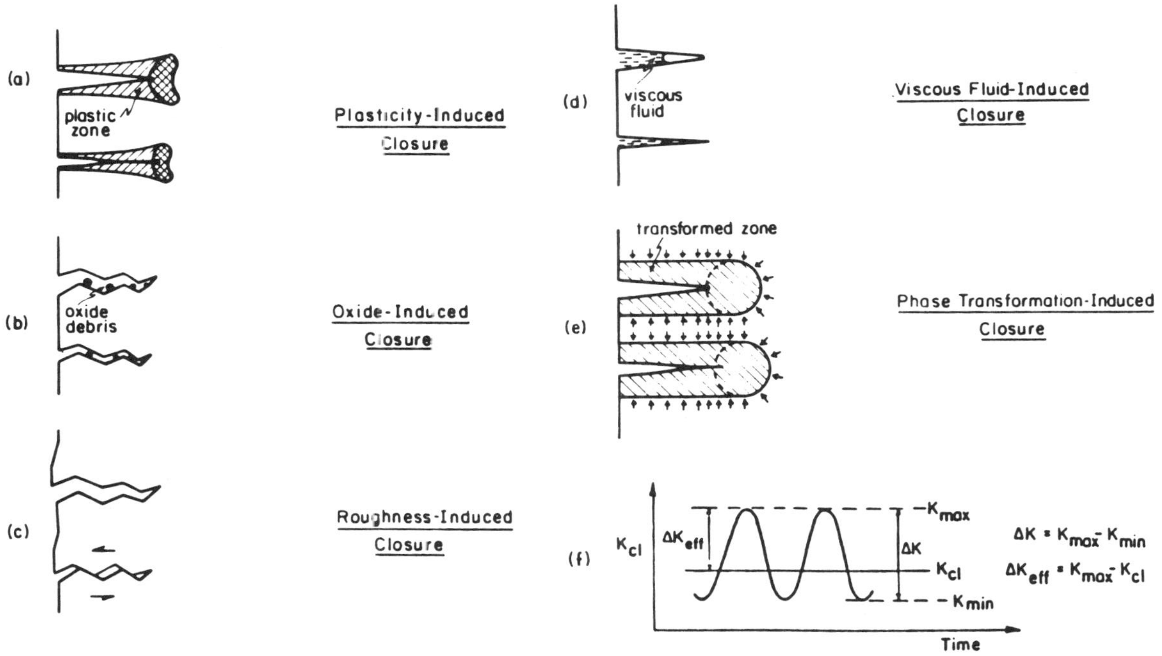

FIG. 1—*Schematic illustration of the primary mechanisms of fatigue crack closure and corresponding nomenclature* [6].

TABLE 1—*Nominal chemical compositions, weight %.*

	Si	C	Cu	Mn	P	S	Ni	Nb	Mg	Zn	Ti	Zr	Cr	Al	Fe
7150 aluminum	0.07	...	2.10	...	...	...	...	...	2.16	6.16	0.02	0.13	...	balance	0.11
7475 aluminum	0.05	...	1.60	...	...	...	...	...	2.00	5.50	...	...	0.20	balance	...
17-4 PH stainless steel	1.00	0.05	4.00	1.00	0.025	0.025	4.00	0.03	...	...	...	...	16.5	...	balance

spectra which contain compressive overload cycles?" Since this question is of obvious importance to aerospace applications, experiments are based on a PH stainless steel and a series of aluminum alloys, representing the current generation of commercial airframe materials. The experiments involve the application of single compression cycles on fatigue cracks arrested at the threshold, ΔK_{TH}, and the close monitoring of any subsequent changes in crack extension and crack closure behavior. The results clearly show a recommencement in crack growth at ΔK_{TH}, following the compression cycle, associated with a reduction in closure—observations which strongly question the use of fatigue thresholds in design or service applications involving variable amplitude loading.

Experimental Procedures

Materials

Experiments were conducted on two conventionally-cast I/M 7150 and 7475 aluminum alloys, heat-treated over a range of aging conditions, and a 17-4 precipitation hardened (PH) stainless steel. Compositions, in weight %, are listed in Table 1. The 7150 alloy was supplied as 25-mm-thick plate in the solution-treated and 2% stretched condition. Blanks, machined from quarter and three-quarter plate thickness locations, were tempered to yield underaged, peak aged, and overaged microstructures, hereafter referred to as 7150-UA, 7150-T6, and 7150-T7, respectively. Underaged structures were hardened by small coherent GP zones, roughly 4 to 8 nm in diameter, compared to hardening by semicoherent η' precipitates in T6 structures. T7 structures, conversely, showed evidence of coarsened η' in the matrix with primarily incoherent η in both matrix and grain boundaries. Grains were pancake-shaped, that is, roughly 15 by 5 μm, with their length aligned along the rolling direction. The 7475 alloy was cold-formed, solution-treated, and stabilized for optimum resistance to stress corrosion cracking, with associated stress relieving by stretching. It is referred to hereafter by its temper designation 7475-T7351. Its grain structure similarly was pancake-shaped, yet far coarser (that is, ~100 by 30 μm) than in the 7150 alloy. Details of the specific heat-treatment schedules, together with the resulting room temperature mechanical properties, are given in Tables 2 and 3, respectively.

Stainless steel specimens were machined from 100-mm-diameter bar and heat-treated to achieve the mechanical properties shown in Table 3. The microstructure consisted of tempered martensite with a prior austenite grain size of 30 to 40 μm.

Fatigue Tests

Fatigue crack propagation tests were performed on 6.4-mm-thick compact (CT) specimens of 7150, 6-mm-thick CT specimens (manufactured from 10-mm-thick plate) of 7475, and 10-

TABLE 2—*Heat treatments utilized for tests on 7150 and 7475 aluminum alloys.*

Alloy	Temper	Heat Treatment
7150-UA	underaged	ST[a] + 1½ h at 121°C
7150-T6	peak aged	ST[a] + 100 h at 121°C
7150-T7	overaged	ST[a] + 24 h at 121°C + 40 h at 163°C
7475-T7351	overaged	ST[a] + 24 h at 121°C + 21 h at 163°C

[a] ST = solution treated, quenched and stretched 2% (W51 condition).

TABLE 3—*Room temperature mechanical properties of 7150 and 7475 aluminum alloys and 17-4 PH stainless steel.*

Alloy	Yield Strength, MPa	Ultimate Tensile Strength, MPa	Elongation,[a] %	Reduction in Area, %	Hardening Exponent	Fracture Toughness, K_Q, MPa $\sqrt{m}$
Aluminum, 7150-UA	371	485	6.8	12.1	0.055	...
Aluminum, 7150-T6	404	480	6.0	10.3	0.046	...
Aluminum, 7150-T7	372	478	7.1	12.5	0.058	...
Aluminum, 7475-T7351	410	495	12.5	...	...	70
Stainless steel, 17-4 PH	1105	1135	14.5	...	...	120

[a] On 32-mm gage length.

mm-thick CT specimens of 17-4 PH steel; the width of all specimen types was 50 to 60 mm. Tests were conducted under load control on electro-servo-hydraulic testing machines (Instron, MTS, and Schenk) at frequencies between 40 and 80 Hz in room temperature laboratory air with load ratios between 0.05 and 0.50. D-c electrical potential, elastic compliance, and visual observation techniques were used to monitor crack length in the compact test specimens. Growth rates (da/dN) were then determined numerically using standard finite difference and incremental polynomial procedures. Determination of crack closure loads to define K_{cl} values was achieved through *in situ* unloading compliance procedures, involving both crack opening displacement and back-face strain measurements [*32*], with an estimated relative error of less than 10%. Full experimental details of the crack extension and closure measurements have been described elsewhere [*2,32,33*].

Compression overload experiments were performed by first determining the long crack threshold under manual load shedding (decreasing ΔK) conditions. As described elsewhere [*2*], the value of the threshold was operationally defined as the highest stress intensity range giving growth rates less than 10^{-11} m/cycle, with the total length of the threshold crack (including initial notch) being of the order of 23 to 25 mm. Single (spike) compression cycles, of magnitude 1 to 5 times the maximum tensile load in the cycle (termed 100% to 500% compression overloads), were then applied to arrested cracks at ΔK_{TH} and any subsequent changes in crack extension and closure behavior were carefully monitored (Fig. 2).

Fracture surfaces were examined by several techniques. In addition to fractographic studies using scanning electron microscopy, the extent of corrosion debris was quantified using Ar^+ sputtering in Auger spectrometry (for example, Ref *9*) and the degree of fracture surface roughness was assessed in terms of the lineal roughness parameter, that is, the ratio of total crack length (a_T) to projected crack length on the plane of maximum tensile stress (a_p). Values of a_T were obtained by taking a series of scanning electron micrographs along a metallographic section through the crack and digitizing the image to derive a total length of crack path.

Results and Discussion

Constant-Amplitude Results

Constant-amplitude fatigue crack propagation results (at R = 0.05 to 0.10) for the four aluminum microstructures, plotted in terms of crack growth rate, da/dN, as a function of the nominal stress intensity range, ΔK, are shown in Fig. 3. For the 7150 alloy, behavior is similar in the UA, T6, and T7 microstructures above ~10^{-9} m/cycle, whereas, at lower

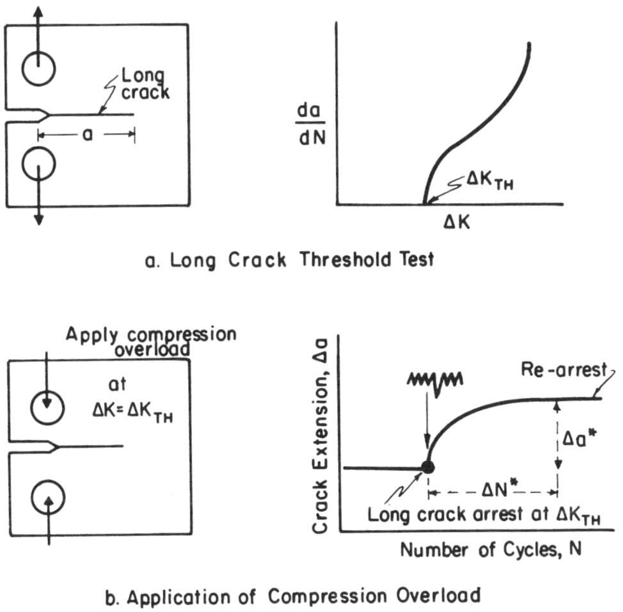

FIG. 2—*Schematic illustration of experimental procedures for applying single compression overloads.*

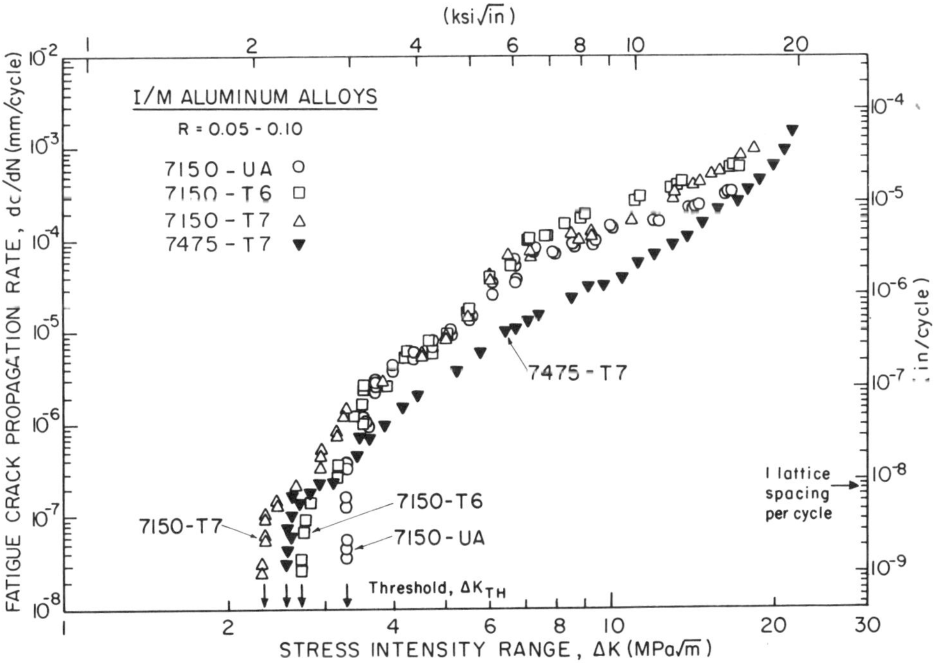

FIG. 3—*Variation in steady-state (constant amplitude) fatigue crack growth rate* (da/dN) *with nominal stress intensity range* (ΔK) *for I/M 7150 and 7475 aluminum alloys tested at* R = *0.05–0.10 in controlled moist air.*

TABLE 4—*Threshold data for 7150 and 7475 aluminum alloys and 17-4 PH stainless steel at* R = *0.05 to 0.10.*

Alloy	ΔK_{TH}, MPa $\sqrt{m}$	K_{cl}/K_{max}	Plastic Zone Size[a] Cyclic, μm	Plastic Zone Size[a] Maximum, μm	Oxide Thickness, nm	Lineal Roughness,[b] (a_T/a_p)
Aluminum, 7150-UA	3.05 to 3.31	0.83	2.9	14.5	~3	1.26
Aluminum, 7150-T6	2.44 to 2.94	0.82	1.8	8.8	~3	1.21
Aluminum, 7150-T7	2.17 to 2.33	0.74	1.5	7.4	~3	1.06
Aluminum, 7475-T7351	2.30 to 2.50	0.65	1.4	6.0	...	1.02
Stainless steel, 17-4 PH	5.30 to 5.75	0.76	1.0	4.4	...	...

[a] Computed from $r_\Delta = \frac{1}{2\pi}(\Delta K/2\sigma_y)^2$ and $r_{max} = \frac{1}{2\pi}(K_{max}/\sigma_y)^2$ where r_Δ and r_{max} are the cyclic and maximum plastic zone sizes, respectively, and σ_y is the yield strength [*34*].
[b] Ratio of total crack length (a_T) to projected length (a_P) on plane of maximum tensile stress.

growth rates, threshold ΔK_{TH} values are lower and crack propagation rates higher with increasing aging. Compared to the correspondingly overaged 7150-T7 alloy, growth rates in 7475-T7351 are somewhat slower over the entire range of ΔK. Threshold stress intensity values, together with associated data on crack closure, crack surface oxide thickness, and fracture surface roughness, are listed in Table 4.

As discussed in detail elsewhere [*35*], the observed trend of decreasing threshold ΔK_{TH} values with increased aging in the aluminum alloys is accompanied by a reduced contribution from crack closure. This is shown for 7150 in Fig. 4 where experimental closure data, in the

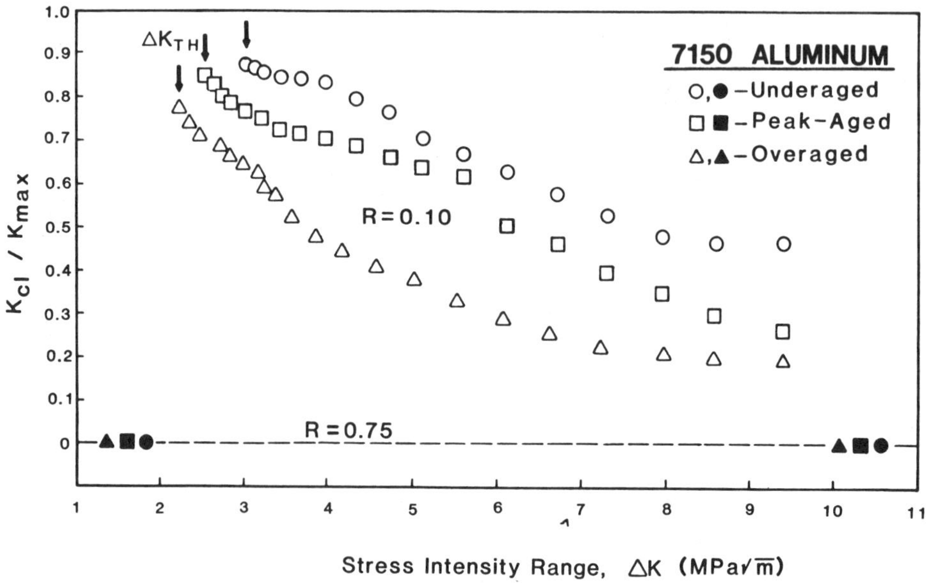

FIG. 4—*Variation in crack closure, in terms of the ratio of closure to maximum stress intensity,* K_{cl}/K_{max} *as a function of* ΔK, *for 7150 aluminum alloy in the underaged (UA), peak-aged (T6), and overaged (T7) conditions. Data at* R = *0.10 correspond to constant-amplitude growth rate results for 7150, shown in Fig. 3* [35].

form of K_{cl}/K_{max} values, are plotted as a function of ΔK. Although the degree of closure at low R increases in all structures as the threshold is approached, that is, $K_{cl}/K_{max} \rightarrow 1$ as $\Delta K \rightarrow \Delta K_{TH}$, overall levels are highest in the underaged condition. Such behavior does not appear to rely on the mechanism of oxide-induced closure, as is commonly observed in lower-strength steels [*15–17*], since Auger spectroscopy measurements [*35*] of the crack surface corrosion deposits at ΔK_{TH} show little accumulation or variation with aging treatment (Table 4). However, as reported for a number of precipitation-hardened aluminum alloys [*35–39*], the coherent particle-hardening mechanism associated with underaged microstructures, which promotes planar slip, generally leads to rougher fracture surfaces (Fig. 5), often termed faceted or crystallographic, from the deflection of the crack path at grain boundaries. This in turn results in a reduction in the near-tip crack driving force, that is, lower ΔK_{eff} values, both from crack deflection mechanics [*40*] and an increased contribution from roughness-induced closure. The association of higher thresholds and increased closure with rougher crack path morphologies can be seen for the present alloys in Table 4, where the degree of fracture surface roughness is assessed in terms of the lineal roughness parameter (a_T/a_p).

Corresponding crack growth rate and closure data for 17-4 PH stainless steel (at $R = 0.05$ and 0.50) are shown in Figs. 6 and 7, respectively.

Variable-Amplitude Results

Following arrest at the threshold, cracks in each microstructure were subjected to single compression overloads and subsequent crack extension and closure behavior monitored under constant $\Delta K = \Delta K_{TH}$ conditions ($R = 0.05$ to 0.10). Small compressive loads, of order 1 to 3 times the maximum tensile load, were found to have little detectable effect. However, the application of single 500% compression overloads led to immediate growth of the arrested cracks in all microstructures, with initial growth rates approaching 10^{-8} m/cycle, even though the nominal ΔK remained at threshold. This is shown in Figs. 8 and 9 in terms of data giving crack extension, Δa, versus number of cycles, ΔN, following the overload. The recommencement of growth following the compression cycle in every case occurred in conjunction with a measured reduction in crack closure. Such closure values, together with the estimated increase in ΔK_{eff}, are listed in Table 5 for the aluminum alloys. No effect was seen for similar experiments at high load ratios, that is, at $R = 0.50$ (Fig. 9) and $R = 0.75$ [see also Ref *41*]. After the initial acceleration, however, post-overload crack growth rates were observed to progressively slow down until rearrest occurred within 5×10^5 cycles (Figs. 8 and 9). This behavior was associated with a measured redevelopment in closure with crack extension, until crack arrest at a K_{cl} value approximately equal to the original (pre-overload) threshold level. The variation in K_{cl}, and hence ΔK_{eff}, for this sequence of events is shown schematically in Fig. 10.

Rearrest was seen to occur over increasingly larger crack extensions, Δa^*, in the more heavily aged microstructures (Fig. 8). Whereas in the underaged 7150 aluminum microstructure, cracks rearrested within ~60 μm, it required ~230 μm of further crack extension to rearrest threshold cracks in the 7475-T7 microstructure (Table 5). Plotted in terms of ΔK_{eff}, however, both constant-amplitude and post-overload growth rate data for all aluminum microstructures fall within a single band (Fig. 11). The scatter within this band is not small, presumably because of intrinsic differences in the resistance to crack extension in the four microstructures and uncertainties in the closure measurements.

As modeled recently by Marissen et al. [*30*], the increase in near-threshold crack growth rates following the application of a compressive load is consistent with a reduction in the residual compressive stresses in the reversed plastic zone directly ahead of the crack tip. Depending upon the relative magnitudes of the applied compressive stress, σ_c, and the yield

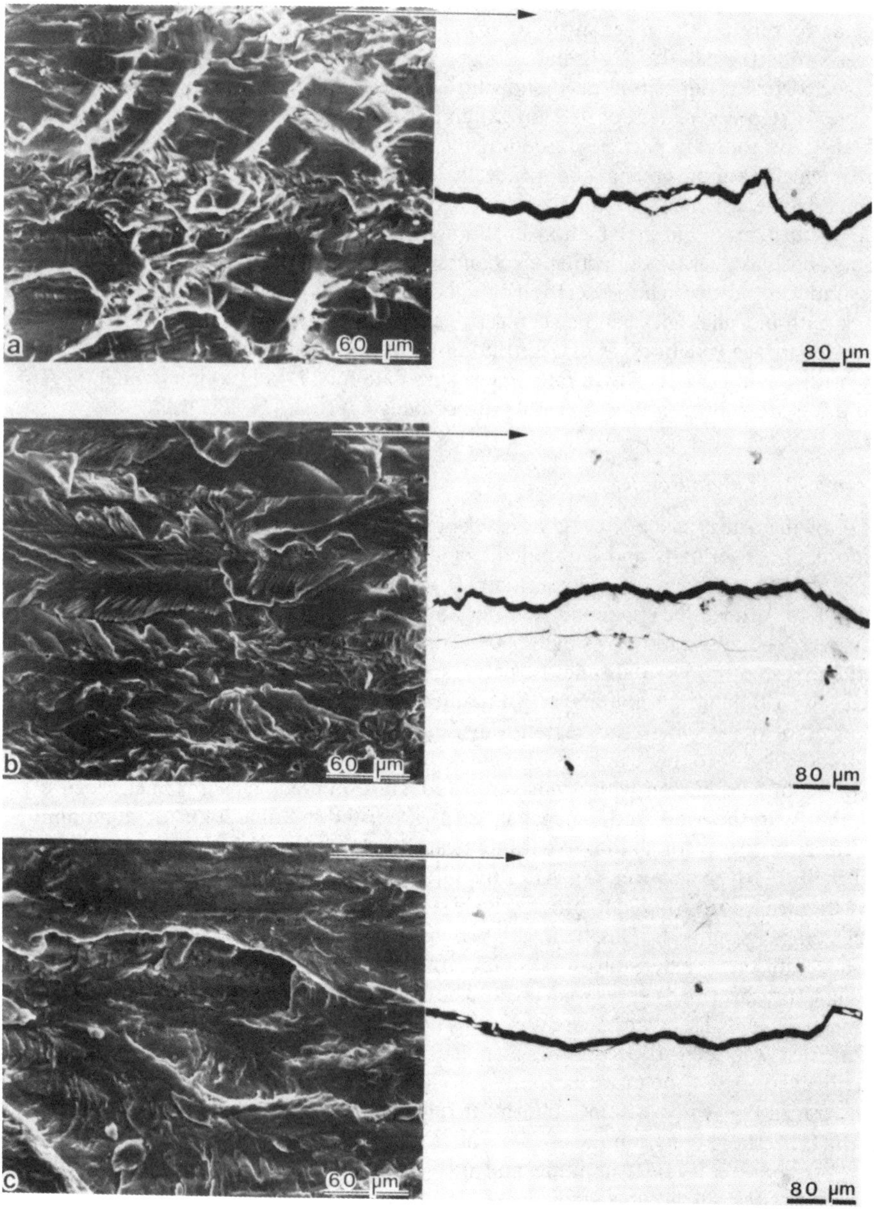

FIG. 5—*Fracture surface and crack path morphologies of near-threshold fatigue crack growth in 7150 aluminum alloy in the* (a) *underaged* (*UA*), (b) *peak-aged* (*T6*), *and* (c) *overaged* (*T7*) *conditions* [35].

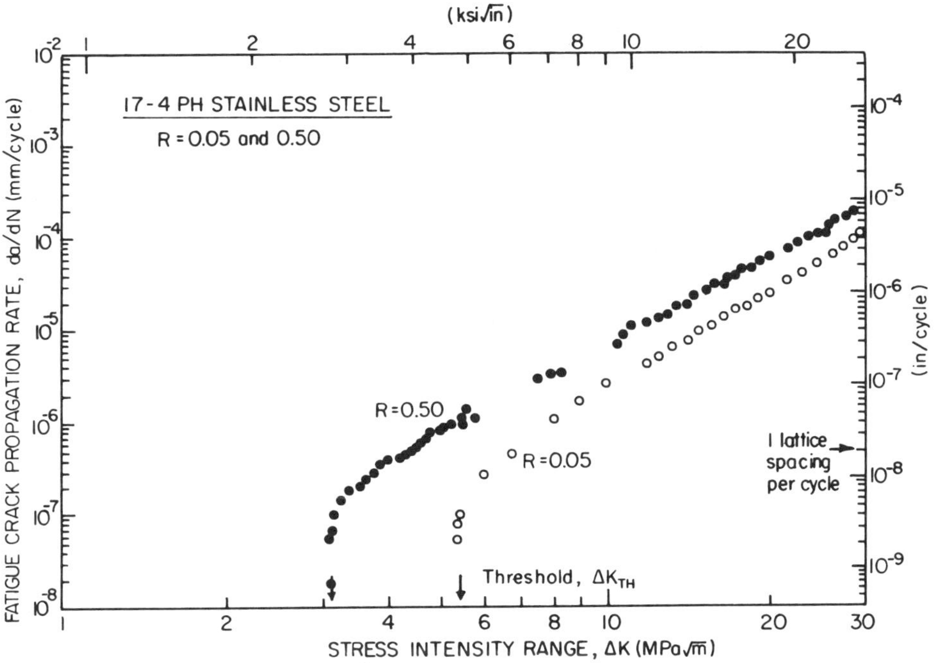

FIG. 6—*Variation in steady-state (constant amplitude) fatigue crack growth rate* (da/dN) *with nominal* ΔK *for 17-4 PH stainless steel tested at* R = *0.05 and 0.50 in controlled moist air.*

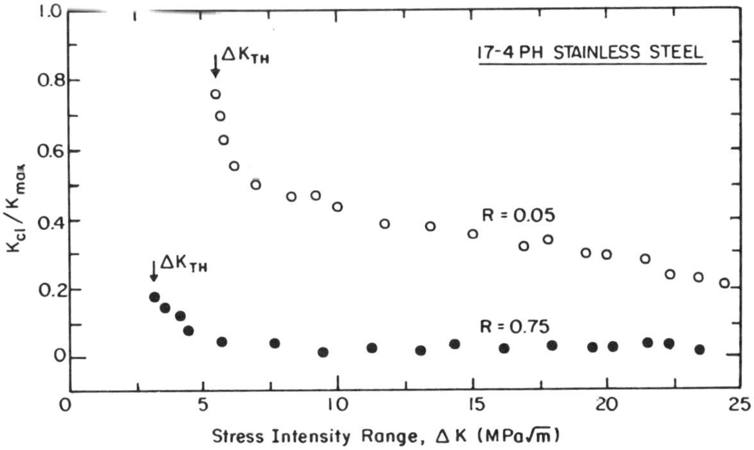

FIG. 7—*Variation in crack closure, in terms of the ratio* K_{cl}/K_{max} *as a function of* ΔK, *for 17-4 PH stainless steel at* R = *0.05 and 0.75. Data at* R = *0.05 correspond to constant-amplitude growth rate results, shown in Fig. 6* [33].

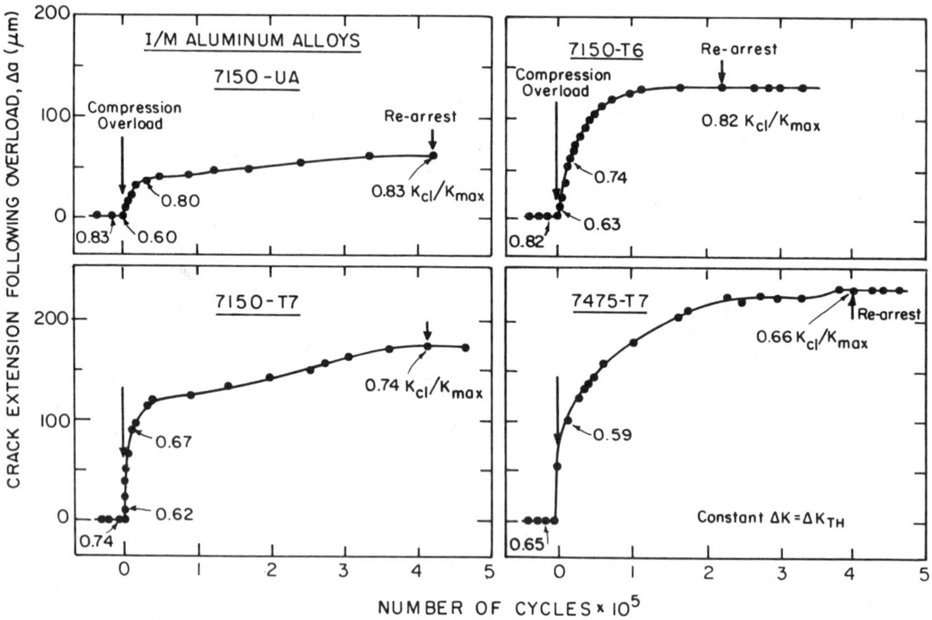

FIG. 8—*Fatigue crack extension* (Δa) *as a function of number of cycles following the application of 500% compression overloads on arrested threshold cracks at* R = *0.05 to 0.10. Data obtained under constant* $\Delta K = \Delta K_{TH}$ *for 7150 and 7475 aluminum alloys.* K_{cl}/K_{max} *closure data are listed along each curve.*

stress in compression, σ_{yc}, such changes can have a strong effect on the subsequent crack opening level. Based on an analysis of plasticity-induced closure, these authors proposed that [*30*]

$$K_{cl}/K_{max} = c[1 - (\sigma_c/\sigma_{yc})^2] \quad (1)$$

where *c* is approximately 0.5. Although the present closure data, which show a reduction

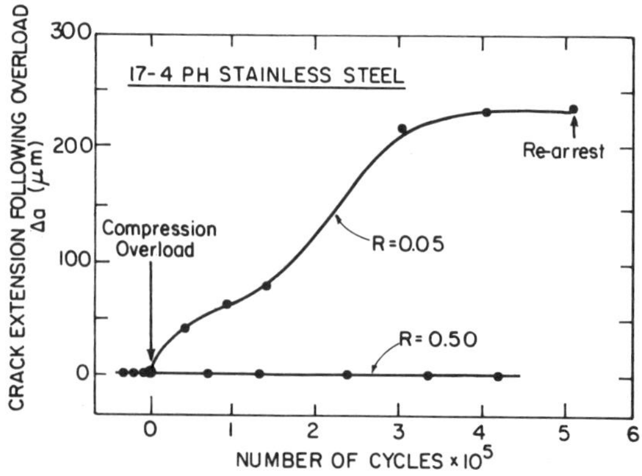

FIG. 9—*Fatigue crack extension* (Δa) *as a function of number of cycles following the application of 500% compression overloads on arrested threshold cracks at* R = *0.05 and 0.50. Data obtained under constant* $\Delta K = \Delta K_{TH}$ *for 17-4 PH stainless steel.*

TABLE 5—*Crack extension and closure data for aluminum alloys at ΔK_{TH} before and after the application of a 500% compression overload.*

Alloy	ΔK_{TH}, MPa $\sqrt{m}$	At Initial Arrest		Following Overload		At Rearrest		Crack Extension to Rearrest	Cycles to Rearrest
		K_{cl}/K_{max}	ΔK_{eff}, MPa $\sqrt{m}$	K_{cl}/K_{max}	ΔK_{eff}, MPa $\sqrt{m}$	K_{cl}/K_{max}	ΔK_{eff}, MPa $\sqrt{m}$	Δa^*, μm	ΔN^*, cycles
7150-UA	3.14	0.83	0.59	0.60	1.40	0.83	0.59	60	4×10^5
7150-T6	2.81	0.82	0.56	0.63	1.15	0.82	0.56	130	2.5×10^5
7150-T7	2.17	0.74	0.63	0.62	0.92	0.74	0.63	170	4.5×10^5
7475-T7351	2.40	0.65	0.88	0.52	1.21	0.66	0.86	230	4×10^5

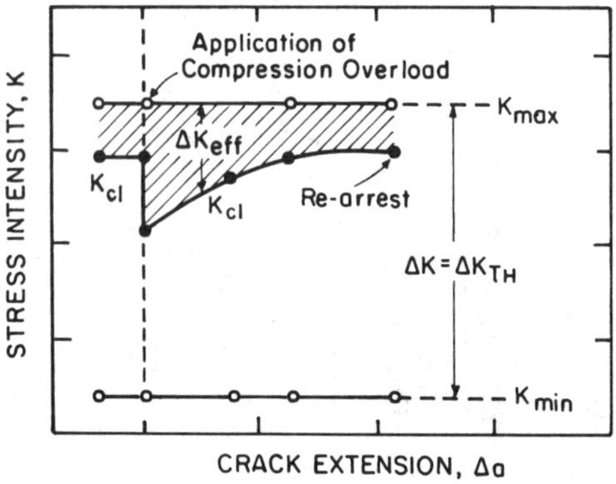

FIG. 10—*Schematic illustration of the variation in* ΔK *and* ΔK_{eff} *following the application of a single compression overload.*

in K_{cl}/K_{max} following the compression cycle (Table 5), essentially are consistent with this equation, it is difficult to evaluate it quantitatively with the current experiments as it is not feasible to calculate accurately the applied compressive stresses in the compact specimen geometry. However, variations in closure and arrest behavior observed in the four different microstructures cannot be rationalized by Eq 1. For example, the underaged and overaged

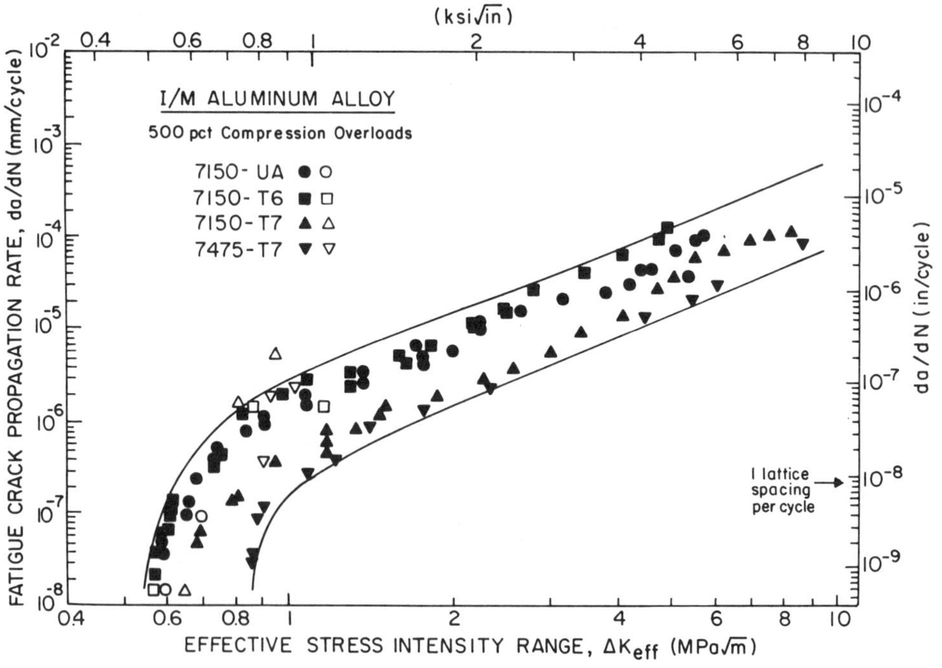

FIG. 11—*Fatigue crack growth rates as a function of effective stress intensity range,* ΔK_{eff}*, for constant amplitude (closed symbols) and following application of 500% compression overloads (open symbols). Data for 7150 and 7475 aluminum alloys at* R = *0.05 to 0.10, with* ΔK_{eff} *calculations based on experimental* K_{cl} *measurements.*

7150 structures are of identical yield strength, yet rearrest occurs after a post-overload crack extension of only 60 μm in the UA condition compared to ~170 μm in the T7 condition. These distances are large compared to the plastic zone sizes at threshold (Table 4), yet are consistent with the primary source of crack closure in these alloys [*35,39*] being from the roughness-induced mechanism resulting from deflections in crack path. Thus, crack closure levels can redevelop with crack extension (causing rearrest) far more efficiently in the less heavily aged microstructures because of the more tortuous nature of the crack paths in these structures (for example, Fig. 5). This is illustrated in Fig. 12 where the extent of crack growth before rearrest, Δa^*, is shown to be inversely proportional to the degree of fracture surface roughness in the aluminum alloys. Moreover, since this implies closure from the wedging action of fracture surface asperities, the reduction in closure following the compression cycle additionally can be attributed to crushing of such asperities. Figure 13, which shows the near-threshold fractography of 7150-UA before and immediately following the overload, provides evidence for this notion. The sharp, faceted profile of fracture surface at ΔK_{TH} (Fig. 13*a*) can be seen to become obscured following the compression cycle, with clear indications of abrasion, compacted fretting oxide debris (Fig. 13*b*) and the flattening and cracking of asperities (Fig. 13*c,d*) in the immediate vicinity of the crack tip.

Such results are consistent with constant-amplitude data [*42*] showing lower thresholds at $R = -1$ compared to $R = 0.05$, which were attributed to lower closure loads resulting from the tension/compression cycling.

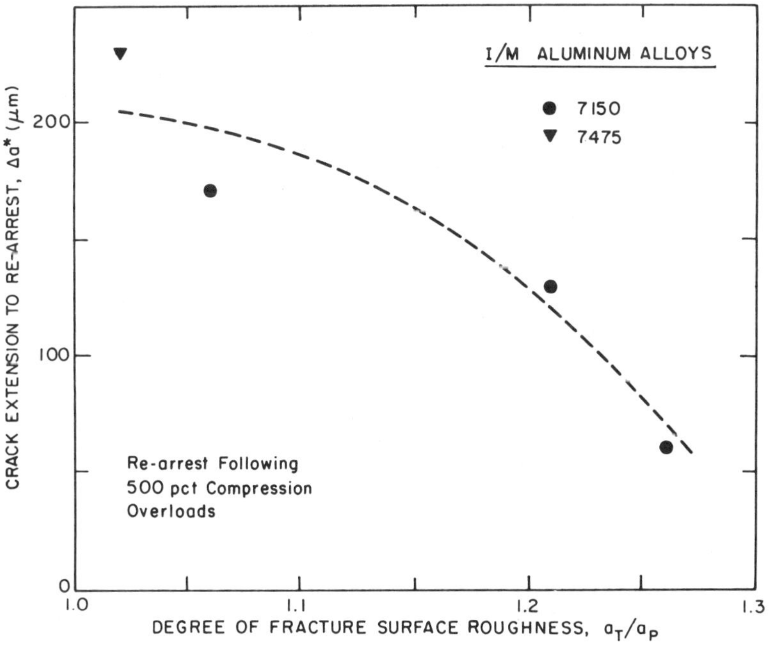

FIG. 12—*Variation of crack extension until rearrest* (Δa^*), *following application of 500% compression overloads, as a function of the degree of near-threshold fracture surface roughness for 7150 and 7475 aluminum alloys at* $R = 0.05$ *to* 0.10. *Surface roughness is assessed in terms of the lineal roughness parameter, that is, the ratio of total crack length to projected crack length on the plane of maximum tensile stress,* (a_T/a_p).

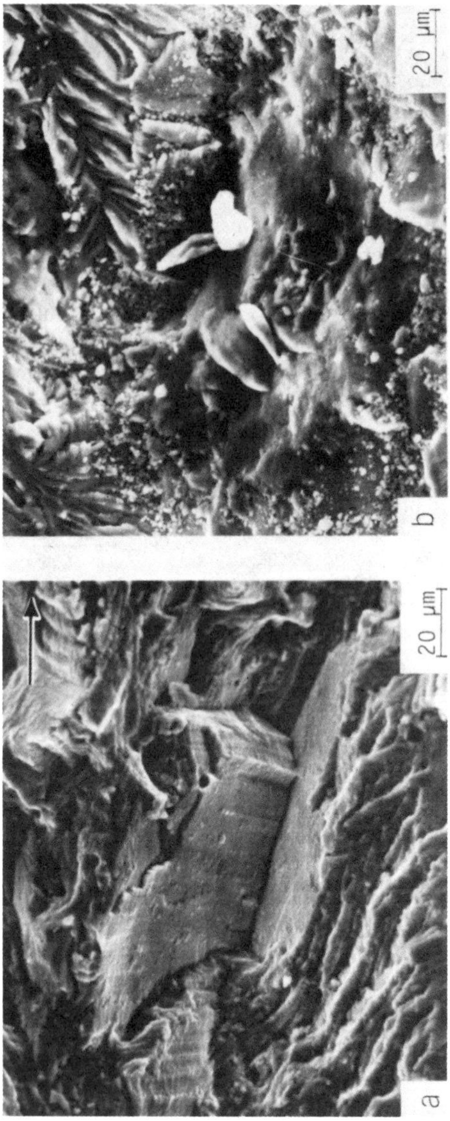
b
20 μm
a
20 μm

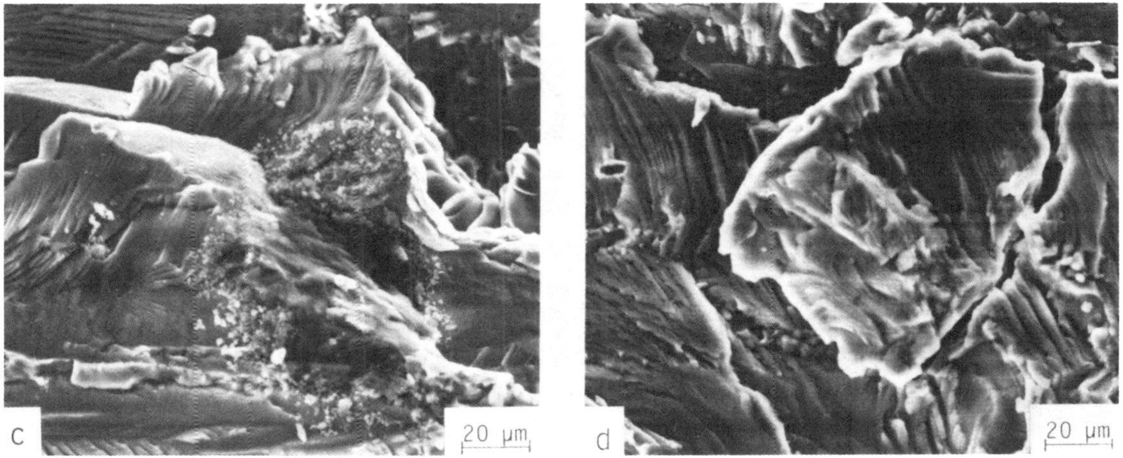

FIG. 13—*Scanning electron micrographs of fatigue fracture morphology in underaged 7150 aluminum alloy directly behind the crack tip at the threshold* ($\Delta K \approx 3.2$ *MPa* $\sqrt{m}$), *showing* (a) *well-defined facets before the application of the compression overload,* (b) *compacted fretting oxide debris,* (c) *asperity flattening, and* (d) *asperity cracking after the application of the overload. Arrow indicates general direction of crack growth.*

Implications

In answer to the original question as to the significance of the fatigue threshold concept in the presence of compression cycles, it is clear that nominal ΔK_{TH} values, as measured by conventional load-shedding procedures currently proposed for ASTM standardization, have little meaning in defining the dormancy of a fatigue crack. This has been explained simply in terms of the immediate reduction in closure following the compression overload, in the present case of 7150 and 7475 aluminum alloys and 17-4 PH stainless steel from the flattening and cracking of fracture surface asperities. This results in a local increase in the near-tip driving force, ΔK_{eff}, *to cause crack propagation at the threshold.* Such results are consistent with the recent studies of Topper and co-workers [*28,29*] who documented a linear decrease in threshold values in both steels and aluminum with increasing compressive peak stress.

Such observations in general highlight the inherent fallacy of employing fatigue threshold values at low load ratios to predict the absence of fatigue cracking. It is now clear that the existence of a threshold is controlled largely by the magnitude of the closure, but this is not incorporated into the computation of the nominal stress intensity range. Thus, the potential exists for accelerated, and sometimes nonunique, crack growth *at or below the low load-ratio threshold* in situations where crack closure may be restricted, such as with small flaws or cracks at notches, with the superposition of steady tensile stresses (that is, at high R), or in the present case of the application of compressive overloads.

The results further question the physical rationale behind many of the existing computational models for fatigue life prediction under spectrum loading. Such models [*43–46*] rely on constant-amplitude data and in general incorporate plasticity effects only. Clearly, at low stress intensity levels, crack growth behavior can be dominated by microstructural influences (compare the behavior in 7150-UA and 7150-T7, which have identical strengths) and these effects are simply not reflected in current durability and damage-tolerant analyses.

Conclusions

Based on a study of near-threshold fatigue crack growth and closure behavior at $R = 0.05$ to 0.50 in 17-4 PH stainless steel, and I/M 7150 and 7475 aluminum alloys in a range of temper conditions, the following conclusions can be made:

1. In response to the question "Is the concept of a fatigue threshold meaningful in the presence of compression cycles?," the present results clearly show that cracks do not remain dormant at ΔK_{TH} following the application of large compression overloads. However, a threshold can be defined locally in terms of ΔK_{eff} where crack growth does arrest.
2. The reinitiation of growth of a crack arrested at ΔK_{TH} due to the application of the compressive cycle was found to be associated with a measured reduction in crack closure. This was attributed principally to a smaller contribution from roughness-induced closure, arising from the compacting and cracking of fracture surfaces asperities close behind the crack tip.
3. Following such reinitiation of crack growth, cracks were found to decelerate progressively until rearrest, consistent with a measured redevelopment of closure over crack extensions ranging from ~60 μm in underaged 7150 to ~230 μm in 7475-T7351. In the aluminum alloys, the more efficient generation of closure with crack extension at ΔK_{TH} in the coherent particle-hardened underaged alloy was attributed to its planar-slip mode of deformation which promotes rougher, more faceted crack paths, thereby enhancing the contribution from the roughness-induced mechanism from the formation of larger fracture surface asperities.

Acknowledgments

This work was sponsored partly under Grant No. AFOSR 82-0181 from the U.S. Air Force Office of Scientific Research and partly by the Swedish Defence Materials Administration (Aero Materials). The authors wish to thank Dr. W. Yu, Ms. P. Donehoo, and Mr. H. Bjermert for experimental assistance and Ms. M. Penton for help in preparing the manuscript.

References

[*1*] Paris, P. C. and Erdogan, F., *Transactions,* American Society of Mechanical Engineers, *Journal of Basic Engineering,* Vol. 85, 1963, pp. 528–534.
[*2*] Ritchie, R. O., *International Metals Reviews,* Vol. 20, 1979, pp. 205–230.
[*3*] Beevers, C. J. in *Fatigue Thresholds, Proceedings,* 1st International Conference, J. Bäcklund, A. F. Blom, and C. J. Beevers, Eds., EMAS Ltd., Warley, U.K., Vol. 1, 1982, pp. 257–275.
[*4*] Ritchie, R. O. in *Fatigue Threshold, Proceedings,* 1st International Conference, J. Bäcklund, A. F. Blom, and C. J. Beevers, Eds., EMAS Ltd., Warley, U.K., Vol. 1, 1982, pp. 503–519.
[*5*] Blom, A. F., *Journal of Aeronautical Materials,* Vol. 2, 1982, pp. 17–28.
[*6*] Suresh, S. and Ritchie, R. O. in *Fatigue Crack Growth Threshold Concepts,* D. L. Davidson and S. Suresh, Eds., The Metallurgical Society of the American Institute of Mining, Metallurgical and Petroleum Engineers, Warrendale, PA, 1984, pp. 227–261.
[*7*] Paris, P. C., Bucci, R. J., Wessel, E. T., Clark, W. G., and Mager, T. R. in *Stress Analysis and Growth of Cracks, ASTM STP 513,* American Society for Testing and Materials, Philadelphia, 1972, pp. 141–176.
[*8*] Schmidt, R. A. and Paris, P. C. in *Progress in Flaw Growth and Fracture Toughness Testing, ASTM STP 536,* American Society for Testing and Materials, Philadelphia, 1973, pp. 79–93.
[*9*] Suresh, S., Parks, D. M., and Ritchie, R. O. in *Fatigue Thresholds, Proceedings,* 1st International Conference, J. Bäcklund, A. F. Blom, and C. J. Beevers, Eds., EMAS Ltd., Warley, U.K., Vol. 1, 1982, pp. 391–408.
[*10*] Minakawa, K., Newman, J. C., and McEvily, A. J., *Fatigue of Engineering Materials and Structures,* Vol. 6, 1983, pp. 359–365.
[*11*] Breat, J. L., Mudry, F., and Pineau, A., *Fatigue of Engineering Materials and Structures,* Vol. 6, 1983, pp. 349–358.
[*12*] Zaiken, E. and Ritchie, R. O., *Scripta Metallurgica,* Vol. 18, 1984, pp. 847–850.
[*13*] Zaiken, E. and Ritchie, R. O., *Metallurgical Transactions A,* Vol. 16A, 1985, pp. 1467–1477.
[*14*] Elber, W., *Engineering Fracture Mechanics,* Vol. 2, 1970, pp. 37–45.
[*15*] Ritchie, R. O., Suresh, S., and Moss, C. M., *Transactions,* American Society of Mechanical Engineers, *Journal of Engineering Materials and Technology,* Vol. 102, 1980, pp. 293–299.
[*16*] Stewart, A. T., *Engineering Fracture Mechanics,* Vol. 13, 1980, pp. 463–478.
[*17*] Suresh, S., Zamiski, G. F., and Ritchie, R. O., *Metallurgical Transactions A,* Vol. 12A, 1981, pp. 1435–1443.
[*18*] Walker, N. and Beevers, C. J., *Fatigue of Engineering Materials and Structures,* Vol. 1, 1979, pp. 135–148.
[*19*] Minakawa, K. and McEvily, A. J., *Scripta Metallurgica,* Vol. 15, 1981, pp. 633–636.
[*20*] Suresh, S. and Ritchie, R. O., *Metallurgical Transactions A,* Vol. 13A, 1982, pp. 1627–1631.
[*21*] Suresh, S. and Ritchie, R. O., *International Metals Reviews,* Vol. 29, Dec. 1984, pp. 445–476.
[*22*] Suresh, S., *Engineering Fracture Mechanics,* Vol. 18, 1983, pp. 577–593.
[*23*] Schijve, J. and Broek, D., *Aircraft Engineering,* Vol. 34, 1962, pp. 314–316.
[*24*] Hudson, C. M. and Scardina, J. T., *Engineering Fracture Mechanics,* Vol. 1, 1969, pp. 429–446.
[*25*] Stephens, R. I., Chen, D. K., and Hom, B. W. in *Fatigue Crack Growth Under Spectrum Loads, ASTM STP 595,* American Society for Testing and Materials, Philadelphia, 1976, p. 27.
[*26*] Hsu, T. M. and McGee, W. M. in *Effect of Load Spectrum Variables in Fatigue Crack Initiation and Propagation, ASTM STP 714,* American Society for Testing and Materials, Philadelphia, 1980, pp. 79–90.
[*27*] Gan, D. and Weertman, J., *Engineering Fracture Mechanics,* Vol. 15, 1981, pp. 87–106.
[*28*] Au, P., Topper, T. H., and El Haddad, M. L. in *Behavior of Short Cracks in Airframe Components,* AGARD Conference Proceedings No. 328, AGARD, France, 1983, pp. 11.1–11.17.
[*29*] Yu, M. T., Topper, T. H., and Au, P. in *Fatigue '84, Proceedings,* 2nd International Conference

on Fatigue and Fatigue Thresholds, C. J. Beevers, Ed., EMAS Ltd., Warley, U.K., Vol. 1, 1984, pp. 179–190.

[30] Marissen, R., Trautmann, K. H., and Nowack, H., *Engineering Fracture Mechanics,* Vol. 19, 1984, pp. 863–879.

[31] Zaiken, E. and Ritchie, R. O., *Engineering Fracture Mechanics,* Vol. 21, 1985, pp. 35–48.

[32] Dutta, V. B., Suresh, S., and Ritchie, R. O., *Metallurgical Transactions A,* Vol. 15A, 1984, pp. 1193–1207.

[33] Blom, A. F. in *Fatigue Crack Growth Threshold Concepts,* D. L. Davidson and S. Suresh, Eds., The Metallurgical Society of the American Institute of Mining, Metallurgical and Petroleum Engineers, Warrendale, PA, 1984, pp. 263–279.

[34] Shih, C. F., *Journal of the Mechanics and Physics of Solids,* Vol. 29, 1981, pp. 305–330.

[35] Zaiken, E. and Ritchie, R. O., *Materials Science and Engineering,* Vol. 70, 1985, pp. 151–160.

[36] Lindgkeit, J., Gysler, A., and Lütjering, G., *Metallurgical Transactions A,* Vol. 12A, 1981, pp. 1613–1619.

[37] Lafarie-Frenot, M. C. and Gasc, C., *Fatigue of Engineering Materials and Structures,* Vol. 6, 1983, pp. 329–344.

[38] Suresh, S., Vasudévan, A. K., and Bretz, P. E., *Metallurgical Transactions A,* Vol. 15A, 1984, pp. 369–379.

[39] Carter, R. D., Lee, E. W., Starke, E. A., and Beevers, C. J., *Metallurgical Transactions A,* Vol. 15A, 1984, pp. 555–563.

[40] Suresh, S., *Metallurgical Transactions A,* Vol. 14A, 1983, pp. 2375–2385.

[41] Donehoo, P., Yu, W., and Ritchie, R. O., *Materials Science and Engineering,* Vol. 74, 1985, pp. 11–17.

[42] Blom, A. F., Hadrboletz, A., and Weiss, B. in *Mechanical Behavior of Materials IV, Proceedings,* International Conference, J. Carlsson and N. G. Ohlson, Eds., Pergamon Press, Oxford, U.K., Vol. 2, 1984, pp. 755–762.

[43] Willenborg, J., Engle, R. M., and Wood, H. A., "A Crack Growth Retardation Model Using an Effective Stress Concept," USAF Technical Report TM-71-1-FBR, Air Force Flight Dynamics Laboratory, Wright-Patterson Air Force Base, Ohio, Jan. 1971.

[44] Wheeler, O. E., *Transactions,* American Society of Mechanical Engineers, *Journal of Basic Engineering,* Vol. 94, 1972, pp. 181–186.

[45] Dill, H. D. and Saff, C. F. in *Fatigue Crack Growth Under Spectrum Loads, ASTM STP 595,* American Society for Testing and Materials, Philadelphia, 1976, pp. 306–319.

[46] Newman, J. C., Jr. in *Methods and Models for Predicting Fatigue Crack Growth under Random Loading, ASTM STP 748,* American Society for Testing and Materials, Philadelphia, 1981, pp. 53–84.

Arthur J. McEvily[1] *and Kunori Minakawa*[1]

Crack Closure and Variable-Amplitude Fatigue Crack Growth

REFERENCE: McEvily, A. J. and Minakawa, K., "**Crack Closure and Variable-Amplitude Fatigue Crack Growth,**" *Basic Questions in Fatigue: Volume I, ASTM STP 924,* J. T. Fong and R. J. Fields, Eds., American Society for Testing and Materials, Philadelphia, 1988, pp. 357–376.

ABSTRACT: The prediction of service lifetime and the selection of alloys for maximum fatigue crack growth resistance are made complicated by the effect of prior loading histories on the rate of crack growth. Some research indicates that crack closure can be an important factor in accounting for this history effect. This paper presents a review of the lifetime on load-interaction effects on fatigue crack growth and indicates some of the future research needed to better understand the factors giving rise to history effects in the growth of long and short cracks.

KEY WORDS: crack closure, fatigue crack growth, overloads, variable amplitude loading

The rate of fatigue crack growth can be significantly affected by the magnitude and sequence of the loads applied under variable-amplitude loading conditions. The retardation of the rate of crack growth after the application of a tensile overload is one of the more common manifestations of a history effect influencing subsequent crack growth. In recent years a new aspect of variable-amplitude loading has been recognized, namely that if several alloys are compared and rated in terms of their resistance to fatigue crack growth under constant-amplitude loading conditions the same order of rating apparently may not apply under variable amplitude loading conditions. This circumstance has been observed, for example, in comparing high-strength aluminum powder metallurgy (P/M) and ingot metallurgy (I/M) products. Such results can have an important influence on material selection for particular applications as well as on alloy designs aimed at optimizing resistance to fatigue crack growth. As yet, the underlying factors responsible for different relative responses of alloys to constant-amplitude or variable-amplitude loading have not been elucidated. This paper is to review the role of crack closure in variable-amplitude fatigue crack growth and to indicate related research areas in need of further study.

Review

Phenomenological

When fatigue cracks are grown under variable-amplitude loading it is possible for the deformation induced at one amplitude to affect the growth rate at another amplitude. In such a case a simple summation of calculated crack growth increments based on constant-

[1] Professor and assistant professor, respectively, Metallurgy Department of Materials Science, University of Connecticut, Storrs, CT 06268. The coauthor is now with NKK Technology Research Center, Kawasaki, Japan.

amplitude loading will not lead to an accurate prediction of the total number of cycles required to grow a crack from one crack length to another. This history effect in crack growth is often investigated by applying a single overload and studying subsequent crack growth behavior with typical results as shown in Fig. 1. The overload retards subsequent growth until the crack has propagated out of the region of influence associated with the overload. The number of delay cycles involved can be defined as indicated in Fig. 1. The distance over which delay occurs, designated as a^*, is often related to the size of the plastic zone which results from the overload, but here it seems that there is an element of ambiguity. For example, if the extent of the overload zone ahead of the crack is taken to be $K_{OL}^2/2\pi\sigma_y^2$, where K_{OL} is the stress intensity factor associated with the overload and σ_y is the monotonic yield strength, one may find that the extent of the affected region exceeds the overload zone size [*1*]. On the other hand if the expression $K_{OL}^2/\pi\sigma_y^2$ is used to define the overload zone, then one may conclude that the region of influence of the overload is less than the overload zone size [*2*].

After a single overload the crack growth rate can vary as the crack traverses the influenced region as shown in Fig. 2. It is noted that in this case the lowest crack growth rate is obtained only after the crack penetrates part way through the affected region, a feature referred to as "delayed retardation." If a series of overloads is applied rather than just a single overload, this delay in reaching the minimum crack growth rate may not be present [*3*].

In the case of tensile overloads the amount of retardation in a given material will depend upon the degree of overload. As shown in Fig. 3, for overload ratios of up to 1.4 there is little effect of the overload on subsequent crack growth rate [*4*]. This figure also illustrates that the relative resistance of two different tempers of the same alloy to crack growth can be dependent upon the magnitude of the overload. For loading conditions wherein the amplitudes do not vary significantly and history effects are minimal as in Fig. 3 for overload ratios below 1.4, it may be possible to devise a simple scheme for the calculation of the

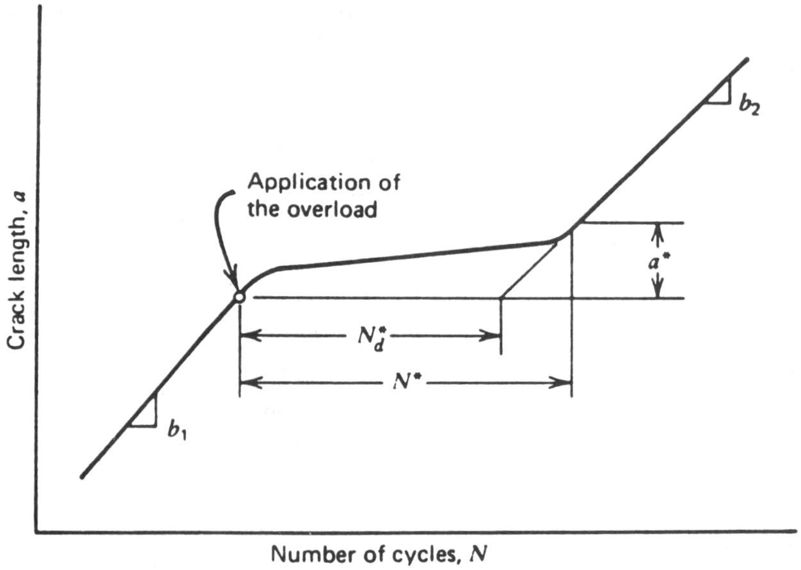

FIG. 1—*Crack growth rate plot illustrating effect of single-peak overload* [3].

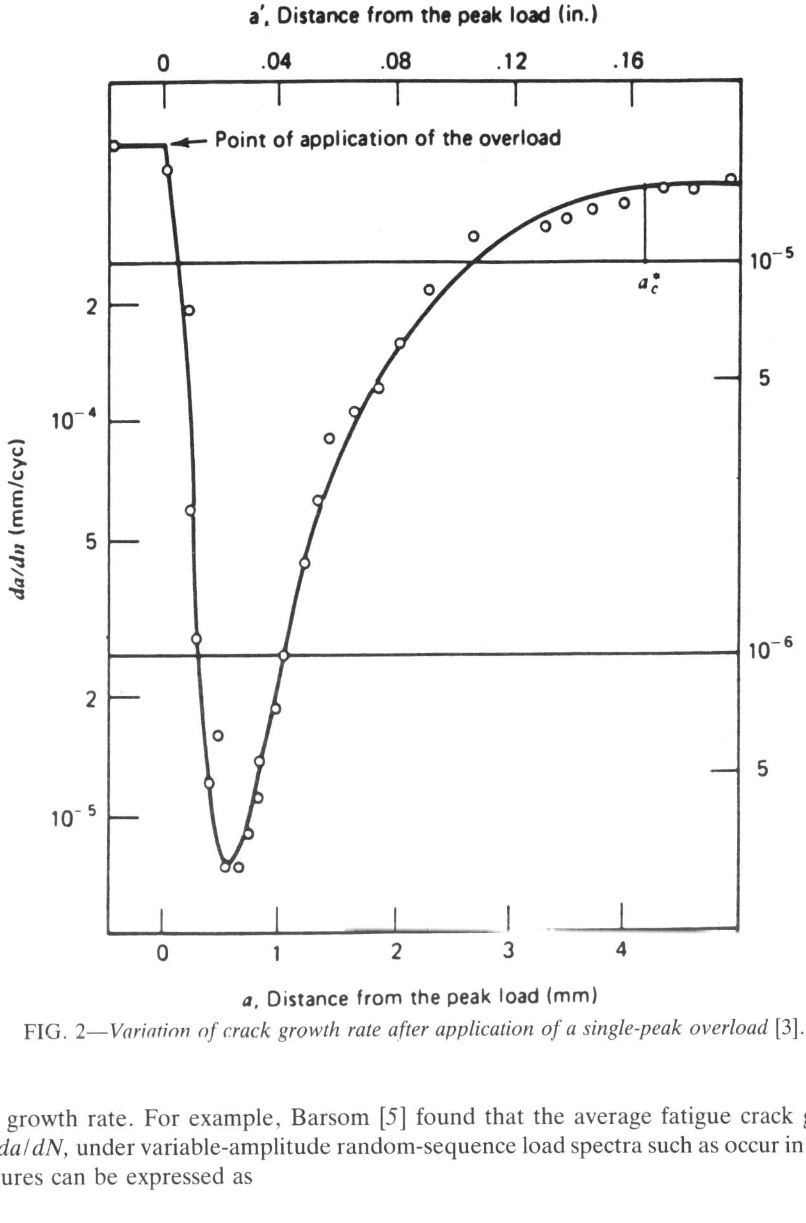

FIG. 2—*Variation of crack growth rate after application of a single-peak overload* [3].

crack growth rate. For example, Barsom [5] found that the average fatigue crack growth rate, da/dN, under variable-amplitude random-sequence load spectra such as occur in bridge structures can be expressed as

$$\frac{da}{dN} = A(\Delta K_{rms})^n \tag{1}$$

where ΔK_{rms} is the root-mean-square stress intensity factor range, and A and n are constants for a given material. It is noted that the rms approach has been shown to be equivalent to a second power of ΔK law, which in turn can be related to the crack-tip opening displacement (CTOD) [6]. Where history effects are not significant, Elber [7] has proposed that ΔK_{eff},

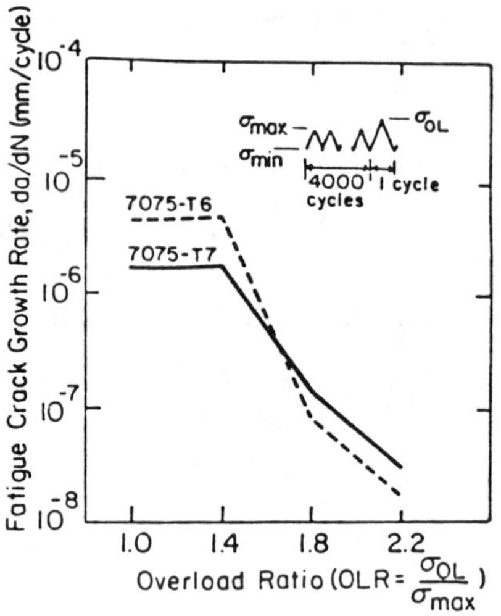

FIG. 3—*Variation of crack growth rates with overload ratio in peak-aged and overaged 7075 aluminum alloys. Tensile overloads were applied after every 4000 constant-amplitude cycles* [4].

determined from constant-amplitude tests, be used in place of ΔK_{rms} in Eq 1 ($\Delta K_{eff} = K_{max} - K_{op}$, where K_{op} is the stress intensity factor at the crack opening level). It may also be of interest to note that if the crack growth relationship is expressed by the Paris law, that is

$$\frac{da}{dN} = A(\Delta K)^n \tag{2}$$

then for variable-amplitude loading the sequence in which the loads are applied has no effect on the calculated number of cycles required to propagate a crack from one size to another, for there is no history effect built into the Paris law. This sometimes comes as a surprise, but is useful to know when setting up computer programs for crack growth analyses where history effects are not important. For example, if $\Delta K = \Delta\sigma\sqrt{\pi a}$ the following expression can be used to determine the crack length, a_f, developed from an initial length, a_0

$$a_f = \left\{ a_0^{-n/2+1} + A\left(1 - \frac{n}{2}\right) \pi^{n/2} \sum_{i=1}^{i=i} \Delta\sigma_i N_i \right\}^{1/1-n/2} \quad (n \neq 2) \tag{3}$$

If the Paris law is used where history effects are known to be important (as in the analysis of crack growth after proof-testing the F-111 aircraft), the exponent n can be modified empirically to account for retardation as in the Wheeler method of analysis [8].

Crack Closure

There is evidence to suggest that the retardation effect associated with a tensile overload is due to increased crack closure and a consequent reduction in ΔK_{eff} as proposed by Elber [9]. The concept of closure was first introduced by Elber, who attributed this phenomenon to residual plastic deformation in the wake of the crack. However, the extent of residual plastic deformation in the wake of a crack differs in plane stress as compared with plane strain. For example, the use of side grooves in low-alloy steel specimens to reduce the plane-stress contribution to closure reduced the crack opening stress intensity factor value from 8 to 4 MPa$\sqrt{m}$ [*10*]. We also now know that in the near-threshold region in which deformation is primarily in plane strain, combined Mode I and Mode II growth mechanisms give rise to fracture surface roughness, which enhances closure [*11–13*]. In addition, closure can be promoted by fretting and oxidation [*14,15*].

The plane-stress zone at the surface is particularly significant with respect to retardation following an overload. We propose that the relaxation of the residual compressive stresses in the overload plastic zone as the fatigue crack penetrates the zone leads to the development of additional closure which is responsible for the retardation. First consider the case of constant amplitude loading at $R = 0$. The amount of plasticity-induced closure for this case can be determined approximately in the following manner. Assume that the extent of the plastic zone perpendicular to the direction of crack propagation is given by $\Delta K^2/\pi\sigma_y^2$, and that the compressive residual stress perpendicular to the direction of propagation at the crack tip is σ_y. As the crack grows, the residual stress behind the crack tip is relaxed and the material in the relaxed zone tries to expand. The amount of this expansion is given by the product of the strain, σ_y/E, and the distance over which the relaxation occurs, $\Delta K^2/\pi\sigma_y^2$, or $\Delta K^2/\pi\sigma_y E$. However, this expansion cannot be accommodated behind the tip at zero load and therefore results in crack closure. The tensile load required to permit the expansion without crack closure is the opening load, and the corresponding stress intensity factor is K_{op}. By a process of superposition, we assume that at the opening load the crack tip opening displacement given as $\Delta K_{op}^2/\sigma_y E$ equals the relaxed displacement, or $\Delta K_{op}^2/\sigma_y E = \Delta K^2/\pi\sigma_y E$. This leads to a ratio of $\Delta K_{op}/\Delta K$ of 0.56, a value close to that reported by Elber. In the case of a 100% overload, the relaxed displacement would be a maximum half-way through the overload zone and is given by which is actually larger than the CTOD which now is given by $4\Delta K^2/\pi\sigma_y E$. The conclusion is that K_{eff} can be greatly reduced as a result of the overload, but probably not completely to the extent indicated in this simple approximation. It is noted that this model is consistent with the delay in retardation observed for a single overload. Further, if the residual extension in the wake of the crack tip is sufficiently large, it might be possible for the retardation effect to persist beyond the extent of the overload plastic zone. Another factor of importance is the shape of the crack front following the overload, particularly if a pop-in event has occurred during the overload. However, this displacement cannot be accommodated behind the tip until the CTOD is about $1/\pi$ of maximum CTOD.

That the plane-stress surface region is indeed important in causing retardation after an overload has been shown by the following experiments [*16*].

A fatigue crack was grown in a 6061-T6 compact specimen and subjected to a 100% overload. Delayed retardation was observed as shown in Fig. 4. A crack was grown in a second specimen in the same manner and also overloaded. However, before the test was resumed the surface layers on each side were machined away to remove the plane-stress regions. Upon resumption of cycling the retardation effect was found to be virtually eliminated. Additional experiments on these two specimens were carried out with results as shown in Fig. 4. Note that in overloading the thinner specimen no delay in retardation was observed.

Experiments with a titanium alloy [*17*] shed further light on the nature of the closure process associated with overloads. As shown in Fig. 5, the crack as it traverses the overload zone is tightly closed in the off-loaded condition. For constant-volume plastic deformation the extension in the loading direction is balanced by contraction in the thickness direction.

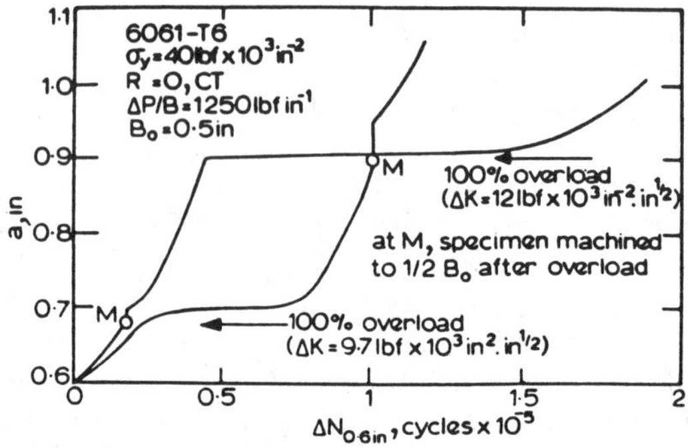

FIG. 4—*Effect of overloads and surface removal on crack length* a *as function of number of cycles to propagate a crack beyond a length of 15 mm 0.6 in.* [16].

The region of contraction, and incidently the shape of the overload plastic zone, can be revealed simply by rubbing the specimen surface on a flat abrasive surface, with results as shown in Fig. 5. By polishing away the surface layers the closure effect is seen to disappear, and we conclude again that the contribution of the plane-strain region to retardation is minimal. From such results we expect that even in the near-threshold region, which normally is dominated by plane-strain processes, plane-stress effects are important when overloads are applied. Further, because of the steepness of the da/dN-versus-ΔK plot in this region it may not take much of a plane-stress overload zone to reduce ΔK_{eff} slightly and to develop a complete arrest or significant number of delay cycles in the near-threshold region.

Further indications of the relative importance of surface regions as opposed to interior regions can be obtained by comparing the output of a strain gage mounted across a crack as employed by Schmidt and Paris [*18*] and a crack mouth opening displacement gage. Both methods are indicated in Fig. 6. The strain gage is particularly sensitive to surface strains whereas the displacement gage averages contributions from both surface and interior regions. The rate of fatigue crack growth in a 2219-T87 aluminum alloy is shown in Fig. 7 for the case of a 100% overload at $R = 0.05$ [*19*]. Upon overloading, a pop-in event occurred in the interior of the specimen which increased the crack length in the mid-thickness. This increment of "tensile crack growth" is indicated in Fig. 7. The fatigue crack growth rate as a function of crack length was determined by measuring the striation spacings for the interior crack and the crack length as a function of the number of cycles optically for the surface crack. Note that as is sometimes observed, the surface crack rate initially accelerated after the overload before entering the region of retardation. Figure 8 shows the crack opening behavior as determined by a surface strain gage and a displacement (clip) gage. In the interior the crack opening level actually decreased after the overload while the crack was within the affected region. (The value of a^* in this case was less than $K^2/\pi\sigma_y^2$.) At the surface, however, the crack-tip opening increased to a maximum at a crack increment which corresponded to the slowest rate of propagation. Beyond this point the CTOD decreased as the crack propagated through the remaining portion of the affected region. These results also indicate that the out-of-plane (z-direction) plastic deformation in the plane-stress region is important in creating closure. The in-plane (y-direction) plastic deformation in the plane-strain region apparently contributes little to closure.

In order to develop a quantitative analysis of crack growth after an overload, a precise determination of the K_{op} level is needed. For this purpose even the surface strain-gage technique may underestimate the opening level. For example, the minimum crack growth rate in Fig. 7 is 5×10^{-5} mm/cycle. For this alloy this would correspond to a ΔK_{eff} of 7 to

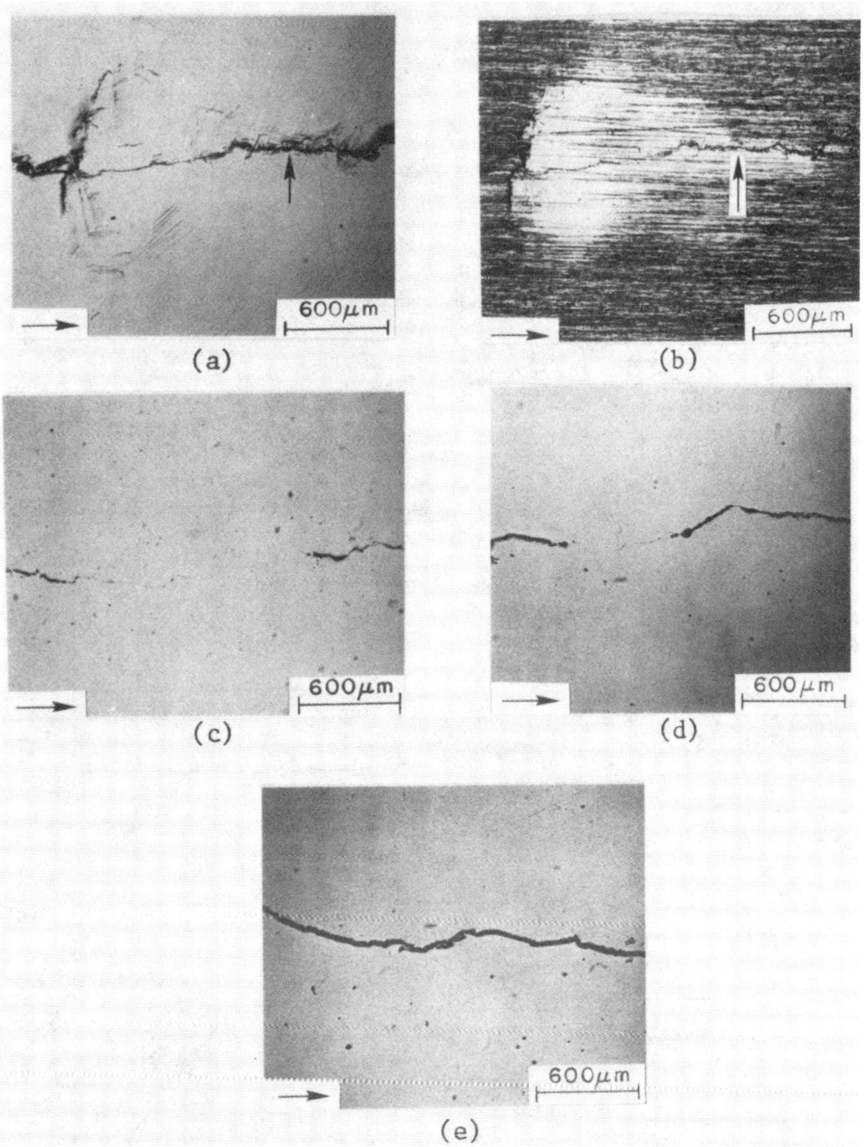

FIG. 5—*Micrographs illustrating* (a) *a tightly closed crack within the overload plastic zone area followed by a more open crack after returning to the baseline crack growth rate* (*arrow*); (b) *lateral contraction* (*necking*) *in the overload plastic zone area;* (c), (d) *and* (e) *appearance of crack at a depth of 0.25, 1.0, and 1.5 mm, respectively* [17].

8 MPa $\sqrt{m}$. However the ΔK_{eff}-value determined from the surface strain gage is 12.2 MPa $\sqrt{m}$, too high a value for correlation. To study the crack-tip opening behavior in greater detail replicas of the crack tip were taken with the specimen under load. The replicas were then examined by scanning electron microscope (SEM). Use of this procedure indicated that the opening level was higher than determined by the strain-gage method. The value of

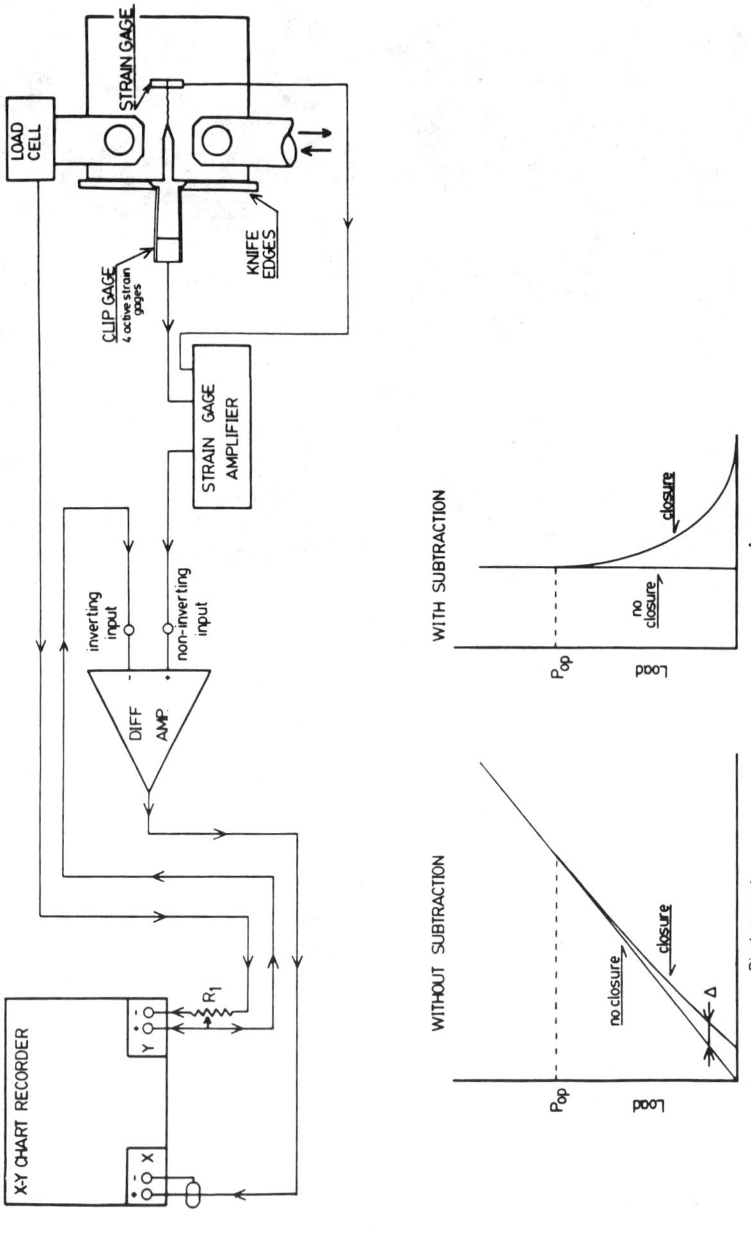

FIG. 6—*Schematic diagram of the compliance method for crack closure measurement.*

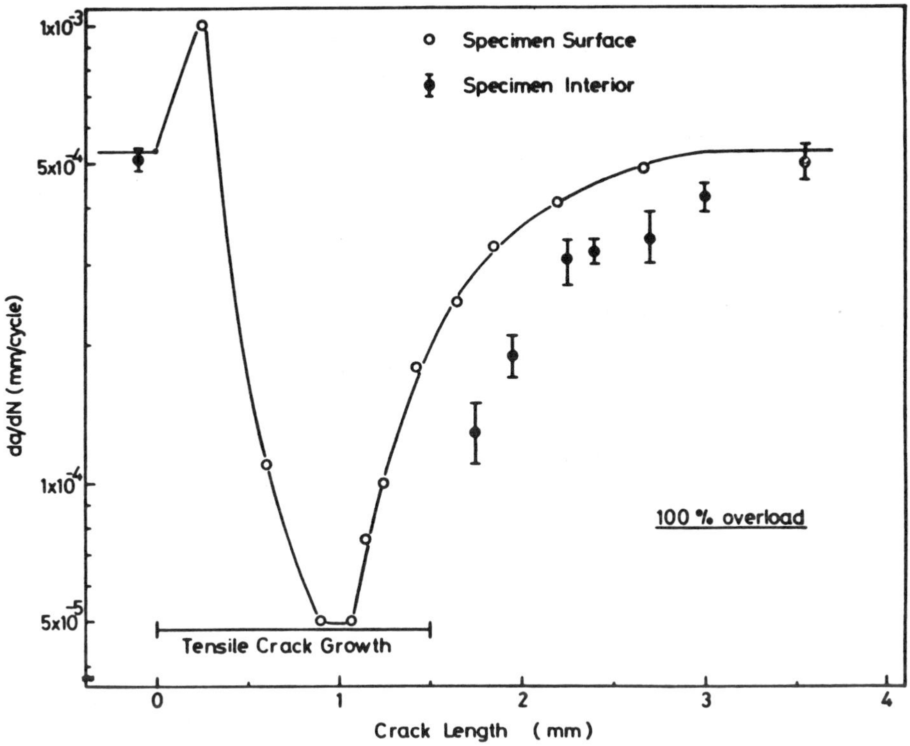

FIG. 7—*Variation of crack growth rate in 2219-T87 aluminum alloy after a single-peak overload* [19].

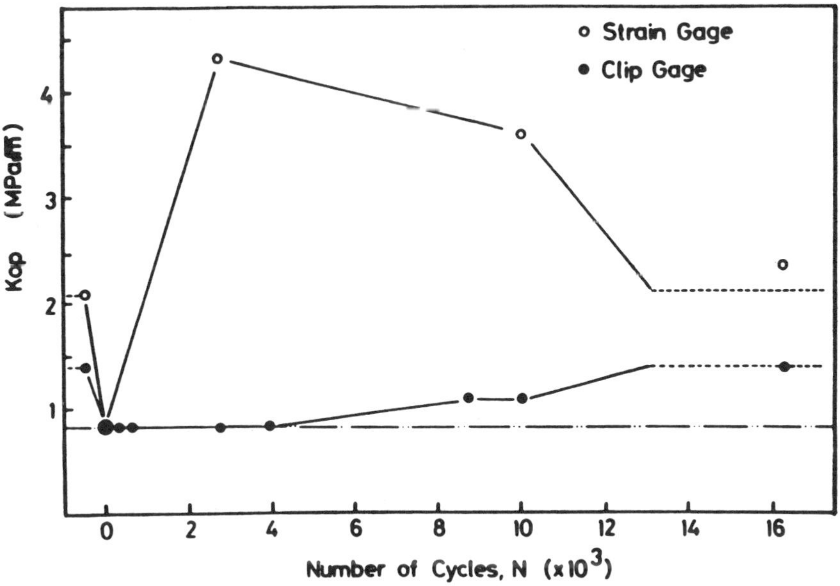

FIG. 8—*Variation of* K_{op} *level in 2219-T87 aluminum alloy after a single-peak overload* [19].

ΔK_{eff} was 7.1 MPa $\sqrt{m}$, about the right value for the rate of surface crack propagation observed. (Incidently, this may be the first quantitative demonstration of this correlation.)

Clearly the precise determination of the opening load is an important consideration. The direct observations utilizing SEM by Lankford and Davidson [*20*] for the aluminum alloys 7075, 2024, and 6061 are therefore of considerable interest. They observed that the tip itself did not open until a crack-opening load which depended upon prior history had been reached. For constant-amplitude loading and for a 100% overload the opening characteristics given in Table 1 were observed. The effect of an overload on ΔK_{eff} is clearly seen. The corresponding crack-tip opening displacements were also determined and a degree of correlation with calculated plane-stress zone sizes was found.

Since an overload can lead to a reduction of ΔK_{eff} in the surface regions, the associated crack growth rate will be reduced, with the slower crack growth rate at the surface retarding that in the interior and with consequent effects on the shape of the crack front. We expect both the overall rate following an overload as well as the crack front shape to be thickness dependent, and the ΔK_{eff} concept as determined by surface measurements needs to be carefully considered. For example, if the effective value of ΔK at the surface were 20 MPa $\sqrt{m}$ before an overload and 10 MPa $\sqrt{m}$ afterwards, one might expect the growth rate to correspond to the lower value based upon the surface measurements, yet what about the interior? There, ΔK_{eff} may have changed from 22 to 25 MPa $\sqrt{m}$ so that the entire specimen is not behaving as if at the lower level of ΔK_{eff}, and considerable tunneling may occur as a result, as has been observed [*21*]. Nevertheless the growth process in the surface layers, which are of course most easily observed, may correspond to a greatly reduced value of ΔK. If this value is low enough, complete arrest or near-threshold crack growth behavior with attendant Mode II induced faceting may occur, as discussed by Suresh [*22*]. If a zigzag path (crack deflection) is followed, a longer total crack growth path will result. In addition, there will be a reduction in the Mode I component since the crack facet may be at an angle to the loading direction. Both of these factors combine to result in an apparent lowering of the macroscopic crack growth rate which in turn adds to the number of cycles spent in traversing the overload zone. We have noted that in overload studies with the aluminum alloy 2219 a marked zigzag pattern of growth can occur even though the retarded crack growth rate is above the near-threshold range. The reason for this behavior needs to be clarified.

Additional Considerations

The influence of the ΔK level at which the overload is applied as well as that of the specimen thickness is shown in Fig. 9 [*23*]. At low ΔK-values or ratios of small plastic zone size to thickness where plane-strain conditions prevail, there is a marked rise in the number of delay cycles with decrease in ΔK level, but the extent of retardation decreases with

TABLE 1—*Crack-opening characteristics.*

	Pre-Overload ΔK = 10 MPa $\sqrt{m}$, R = 0.2		Post-Overload (at Minimum da/dN) (100% Overload)	
	K_{op}, MPa $\sqrt{m}$	ΔK_{eff}, MPa $\sqrt{m}$	K_{op}, MPa $\sqrt{m}$	ΔK_{eff}, MPa $\sqrt{m}$
6061-T6	4.8	5.2	7.8	2.2
2024-T4	5.7	4.3	9.7	0.3
7075-T6	6.3	3.7	8.7	1.3

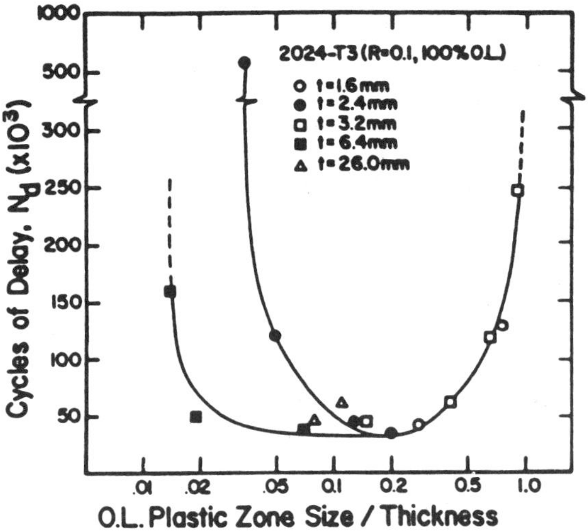

FIG. 9—*Number of cycles of delay as a function of the overload plastic zone size to sheet thickness ratio in 2024-T3 alloy* [23].

increase in thickness. At high ΔK-values where plane-stress conditions prevail there is an increase in the number of delay cycles with ΔK, and a single-valued dependency of the number of delay cycles on the overload plastic zone size per unit of thickness was obtained.

The overload effect is sensitive to the base R-ratio as illustrated in Fig. 10 [*24*] for 2024 where R is the ratio of the minimum to the maximum stress in a cycle. In this case for the same magnitude of the overload subsequent crack growth behavior depended on the R-ratio used. Compressive loading is seen to greatly reduce the overload effect. Crack growth behavior for a crack initially grown in 7075 at $R = 0$ and overloaded by a factor of 2.5 is shown in Fig. 11 [*24*]. Again an influence of the R-ratio used in subsequent testing is seen, and in particular it is noted that crack arrest was observed for $R = 0$ subsequent test

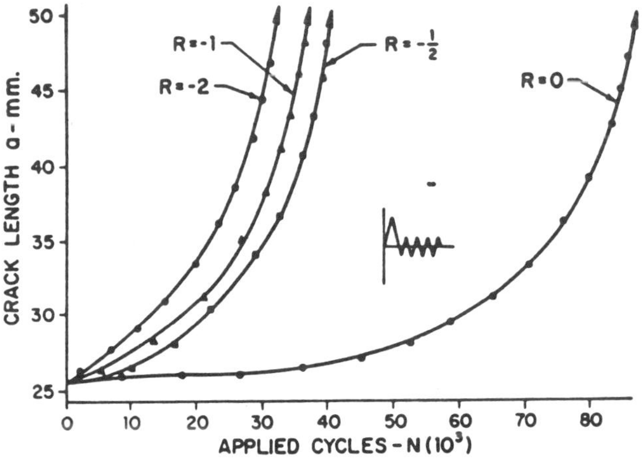

FIG. 10—*Crack growth at different* R *levels following single tensile overload in 2024-T3 aluminum alloy, overload ratio = 2.0* [24].

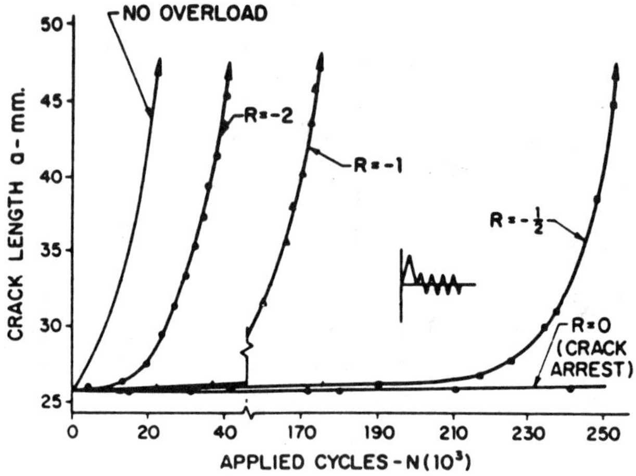

FIG. 11—*Crack growth following single tensile overload in 7075-T6 aluminum alloy, overload ratio = 2.5* [24].

conditions. The influence of compressive overloads or various combinations of compression-tension overloads is generally of lesser significance than a single tensile overload, an effect which can be related to a reduction of the crack opening level as the result of a compressive cycle.

There is limited evidence for the effect of overloads where high-baseline *R*-ratios are involved. Such tests are of interest since closure effects would be less pronounced than at lower *R*-ratios, and one might then determine the significance of other processes such as particle decohesion (to be discussed). Knott and Picard [*25*] did carry out some experimental work with an aluminum-zinc-magnesium alloy at different temper conditions at a baseline *R*-value of 0.5, but applied a series of overloads at an $R = 0.2$ rather than a single tensile overload. Retardation was noted, particularly in an underaged temper, but this retardation could have been due to closure induced by the 0.2 *R* overload cycling. It may be significant that this temper exhibited the highest threshold level, an indication of a propensity of an increased planar glide in the underaged temper utility to promote the development of fracture roughness.

Brown and Weertman [*26*] studied the effects of overloads in the aluminum alloy 7050-T76 at a base *R*-ratio of 0.5. The specimen was a center-cracked panel type, and closure levels were determined by means of a clip gage mounted on the surface. This technique was able to determine that closure occurred at lower *R*-ratios, but none was determined at an *R*-ratio of 0.5, even after a 180% tensile overload. On the other hand they found that the crack growth rate after the overload was significantly retarded. The absence of detectable closure led them to conclude that closure was not the only process involved in retardation. They suggested that the residual compressive stress field induced by the overload as well as crack-tip blunting might be contributing factors. However, we have seen that the determination of the opening load can be sensitive to the detection method employed, for example, in comparing results obtained with a strain gage with results obtained by the replica method, and therefore suggest that the experiments of Brown and Weertman be repeated using a more sensitive method for the determination of the crack opening level such as direct observation by SEM or replicas.

In certain alloys, particularly those of high toughness, a stretched zone on the fracture surface due to an overload as discussed by Fleck [*27*] may also contribute to retardation.

Influence of Material

One of the most interesting and important aspects of variable-amplitude loading is that the relative ratings of materials can change. An example of this is shown in Fig. 3. Bucci [*28*] has noted that there can be important interactions between microstructure and crack growth under variable-amplitude loading which are not accounted for by the constant-amplitude test. For example, the aluminum alloy 7050-T736 exhibited particularly good resistance to fatigue crack propagation under a variable load history, a fact not apparent from a comparison of constant-amplitude results. The favorable response of the 7050 alloy was attributed to an improved crack-growth retardation capability, a characteristic also not revealed by the constant-amplitude tests. More recently the work of Chanani (private communication) has shown that high-strength P/M aluminum alloys can be superior to comparable ingot aluminum alloys when tested under spectrum loading conditions, and yet be inferior under constant-amplitude conditions.

What are the characteristics of an alloy that contribute to superior resistance to crack growth under variable-amplitude loading? One aspect of the answer may relate to comparisons made under constant-amplitude loading conditions. The superiority of one alloy over another may hold true only in the intermediate crack growth range where most data traditionally have been obtained. The same superiority may not hold in the near-threshold range where the crack growth curves may cross as indicated in Fig. 12 [*29*]. If after an overload the retarded rate of crack growth is in the near-threshold range, then the material of higher threshold may exhibit superior behavior under variable-amplitude loading, and comparisons made in the intermediate growth range may be less important. Further, Suresh and Vasudevan [*30*] have noted that the low crack growth rates associated with retardation can take on the characteristics of near-threshold crack growth behavior.

With respect to the material characteristics which promote retardation, there is some lack of agreement. In some cases the yield strength is considered to be important [*16*] in that considerable retardation was found in the case of the low-strength (280 MPa) 6061 aluminum alloy in contrast to the limited amount of retardation found in the high-strength (490 MPa) 7075 alloy. The results of Hertzberg [*23*], Fig. 9, also suggests that the yield strength is important. Similarly Bucci [*28*] has noted a general trend of increased life with decreasing yield strength for some aluminum alloys tested under spectrum loading, and overaging the aluminum alloy 7475 to reduce the yield strength from 482 to 427 MPa resulted in longer spectrum fatigue life due presumably to better crack-retardation characteristics at the lower strength level. Further, Gallagher and Hughes [*31*] reported a similar effect of strength level on the fatigue crack growth life of a 4340 low-alloy steel subjected to tensile overloads. However, it has also been observed [*27*] that the fatigue crack propagation resistance to spectrum loading was better for 7050-T73651 (yield strength 440 MPa) compared with 7075 T7351 (yield strength 382 MPa). Therefore there are exceptions to the simple view that a lower yield strength leads to improved resistance to crack growth under variable-amplitude loading conditions. Other factors have been proposed. For example, Lankford and Davidson [*20*] have suggested that increased monotonic strain hardening capacity may result in increased retardation, and Knott and Pickard [*25*] have suggested that a high cyclic work hardening rate may be important.

In addition to the above, another possible explanation relates to the extent of closure present prior to an overload. For example, our studies with the P/M alloy 9021-T4 have shown that because of its extremely small grain size (0.2 to 1.0 μm) closure could not be detected at any ΔK range, even in the near-threshold region. Similarly, for the 7090-T6, no closure was detected by the clip-gage method in the intermediate range of crack-growth rates. On the other hand, for 7075-T76, K_{op}/K_{max} levels of 0.25 were determined in the

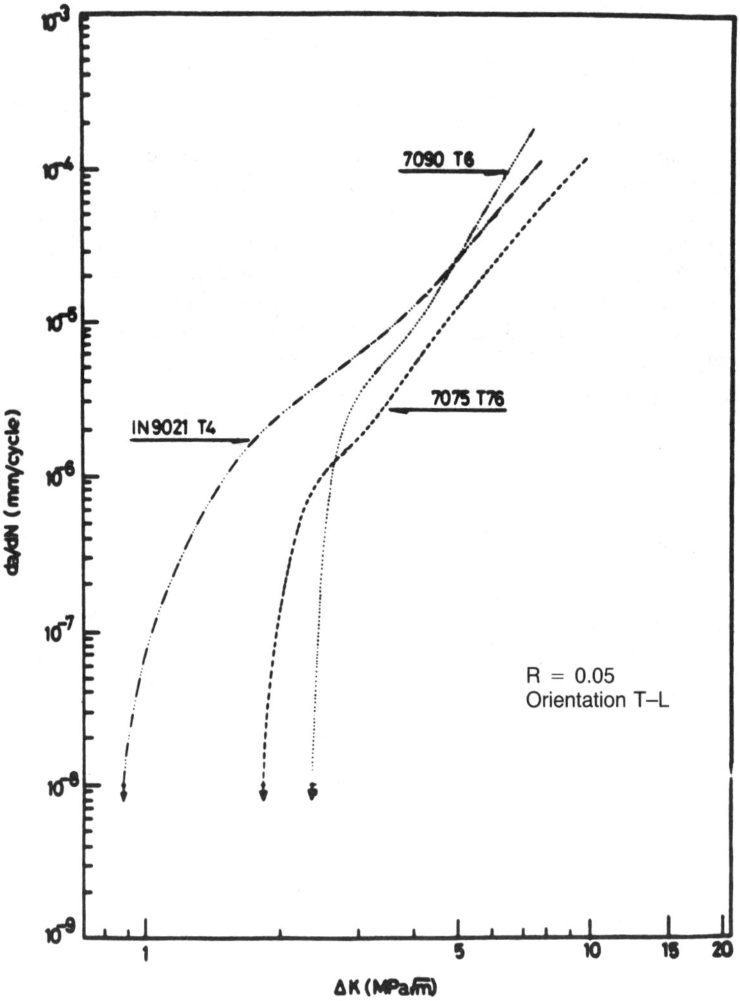

FIG. 12—*Crack growth rates* da/dN *as a function of* ΔK *at* R = *0.05 for three types of high-strength aluminum alloys: P/M 7090-T6, P/M IN9021-T4, and I/M 7075-T76* [29].

intermediate range at $R = 0.05$, and in this range the crack growth rate of 7075-T76 was lower than for either of the other two alloys, Fig. 12. Now if a 100% overload relative to a base ΔK level were to be applied to these materials the overload plastic zone size would be proportional to $4K^2/\sigma_y^2$ for the P/M alloys but only $(2\text{-}0.25)^2\ K^2/\sigma_y^2$ or $3.06K^2/\sigma_y^2$ for the I/M alloy. The yield strengths of these alloys are: 7090, 650 MPa; 9021, 535 MPa; and 7075, 470 MPa. The corresponding overload plastic zone sizes vary as: 7090, $9.4 \times 10^{-6}\ K^2$; 9021, $1.4 \times 10^{-5}\ K^2$; and 7075, $1.4 \times 10^{-5}\ K^2$. Therefore, differences in the degree of retardation based upon differences in the base closure level do not appear to offer an obvious explanation for changes in ranking that may occur after overloads. However, this possibility should be examined more carefully in the course of future test programs. In the case of 9021 we have determined by means of the crack mouth displacement gage that an overload can lead to the development of closure where none could be detected in the absence of the overload. It would be of interest to determine if closure in the surface region can be detected by strain gages mounted on the surface as in Fig. 6.

After an overload, shear localization and the tendency for deflected crack growth in certain alloy conditions may contribute to retardation as discussed by Suresh and Vasudevan [*30*] and might affect alloy comparisons. The contribution of such factors to the overall amount of retardation should be evaluated.

Another factor to be considered is the relative cyclic fracture toughness of the materials being compared, for a material of lower fracture toughness might exhibit greater crack advance due to pop-in events and this might affect the relative rankings in terms of the number of flights to advance a crack from one size to another. However, this does not appear to provide an explanation for any superiority the P/M alloys shown in Fig. 12 might have over the I/M alloy, since the fracture toughness of the P/M alloy is lower than that of the I/M alloy. Nevertheless, in spectrum loading with high tensile overloads the fracture toughness may be of influence according to Sanders and Staley [*32*]. They concluded that at low-baseline mean stress intensity levels the residual stresses induced by tensile overloads reduce the effective stress intensity at the crack tip and thus retard growth during subsequent lower-amplitude cycles. The influence of yield strength in this case is as depicted in Region A of Fig. 13. When the mean stress intensity is higher, Region B of Fig. 13, tensile overloads may become high enough to cause coarse intermetallic particles ahead of the crack to separate from the matrix. Sanders and Staley proposed that these incipient cracks reduce the stress intensity at the main crack tip and thus increase the magnitude of retardation. The severity of cracks induced decreases with increasing toughness (decreasing yield strength). Consequently, fatigue cracks grow slower in material aged to higher strength tempers when overloads induce particle-matrix decohesion. Thus the ranking of resistance is the reverse of the ranking under constant-amplitude loading in this regime. At higher stress intensities, Region C of Fig. 13, the crack advances significantly during tensile overloads, particularly in the material of lower toughness (higher strength). As a consequence the ranking of different tempers can again change in this range. Experimental results for the number of cycles required to grow a crack from one length to another as a function of yield strength

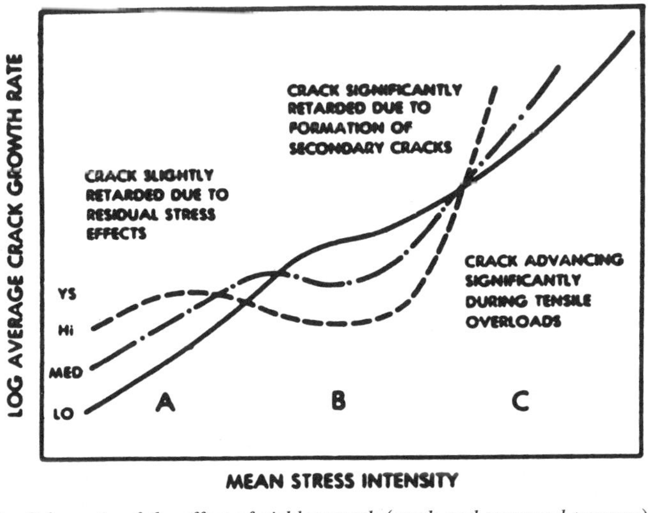

FIG. 13—*Schematic of the effect of yield strength (peak and overaged tempers) on fatigue crack growth rate under spectrum loading* [32].

were presented which exhibited the same trends as to be expected from Fig. 13. Unfortunately, no crack closure data were presented.

In the attempt to develop a quantitative treatment of the rate of fatigue crack growth, the following expression based upon crack opening displacement considerations has been proposed [33]

$$\frac{da}{dN} = \frac{A}{E^2}(\Delta K_{eff} - \Delta K_{TH.eff})^2 \left(1 + \frac{\Delta K_{eff}}{K_{cc} - \Delta K_{max}}\right) \quad (4)$$

This expression can provide reasonable agreement with crack growth data at room and elevated temperatures provided that environmental effects are not overwhelming. Consideration should be given to modifying this expression to account for changes in ΔK_{eff} as more is learned about interaction effects.

Short Cracks and Tensile Overloads

Curves A, B, and C in Fig. 14 indicate the anomolous behavior of short cracks. Much of their behavior can be explained on the basis of crack closure, for when a crack is first formed it has no wake and it takes a distance of the order of 0.5 mm for closure to develop and change the short crack behavior to long crack behavior. Until this closure is developed the ΔK_{eff} will be higher than might be expected and much higher rates of crack growth can occur. Load interaction effects can occur in this range which can have an effect on crack growth behavior, but this aspect of the crack growth behavior has not received attention.

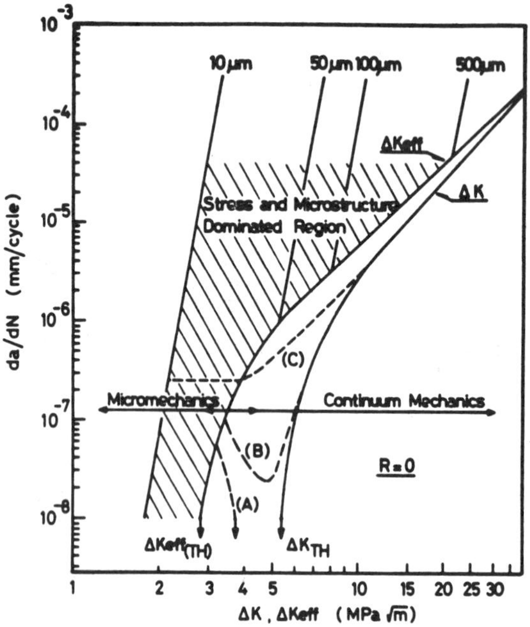

FIG. 14—*Rate of fatigue crack growth in a ferritic steel. Regions of applicability of micromechanics and continuum mechanics are indicated as is the microcrack region dominated by stress and microstructure. Microcrack sizes are indicated. Curves* (A), (B), *and* (C) *indicate anomalous crack growth behavior during development of closure* [32].

In view of the fact that the U.S. Air Force will soon be embarking on Retirement for Cause procedures for turbine disks wherein a 0.38-mm (15 mil) crack is of significance [*34*], it seems all the more important to understand the basic factors involved in affecting the rate of growth of short cracks under variable-amplitude loading conditions. As a first step in the development of this understanding the rate of development of crack closure in the wake of a short crack under constant-amplitude conditions should be determined. A start has already been made in a study of closure development in a ferritic steel [*35*]. The procedure involved the growth of a long fatigue crack, the removal of the wake of the crack to within a millimetre of the crack tip, and heat treatment in vacuo to reduce the residual closure level to below the level of detection. Fatigue cycling was then resumed and the closure level determined in the first millimetre of growth. As expected, the closure level built up rapidly to approach that of a long crack. This work was carried out in the near-threshold region where plane-strain conditions prevailed. Similar experiments could be carried out to determine the effect of an overload on the crack closure development in the short crack range and relate this to the rate of fatigue crack growth. Such overloads might lead to a more rapid buildup of closure and therefore be effective in arresting short crack growth.

Since it seems clear that the faster the rate of development of closure the less there will be of rapid, anomolous crack growth, it is important to understand the factors influencing the rate of development of closure. Thus far our results have indicated that large facets on the fracture surface, which are associated with a larger grain size as well as planar glide, are important in steel and aluminum alloys. Exploratory work should be carried out to modify the microstructure of an aluminum alloy to optimize these characteristics and determine their influence on crack closure. It should of course be recognized that these closure effects are most pronounced at low R-ratios and are of much lesser significance at high R-ratios.

Long Crack Growth Under Spectrum Loading

A number of research programs are aimed at the study of fatigue crack growth under spectrum loading. For example, we have grown cracks in an aluminum alloy under spectrum loading conditions, using a spectrum known as Mini-Twist, which represents the loads experienced by a transport aircraft wing structure as sampled in 4000 flights. The sequence of flights can be randomized, but the sequence of loading within each flight is not varied. The application of all 4000 flights constitutes one loading block, and a number of repeated blocks may be applied in the course of a test. Figure 15 shows some early results. It would be of interest to study crack growth behavior of the alloys and to make crack closure measurements at regular intervals to determine if different rates of growth correspond to differences in closure behavior. These studies should be carried out as a function of mean stress level to determine the influence of this variable. An objective would be to determine if there is a correlation in response of alloys tested under spectrum loading with behavior observed under just tensile overloads. Additionally, Eq 4 should be modified to account for load interaction effects. Linear superposition will generally provide an upper bound to the crack growth rate [*36*], and the important effects resulting in retardation may be associated with only the most severe of the loading cycles.

Short Cracks and Spectrum Loading

The considerations under this topic heading are similar to those outlined above with the exception that it would be of interest to determine the effect of spectrum loading on the development of closure and the resultant crack growth rate of short cracks.

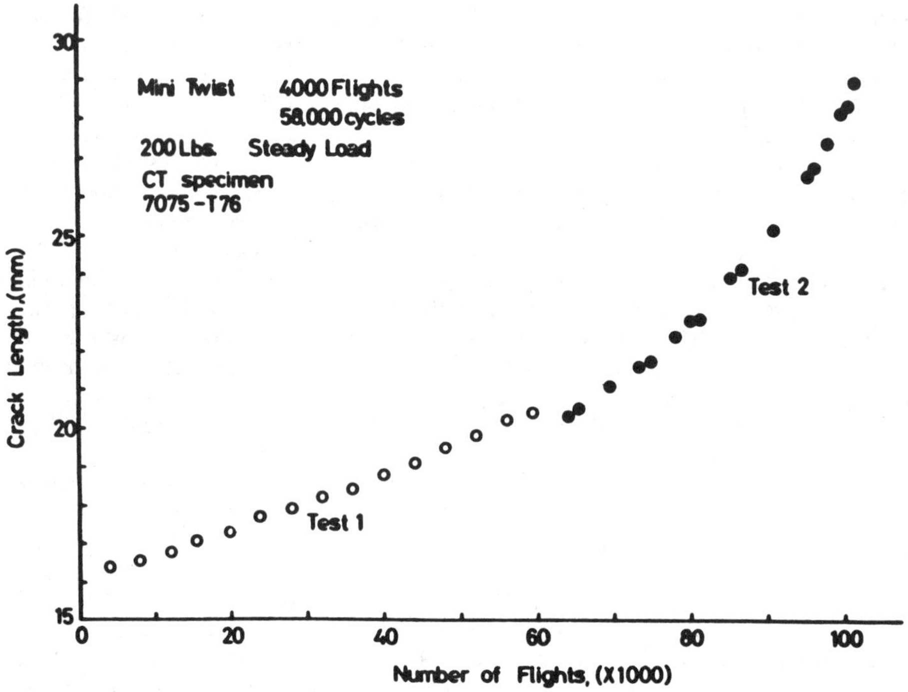

FIG. 15—*Crack length as a function of number of flight under the Mini-Twist loading program.*

Research Needs

The previous section has provided a brief review of the state of the art concerning load-interaction effects in fatigue crack propagation. It seems clear that our understanding of the processes involved in load interaction and the ability to make use of this understanding as a basis for quantitative analysis is far from complete. In particular the cause for the change in ranking of certain alloys in going from constant-amplitude to variable-amplitude loading is an important matter in need of clarification. In this section some areas in need of further research are discussed. Some questions to be answered by this research are:

1. When an overload is applied in the near-threshold region, what is the mechanism of retardation?
2. Does the mechanism of retardation differ as a function of ΔK and the extent of the overload?
3. What is the influence of the base R-ratio on overload effects?
4. What is the effect of thickness on retardation?
5. How does microstructure influence load-interaction effects?
6. Can a quantitative model for load-interaction effects be developed which is based on observed mechanisms?
7. How does an overload affect the propagation of a short crack?
8. How does spectrum loading affect the propagation of long and short cracks?

To answer such questions the following research tasks should be carried out:

(*a*) The determination of the effect of tensile overloads on subsequent crack growth as a function ΔK level by the use of the replica technique or the direct observation of the crack tip by SEM to determine crack-tip opening behavior as in the work of Lankford and Davidson [*20*]. The variation in opening level as the crack traverses the overload zone would be of particular interest. A few underload experiments should be carried out for purposes of comparison.

(*b*) The use of surface removal procedures as in Fig. 4 to observe the nature and variation with depth of the closure process by the SEM technique with and without overloads.

(*c*) The use of a clip gage and a strain gage as in Fig. 6 to determine if correlations exist between the crack-tip opening levels determined by these techniques and by the replica and direct SEM methods. We have already shown that examination of replicas by SEM gives a higher opening level than does the strain-gage technique in the case of an overload. It will be of interest to check for the generality of this result. We expect that better agreement between the strain-gage technique and the replica technique may be found in the case of constant-amplitude loading.

(*d*) The use of a range of specimen thicknesses to evaluate the effect of thickness on interaction effects as a function of ΔK. It would be of interest to determine for a given overload if the surface tip opening level is influenced by specimen thickness.

(*e*) The application of overloads at various *R*-ratios in order to assess the influence of *R*-ratio on retardation. The replica or direct technique should be valuable in determining whether or not closure exists at high *R*-ratios.

(*f*) the use of metallographic examination to determine if decohesion results after an overload.

(*g*) The study of short crack growth behavior is influenced by overloads.

(*h*) The study of long and short crack behavior under spectrum loading conditions.

Closing Remarks

The research outlined herein should provide the desired further insight into the factors responsible for load-interaction effects. This insight in turn should lead to sounder methods for the prediction of crack growth rates where load-interaction effects are important. In addition, it should also serve to identify those microstructural features which are important in developing closure and the retardation of crack growth under spectrum loading. Furthermore, it should identify those factors responsible for changes in rankings of alloys when compared under constant-amplitude and variable-amplitude loading conditions.

Acknowledgment

The financial support for this study provided by the U.S. Air Force Office of Scientific Research (AFOSR 81-0046) is gratefully acknowledged.

References

[*1*] Jacoby, G. H., Nowack, H., and Van Lipzig, H. T. M. in *Fatigue Crack Growth Under Spectrum Loads, ASTM STP 595,* American Society for Testing and Materials, Philadelphia, 1976, p. 172.

[*2*] Bernard, P. J., Lindley, T. C., and Richards, C. E. in *Fatigue Crack Growth Under Spectrum Loads, ASTM STP 595,* American Society for Testing and Materials, Philadelphia, 1976, p. 78.

[3] von Euw, E. F. J., Hertzberg, R. W., and Roberts, R. in *Stress Analysis and Growth of Cracks, ASTM STP 513,* American Society for Testing and Materials, Philadelphia, 1972, p. 230.
[4] Bucci, R. J., Thakker, A. B., Sanders, T. H., Sawtell, R. R., and Staley, J. T. in *Effect of Load Variables on Fatigue Crack Initiation and Propagation, ASTM STP 714,* American Society for Testing and Materials, Philadelphia, 1980, p. 41.
[5] Barsom, J. M. in *Fatigue Crack Growth Under Spectrum Loads, ASTM STP 595,* American Society for Testing and Materials, Philadelphia, 1976, p. 217.
[6] McEvily, A. J., Discussion of Ref 5.
[7] Elber, W. in *Fatigue Crack Growth Under Spectrum Loads, ASTM STP 595,* American Society for Testing and Materials, Philadelphia, 1976, p. 236.
[8] Wheeler, O. E., *Journal of Basic Engineering, Transactions,* American Society of Mechanical Engineers, Series D, Vol. 94, 1972, p. 181.
[9] Elber, W. in *Damage Tolerance in Aircraft Structure, ASTM STP 486,* American Society for Testing and Materials, Philadelphia, 1971, p. 230.
[10] Jono, M., Kanaya, T., Sugeta, A., and Kikukawa, M., *Journal of the Society of Material Science,* Japan, Vol. 31, No. 344, 1982, p. 483.
[11] Minakawa, K. and McEvily, A. J., *Scripta Metallurgica,* Vol. 15, 1981, p. 633.
[12] Walker, N. and Beevers, C. J., *Fatigue of Engineering Material Structures,* Vol. 1, 1979, p. 135.
[13] Suresh, S. and Ritchie, R. O., *Metallurgical Transactions,* Series A, Vol. 13, 1982, p. 1627.
[14] Ritchie, R. O., Suresh, S., and Moss, C. M., *Journal of Engineering Materials Technology,* Vol. 102, 1980, p. 293.
[15] Vasudevan, A. K. and Suresh, S., *Metallurgical Transactions,* Series A, Vol. 13, 1982, p. 2271.
[16] McEvily, A. J., *Metal Science,* Vol. 11, 1977, p. 274.
[17] Ruppen, J. A. and McEvily, A. J. in *"Titanium 80," Proceedings,* American Institute of Mining, Metallurgical, and Petroleum Engineers, Fourth International Conference on Titanium, Kyoto, Japan, Vol. 3, 1980, p. 1738.
[18] Schmidt, R. A. and Paris, P. C. in *Progress in Flaw Growth and Fracture Toughness Testing, ASTM STP 536,* American Society for Testing and Materials, Philadelphia, 1973, p. 79.
[19] Minakawa, K. and McEvily, A. J., to be published.
[20] Lankford, J. and Davidson, D. L. in *"Advances in Fracture Research," Proceedings,* Fifth International Conference on Fracture, Cannes, France, Pergamon Press, New York, Vol. 2, 1981, p. 899.
[21] Hardrath, H. F. and McEvily, A. J. in *Proceedings,* Crack Propagation Symposium, Cranfield, England, Vol. 1, 1961, p. 231.
[22] Suresh, S., *Metallurgical Transactions,* Series A, Vol. 14, 1983, p. 2375.
[23] Vecchio, R. S., Hertzberg, R. W., and Jaccard, R., *Scripta Metallurgica,* Vol. 17, 1983, p. 343.
[24] Stephens, R. I., Chen, D. K., and Hom, B. W. in *Fatigue Crack Growth Under Spectrum Loads, ASTM STP 595,* American Society for Testing and Materials, Philadelphia, 1976, p. 27.
[25] Knott, J. F. and Pickard, A. C., *Metal Science,* Vol. 11, 1977, p. 399.
[26] Brown, R. D. and Weertmen, J., *Engineering Fracture Mechanics,* Vol. 10, 1978, p. 867.
[27] Fleck, N. A., this publication, pp. 157–183.
[28] Bucci, R. J. in *Flaw Growth and Fracture, ASTM STP 631,* American Society for Testing and Materials, Philadelphia, 1977, p. 388.
[29] Minakawa, K., Levan, G., and McEvily, A. J. *Metallurgical Transactions,* Vol. 17A, 1986, p. 1787.
[30] Suresh, S. and Vasudevan, A. K. in *Proceedings,* American Institute of Mining, Metallurgical, and Petroleum Engineers, International Symposium on Fatigue Crack Growth Threshold Concepts, 1984, p. 361.
[31] Gallagher, J. P. and Hughes, T. F., Technical Report AFFDL-TR-74-27, Air Force Flight Dynamics Laboratory, Wright Patterson Air Force Base, Ohio, 1974.
[32] Sanders, T. H., Jr., and Staley, J. T. in *Fatigue and Microstructure,* American Society for Metals, Metals Park, OH, 1978, p. 467.
[33] McEvily, A. J., Minakawa, K., and Nakamura, H. in *Proceedings,* American Institute of Mining, Metallurgical and Petroleum Engineers Symposium on Fracture: Interactions of Microstructure, Mechanisms and Mechanics, Los Angeles, 1984, p. 215.
[34] Larsen, J. M. and Nicholas, T., presented at the 63rd (B) Specialists' Meeting of the Propulsion and Energetics Panel on Engine Cyclic Durability by Analysis and Testing, Lisse, the Netherlands, 30 May–1 June 1984.
[35] Minakawa, K. Nakamura, H., and McEvily, A. J., *Scripta Metallurgica,* Vol. 18, 1984, p. 1371.
[36] Katcher, M., *Engineering Fracture Mechanics,* Vol. 5, 1973, p. 793.

Author Index

T

W

Y

Z

Subject Index